Jean-Raymond Abrial   Egon Börger
Hans Langmaack   (Eds.)

Formal Methods for Industrial Applications

Specifying and Programming the Steam Boiler Control

Springer

Series Editors

Gerhard Goos, Karlsruhe University, Germany
Juris Hartmanis, Cornell University, NY, USA
Jan van Leeuwen, Utrecht University, The Netherlands

Volume Editors

Jean-Raymond Abrial
26, Rue des Plantes, F-75014 Paris, France

Egon Börger
Dipartimento di Informatica, Università di Pisa
Corso Italia 40, I-56125 Pisa, Italy

Hans Langmaack
Institut für Informatik und Praktische Mathematik, Universität Kiel
Preusserstr. 1-9, D-24105 Kiel, Germany

Cataloging-in-Publication data applied for

Die Deutsche Bibliothek – CIP-Einheitsaufnahme

Formal methods for industrial applications : specifying and programming the steam boiler control / Jean-Raymond Abrial ... (ed.). - Berlin ; Heidelberg ; New York ; Barcelona ; Budapest ; Hong Kong ; London ; Milan ; Paris ; Santa Clara ; Singapore ; Tokyo : Springer, 1996
  (Lecture notes in computer science ; Vol. 1165)
  ISBN 3-540-61929-1
NE: Abrial, Jean-Raymond [Hrsg.]; GT

CR Subject Classification (1991): D.1, D.2, D.3.1, F.3.1, K.6, C.2.4

ISSN 0302-9743
ISBN 3-540-61929-1 Springer-Verlag Berlin Heidelberg New York

This work is subject to copyright. All rights are reserved, whether the whole or part of the material is concerned, specifically the rights of translation, reprinting, re-use of illustrations, recitation, broadcasting, reproduction on microfilms or in any other way, and storage in data banks. Duplication of this publication or parts thereof is permitted only under the provisions of the German Copyright Law of September 9, 1965, in its current version, and permission for use must always be obtained from Springer-Verlag. Violations are liable for prosecution under the German Copyright Law.

© Springer-Verlag Berlin Heidelberg 1996
Printed in Germany

Typesetting: Camera-ready by author
SPIN 10549218   06/3142 - 5 4 3 2 1 0   Printed on acid-free paper

Preface

This book is the result of the collaborative effort of many persons and institutions, willing to contribute to a solution of the problem of finding means for a realistic evaluation of the practicality of formal methods for usage under industrial constraints. We would like to thank them all and will name them individually, in the chronological order of the help obtained.

We thank Schloß Dagstuhl for the opportunity to organize the workshop "Methods for Semantics and Specification" in June 1995, which was the starting point for our enterprise.

We thank Lt.-Col. J. C. Bauer from the Institute for Risk Research of the University of Waterloo, Ontario, Canada, for allowing us to use and modify his steam-boiler problem description for our own problem formulation, which was the basis for the specification competition reported in this book. We also thank the Institut de Protection et de Sureté Nucléaire, Fontenay-aux-Roses, France, through which we obtained J. C. Bauer's text.

We thank Annette Lötzbeyer for the Karlsruhe steam-boiler simulator which allowed the participants to run and test their steam-boiler implementations through the internet.

We thank the authors of the submitted solutions for having accepted the challenge to work on a concrete, non-trivial, and non-academic specification problem coming from an application domain, and in particular for participating under the condition that during the whole competition period the full documentation of all the contributions be completely open to everybody who wanted to participate in the comparison effort.

We thank the participants of the Dagstuhl workshop for their open criticism in a friendly atmosphere which helped considerably to bring researchers from different communities together for a fruitful exchange of ideas.

We thank Martin Fränzle for the creation and maintenance of the steam-boiler www site at the University of Kiel, which made the transparency of the competition possible and added a world-wide dimension to the comparative research effort reported in this book. We thank Martin Fränzle also for his valuable technical assistance during the preparation of this volume.

We thank the following reviewers for their detailed study, in a relatively short period of time, of the 33 steam-boiler problem solutions submitted for this book: Sten Agerholm, Christoph Beierle, Jonathan Bowen, Reinhard Budde, Holger Busch, Jorge Cuellar, Werner Damm, Igor Đurđanović, Martin Fränzle, Uwe Glässer, Mats Heimdahl, Friedrich von Henke, Tom Henzinger, Mike Hinchey, Leszek Holenderski, Jozef Hooman, Martin Huber, Klaus Indermark, Burghard von Karger, Christoph Kreitz, Peter Gorm Larsen, Thomas Lindner, Peter Lucas, Andrea Masini, Jose Meseguer, Tobias Nipkow, Peter Päppinghaus, Anders P. Ravn, Elvinia Riccobene, Hans Rischel, Peter Ryan, Michael Schenke, Fritz

Vogt, Isolde Wildgruber, Thomas Wilke, Martin Wirsing, Howard Wong-Toi, Wolf Zimmermann.

We thank Springer-Verlag, the editors of the LNCS, and in particular Alfred Hofmann from Springer-Verlag for publishing this book in the series and for offering the experiment of the first LNCS volume to come out with a CD annex containing full details of all the contributions, including the executable programs and their documentation which could not be reported in print.

Our thanks also go to Vicky Hartonas-Garmhausen for commenting on the introduction.

We are confident that the collaborative effort of the persons and institutions cited above will be rewarded by the positive contribution the material presented in this book will make to the discussion on the use of formal methods in practical applications.

September 1996

Jean-Raymond Abrial
Egon Börger
Hans Langmaack

Table of Contents

ABL: The Steam Boiler Case Study: Competition of Formal Program Specification and Development Methods 1
    Jean-Raymond Abrial, Egon Börger, Hans Langmaack

AT: Structural Synthesis of Programs from Refined User Requirements (Programming Boiler Control in NUT) 13
    Mattin Addibpour, Enn Tyugu

AL: Using FOCUS, LUSTRE, and Probability Theory for the Design of a Reliable Control Program 35
    Christoph Andriessens, Thomas Lindner

BBDGR: Refining Abstract Machine Specifications of the Steam Boiler Control to Well Documented Executable Code 52
    Christoph Beierle, Egon Börger, Igor Đurđanović, Uwe Glässer, Elvinia Riccobene

BCPR: An Algebraic Specification of the Steam-Boiler Control System .. 79
    Michel Bidoit, Claude Chevenier, Christine Pellen, Jérôme Ryckbosch

BW: A Steam-Boiler Control Specification with Statecharts and Z 109
    Robert Büssow, Matthias Weber

BSS: An Action System Approach to the Steam Boiler Problem 129
    Michael Butler, Emil Sekerinski, Kaisa Sere

CD: The Steam-Boiler Problem in Lustre 149
    Thierry Cattel, Gregory Duval

CW1: The Steam Boiler Problem - A TLT Solution 165
    Jorge Cuéllar, Isolde Wildgruber

CW2: The Real-Time Behavior of the Steam Boiler 184
    Jorge Cuéllar, Isolde Wildgruber

DC: Specifying and Verifying the Steam-Boiler Problem with SPIN 203
    Gregory Duval, Thierry Cattel

GM: TRIO Specification of a Steam Boiler Controller 218
    Angelo Gargantini, Angelo Morzenti

GDK: A Formal Specification of the Steam-Boiler Control Problem by Algebraic Specifications with Implicit State 233
    Marie-Claude Gaudel, Pierre Dauchy, Carole Khoury

HW: Using HyTech to Synthesize Control Parameters for a Steam Boiler . 265
 Thomas A. Henzinger, Howard Wong-Toi

LP: A VDM Specification of the Steam-Boiler Problem 283
 Yves Ledru, Marie-Laure Potet

LL: Proving Safety Properties of the Steam Boiler Controller 318
 Gunter Leeb, Nancy Lynch

LM: Steam Boiler Control Specification Problem: A TLA Solution 339
 Frank Lesske, Stephan Merz

LW: Specifying Optimal Design of a Steam-Boiler System 359
 Li XiaoShan, Wang JuAn

OKW: An Object-Oriented Algebraic Steam-Boiler Control Specification . 379
 Peter Csaba Ölveczky, Piotr Kosiuczenko, Martin Wirsing

SR: Refinement from a Control Problem to Programs 403
 Michael Schenke, Anders P. Ravn

S: VDM Specification of the Steam-Boiler Control Using RSL Notation . . 428
 Christian P. Schinagl

VH: Assertional Specification and Verification Using PVS of the Steam
Boiler Control System 453
 Jan Vitt, Jozef Hooman

WS: Specifying and Verifying the Steam-Boiler Control System with Time
Extended LOTOS 473
 Andreas Willig, Ina Schieferdecker

L: Simulation of a Steam-Boiler 493
 Annette Lötzbeyer

A: Steam-Boiler Control Specification Problem 500
 Jean-Raymond Abrial

Author Index 511

The Steam Boiler Case Study: Competition of Formal Program Specification and Development Methods

Jean-Raymond Abrial, Egon Börger, Hans Langmaack

This book wants to contribute to a realistic comparison, from the point of view of practicality for applications under industrial constraints, of the major techniques which are currently available for formally supported specification, design and verification of large programs and complex systems.

The contributions appearing here are the result of a world-wide competition which started in the Fall of 1994 by asking the researchers of all the formal specification approaches to apply their method to a non trivial and non academic case study, namely the design of a steam boiler control, which realizes the informal specification handed out to the participants of the competition and reprinted in Chapter A below. The competition has been conducted on a universal scale using the new communication means offered by the world wide web. The solutions which appear in this book are documented in detail in the appendices' which are stored on the accompanying CD.

In the rest of this introduction we describe in more details the intentions, the conditions and the realization of the competition and then survey briefly its result by analyzing the main features of the single contributions to this book. This survey is condensed into three tables at the end of the introduction. We hope that this concrete case study will help to assess which formal methods have a chance of entering the dimension of realistic practical applicability.

1 The Competition Project

In the Spring of 1994 we started to think about how to realize a competition among the great variety of formal approaches in the area of semantics and formal specification. The competition should focus on the feasibility of the methods for programming complex (in particular distributed embedded) systems under industrial constraints. We thought about a workshop with the goal of bringing together

– researchers representing the major formal methods and
– representative potential industrial users of such methods

for a critical comparison of the advantages and drawbacks of such methods for practical applications.

Schloß Dagstuhl generously offered its hospitality. This allowed us to invite, in the Fall of 1994, more than 75 researchers, representing all the approaches to formal specification, to a workshop which was held from June 5-9, 1995 with 49 participants from accademia and industry. In order to have a concrete basis for the intended evaluation of the proposed methods we asked the participants to prepare a solution to a problem which was large enough to exhibit features of a real-life problem, but small enough to be completed in a reasonable amount of time. Since we wanted to evaluate the use of formal methods for practical applications we chose a problem which comes from a non-computer-science application domain in a form that is typical for industrial informal requirement specifications.

The proposed problem and its specification is derived from an original text written by LtCol. J. C. Bauer for the Institute for Risk Research of the University of Waterloo, Ontario, Canada. The original text had been submitted as a competition problem to be solved by the participants of the International Software Safety Symposium organized by the Institute for Risk Research. It was given to us by the Institut de Protection et de Sureté Nucléaire, Fontenay-aux-Roses, France.

The steam boiler-control specification problem was sent out to the participants nine months before the workshop. Four months before the workshop some additional requirements for the physical behaviour of the components of the steam boiler were sent to the participants for possible inclusion into the formal model and as a possible test for the adaptability of the used method to requirement modifications (see chapter A in this book). One month before the workshop A. Lötzbeyer from the FZI in Karlsruhe provided the participants with an environment in which to run and test the implementations of proposed solutions (see chapter L in this book for a short description of this simulator).

We deliberately abstained from imposing any specific constraints on the expected solution. The idea was not to exclude any approach and to permit each method to be shown at its best, be it by providing a formal requirement specification, an architectural design, a sequence of stepwise refinements, an executable program or an analysis and proof of behavioural properties one wants to guarantee for the system. We hope that this decision to keep the specific task as free as possible neutralizes somehow the difficulty that the authors had no possibility of interaction with the virtual customer.

The workshop in Dagstuhl brought a fruitful, vivid, frank and critical discussion of a wide spectrum of different solutions, including a review at the end of the week by T. Lindner, J. Loeckx, P. Y. A. Ryan and M. Sintzoff. A list of evaluation criteria emerged by which various features of specification methods can be compared on objective grounds (see the Dagstuhl-Seminar-Report *Methods for Semantics and Specification*, no.117). The unexpectedly great success of the workshop encouraged us to go one step further; we invited the participants and other interested researchers to elaborate with detail a solution of the steam boiler problem, in the light of the critical discussion in Dagstuhl, and to submit their solution to be included in this book after a regular reviewing process. To promote the interaction among the different approaches, we asked every participant in the competition to study the material made available by the other groups.

After the Dagstuhl workshop Martin Fränzle established and since then administrated a steam boiler **www** page at the University of Kiel through which the competition received a truly universal dimension. The participants of the competition were asked to deposit at this **www** site the complete documentation of their contribution. This had an unexpected effect. At the end of 1995 we received a total of 33 submissions for this book, only 13 of which had been presented in preliminary form at the Dagstuhl workshop. (For accidental reasons, two of the solutions presented at the Dagstuhl workshop have not been submitted, another one was published elsewhere but is referenced in some of the solutions appearing in this volume, a fourth one was not ready at the time the typoscripts had to be sent to the publisher). 21 of these submissions appear here as the result of a thorough reviewing process; each solution has been reviewed by at least two referees.

The number of accesses to the steam boiler **www** site in Kiel demonstrates the intensive interaction, which through the steam boiler case study, has taken place among the different research groups in the area of specification methods. Above all, a great number of specific comparisons of the different approaches has become possible and is reflected in the contributions to this book. We hope that the fruitful world-wide research effort made possible by the **www** medium will continue; the steam boiler **www** page in Kiel accessed by:

`http://www.informatik.uni-kiel.de/ procos/dag9523/dag9523.html`

will stay active. It will contain contributions to the problem not included in this book, or those received after the deadline for this volume. Potential authors of new contributions to the steam boiler problem are invited to send their material to the above **www** site. We are glad to see that our idea of a concrete competition and evaluation of different formal specification methods has been well received and points into the future.

This book comes with a CD which contains the printed text and a series of appendices, where the authors had the possibility to insert all the details they consider necessary for a full documentation of their solution. This includes executable code, full definitions of all the parts of the specification, and where needed a detailed description of the mathematical foundation of the used method. As far as we know this is the first time that a representative group of formal methods researchers have got the chance to exhibit their approach by presenting in full detail their solution of a common concrete application problem, without any space restrictions. We believe that this particular feature of the present combined paper and digital publication will contribute to a realistic comparison of the merits and drawbacks of the currently available methods for formally supported design and verification of systems.

Before proceeding in the next section to a short survey of the contributions to this book, we conclude the description of the rationale and the history of our project by listing those questions which we asked the authors to answer in order to provide the reader with a guideline that is not affected by our personal evaluation.

1. Which parts of the system are specified, at which level, and in which way (verbally, rigorously, formally)?
2. (a) Does the solution comprise an implementation of a control program?
   (b) Has the control program been linked to the FZI Karlsruhe simulator?
   (c) Has some experimentation been done with the control program?
3. What are the comparable other solutions and what are the major differences with respect to them? What other solutions complement the given solution and in which respect?
4. (a) How much time has been spent on producing the solution? Give the number of person months, if possible, for the various parts of the solution.
   (b) How much preparation is needed to become sufficiently expert of the used specification framework in order to be able to produce a solution to such a problem in that framework? Indicate the number of weeks of training which you believe is needed for an average programmer to learn the method.
5. What are the premises for a good understanding of the proposed solution?
   (a) Is a detailed knowledge of the used formalism needed?
   (b) Can an average programmer understand the solution?
   (c) How much time do you believe is necessary for the average programmer without knowledge of the used specification method to learn what is needed to be able to understand the solution?

The reader can find the answers to these questions at the end of each contribution to this book.

2 Typology of the Contributions to this Book

A general pattern of the solutions in this book is that independently of the particular framework used for the requirement specification or for the implementation, almost all the solutions use some form of *modularization* (decomposition into components of manageable size) and of *refinement* (incremental introduction of further details on data and actions).

Another general phenomenon is that in most of the contributions, the use of formal methods helped to discover inconsistencies, loose ends, incomplete or ambiguous parts, and implicit assumptions that are hidden in the informal system description. The high degree of formality most of the authors have imposed upon themselves improved the detailed knowledge of the system suggesting clear modularization and refinement principles for the implementation of the system. This observation corresponds to the experience reported in [?] that the use of formal languages or methods at the specification stage is widely felt as a significant improvement of the traditional and mostly still current specification and program development practice.

Another observation is that most of the contributions that used tool based support (in particular theorem proving systems) had to simplify somehow the informally described steam boiler system in order to avoid the combinatorial or state explosion problem. It seems that at present, if the system to be analysed gets really complex, the traditional form of mathematical abstraction and reasoning activity can be replaced only partially (but obviously be supported) by machines.

The contributions to this book come with a rich variety of different tasks which have been accomplished: formal requirement specifications (of the controller or of the physical environment), intermediate refined models, analysis of system properties (failure analysis with or without fault tolerance, performance analysis, proofs (using also machine based theorem proving systems) for correctness, safety, or liveness), automated synthesis of conditions implying safety for parameters of the controller, design, or generation of executable code linked to the Karlsruhe simulator. In an attempt to differentiate the multitude of solutions, in the following we consider mainly the extent of the formalizations and the basic method adopted for the formal descriptions and the analysis.

In this book six out of the twenty one contributions come together with an executable program, namely AT, AL, BBDGR, BSS, CD, DC.[1] All of these programs except the one in AT have been linked for experimentation to the Karlsruhe simulator. At the Dagstuhl workshop only two solutions (AT, BBDGR) were shown to lead from a requirement specification to executable code which came as part of the proposed complete solution of the problem; the C++ program developed in BBDGR was presented in a demo in Dagstuhl and shown to successfully control the Karlsruhe simulator. One solution (SR) is refined to the level of an Occam-like program (which has not been compiled nor linked to the Karlsruhe simulator). All these solutions use some form of state based abstract specification of the steam boiler: AL formalizes the controller (for one pump and one pump controller) directly in LUSTRE, BBDGR derives C++ code from explicitly defined abstract machines, CD and DC start from temporal logic formalizations of the next-step relation of the steam boiler control, BSS starts from an action system description and AT from a direct formalization of the informal specification by restricted formulae of the intuitionistic propositional calculus. The remaining contributions restrict their attention to a formal specification and mathematical analysis of the desired properties of possible solutions of the steam boiler problem.

Most solutions in this book follow a suggestion, which is implicitly contained in the informal problem description, namely not to formalize the continuous system behaviour and to specify a discrete controller. Obviously all the programs that have been linked to the Karlsruhe simulator had to take into account the assumptions made by Annette Lötzbeyer for the realization of the continuous physical system behaviour. Some of the contributions which focus on the requirement specification and the mathematical analysis of the expected properties of the steam boiler system formalize and investigate to some extent the continuous system behaviour (in particular CW2, HW, LL, LW, SR, VH), partly at the expense of a simplification of the original problem dictated by the resulting state explosion problem for the adopted method. The solutions can be classified fur-

---

[1] We refer to the single chapters by the sequence of first capital letters of their authors.

thermore in terms of the specific language used for the abstract specification, the main categories being algebraic (BCPR, GDK, OKW), logical and operational languages. The logical approaches can be further subdivided according to the specific underlying logic, namely temporal logics (CD, DC, CW, GM, LM), real-time interval logic (LW, SR), intuitionistic propositional logic (AT) and higher order logic (VH). The operational approaches can be differentiated according to the adopted underlying computational paradigm, namely abstract machines (BBDGR), hybrid automata (HW), Lynch-Vaandrager timed automata (LL), statecharts and Z (BW), VDM (LP,S) and time-extended LOTOS (WS). Some contributions pay particular attention to (mathematical or formalized) proofs of expected system properties (in particular BCPR, CD,DC,CW, GM, HW, LL, LM, LW, OKW, SR, VH, WS, for illustration purposes also BBDGR).

In the following we characterize shortly some specific features that distinguish the single contributions in the two groups.

AT presents an automated implementation of a simulator and of a control program out of a formalized requirement specification, written in a restricted form of the language of intuitionistic propositional logic. The program synthesis tool which is used for this purpose is based on automatic proof search in an intuitionistic propositional calculus. The correctness of the program depends on the correctness of the synthesis tool and on the correctness of the propositional logic formalization.

AL splits the design task, on the basis of an architectural specification, into the construction of three modules, namely for the communication, the failure handling, and the pump controller. The communication module is designed using LUSTRE, the reliability of the failure manager and of the pump controller is discussed in probabilistic terms. The contribution argues against a uniform approach to code development; nothing is said however about how the C++ code is derived from and related to the formal specification.

BBDGR derives C++ code by stepwise refining abstract machine models of the steam boiler control that document the various design decisions as well as the structure and the functionality of the code. This makes the executable code inspectable by formal methods—an appendix illustrates the possibility of mathematical model analysis through proofs for some simple system properties—and easily modifiable. Care is taken that the starting model of this refinement hierarchy, i.e. the formal requirements specification, is stated in terms which directly formalize the basic objects and notions of the application domain so that the correctness of this specification can be checked by the application domain expert through a direct comparison with the informal task desciption. The correctness of the C++-code is related to the correctness of the abstract models via the refinement steps.

BSS starts with requirement specifications of both the control and the physical environment using action systems as defined by Back and Kurki-Suonio. A predicate transformer based refinement calculus is adopted to transform the abstract action system for the controller and for the physical environment into a more detailed action system closely resembling the final PASCAL implementation. The refinement calculus is also used as framework for the reasoning about these specifications.

CD and DC are based on synchronous programming. In CD, C-code has been generated automatically from a requirement specification in LUSTRE. The correctness of the code therefore depends on the correctness of the code generator and on the correctness of the LUSTRE specification. The architecture of the LUSTRE model for the controller is essentially the same as in DC; it is given by a graphical data-flow diagram and has been inspired by the temporal logic formalization presented in (a previous version of) LM. Safety properties are expressed in linear temporal logic and then coded into LUSTRE so that they can be checked using the LUSTRE model checker Lesar. In DC also liveness properties are considered for various abstract models of the controller; they are expressed by temporal logic formulae which are then verified by the SPIN modelchecker. The executable code in DC is written in a concurrent extension of C++. Nothing is said about how this code (and therefore its correctness) is related to the formal abstract specification.

CW presents a temporal logic formalization of (the next-step relation of) the discrete control program which is coupled to a rigorous model of the dynamics of the physical components of the system determining the continuous environment in which the control program is supposed to act. Several positive and impossibility results about the expected properties of the system are proved on the basis of the temporal-logic specification, a previous version of which was implemented in C (apparently without considering how the formal models are related to the code).

SR uses the Duration Calculus for a requirement specification (including safety and liveness requirements) and for its refinement to a functional design. The specifications of the components of the distributed architecture as communicating sequential processes are obtained by syntax directed transformation rules and form the basis for a further transformation to occam-like programs. The paper analyses both the controller and the environment and concentrates on a detailed investigation of fault detection and tolerance mechanisms (for a simplified steam boiler system).

BCPR presents a requirement specification and analysis in classical algebraic style, coming as axiomatic first-order description of the desired system properties and avoiding any notion of state (which is conceived as belonging to the implementation view of a system). The contribution uses Bidoit's algebraic specification language PLUSS and the interactive many-sorted first-order logic theorem prover LARCH for discharging the proof obligations introduced by the different refinement steps of the specification and for proving some of its logical consequences. Particular attention is paid to an analysis of the various failure situations. In GDK a variant of the purely algebraic approach is proposed by introducing an implicit notion of state through so-called modifiers (a formalization of function updates in the algebraic specification framework). A similar operational algebraic formalization (which uses also some object-oriented features) appears in OKW where the static system parts are described by equational specifications and the behaviour of processes by timed term rewriting, i.e. term rewriting taking place in explicitly prescribed periods of time. In the underlying timed rewrite logic which is an extension of Meseguer's rewriting logic, processes are encoded as terms and process behaviours as proofs.

The temporal logic based requirement specification GM of the controller incorporates some object-oriented language extension. LM is based upon Lamport's linear temporal logic TLA and uses stepwise refinement and grouping of related requirements into "modules" for the specification of both the controller and the physical environment. It contains also a "data flow analysis" in terms of input/output dependencies between the different "modules". A previous version of this specification has been used in Chapters CD and DC as starting point for the implementation. The real-time interval logic based specification of a simplified version of the steam boiler in LW studies some performance and optimization features. VH presents (for a simplified steam boiler system) a declarative extended Hoare logic specification apt for checking proofs of the desired system properties using PVS. It proceeds in three successive phases, starting with a formalization of a real-time control program without communication or component failures. This description is then extended to cope with communication failures and finally with failure situations for single components.

The hybrid automata model in HW describes the continuous (simplified steam boiler) behaviour without discretization. The symbolic model checker HYTECH is used to automatically synthesize control parameter constraints that provide values for these parameters for which the steam boiler is correct. The Lynch-Vaandrager timed automata model in LL includes a detailed failure model and is geared to the proof of desired safety properties of both the continuous steam boiler and the discrete controller. The proof for the controller model tolerating sensor faults is obtained incrementally from the proof for a simple non fault-tolerant controller by providing consistent transitions between different abstraction layers.

BW advocates the loose use of formal elements for the description of different views and parts of a complex system. An architectural model of the steam boiler is described in an object-oriented style using class diagrams to depict classes and their structural relationships in the system configuration; the reactive model for the interaction between the components and for the timing control is formalized using a timed variant of Harel's statecharts; the functional behaviour of single components—i.e. their local state transformations—is defined using the Z notation.

The contribution S contains a classical VDM style description, using the RAISE specification language RSL, of all the steam boiler components and of the behaviour of the controller. The author states that he has abstained from proving system properties of the resulting formal model because the complexity of this model—with the same level of abstraction as the informal specification—makes such proofs too difficult. LP provides another model-based formal specification using the VDM-SL language, the modular extensions contained in the IFAD tools, and refinements of data and control constructs. It incorporates into

VDM-LS reactive, real-time, and parallel elements needed for the formalization of the physical steam boiler, of the interface between the physical boiler and the control system and of the controller itself. The specification is defined in an *environment-based* way. It starts from a model for the environment, i.e. the physical uncontrolled boiler whose state space is characterized by (mostly continuous) external variables which are governed by physical laws and only monitored by the controller. Then it introduces the interfaces with the control system, more precisely discrete time and "actuators" (variables which may be written by the controller and have an impact on the environment). Finally the behaviour of the controller is specified by a controller architecture and a module, which defines the controller's use of the monitored variables.

WS extends LOTOS by features that formalize the timing constraints imposed for stating correctness and safety properties of the steam boiler system. The purely functional design step which does not contain any timed behaviour has been verified and tested using LOTOS tools assuming non faulty system and communication behaviour. The generation of the underlying extended finite state machine took several days and was therefore interrupted before it finished. In a third step the data dependencies of the steam boiler system are specified in the functional data type language GOFER.

The richness and variety of the different tasks and the specific conditions of this competition—including our desire not to be influenced by any a priori bias and the necessity to accept the historical fact that at present the development of formal methods is mainly driven by accademic research—prevented us from coming up with a list of clear-cut and uniform evaluation criteria. A future edition of such a competition of methods should probably start with a sharper catalogue of specific tasks whose solution is made mandatory. Nevertheless, we hope to provide enough and representative material with this book which allows the reader, in particular the practitioner, to guide and support his personal selection. The reader is therefore invited to judge for himself the different conceptual and methodical efforts which are needed to apply the proposed specification techniques for a well documented and formally supported practical program design and analysis.

References

1. D. Craigen, S. Gerhart, and T. Ralston. An international survey of industrial applications of formal methods. Technical Report NISTGCR 93/626, U.S. National Institute of Standards and Technology, 1993.

	(requirement) specification	exec. code	linked to FZI simul.	proofs of syst.propties	analys. of contin.env.
AT	in intuition. propos.logic	+			for the simulator
AL		C++	+	+	for the code
BBDGR	as abstract machines	C++	+	+	for the code
BCPR	by 1st-order logic axioms			refinement proofs	
BW					
BSS	as action system	PASCAL	+	refinement proofs	for the code
CD	as LUSTRE program	C	+	safety model-checked	for the code
CW1-2	by temporal logic axioms			+	+
DC	taken from Chapter LM	synchr. C++	+	safety/liven. SPIN verified	for the code
GM	in temp.logic				
GDK	in pred.logic				
HW	as lin. hybr. automata			+	+
LP	in model-based VDM-SL			safety and consistency	+
LL	as timed autom.			safety	+
LM	in temp.logic			+	+
LW				+	+
OKW	in Timed Maude			+	for simple environment
SR	in real-time interv. log.			+	+
S	in RAISE-SL				+
VH	as PVS spec			+	+
WS	in TE-LOTOS			+	

	main used language or method	particular features	remark	country
AT	intuitionistic propos. logic /NUT system	automated program synthesis from formal req spec		S
AL	LUSTRE stochastics	probabilistic failure analysis	restriction to 1 pump	D
BBDGR	abstract machines	req spec near to applic., uniform refinement to code		D, I
BCPR	PLUSS alg spec lg LARCH prover	thm proving for stepw. refinement/validation	initialization not specified	F
BW	Z,diagrams,timed statecharts	loose combination of various formalisms		D
BSS	Back/Kurki-Suonio action systems	formalizes environment and controller		GB,SF
CD	temporal logic and LUSTRE	specification taken from LM	safety stated in temp.logic	CH
CW1-2	temporal logic TLT	analysis of real-time behaviour		D
DC	PROMELA/SPIN	arch.model reused in CD	no transm. failure	CH
GM	temp.logic/TRIO			I
GDK	PLUSS algebraic spec language	algebraic spec. with implicit states		F
HW	linear hybrid automata,HYTECH	automatic synthesis of safe controller params constraints	only 2 pumps in norm. mode	USA
LP	VDM/IFAD tools	env-based approach	fail.,comm., initial. sketchy	F
LL	Lynch-Vaandrager timed automata	detailed failure model for fault-tolerant controller		AU, USA
LM	temp. logic TLA	spec used in CD, DC	fail.det. sketchy	D
LW	real-time interval logic	optimal design for broken water sensor discussed	simplified boiler	Macau
OKW	Timed rewriting logic TRL	equational alg. spec with timed rewriting		N, PL, D
SR	duration calculus	uses real-time refinement calculus	fault det./toler. for simplif.boiler	D, DK
S	RAISE lg/tools	abstraction level of inf.spec.		AU
VH	PVS	PVS proofs for refinem.steps	simplif.boiler	D, NL
WS	TE-LOTOS			D

	time to find the solution (months)	preparation needed to understand the solution (hours, days,weeks)	training needed to learn the method (weeks)
AT	3	knowledge of logic + 1w	4
AL	4	knowledge of probability theory	?
BBDGR	2	0	0.5
BCPR	1	knowledge of 1st-order logic	8
BW	1.25	2w	8
BSS	4	1h	2
CD	1 with DC reused	4d	2
CW1-2	3	knowledge of logic + 1h	logic + automata theory + 1w
DC	1 using LM	6d	12
GM	0.75	logic + TRIO	math.backgrd.
GDK	1.5	1st-order logic + 2w	8
HW	?	?	3
LP	1.5	1w	several months
LL	2	2w	8 months
LM	1.5	a few days	a few days
LW	4	4h	4
OKW	a couple of weeks	?	?
SR	6	?	work amount of a master thesis
S	1	2w	6
VH	9	knowledge of logic	logic + a few d
WS	2.75	fctal pgg + 2w	8

Structural Synthesis of Programs from Refined User Requirements
(Programming Boiler Control in NUT)

Mattin Addibpour and Enn Tyugu

[mattin, tyugu]@it.kth.se
Department of Teleinformatics, Royal Institute of Technology Electrum-204 KTH,
164 40 Kista, Sweden

Abstract. The aim of this work is to demonstrate the feasibility of using a declarative language as a tool for automated implementation of requirements written in a semiformal manner. The technique of structural synthesis of programs based on automatic proof search in intuitionistic propositional calculus implemented in the NUT system is used for solving the steam-boiler problem. The goal of the experiment is to bridge the gap between the language of requirements and an implementation. An appropriate set of concepts is developed to represent the problem, i.e. the requirements are written in a form understandable by the program synthesis tools. An implementation, including simulator of the actual steam-boiler and control panel, is written in NUT. The program synthesizer is a part of the specification language compiler and it is completely hidden from the user.

1 Introduction

This paper concerns automatic application of formal methods in implementation of industrial control systems. The idea is to use an extensible very high level language, extending it with a set of concepts sufficient for writing requirements specifications in a form close to a selected formal specification language. It is assumed that the specifications are already refined and verified, so we have to take responsibility only for the correctness of the implementation. We considered the TLT specification language as a possible candidate for the requirements specification language, but have decided to take the specification directly as it has been presented in [AS] (see Chapter AS, this book), although several other candidates (TLT [CW1] (see Chapter CW1, this book), Z [BW] (see Chapter BW, this book)) were also possible.

In order to solve the problem, we used a technique of deductive synthesis of programs based on automatic proof search in intuitionistic propositional calculus, implemented in the NUT system [Tyu]. In the NUT system, an appropriate set of concepts represented as classes has to be developed for each kind of problems. The concepts for simulating the steam-boiler and programming a boiler model are: device, boiler, pump, pump controller, level meter, flow meter, etc. These concepts are represented as sets of equations taken directly from the requirements specification.

Another set of concepts is required for implementing a control algorithm: propositions, rules and actions. The control algorithm is represented as a collection of rules written as relations in NUT. The rules are triggered by conjunctions of propositions describing a state of the system at an observable time moment. These rules are also derived directly from the requirements specification after introducing a proper collection of propositions and predicates which are implicitly present in the text of the requirements. With the present experiment, we demonstrate that the presentation of requirements, as given in the steam-boiler example, enables us to extract rules immediately from the requirements text. Formally, the steam-boiler control system can be considered as an attributed automaton, whose attribute models are represented as classes, and transitions are specified by rules. This technique has been used earlier, in particular, for protocol simulation in NUT [Pen].

As we were concerned with the method of implementation and not with the completeness of the given initial specifications, we took the specifications as given in [AS] (see Chapter AS, this book), including the assumptions about the fixed time cycle and the transmission system.

2 Theory and Tools

2.1 Structural Synthesis of Programs

We are using a completely automatic deductive program synthesis as an implementation technique for the semantics of the specifications. Therefore, we need a method which is both complete and efficient. This is a very strong requirement, and it can be achieved only at the price of restricting the generality of a language. We have chosen a restricted fragment of the intuitionistic propositional calculus (IPC) which is still universal with respect to the whole IPC. Transformation of an arbitrary formula of IPC into the fragment we use has been described in [MT]. We use this logical language, hidden from the user, only for internal representation of specifications. We call our method structural synthesis of programs (SSP).

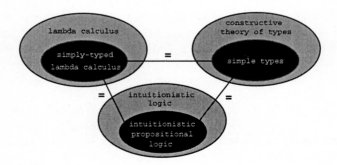

Fig. 1. Three formalisms for representing structural synthesis of programs.

The SSP can be explained in terms of lambda calculus or type theory as well. The fragment of logic on which the SSP is based, covers the largest fragment of intuitionistic logic, lambda calculus or type theory which we are still able to use efficiently. The relations between different representations of the SSP are shown in Fig. 1. The equivalences shown in the figure can be proved according to Curry-Howard correspondence between intuitionistic logic and typed lambda calculus. An interested reader can find the theory of SSP in [MT] and [Uus].

The basic building block of formulas of the SSP is an implication:

$$A_1 \wedge A_2 \wedge ... \wedge A_n \rightarrow B$$

which has a computational interpretation as a function f for computing b from $a_1, a_2,...,a_n$ written in a usual way as follows:

$$b = f(a_1, a_2, ..., a_n)$$

The propositions $A_1, A_2, ..., A_n$ can be assumed to have the meaning "a_1 is computable", "a_2 is computable" etc. This gives the following meaning of the implication: "b is computable if a_1 is computable and a_2 is computable and ... and a_n is computable". Implications given above can be combined into a new implication as follows:

$$(A_1 \wedge A_2 \wedge ... \wedge A_n \rightarrow B) \wedge ... \wedge (C_1 \wedge C_2 \wedge ... \wedge C_k \rightarrow D) \rightarrow (X_1 \wedge X_2 \wedge ... \wedge X_m \rightarrow Y)$$

The meaning of the nested implication is "If b is computable from a-s, and ... and d is computable from c-s, then y is computable from x-s". This nested implication combines computabilities, which gives us sufficient expressive power for representing arbitrary algorithms by means of a few preprogrammed constants. Indeed, a computational interpretation of the nested implication from above can be written as follows:

$$y = f(x_1, ..., x_m)$$

where, first, $f = F(g_b, ..., g_d)$ and, second, $g_b,...,g_d$ are respective interpretations of the implications $(A_1 \wedge A_2 \wedge ... \wedge A_n \rightarrow B), ..., (C_1 \wedge C_2 \wedge ... \wedge C_k \rightarrow D)$, i.e. they are functions which have to be synthesized. One can immediately detect a similarity between this notation and functional programming languages.

Inference rules of the SSP together with the program derivation rules are given below:

Introduction of implication: **Elimination of implication:**

$$\begin{array}{cc} \begin{array}{c} F \\ \vdots \\ G \\ \hline F \rightarrow G \end{array} & \begin{array}{c} x \\ \vdots \\ g(x) \\ \hline \lambda x.g \end{array} \end{array} \qquad \dfrac{F \quad F \rightarrow G}{G} \quad \dfrac{f \quad \lambda x.g}{g(f)}$$

Introduction of conjunction: **Elimination of conjunction:**

$$\dfrac{F \quad G}{F \wedge G} \quad \dfrac{f \quad g}{(f,g)} \qquad \dfrac{F \wedge G}{F} \quad \dfrac{(f,g)}{f} \quad \dfrac{F \wedge G}{G} \quad \dfrac{(f,g)}{g}$$

The proof of a formula is a derivation tree whose leaves are axioms and assumptions, and any other node represents a derivation step where one of the inference rules is applied and a new formula is derived. Text in lower-case represents programs derived in parallel with a proof. Axioms have predefined realizations in the form of programs, equations and structural relations written in NUT. Assumptions give variables which are bound in lambda expressions. A program emerges gradually as a realization of derived formulas, and the complete program is obtained as a realization of the formula derived last, i.e. the goal. The axioms are derived automatically from class specifications and the proof search is completely hidden from the user. So are the synthesized programs which exist only during the computations. They are visualized only for debugging purposes. Appendix D (see CD-ROM Annex AT.D) contains an example of a synthesized algorithm which calculates the next state of a boiler from the current state and is synthesized on the boiler model. Finally, propositions of our logical language can be interpreted arbitrarily, and implications used in a conventional constructive way. This gives us another possibility to represent the logic of the requirements specification as a set of rules in the same language which is used for the SSP, and to use the theorem prover of the SSP for making inferences about the actions to be taken for controlling the boiler. This will be used in the Sect. 4 for implementing the control algorithm. In this case, the program synthesizer is called repeatedly, for selecting an applicable rule at each step of the computation.

2.2 The NUT Language

Here we introduce the NUT[1] language to the extent needed for understanding the implementation descriptions in the following sections. The NUT language is an object-oriented language with multiple inheritance, parametric polymorphism, and a flexible mechanism for message passing. It has predefined classes: *num, bool, array, program, text* and *any*. Specifying the user-defined classes is the main activity in an implementation process. In the NUT language, classes serve the following two purposes:

1. As is usual in an object-oriented system, they are specifications of behaviour of objects. They contain specifications of variables, called in this case components of objects, and specifications of methods written in an imperative language.
2. Classes serve also as specifications which can be used for automatic construction of programs. In particular, a method called *compute* is available by default in every class which can be sent to any object x as a request to compute some of its components, let us say, y. The expression $x.compute(y)$ returns a value of y, if the class specification of x contains sufficient information to compute this value.

---

[1] The NUT system together with its documentation is available via www under http://www.it.kth.se/labs/se/nut.html

The usage of classes as specifications for program synthesis is an unusual feature of an object-oriented language which distinguishes NUT from other object-oriented systems. The program synthesis in NUT is completely automatic, and syntactically fits into the object-oriented framework, although semantically, it enforces a different encapsulation border on the classes of the language. There exist systems supporting semi-automatic program synthesis like the KIDS system which has been used for constructing impressive algorithms [Smi].

The overview of the whole specification language is given in appendix A (see CD-ROM Annex AT.A). Below, we concentrate on the interplay between data and program modules (relations). Thus only two parts of a class specification are considered, which are directly related to the program synthesis. These two parts are specifications of state variables, also called *components* of a class, and specifications of methods, also called *relations*, because of their usage in automatic construction of programs. They are distinguished by the keywords *var* and *rel*, as shown in the following simple class specification:

```
class complex
 var
   re, im: num;
   mo, arg: num;
 rel
   re2mo: re,im -> mo,arg {mo:=sqrt(re^2+im^2); arg:=acos(re/mo)};
   mo2re: mo,arg -> re,im {re:=mo*cos(arg); im:=mo*sin(arg)}
```

The relations *(re2mo, mo2re)* of the class *complex* are specified by the following two parts. First, they have external views (called also axioms) which specify their input and output parameters and give some type information:

```
   re, im -> mo, arg
   mo, arg -> re, im
```

Second, they contain programs written in an imperative language which is a part of the NUT. These programs implement the methods. This is the semantics of the specifications used in object-oriented computations. An input parameter of a relation may be a function, which in this case will be specified implicitly, by giving its type, as shown in the following example:

```
   rr: [u -> v]   x -> y {<program>};
```

This example describes a method with two input parameters. The first is a function for computing v from u, the second is x. Functions as input parameters of methods play an important role in program synthesis. The axiom of the example will be translated into the following formula of the internal language:

$$(U \to V) \to (X \to Y)$$

where U, V, X and Y denote computability of u, v, x and y respectively.

Arithmetic equations, together with the default typing of numeric variables, provide a possibility to give equational specifications of concepts in the form of classes. For instance the class complex can in a declarative and more expressive way be specified as follows:

```
class complex
 var
   re, im: num;
   mo, arg: num;
 rel
   mo^2 = re^2 + im^2;
   re = mo * cos (arg);
   im = mo * sin (arg);
```

In the same manner, equational specifications can be used for specifying an essential part of the physical behaviour of the boiler by writing only equations as they are presented in the specifications text:

```
class simpleboiler
 rel
   qc1 = qa1-va2*Dt-0.5*U1*Dt^2+pa1*Dt;
   qc2 = qa2-va1*Dt-0.5*U2*Dt^2+pa2*Dt;
   vc1 = va1-U2*Dt;
   vc2 = va2+U1*Dt;
```

Equations are considered as sources of methods for computing variables bound by the equations. The equation vc1=va1-U2*Dt in the class simpleboiler gives the following four methods:

```
va1,U2,Dt   -> vc1{vc1:=va1-U2*Dt};
vc1,U2,Dt   -> va1{va1:=vc1+U2*Dt};
va1,vc1,Dt  -> U2{U2:=(va1-vc1)/Dt};
va1,vc1,U2  -> Dt{Dt:=(va1-vc1)/U2};
```

which have no explicit message names for invoking them. Each of the other three equations also gives us a number of methods. These methods can be used automatically in computations, i.e. without explicitly sending messages to them, for instance, by sending the compute message to an object of the class simpleboiler.

Prototype is any class which is used as a type specifier in the construction

```
        <identifier> : <name of class>;
```

where *<identifier>* is a name for a new component of a class which acquires the type of the class with the name *<name of class>*. This feature of the specification language is essential, and it is very much used in the applications where automatic synthesis of programs is present. It enables one to put together large classes which represent problem specifications for automatic synthesis of programs.

Any existing class can be used as a *superclass* for new classes. We can specify various boilers, using the class *simpleboiler* as a superclass and extending classes with new components: pumps, valves etc. The following example demonstrates usage of a superclass and prototypes:

```
class boiler_with_pumps
 super simpleboiler;
 var
```

```
  P1, P2, P3, P4: Pump P=10;
rel
  pc1 = P1.pc1+P2.pc1+P3.pc1+P4.pc1;
  pc2 = P1.pc2+P2.pc2+P3.pc2+P4.pc2;
alias
  Pumps = (P1,P2,P3,P4);
```

The last two lines of this class demonstrate one more feature of the language - aliasing. If it is desirable to refer to several objects by one name, or just to use another name for an object, we can introduce a new name by using *alias* as shown in the example above. There are two more features of the language we have to introduce here: *bindings* and *polymorphism*.

Bindings are amendments to types introduced by the prototypes. Bindings enable us to write specifications where components of components given by prototypes are bound to values or other known objects.

The type *any* provides an interesting form of *polymorphism* to the language, which can be used in object-oriented computations as parametric polymorphism, but has a more general interpretation in specifications - components of an object of type *any* can be referred to before the type of this object is concretely specified. We can rewrite the class of *boiler_with_pumps* in the following way:

```
class boiler_with_pumps
super
  simpleboiler;
var
  P1, P2, P3, P4: any P=10;
rel
  pc1 = P1.pc1+P2.pc1+P3.pc1+P4.pc1;
  pc2 = P1.pc2+P2.pc2+P3.pc2+P4.pc2;
alias
  Pumps = (P1,P2,P3,P4);
```

This specification defines essentially the same class, except that the class of pumps is still *any*, i.e. not precisely defined yet. This concludes our brief introduction of the NUT language. Appendix A (see CD-ROM Annex AT.A) contains an informal description of the whole specification language of NUT, which can be used as a reference when reading specifications of the steam-boiler control system in the following sections.

3 Overall Structure of the Steam-Boiler Control System

Using the NUT system as a tool serves the purpose of implementing the steam-boiler control system simply and reliably. Our intention is to use the requirements specification as directly as possible to obtain executable specifications. The design phase which precedes the implementation has to support this goal, preserving the overall structure of the system outlined in the requirements specification. In this section, we present the structure of the boiler control system as it appears after the design phase.

A design decision we have made is to include a system simulator into the program for testing and demonstration purposes. As a consequence, the boiler control system consists conceptually of the following three parts: system simulator, control program and message switch. These are implemented as five global objects in NUT as shown in Fig. 2.

The system simulator consists of the following components:

- *SysSim:* For simulating the steam-boiler system,
- *MsgPanel:* For simulating the control panel used for generating events like breakdowns and failures.

We have developed our own simulator in NUT by using the NUT scheme editor for visual specification of the system and the NUT graphics for visualization of simulation, but our control program can also be connected to the FZI simulator.

The steam-boiler control program consists of two main components:

- *Controller*: For specifying the control algorithm, described in detail in Sect. 4,
- *SysModel*: The control program's own simulation of the system for predicting the state of the steam-boiler in the case of failure, described in Sect. 5.

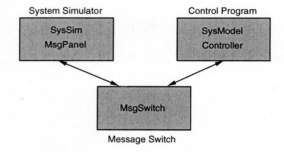

Fig. 2. Global objects (structure of the steam-boiler control system).

Thus we have two very similar models of the steam-boiler, one *(SysSim)* as a substitute of an actual boiler and other devices for making experiments with the control program, and another *(SysModel)* as a part of the control program for making predictions about behaviour of the steam-boiler.

The system simulator and the control program communicate through a message switch:

- *MsgSwitch*: For simulating the transmission system between the control program and the system simulator as well as for detecting the transmission failure. *MsgSwitch* can act internally in NUT (using our own simulator) or can read commands from stdin in the case of using the FZI simulator.

As we mentioned above, the most complicated classes *SysSim* and *SysModel* are quite similar to one another. Figure 3 shows a screen snapshot of the scheme editor window with a graphical specification of the *SysModel* class. A precise textual representation of this class is obtained completely automatically from this scheme by means of the scheme editor. In this scheme, we see four pumps with pump controllers, as well as a boiler, a flow-meter, a level-meter and a valve. These components of the system are properly connected according to the schema. A class exists behind each of the graphical icons in the scheme. Some of these classes will be presented in Sect. 5 and appendix C (see CD-ROM Annex AT.C). Graphical specification of the *SysSim* is quite similar, and is not presented here.

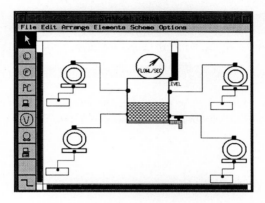

Fig. 3. Graphical specification of the SysModel class.

4   Implementing the Control Algorithm

The control program seen in more detail consists of the following parts (see Fig. 4):

- *SysModel* : A system model for predicting the next state of the steam-boiler needed for estimating water level in the case of failure of the water level measuring unit; furthermore a failure of the steam measuring unit can be detected by comparing a predicted value with a measured value. (We give specifications of physical devices later in Sect. 5 and appendix C (see CD-ROM Annex AT.C). This allows us to show the place of the physical model *SysModel* in the whole picture before its specifications. We shall focus on the control algorithm represented mainly as a set of rules.)
- *Controller*: Implementation of the control algorithm consists of the following parts:
    - *Analyser*: A program for analysing the incoming messages from the physical part, and producing necessary propositions used in the *Rules* and other parts of the *Controller*. The *Analyser* class is described in 4.2.

- *Rules:* A heap of rules, which represent the decision making algorithm in the *Controller*. Here we use an easily extendable set of rules each of which can be fired as soon as its preconditions are satisfied. This mechanism is described in detail in 4.3.
- *Actions:* Once actions have been selected by firing the rules, the *Actions* part translates them into proper messages sent to the physical units, described in 4.4.

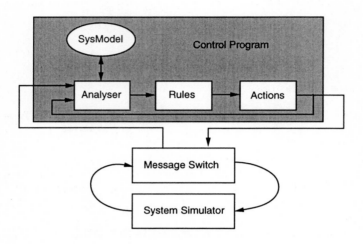

Fig. 4. Structure of the control program.

4.1 Propositions

Each condition which appears in the text of requirements is represented as a proposition (a boolean variable) in the *Controller*. The propositions represent either conditions directly obtained from the incoming messages (external propositions), or conditions calculated and used by the control program internally (internal propositions). We have tried to keep names of propositions reasonably close to the text of the requirements.

External propositions The following propositions are obtained directly from the incoming signals:
- STOP, STEAM_BOILER_WAITING, PHYSICAL_UNITS_READY, LEVEL_REPAIRED, STEAM_REPAIRED, LEVEL_FAILURE_ACKNOWLEDGEMENT, STEAM_OUTCOME_FAILURE_ACKNOWLEDGEMENT.

On receiving a message of this kind, the *MsgSwitch* will assign true to the corresponding proposition. If a message is not received, the value *nil* will be assigned to the related proposition. The following are arrays of external propositions indexed from 1 to 4 by the number of a pump:

- PUMP_STATE, PUMP_CONTROL_STATE, PUMP_REPAIRED, PUMP_CONTROL_REPAIRED, PUMP_FAILURE_ACKNOWLEDGEMENT, PUMP_CONTROL_FAILURE_ACKNOWLEDGMENT.

On receiving a message of this kind, the *MsgSwitch* will assign *true* to the corresponding indexed proposition. If a message is not received, the related indexed proposition will be *false*.

Incoming numeric messages Besides the external propositions, the *Controller* receives two messages containing numerical values:
- LEVEL, STEAM.

Internal propositions These propositions are produced by rules or by small programs given in the *Analyser*.

- The *MsgSwitch* can detect the transmission failure. This will be done by using an automaton for detecting a failure in transmission by checking the possible states:
  - TRANS_FAIL. (either true or nil)
- Some messages are filtered and analysed by the *Analyser*; the external proposition STOP is transformed into an internal proposition STOP_DECISION (either *true* or *nil*), which indicates that STOP has been received three times in a row.
- The *Analyser* can detect failures in pumps, pump controllers, level measuring unit and steam measuring unit, according to Sect. 7 of [AS] (see Chapter AS, this book). These failures are represented by the following internal propositions:
  - PUMP_FAILURE, PUMP_CONTROL_FAILURE, PUMP_FAIL_NOT_ACK, PUMP_CONTROL_FAIL_NOT_ACK. (array of true or false)

  ALL_OK, NOT_ALL_OK, ALL_PUMP_OK, SOME_PUMP_FAIL, ALL_PUMP_CONTROL_OK, SOME_PUMP_CONTROL_FAIL, LEVEL_OK, LEVEL_FAIL, LEVEL_FAIL_NOT_ACK, STEAM_OK, STEAM_FAIL, STEAM_FAIL_NOT_ACK. (either true or nil)
- The *Analyser* converts numerical values of STEAM and LEVEL into the following internal propositions:
  - STEAM_ZERO, STEAM_NOT_ZERO, LEVEL_N1 ($level < N_1$), LEVEL_N2 ($N_2 < level$), LEVEL_M1 ($level < M_1$), LEVEL_M2 ($level > M_2$), LEVEL_N12 ($N_1 < level < N_2$). (either true or nil)
- The *Analyser* generates the following propositions on the basis of the information obtained from the *SysModel*. These propositions are used in the rescue mode of operation [AS] (see Chapter AS, this book):
  - LEVEL_CASE_12 (case 1 and 2),
  - LEVEL_CASE_3 (case 3: shutdown decision),
  - LEVEL_CASE_4 (case 4: do nothing),
  - LEVEL_CASE_56 (case 5 and 6). (either true or nil)
- The *Actions* class also informs other parts of the *Controller* about opening or closing pumps by means of messages which are sent to the pumps:

- OPEN_PUMPS. (array of true or false)
- The *SysModel* imitates the behaviour of pumps, and generates PUMP_STARTED, which indicates those pumps which have been effectively started.

4.2 Analyser

The *Analyser* filters all the numerical values giving them a logical meaning by expanding them into a set of propositions (this can be done directly by following the specification text), which can be used internally for synthesis of the behaviour of the system. The *Analyser* detects failures in the physical units. This is a direct implementation of Sect. 7 of [AS] (see Chapter AS, this book). The *Analyser* detects erroneous behaviour of physical units by observing the messages sent to the program and by saving and comparing them with previous states of the system. The failure detection messages are directly passed to the *Actions*.

Below, we see the class specification of the *Analyser*. Most of the relations are easy to follow from the text. After declaring the internal propositions needed by the different parts of the *Controller*, it is easy to follow the specifications text and specify the relations. It is also possible to use some other form of refined specifications, and implement them in the same manner. Here we present the complete specification for analysing the water level and water-level-measuring unit, steam level and steam-level-measuring unit and also the external STOP message causing the program to enter emergency stop. Note that the text written in Italics is directly copied from the control specifications given in [AS] (see Chapter AS, this book), our comments are written in the brackets.

```
Class Analyser
 super MsgIn; %MsgIn contains all the propositions
 var
    stop_counter, last_stop_time, current_time, Dt: num;
 rel
```

- Analysing water level and water-level-measuring unit:
    - *If the program detects that the unit indicates a value which is out of the valid static limits - i.e. between 0 and C or the program detects [by using the SysModel] that the unit indicates a value which is incompatible with the dynamics of the system...*
      ```
      LEVEL , LEVEL_OK -> LEVEL_FAIL, LEVEL_FAIL_NOT_ACK
         {if LEVEL<0 | LEVEL>BoilerPar.C | ~SysModel.LEVEL_OK() ->
              LEVEL_FAIL:=true; LEVEL_OK:=nil;
              LEVEL_FAIL_NOT_ACK:=true fi};
      ```
    - *To indicate to the physical units that the program has detected a failure of the physical unit which measures the outcome of water, LEVEL_FAILURE_DETECTION is sent*
      ```
      LEVEL_FAIL, LEVEL_FAIL_NOT_ACK ->
              {LEVEL_FAILURE_DETECTION()};
      ```
    - (until receipt of the corresponding acknowledgment).
      ```
      LEVEL_FAIL, LEVEL_FAILURE_ACKNOWLEDGEMENT
              -> LEVEL_FAIL_NOT_ACK {LEVEL_FAIL_NOT_ACK:=nil};
      ```

- *LEVEL_REPAIRED indicates that the unit, which measures the outcome of steam, has been repaired. It is sent by the physical units until a corresponding acknowledgment message has been sent by the program.*
  ```
  LEVEL_REPAIRED, LEVEL_FAIL ->  LEVEL_OK
    {LEVEL_REPAIRED_ACKNOWLEDGEMENT();
     LEVEL_FAIL := nil; LEVEL_OK := true};
  ```
  The following internal propositions are simply generated by analysing the water level in the case of correct operation of devices:
  ```
  LEVEL , LEVEL_OK
      -> LEVEL_N1, LEVEL_N2, LEVEL_M1, LEVEL_M2, LEVEL_N1N2
   {if LEVEL < BoilerPar.M1 -> LEVEL_M1:=true
    || LEVEL > BoilerPar.M2 -> LEVEL_M2:=true fi;
    if LEVEL < BoilerPar.N1 -> LEVEL_N1:=true
    || LEVEL > BoilerPar.N2 -> LEVEL_N2:=true
    || LEVEL > 0 & LEVEL<BoilerPar.C -> LEVEL_N12:=true fi};
  ```
  and if the water level measuring unit is defective, then the water level is estimated by the *SysModel* and the following four cases are decided (used in the rescue mode as written in [AS] (see Chapter AS, this book):
  ```
  LEVEL_FAIL, SysModel.qa1, SysModel.qa2 ->
    LEVEL_CASE_12, LEVEL_CASE_3, LEVEL_CASE_4, LEVEL_CASE_56
   {if (SysModel.qa1 < BoilerPar.N1 &
        SysModel.qa2 < BoilerPar.N2) -> LEVEL_CASE_12:=true
    || (SysModel.qa1 < BoilerPar.N1 &
        SysModel.qa2 > BoilerPar.N2 &
        (SysModel.qa1 <= BoilerPar.M1 |
         SysModel.qa2 >= BoilerPar.M2 |
         SysModel.qc1 <= BoilerPar.M1 |
         SysModel.qc2 >= BoilerPar.M2)) -> LEVEL_CASE_3:=true
    || (SysModel.qa1 > BoilerPar.N1 &
        SysModel.qa2 > BoilerPar.N2)  -> LEVEL_CASE_56:=true
    ||      true     -> LEVEL_CASE_4 :=true fi};
  ```
- Analysing steam level and steam-level-measuring-unit:
  - *If the program detects that the unit indicates a value which is out of the valid static limits - i.e. between 0 and W or the program detects [by using the SysModel] that the unit indicates a value which is incompatible with the dynamics of the system...*
    ```
    STEAM, STEAM_OK ->  STEAM_FAIL, STEAM_FAIL_NOT_ACK
     {if STEAM < 0 | STEAM >BoilerPar.W | ~SysModel.STEAM_OK()
         -> STEAM_FAIL:=true; STEAM_OK:=nil;
            STEAM_FAIL_NOT_ACK:=true fi};
    ```
  - *To indicate to the physical units that the program has detected a failure of the physical unit which measures the outcome of steam, STEAM_FAILURE_DETECTION is sent*
    ```
    STEAM_FAIL, STEAM_FAIL_NOT_ACK ->{STEAM_FAILURE_DETECTION()};
    ```
  - *(until receipt of the corresponding acknowledgment).*
    ```
    STEAM_FAIL, STEAM_OUTCOME_FAILURE_ACKNOWLEDGEMENT
           -> STEAM_FAIL_NOT_ACK{ STEAM_FAIL_NOT_ACK:=nil };
    ```

- *STEAM_REPAIRED indicates that the unit which measures the outcome of steam has been repaired. It is sent by the physical units until a corresponding acknowledgment message has been sent by the program.*
  ```
  STEAM_REPAIRED, STEAM_FAIL -> STEAM_OK
     {STEAM_REPAIRED_ACKNOWLEDGEMENT(); STEAM_FAIL := nil;
      STEAM_OK := true};
  ```
  The internal propositions STEAM_ZERO and STEAM_NOT_ZERO are simply generated from the following:
  ```
  STEAM, STEAM_OK -> STEAM_NOT_ZERO, STEAM_ZERO
     {if STEAM==0 -> STEAM_ZERO:=true
     || STEAM >0  -> STEAM_NOT_ZERO:=true fi};
  ```
- Analysing the STOP message from physical units:
  - *When the STOP message has been received three times in a row by the program, the program must go into emergency stop.*
  ```
  STOP, current_time, last_stop_time, stop_counter, Dt
                                              -> STOP_DECISION
     {if current_time-last_stop_time==Dt & stop_counter==3
         -> STOP_DECISION:=true
     ||  current_time-last_stop_time==Dt
         -> stop_counter:= stop_counter + 1
     ||  true
         -> last_stop_time:=current_time; stop_counter:= 1 fi}
  ```

4.3 Rules

The overall behaviour of the control program, more specifically, the interpretation of Sect. 4 of [AS] (see Chapter AS, this book), is specified by a collection of rules. These rules are written directly in NUT, and are fired by performing the *compute* command repeatedly until a special rule fires notifying that no other rule can be fired. The next mode of the operation and the necessary actions to be taken are decided by firing these rules. Each rule has the form:

$$P_1 \& P_2 \& ... \& P_n \rightarrow mode\ \{< actions >\};$$

where P_1, P_2,..., P_n are preconditions, i.e. propositions which must have a value different from *nil* for firing the rule. Rules are fired by automatically testing the preconditions in each axiom using the automatic synthesis feature of the NUT system. Encoding a condition C given in requirements, we represent the condition and its negation by two separate propositions: "C is *true*" and "C is *false*". This, together with the logic of the NUT synthesizer, provides a constructive logic for our rule-based system.

Actions of rules affect the steam-boiler state by sending messages to the system, and the *Controller* may enter a new operational mode after firing a rule. The set of rules given in this section has been optimized, in such a way that a set of general rules is fired first and, thereafter, each transition between two states of the system has been specified by a single rule. Since in this approach, only one rule is expected to be fired after the general rules, any unexpected transition can be easily detected. The control part of the *Rules* class is not discussed here, and the focus will be on the rules themselves.

Preconditions of rules are also set by incoming messages from the boiler after being filtered by the *Analyser*. Actions are internal messages between the *Rules* and the *Actions*. These messages have been abstracted, and a message may require further analysis by the *Actions*. For instance, closing a pump or opening the valve is represented by one and the same internal message DECREASE_WATER which leads to the reduction of the water level. The *Actions* then takes the proper decision on the basis of the state of the pumps and of the valve. This abstraction of both incoming (from the *Analyser*) and outgoing (to the *Actions*) messages separates the rules from the physical parts of the system and makes the control more flexible. The only state saved in the *Rules* is the operational mode of the *Controller*. States of the devices are saved in the *SysModel*, *Analyser* and *Actions*.

The translation of input messages is performed separately from the *Rules* (by converting each measured value into a set of propositions) which makes the control more flexible and easily changeable in order to be able to respond to changes in the system and to further refinements. Separating outgoing messages from the Rules plays the same role in the output interface. In the following, we see that Sect. 4 of [AS] (see Chapter AS, this book) has been implemented by a set of rules directly taken from the requirements specification text and has been written almost sentence by sentence from the text. One can follow the axioms and compare them directly with the requirements specification. Note that the text written in Italics is directly copied from the requirements specification, our comments are written in the brackets.

```
Class Rules
 super Mode;
 rel
```

- General rules:
    - *The program sends, at each cycle, its current mode of operation to the physical units.*
      `true -> Mode {MODE(CurrentMode())}`
    - *If the transmission line fails go to emergency mode!*
      `TRANS_FAIL -> Mode {ChMode('EMER')};`
    - *If the program receives the stop message three times [counted by the Analyser] go to emergency mode!*
      `STOP_DECISION -> Mode {ChMode('EMER')};`
- Initialization mode:
    - *As soon as message STEAM_BOILER_WAITING has been received, the program checks whether the quantity of steam coming out of the steam-boiler is really zero; if not, the program enters the emergency stop mode.*
      `INIT, STEAM_NOT_ZERO, STEAM_BOILER_WAITING -> Mode`
      `    {ChMode('EMER')};`
    - *If the quantity of water in the steam-boiler is above N2 the program activates the valve of the steam-boiler in order to empty it.*
      `INIT, LEVEL_N2 -> Mode {DECREASE_LEVEL('INIT')};`

- *If the quantity of water in the steam-boiler is below $N1$ then the program activates a pump to fill the steam-boiler.*
  ```
  INIT, LEVEL_N1 -> Mode {INCREASE_LEVEL('INIT')};
  ```
- *As soon as a level of water between $N1$ and $N2$ has been reached the program can send continuously the signal PROGRAM_READY to the physical units, until it receives the signal PHYSICAL_UNITS_READY.*
  ```
  INIT, LEVEL_N1N2 -> Mode {PROGRAM_READY()};
  ```
- *As soon as the signal PHYSICAL_UNITS_READY has been received the program enters either the mode normal if all the physical units operate correctly,*
  ```
  INIT, LEVEL_N1N2, ALL_OK, PHYSICAL_UNITS_READY -> Mode
        {ChMode('NORM')};
  ```
- *or the mode degraded if any physical unit is defective.*
  ```
  INIT, LEVEL_N1N2, NOT_ALL_OK, PHYSICAL_UNITS_READY -> Mode
        {ChMode('DEGR')};
  ```

Rules of other modes are presented in appendix B (see CD-ROM Annex AT.B).

4.4 Actions

The decisions made in the *Rules* are not concrete messages to the physical units, but abstracted ones which must be translated into concrete messages taking into account the state of the steam-boiler. The *Rules* trigger the *Actions*, which in turn sends the proper messages to the steam-boiler. The message DECREASE_WATER_LEVEL has the following specification:

```
DECREASE_WATER_LEVEL: mode, OPEN_PUMPS ->
  {if mode == 'INIT' -> MsgSwitch.VALVE()
  || true -> for i to 4 do
                 if OPEN_PUMPS[i] ->
                     MsgSwitch.CLOSE_PUMP(i);OPEN_PUMPS[i] := false;
                 exit fi od fi};
```

All other messages like PROGRAM_READY, LEVEL_FAILURE_DETECTION have very simple specifications, containing only a message-sending statement:

```
PROGRAM_READY: -> { MsgSwitch.PROGRAM_READY() };
LEVEL_FAILURE_DETECTION: ->
       { MsgSwitch.LEVEL_FAILURE_DETECTION() };
```

4.5 Main Controller Program

The *Controller* has a method RUN which is executed in every cycle and is the main controller program:

```
RUN:  ->{ SysModelUpDate();   % Update the state of the
                              % modelled system (in the SysMod),
          AnalyseMessages();  % Analyse the incoming messages
                              % using the programs in the Analyser,
          FireRules() }       % Fire rules (in the class Rules).
```

The three programs called in the body of RUN are simple relations written in classes *Analyser*, *Rules* and *Actions*.

5 Specifications of Physical Behaviour

5.1 Model of the Steam-Boiler System

The *SysModel* simulates the behaviour of the steam-boiler system as a whole, partly on the basis of the information delivered by the *Analyser* and partly by reflecting the actions taken by the *Actions*. It compares the measured steam flow and water level with their predicted values to detect the failure of level and steam measuring units. It delivers the adjusted values to the *Analyser* which considers the decisions needed in the rescue mode on the failure of level measuring unit. The *SysModel* is designed visually by using the scheme editor of the NUT system as shown in Fig. 3. It has the following specification (generated automatically by the scheme editor):

```
class SysModel
super ModelComp;
% scheme begin
var
    PumpMod1: PumpMod;
    PumpMod2: PumpMod;
    PumpMod3: PumpMod;
    PumpMod4: PumpMod;
    PumpControlMod5: PumpControlMod;
    PumpControlMod6: PumpControlMod;
    PumpControlMod7: PumpControlMod;
    PumpControlMod8: PumpControlMod;
    LevelMeterMod9: LevelMeterMod;
    SteamMeterMod10: SteamMeterMod;
    Valve11: Valve;
    BoilerMod12: BoilerMod;
rel
    PumpMod1 = BoilerMod12.P1;
    PumpMod2 = BoilerMod12.P2;
    PumpMod3 = BoilerMod12.P3;
    PumpMod4 = BoilerMod12.P4;
    PumpMod1.control = PumpControlMod5;
    PumpMod2.control = PumpControlMod6;
    PumpMod3.control = PumpControlMod7;
    PumpMod4.control = PumpControlMod8;
    Valve11 = BoilerMod12.Valve;
    SteamMeterMod10 = BoilerMod12.S;
    LevelMeterMod9  = BoilerMod12.L;
% scheme end
```

Equalities in the text show bindings of devices. The equality *PumpMod1=BoilerMod12.P1* shows that the slot *P1* for a pump in boiler *BoilerMod12* is filled by *PumpMod1*, etc. The presented specification is complete in the sense that a program for simulating is synthesized automatically from this specification when executing the relation

```
compute_next: BoilerMod12 ->
   {BoilerMod12.old_adjusted:=adjusted;
    BoilerMod12.adjusted:=nil;
    BoilerMod12.compute(adjusted)}
```

given in the class *ModelComp*. Now a message *SysModel.compute_next()* will cause the synthesis of the program, which consequently computes all new values.

5.2 Classes of Physical Units

Two different classes have been developed for each physical unit. One is used for imitating the devices, and in particular, for imitating their failures. The other set of classes is developed to simulate the behaviour of the devices on the basis of informations filtered by the *Analyser*. These classes are used in the *SysModel* for keeping track of the actual state and the dynamic of the system and for predicting the state values in the case of failure in measuring units. Classes of both sets inherit a class *Device*, as shown in Fig. 5.

```
class Device
   var
      OK: bool;       % true, if operating correctly
      name: text;     % name of a device
      mode: text;     % operation mode
      time: num;      % Current time
      Dt: num;        % time step
   init
      Dt := 5;
      OK := true;
```

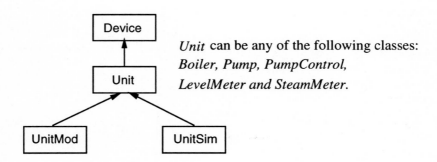

Unit can be any of the following classes: *Boiler, Pump, PumpControl, LevelMeter* and *SteamMeter*.

Fig. 5. Classes of physical units inherit from Device.

Boiler models. As we can see in Fig. 2, we need two different specifications of the boiler, one for the system simulator (*SysSim*) and another for the system model (*SysModel*) used in the control program. The former represents only the purely physical behaviour of the boiler. The latter also contains relations for making predictions about the behaviour of the boiler in the case of uncertainty about the correctness of signals.

Boiler parameters. Both boiler models inherit specifications of variables representing boiler parameters according to [AS] (see Chapter AS, this book). The class *BoilerPar* specifies these parameters:

```
class BoilerPar
  var
    C: num;        % max capacity in liters,
    M1, M2: num;   % min and max limits in liters,
    N1, N2: num;   % min and max normal in liter/sec*sec,
    U1, U2: num;   % max gradients of increase and decrease of
                   % steam flow in liter/sec,
    W: num;        % max quantity ,
    P: num;        % nominal capacity .
```

Bioler. The class *Boiler* shows the configuration of the boiler environment, including a valve and four pumps (being either of class *PumpMod* or *PumpSim*). Pumps and valve are represented in both boiler models, because values of their parameters are needed in equations:

```
class Boiler
  super Device; BoilerPar;
  var
    valve: Valve;
    P1, P2, P3, P4: any;
    p, q, v: num;
  alias
    Pumps = (P1,P2,P3,P4);
```

Boiler model. Here begins a specification of the class *BoilerMod* which predicts the dynamics of the boiler:

```
super
    Boiler;
```

The following is a description of variables as listed in the requirements specification:

```
var
  qa1,qa2: num; % min and max adjusted water quantities
  qc1,qc2: num  % min and max calculated water quantities
  pc1,pc2: num; % min and max calculated pump throughputs
  va1,va2: num; % min and max adjusted steam flows
  vc1,vc2: num; % min and max calculated steam flows
```

The classes LevelMeterMod and SteamMeterMod are needed to obtain the new values:

```
    L: LevelMeterMod;
    S: SteamMeterMod;
```

New names for denoting tuples of values (objects) used in computations: raw, calculated, adjusted, newval, Pumps are introduced by means of aliasing:

```
alias
  raw=(q,q,v,v,p,p);
  calculated=(qc1,qc2,vc1,vc2,pc1,pc2);
  adjusted=(qa1,qa2,va1,va2,pa1,pa2);
  old_adjusted=(qo1,qo2,vo1,vo2,po1,po2);
```

Raw, adjusted and calculated values are specified as follows:

```
rel
   v = S.v; %get the steam value from the steam meter!
   q = L.q; %get the water level from the level meter!
   p = P1.p+P2.p+P3.p+P4.p-valve.v; % get the throughputs of the
                                   % pumps, valve is also considered!
   adjust: raw, calculated, L.OK, S.OK-> adjusted
     {adjusted:=raw;
      if ~L.OK -> adjusted[1]:=calculated[1];
                  adjusted[2]:=calculated[2]
      || ~S.OK -> adjusted[3]:=calculated[3];
                  adjusted[4]:=calculated[4]
      fi;         calculated:=nil};
% new calculated values:
   throughputs: Pumps -> pc1, pc2
      {pc1:=0; pc2:=0;
       for i to 4 do
          pc1:=pc1+Pumps[i].pc1; pc2:=pc2+Pumps[i].pc2;
       od};
```

Equations are written precisely as in the requirements specification:

```
   qc1 = qo1-vo2*Dt-0.5*U1*Dt^2+po1*Dt;
   qc2 = qo2-vo1*Dt+0.5*U2*Dt^2+po2*Dt;
   vc1 = vo1-U2*Dt;
   vc2 = vo2+U1*Dt;
```

This is a complete specification of the boiler as a physical device. Classes of other devices are presented in the Appendix C (see CD-ROM Annex AT.C).

6 Evaluation and Comparison

1. We have shown by the present work the feasibility of using a declarative language as a formal tool for automated implementation of requirements taken directly from a sufficiently complete informal specification.
2. (a) Our solution comprises an implementation of a control program.
   (b) It is not linked to FZI simulator, but we have developed our own simulator.

(c) The first version of the control system was demonstrated already at the Dagstuhl seminar on the boiler control problem in June 1995. The present version differs from the initial version in the strategy of usage of rules as well as in a number of details of specifications.
3. We have compared the TLT [CW1] (see Chapter CW1, this book) and Z [BW] (see Chapter BW, this book) notations to our specification language and are quite convinced that our synthesis can be used in implementing an essential part of specifications written in TLT or Z. Our intention is to continue this experiment by implementing requirements specifications written in some formal notation, e.g. TLT, Z.
4. (a) The development of the program for the physical parts took 1 person month, and for the control part it took 2 p.m.
   (b) It may take 4 weeks to become sufficiently expert in using the specification framework.
5. (a) Detailed knowledge of the specification language is required.
   (b) Assuming that the average programmer knows a bit about logic and object-orientation, she can understand the solution.
   (c) It takes 1 week to learn what is needed to be able to understand the solution.

7 Conclusion

In the present implementation, we used formal methods for program synthesis, and did so completely automatically. This was possible due to heavy restrictions put on the formal language used as a logical specification language. The SSP method is not new, therefore it was not explained in details. It was important that we had a convenient language for specifications written by hand, and a translator for transforming the specifications into the restricted logical language of SSP. The usage of this formal but user-friendly specification language was demonstrated thoroughly in this paper.

The overall design of the system was done in the object-oriented style, by specifying classes for each physical device and putting them together into a specification of the whole system. This was facilitated by tools for visual specification of structure of systems. We used a formal graphical language for specifying the steam-boiler system. The semantics of elements of this language was described by classes of devices. The semantics of the whole graphical specification was essentially dependent on the usage of the program synthesizer which produced all required control and simulation programs.

The specification of the control algorithm was written in the form of rules in a style well-known in knowledge-based software technology. Developing the rule-based part, we tested two strategies of rule-based programming, first, firing rules one by one, and, second, allowing all acceptable rules to fire at the same step. The first strategy was finally chosen and implemented as the more reliable one, which also allowed us to write rules in a form very close to the initial requirements specification.

8 Acknowledgements

We wish to express our thanks to the colleagues who have helped us to bring the NUT system to the state where it could be applied as a tool for solving the steam-boiler control problem. We express our thanks to Tarmo Uustalu, Jaan Penjam, Mait Harf and Rando Valt. This work has been supported by the Swedish National Board for Industrial and Technical Development (NUTEK) under grant number 9303405-2.

References

[AS] Abrial, J. -R.: Steam-Boiler Control Specification Problem. In this book.

[BW] Buessow, R., Weber, M.: A Steam-Boiler Control Specification with Statecharts and Z. In this book.

[CW1] Cuellar, J., Wildgruber, I.: The Steam Boiler Problem - A TLT Solution. In this book.

[MT] Mints, G., Tyugu, E.: Justification of the structural synthesis of programs. Science of Computer Programming **2 3** (1982) 215–240

[Pen] Penjam, J.: Attributed automata: A formal model for protocol specification. CSLab, Teleinformatics/KTH, TRITA-IT-R **94:30** (1994)

[Smi] Smith, D. R.: KIDS - a semi-automatic program development system. IEEE Transactions on Software Engineering. Special Issue on Formal Methods in Software Engineering **16 9** (1990) 1024–1043

[Tyu] Tyugu, E.: Using classes as specifications for automatic construction of programs in the NUT system. Journal of Automated Software Engineering **1** (1994) 315–334

[Uus] Uustalu, T.: Aspects of Structural Synthesis of Programs. CSLab, Teleinformatics/KTH, Licentiate Thesis TRITA-IT-R **95:09** (1995)

Using FOCUS, LUSTRE and Probability Theory for the Design of a Reliable Control Program

Christoph Andriessens and Thomas Lindner

Forschungszentrum Informatik (FZI)

Abstract. This paper[1] describes a combined formal approach to the Steam Boiler Problem. We show how the problem of specification correctness can be addressed using a careful initial phase which we call system identification. In this phase we use a rigorous refinement approach and apply the formal method Focus to decompose the task of implementing and verifying the control program into three components: First, a Communicator module is identified and designed using the synchronous language Lustre. The correctness of this module, which deals with the message exchange with the console of the system, is proved using symbolic model checking. The second module is the Failure Manager which deals with the detection of equipment failures and the estimation of the water level. We argue that a deterministic model makes no sense for this module and present a probabilistic approach based on Probability Theory. We calculate probablilities for the correctness of the failure detection and thereby quantify its reliability. The third module is the Pump Controller, which must be treated with the same methods as the Failure Manager.

1 Problem and Goal

The steam boiler problem [Abr95b] (see CD-ROM AS.) is presented in Chapter 1 of this volume. It is a control specification problem for a non-reliable piece of hardware which requires a fault-tolerant control program.

Based on our experiences with the Case Study Production Cell [LL94], where plenty of verified control programs failed to control the system correctly, our main objective in this study was to use the most appropriate notation and method at each point during the formal development, and to validate the models as carefully as possible. Emphasis has been put on the early phases of development, where a novel phase called *system identification* was introduced to engineer the requirements and select appropriate models and notations as carefully as possible.

The second objective was to use an approach which was not biased towards one favourite method right from the beginning, and instead profits from the experience from the aforementioned Case Study Production Cell by selecting the method at each step that to the best of our knowledge is the most appropriate.

---

[1] For an extended version please contact the authors: {andriess,lindner}@fzi.de .

Our approach was therefore to first determine the purpose of the formal method — what *must* be proved — and then select the corresponding formalism, rather than taking a formalism and applying it — investigating what *can* be proved.

The rest of the paper is organized as follows: In Section 2 we describe our general approach and introduce the formalisms which were used in the solution. Section 3 describes our ideas about system identification and explains them using the formal method FOCUS. According to our strategy, this is the most elaborate section. Section 4 describes how the first identified component, the *Communicator* module has been implemented and verified using the synchronous programming language LUSTRE. In Section 5, our approach to the second component identified, the *Failure Manager*, is explained. As no deterministic properties can be proved, we use a stochastic model based on Probability Theory. Section 6 concludes with an evaluation.

2 Approach

2.1 System Identification

Formal methods can be characterized as the application of mathematical modelling for the systematic support of the construction of software. The application of mathematical models implies the use of abstraction. Details of the real world phenomenon are neglected to build the model. The abstract, mathematical model must be suitable for its purpose, i.e. it has to model the relevant phenomena adequately (in order to predict their influence on the properties under analysis correctly) and has to abstract away from irrelevant details (in order to make the model managable).

The selection and construction of a suitable model is a difficult task. We believe that in most cases the best solution is achieved by the combination of different models for different aspects or components of the system. Our experiences with the case study "Production Cell" substantiates that this claim can even be true for medium-sized examples. We call the decomposition of the system under study and the selection of appropriate models for different components *system identification*, as it is called in control theory.

In our understanding, system identification is

- the identification of different system components and their interfaces, as well as
- the classification of these components according to the classes of mathematical models which are suitable for the modelling of the single components, according to the purpose which is pursued by this modelling.

The notion of *purpose* in the above definition can have multiple meanings. One of the purposes which is typical for formal methods is the proof of properties, or, more generally, the analysis of the behavior of the system under consideration. We can analyze safety or liveness properties, we can analyze whether an algorithm computes the function it is supposed to compute, or we can look at the mean time between failures of a controlled physical process. For each of these analyses, a different mathematical model will be used; we therefore have to make ourselves clear about the purpose of the models of our system during system identification.

The result of the system identification phase is an initial model of the system. In our approach,

- this initial model is described in a language with formal semantics (to allow for early analyses and consistency checks),
- the initial model is developed using stepwise refinement (to allow for the incremental addition of knowledge), but the refinement steps are in general not formally verified (as verification would not mean very much in this early phase of requirements engineering), and
- system identification prefers to aim to automatically analyzable types of mathematical models, for instance finite state models which can be verified mechanically by model checking.

We have decided to use the formal method FOCUS for system identification. We did this for the following reasons:

- FOCUS supports the stepwise refinement approach.
- The proof rules of FOCUS give excellent support for architectural design. There are a number of rules which incorporate knowledge about system architecture, for instance there are rules for refining a component into two parallel, sequential, or two components cooperating in a master/slave style.
- FOCUS allows to describe the architectural aspects of the specification graphically. This makes the writing and understanding of specification much easier.
- FOCUS allows the developer to think data flow oriented. On a first glance of the system, this seemed to be a helpful paradigm to come up with a first description of the system.

2.2 FOCUS

FOCUS [MBW] is a method for the development of distributed systems. The method is divided into the following three phases: In the *Requirements Specification Phase* the requirements are formalized using streams and stream processing functions. In the *Design Specification*, which is written in the second phase, the specification is refined until it is constructive enough to be easily implemented. This implementation is done in the *Implementation Phase*.

FOCUS' main structuring feature are so-called *agents*. Systems are described by concurrently operating and asynchronously communicating agents. An agent has a finite number of input channels. It receives messages via a buffered, asynchronous channel. The agent then computes a finite number of messages which it sends via output channels.

The basic data type of FOCUS are *streams*. The set of all streams over a set M is defined as the set of all finite and infinite sequences over the set M. The standard operations on sequences are also available for streams: concatenation, head, front, tail and last, length, prefix relation, and a filter operation @, where for a subset A of M and a stream s the stream $A@s$ consists of the stream which remains after deleting all elements of A in s (Example: $\{a,b\}@\langle a,b,c,a,b,c,...\rangle = \langle a,b,a,b,...\rangle$).

Agents can be specified in various ways by *stream processing functions* (continuous functions on streams). We are particularly interested in the assumption-commitment style developed by Ketil Stolen. Here, the agent is specified by using two predicates over stream processing functions. The *assumptions* which are made by an agent S on its inputs are denoted by A_S, and the commitments which S guarantees if the assumptions are fulfilled, are denoted by C_S.

In [KSW], a proof calculus for FOCUS specifications written in the assumption-commitment style is presented. There are refinement rules both for architectural and behavioral refinement.

2.3 LUSTRE

LUSTRE [Hal93] is a declarative language for programming synchronous reactive systems. It is based on the perfect synchronization hypothesis, which assumes that the execution time of the reactive program can be neglected. Therefore, inputs and outputs are assumed to be synchronous.

LUSTRE programs are structured along nodes. A simple node is shown in Figure 3. There, two boolean input signals *low* and *high* exist, and two boolean output signals, called *on* and *off*. The output *on* is set true iff *low* is true and has not been true in the last instance, and *off* is set iff *high* is true and has not been true in the last instance. Both *high* and *low* are initially false (in the first instance of time, the previous value of *high* and *low* does not exist).

LUSTRE programs can be compiled to finite automata. There is a symbolic model checker available [Hal93] which can check safety properties written in a very simple temporal logic directly built into LUSTRE.

3 System Identification Using FOCUS

In the system identifications phase we strive for finding the most appropriate models and modelling methods. We construct an initial model of the system by incrementally adding detail in a stepwise refinement fashion. The purpose of this activity is to avoid specification errors or inappropriate models. We have chosen the FOCUS notation for this phase; the reasons for this decision have been given above.

It would take far too much time to present the whole FOCUS development. For reasons of brevity we skip all formulas and only present the architectural view of the resulting initial system model (cf. the Appendix for more details). We stress that in this phase our emphasis is in exploring the system and not formally verifying refinement relations. The proof calculus of FOCUS would give a developer the possiblity to do formal proofs and thereby gain confidence about difficult refinement steps, but at this stage we have not enough confidence in our abstract specifications to take them as references for more detailed descriptions.

Our primary goal with this section is to show that system identification is well suitable to link the informal to the formal and therefore helps to avoid specification errors. We also want to show that the proof rules of FOCUS are useful as a set of design rules which can be exploited when formalizing the system architecture of a given system. Finally, we want to reason that the components we have identified in the system identification are useful units and that the models and languages we have assigned make sense.

In a first step, the system can be described as an agent which consumes water and produces steam. We call this *System 0* (cf. Fig. 1).

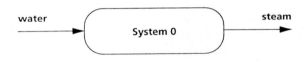

Figure 1. System 0

In this paper, we do not deal with the obvious question how to match the asynchronous communication of FOCUS with the discrete time model of the steam boiler problem, where the system can only be observed every 5 seconds.

As explained in Section 2, system identification is about finding models and their purpose. Therefore, we note that at this stage an obvious observation about the purpose of the system is that it is supposed to produce as much steam as possible. We know that we have to design a control program for the steam boiler. Is it a requirement for the program that it controls the boiler in a way that the latter produces as much steam as possible?

At this point, we cannot give a final answer to this question, but it remains a good candidate for both a requirement on some component (or even the whole system) and will certainly have to be considered when the choice of the mathematical model has been decided. At this time, we further refine System 0. Our intention is to stepwise factor out the environment of the system, because our main interest is in the control software. Obviously, there is something like a console which communicates with the rest of the system, called System 1. The structure of the communication between System 1 and the Console is best described by the master/slave architecture of the FOCUS proof system (cf. Fig. 2).

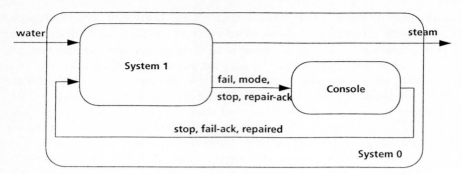

Figure 2. System 1

This informal refinement process is continued until we arrive at a final system model.

Figure 3 shows the final model (in its graphical representation) of the system after the system identification phase. The solid line shows the software boundaries. The arrows crossing the boundary from inside to outside show the actuator commands, the arrows crossing the boundary from outside to inside show the sensor values. The model is sound in the sense that each element of the interface description of the task definition can be found (except the initialization handshaking, see above).

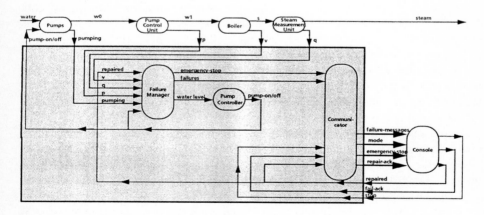

Figure 3. System chart

We have identified three major blocks:

- The *Communicator*. The Communicator can be modelled using a finite state model. We have decided to use the synchronous programming language LUSTRE and its verification environment for its implementation. Key properties

are safety properties which basically state protocol conformance. The design of the Communicator will be described in the next section.
- The *Failure Manager*. We found out that only a probabilistic model gives enough power to analyze interesting properties of the Failure Manager, as deterministic properties are very weak. We do not know of any powerful framework for such problems, and have decided to use basic mathematical stochastic. Careful analysis (cf. Section 5) shows that the physical process under consideration allows the application of Probability Theory. We use this model to design a failure detection algorithm which will then be straightforward to implement. The most important requirement about this failure detection will be its reliability. As this is even mathematically a quite challenging problem, we have decided to provide the Failure Manager with a memory of just length one, i.e. it takes the current and the previous value for its failure detection. The other important requirement is the quality of the water level prediction.
- The *Pump Controller*. For the pump controller, a similar model would be useful. The basic strategy is straightforward: use some constant values, for instance N_1 and N_2 to switch the pumps on or off.

4 Implementing and Verifying the Communicator Module Using LUSTRE

For brevity our implementation works only for one pump and one pump controller, but the extension is straightforward.

The implementation was very easy and finished within half a day. The implementation is 62 lines long.

It was not quite as easy to formulate the requirements for the correctness of the implementation. The simple temporal logic built in into LUSTRE made it very difficult to write down the requirements. It is quite easy to express some of the properties as Symbolic Timing Diagrams [SD93] (SDT), but there is no automatic translator available which would transform SDT or a subset of them into LUSTRE's temporal logic.

After about another half a day, 80 lines of properties had been written and verified. For this verification the symbolic model checker LESAR was used which is available for LUSTRE. Its longest run on a SUN Sparc 10/20 was about 5 minutes with a maximum BDD size of 85,000. There were 6 errors detected, five of them in the specification, and one in the implementation.

LUSTRE proved to be a very useful tool for the implementation. We are sure that there are not very many languages which allow such a concise description of the program. The implementation is very easy to read and write. For verification purposes, LUSTRE's logic is not expressive enough, but its simplicity allows very efficient model checking even with public domain tools.

The implementation of the Communicator was then used for a module test with the LUSTRE simulator XSIMLUS. There were no further errors detected.

5 The Failure Manager

The task of the Failure Manager is the diagnostics of broken components using the messages sent by the other devices. In Section 3 we have decided to use a stochastic model for the design and the validation of the Failure Manager. We now explain this model and the design of the Failure Manager in detail.

After presenting our model, we will show the following results:

- We will prove that the Failure Manager detects the failure of exactly one component, if the probabilities of failure satisfy certain conditions. If not, certain failures cannot be detected.
- More generally, we will present a formula which allows to compute the probability that the diagnostics of the Failure Manager is correct at a given time t depending on the history of the system. This computation can only be done for a specific history.

The first characteristic helps us to evaluate our design of the Failure Manager. Ideally, we would like to show that our Failure Manager is designed such that all failures are detected in all circumstances, i.e. we have implemented an optimal Failure Manager. Unfortunately, this is not feasible.

The second characteristic — the probability of a correct decision of the failure manager at instant t — can be used during the runtime of the system. If it is displayed on the operator console, the operator can decide at which time maintenance should be called and have all devices checked. Afterwards, the Failure Manager can be reset with definite information about the status of the system.

In the System Identification Phase we have decided to design a Failure Manager with a memory of length one, i.e. the failure detection algorithm will only use the sensor values at the current and the previous cycle. This is a deliberate design decision which we take for reasons of simplicity. It is possible to generalize our approach to larger memories, but the stochastic analysis will become increasingly complex.

The rest of this chapter is organized as follows: In 5.1, we present our stochastic model. In 5.2 we introduce the concept of Indicators. These are numbers which are computed from the sensor and actuator values of the current and the previous instant of time, and are subsequently used for the detection of failures. The latter is presented in 5.3. In the last subsection, 5.4, we use the stochastic model to do the aforementioned analysis.

5.1 The Stochastic Model

In order to develop a stochastic model describing the behaviour of the steam boiler, we have to assume that we have information about the reliability of the devices. This is an assumption which does not appear in the task description. We find it reasonable, as in a safety critical environment, it is common to know the mean time between failure of the physical devices. In the following we will assume that such mean times are known.

In this section the i-th pump is identified by P_i, the i-th pump monitor by PM_i, the water level measurement device by WL and the steam output device by SO so that the set $\{P_1, P_2, P_3, P_4, PM_1, PM_2, PM_3, PM_4, WL, SO\}$ is the set of all devices. Additionally, we write 6 for PM_2 and 10 for SO, for instance, to iterate on this set. For reasons of simplicity we write $P_1, \ldots, SO$ to enumerate all devices.

A Model for One Single Device. We distinguish the states *OK* for correct behaviour and *DEF* for failure for all devices, respectively. Instead of writing that a device is *in state OK* or *DEF*, we just write that a device *is OK* or *DEF*. If a device is detected to be broken, its only possible state is *DEF*.

A random variable and its probability distribution used to describe the states of a device should meet the following requirements: Provided that a device K is *OK* at cycle t, the probability for this device to be *DEF* at cycle $t+1$ should be a constant p_K and thus the probability for being *OK* $1 - p_K \, \forall t, K$. If K is *DEF* at cycle t, K should be *OK* at cycle $t+1$ with probability 0 and *DEF* with probability 1.

We assume that the failures of the devices are distributed geometrically and define a probability space $(\Omega_K, P_K, \wp(\Omega_K))$ with $\Omega_K = \{1, 2, \ldots\}$ the set of all natural numbers, P_K the geometric distribution with probability of failure p_K and $\wp(\Omega_K)$ the set of all subsets of Ω_K. Every $\omega \in \Omega_K$ represents the failure of K at cycle ω, so that the event "K *DEF* at cycle t" is described by the set $\{1, \ldots, t\}$, the event "K *OK* at cycle t" by $\{t+1, t+2, \ldots\}$. Formally, we define the random variable $X_{K,t}$ on Ω_K as follows:

$$X_{K,t} : \Omega_K \to \{OK, DEF\}$$
$$\omega \mapsto \begin{cases} OK, & \omega > t \\ DEF, & \omega \leq t \end{cases}$$

We have $P_{X_{K,t}}(OK) = (1 - p_K)^t$ and $P_{X_{K,t}}(DEF) = 1 - (1 - p_K)^t$ for the distribution of $X_{K,t}$. It is easy to compute that for the conditional probabilities the following is valid (cf. Appendix):

$$P_K(X_{K,t} = OK | X_{K,t-1} = OK) = 1 - p_K,$$
$$P_K(X_{K,t} = DEF | X_{K,t-1} = OK) = p_K,$$
$$P_K(X_{K,t} = OK | X_{K,t-1} = DEF) = 0, \text{ and}$$
$$P_K(X_{K,t} = DEF | X_{K,t-1} = DEF) = 1 \ .$$

So the definition and the distribution of $X_{K,t}$ conform to our demands.

Given a mean time between failure (MTBF, measured in numbers of cycles) t_{MTBF} for the component K, we can now calculate p_K by $p_K = 1/t_{\text{MTBF}}$ as we assume the failures to be distributed geometrically.

A Model for the Whole System. Before we present our stochastic model for the whole system, we make another assumption: we assume that the failure of two components is (stochastically) independent. This assumption is reasonable and reflects the behaviour of the system in the real world. Nevertheless, it needs to be checked before system integration as it is possible to build a physical system which does not conform to this hypothesis.

As the probability space describing the whole system we define $(\Omega, P_\Omega, \wp(\Omega))$ with $\Omega := \Omega_{P_1} \times \ldots \times \Omega_{SO}$ and for $\omega = (\omega_{P_1}, \ldots, \omega_{SO}) \in \Omega$

$$P_\Omega(\omega) := P_{P_1}(\omega_{P_1}) \cdot \ldots \cdot P_{SO}(\omega_{SO}) \ .$$

Let X_t be the random variable which is defined by $X_t := (X_{P_1,t}, \ldots, X_{SO,t})$. For $\Omega_X := \{OK, DEF\}^{10}$ and $K \in \{P_1, \ldots, SO\}$ we have

$$X_t : \Omega \to \Omega_X$$

$$\omega_K \mapsto X_{K,t}(\omega_K) = \begin{cases} OK, & \omega_K > t \\ DEF, & \omega_K \leq t \end{cases}$$

The random variable X_t describes the state of all components of the system at cycle t. At the same time, t describes the "age" of all components. In order to allow for various ages of components (caused by repairs of components for example) we define t to be a vector $t := (t_{P_1}, \ldots, t_{SO})$ and generalize our previous definition of X_t by replacing t with t_K where necessary. If t is used with several components, we refer to t as a vector as above, and in all other cases as a natural number.

For the distribution of X_t, the product measure of the P_{X_K, t_K}'s, we have

$$P_{X_t}(x) = \prod_{K=P_1}^{SO} P_{X_K, t_K}(x_K) \ ,$$

Therefore, we have a new probability space $(\Omega_X, P_{X_t}, \wp(\Omega_X))$ which describes the probabilities of all simulataneously possible component states (note that P_{X_t} is dependend on t).

Using the probability space Ω_X it is now possible to compute the probability of the failure of a device at a given cycle, and therefore the probability for any state of the devices of the system at any cycle. This computation, however, would only rely on the failure probabilities of the components and therefore could not really be used to detect failures as we do not use the additional information we have at every cycle, namely the sensor and actuator values of the current and the previous cycle. In the following we describe how we use this information to restrict the choice of possible system states at a given cycle to a subset of Ω_X deterministically by eliminating events which are inconsistent with the current and previous sensor and actuator values. For example, if the messages received about the state of a pump tells us that a command sent to it in the previous cycle has not been executed, we know that the pump must be broken.

In the next subsection we explain the concept of *indicators*, which we use to identify the elements of the restriction of Ω_X at a given cycle.

5.2 Indicators

In the following we denote Adjusted Values (cf. [Abr95b]) (see CD-ROM AS.) just as the upper and lower bounds for the water level and the steam outcome by over- or underlining, respectively. If a variable is neither overlined nor underlined, it denotes an exact value.

We use the same bounds as described in [Abr95a] (see CD-ROM AS.) naming the bounds for the water level at cycle t $\underline{bw}^{(t)}$ and $\overline{bw}^{(t)}$ as well as the bounds for the steam outcome at cycle t $\underline{bo}^{(t)}$ and $\overline{bo}^{(t)}$.

Our aim is to eliminate at a given cycle t as much states from Ω_X that are inconsistent with the current and previous sensor and actuator values as possible. The idea is to define a formal function which measures the distance between the measured system state at cycle t and a computed "ideal" (i.e. error-free) system state for each device. This formal function yields a indicator vector $y^{(t)}$ which contains for each device at least one indicator value (short indicator). We use these values to determine the inconsistent states.

Definition and evaluation of the Indicators. If a component has already been detected to be faulty, the corresponding indicator(s) has (have) the value 0; for a broken pump the indicator for the pump monitor is 0 and vice versa. For devices which have not yet been identified to be broken, the indicators are defined in the following paragraphs. We define one indicator for each device, except for the water level measurement device. For this one we define two indicators named $y_{WL}^{(t)}$ and $y_{WLDIF}^{(t)}$. So far we have $y^{(t)} = (y_{P_1}^{(t)}, \ldots, y_{PM_4}^{(t)}, y_{WL}^{(t)}, y_{SO}^{(t)}, y_{WLDIF}^{(t)})$.

Let $P_i = 0$ denote "Pump i switched off", $P_i = 1$ "Pump i switched on" and analogeously for the pump monitors. Let $q_i^{(t)}$ be the state of P_i or PM_i at cycle t. We have $q_i^{(t)} \in \{0,1\}$. Let the set $C_t := \{0,1\}^4$ denote the commands which the control program gives at cycle t. Here the command 0 denotes "stay in the same state" and 1 "change state". For $c^{(t)} := (c_1^{(t)}, \ldots, c_4^{(t)}) \in C_t$ we can compute $q_i^{(t+1)}$ of P_i at cycle $t+1$ by $q_i^{(t+1)} := q_i^{(t)}$ XOR $c_i^{(t)}$ — if P_i does not fail. If both P_i and PM_i work properly, their messages are equal.

The indicators for the pumps and the pump monitors are defined as follows:

$$y_i^{(t)} := |(q_i^{(t-1)} \text{ XOR } c_i^{(t-1)}) - q_i^{(t)}| \, , \quad P_1 \leq i \leq P_4$$
$$y_i^{(t)} := |(q_{i-4}^{(t-1)} \text{ XOR } c_{i-4}^{(t-1)}) - q_i^{(t)}| \, , \quad PM_1 \leq i \leq PM_4$$

Let $w^{(t)} \in [0, \infty[$ denote the water level at cycle t and let d be the throughput of the pumps. For the first indicator for the water level, $y_{WL}^{(t)}$, we have $y_{WL}^{(t)} \in \{0, \ldots, 5\}$. If the water level is not within the computed bounds, this indicator gives the deviation of the water level from the bounds computed in rounded multiples of the pump capacity d. As there are only four pumps, we can limit $y_{WL}^{(t)}$ to be not bigger than 5.

If $y_{WL}^{(t)} = k$ with $0 < k < 5$, either the water level measurement device is broken or k pumps are defective (there may even be more than k pumps broken, as the failure of two pumps may cancel their effects on the water level mutually, but our model does not reflect this fact). If the indicator has the value 5, the water level measurement device must be broken. Formally:

$$y_{WL}^{(t)} := \begin{cases} 0, & \underline{bw}^{(t)} \leq w^{(t)} \leq \overline{bw}^{(t)} \\ \left\lceil \frac{w^{(t)} - \overline{bw}^{(t)}}{d} \right\rceil, & \overline{bw}^{(t)} < w^{(t)} \leq \overline{bw}^{(t)} + 4d \\ \left\lceil \frac{\underline{bw}^{(t)} - w^{(t)}}{d} \right\rceil, & \underline{bw}^{(t)} > w^{(t)} \geq \underline{bw}^{(t)} - 4d \\ 5, & (\underline{bw}^{(t)} - 4d > w^{(t)}) \vee (\overline{bw}^{(t)} + 4d < w^{(t)}) \end{cases}$$

The second indicator for the water level $y_{WLDIF}^{(t)}$ is a refinement of the first one. This indicator yields, in rounded multiples of d, the difference between the upper and the lower bounds of the water level. Therefore, $y_{WLDIF}^{(t)}$ measures the number of pumps which may fail simultaneously and whose failure can not yet be detected by observing the water level. Pumps whose failures have already been detected do not influence this value, as these failures have been observed when computing the adjusted value for the water level. Formally:

$$y_{WLDIF}^{(t)} := \left\lfloor \frac{\overline{bw}^{(t)} - \underline{bw}^{(t)}}{d} \right\rfloor$$

When evaluating $y_{WLDIF}^{(t)}$, it has to be taken into account that there is always a minimum distance greater than zero between $\overline{bw}^{(t)}$ and $\underline{bw}^{(t)}$. Thus in case $y_{WL}^{(t)} = 0$ is valid, $y_{WLDIF}^{(t)}$ gives no information.

Let $o^{(t)} \in [0, \infty[$ denote the steam outcome at cycle t. Then the indicator $y_{SO}^{(t)}$ for the steam outcome measurement device is defined by the following formula:

$$y_{SO}^{(t)} := \begin{cases} 0, & \underline{bo}^{(t)} \leq o^{(t)} \leq \overline{bo}^{(t)} \\ 1, & \text{otherwise} \end{cases}.$$

This indicator determines whether the amount of outcoming steam is outside the computed bounds. If this is true, $y_{SO}^{(t)} = 1$ and the steam measurement device is certainly broken.

The following tables summarize the evaluation of the indicators for components which are not broken.

$y_{P_i}^{(t)}/y_{PM_i}^{(t)}$	possible states of P_i/PM_i at cycle t
0 / 0	OK / OK $\vee$ DEF / DEF
1 / 0	DEF / OK $\vee$ DEF / DEF
0 / 1	OK / DEF $\vee$ DEF / OK
1 / 1	DEF / OK $\vee$ DEF / DEF

$y_{WL}^{(t)}/y_{WLDIF}^{(t)}$	possible states of WL/ the pumps at cycle t
5 / 0 … 5	DEF / (OK ∨ DEF)
4 / 0	DEF / (OK ∨ DEF) ∨ OK / 4 pumps DEF
4 / 1 … 5	DEF / (OK ∨ DEF)
3 / 0	DEF / (OK ∨ DEF) ∨ OK / 3 pumps DEF
3 / 1	DEF / (OK ∨ DEF) ∨ OK / 4 pumps DEF
3 / 2 … 5	DEF / (OK ∨ DEF)
2 / 0	DEF / (OK ∨ DEF) ∨ OK / 2 pumps DEF
2 / 1	DEF / (OK ∨ DEF) ∨ OK / 3 pumps DEF
2 / 2	DEF / (OK ∨ DEF) ∨ OK / 4 pumps DEF
2 / 3 … 5	DEF / (OK ∨ DEF)
1 / 0	DEF / (OK ∨ DEF) ∨ OK / 1 pumps DEF
1 / 1	DEF / (OK ∨ DEF) ∨ OK / 2 pumps DEF
1 / 2	DEF / (OK ∨ DEF) ∨ OK / 3 pumps DEF
1 / 3	DEF / (OK ∨ DEF) ∨ OK / 4 pumps DEF
1 / 4 … 5	DEF / (OK ∨ DEF)
0 / 0 … 5	(OK ∨ DEF) / (OK ∨ DEF)

$y_{SO}^{(t)}$	possible states of SO at cycle t
0	OK / DEF
1	DEF

In the last table, for example "DEF / (OK ∨ DEF)" means that the water level measurement device is broken and that the pumps can be intact or defective. "OK / 3 pumps DEF" means, that the water level measurement device is intact and 3 pumps are defective.

Let Γ denote the set of all indicator vectors. The subset $M \subseteq \Omega_X$, which describes the set of all possible component state at cycle t is computed as follows[2]: Those $\omega \in \Omega_X$ are selected, that for each $K \in \{P_1, \ldots, SO\}$ ω_K fulfills the conditions yielded by the evaluation of the indicator vector $y^{(t)}$ as described by tables above. Thus, we define a mapping $\sigma : \Gamma \longrightarrow \Omega_X$, which maps every indicator vector $y^{(t)}$ to a subset M_t of Ω_X. We say the indicator vector $y^{(t)}$ *identifies* (by σ) a subset M_t of Ω_X.

Additional Indicators. Our set of indicators can easily be extended. For example we only have an indicator watching the water level for changing too much so far. It should be possible to construct an indicator measuring whether the change of the water level is too small and to define appropriate rules for σ.

5.3 Failure Detection

Our aim was the computation of the probability of the failure of components in order to detect whether a component is broken or not with the highest possible

---
[2] An example is given in the appendix.

probability. We use the restriction of Ω_X that we gain through the use of the indicators to complete this task.

Let be $M_t := \sigma(y^{(t)})$ with $y^{(t)}$ the indicator vector at cycle t. Then we have a new probability measure on M_t by

$$\mathbb{P}_t(A) := P_\Omega(\{X_t \in A\}|\{X_t \in M_t\})$$

for $A \subseteq \Omega_X$.

Let $\mathcal{K}_t$ be the event "device K is OK in cycle t" and $\overline{\mathcal{K}}_t$ the event "device K is DEF in cycle t". We have $\mathcal{K}_t = \{\omega \in \Omega_X : \omega_K = OK\}$ and $\overline{\mathcal{K}}_t = \complement\mathcal{K}_t$. For $\mathbb{P}_t(\overline{\mathcal{K}}_t)$, which is the probability for K to be faulty at instant t provided our knowledge about the system state, we have

$$\mathbb{P}_t\left(\overline{\mathcal{K}}_t\right) := P_\Omega\left(\{X_t \in \{\omega \in \Omega_X : \omega_K = DEF\}\} \mid \{X_t \in M_t\}\right)$$

and for $\mathbb{P}_t(\mathcal{K}_t)$

$$\mathbb{P}_t(\mathcal{K}_t) = 1 - \mathbb{P}_t\left(\overline{\mathcal{K}}_t\right) .$$

Our failure detection now calculates for each $K \in \{P_1, \ldots, SO\}$ these probabilities and selects the event from $\{\mathcal{K}_t, \overline{\mathcal{K}}_t\}$ that has the higher probability. If both events have the probability $\frac{1}{2}$, it assumes the worst case and selects $\overline{\mathcal{K}}_t$. Thus we get a vector $\vartheta \in \{\mathcal{P}_1, \overline{\mathcal{P}}_1\} \times \ldots \times \{\mathcal{SO}, \overline{\mathcal{SO}}\} = \{OK, DEF\}^{10}$ describing the most probable state for each device at cycle t.

5.4 Analysis

In this subsection we investigate the reliability of our failure detection algorithm.

The Reliability of the Failure Detection. In the following lemma and in both theorems we require for all devices that for the probability p_K is valid $p_K > 1 - p_K$. This is reasonable as otherwise one would not use these devices in a safety critical environment. Furthermore we do not distinguish between the failure of a pump and a pump monitor, because both devices form an integrated whole.

All the proofs are dependent on our set of indicators. Thus a lemma or theorem need not to be true any more if the set of indicators would be changed. We start with

Lemma 5.1. *Suppose that at a given instant of time exactly one component is broken. If the failed component is a pump or a pump monitor, we can deterministically detect the failure in this couple.*

Theorem 5.2. *Suppose that a given cycle t exactly one component is broken. If the condition $[y^{(t)} \notin \{y \in \Gamma : (y_{WL} = 0) \vee (y_{SO} = 0)\}]$ is valid and the probability for the water level measuring device being OK if*

$$y^{(t)} \in \{y \in \Gamma : y_{P_1} = \ldots = y_{PM_4} = 0, y_{WL} \in \{1, \ldots, 4\}, y_{WL} + y_{WLDIF} \leq 4\}$$

is equal to or less than $\frac{1}{2}$, then the failure manager detects the failure regardless of which component is broken. If $y^{(t)} \in \{y \in \Gamma : (y_{WL} = 0) \vee (y_{SO} = 0)\}$, the failure is only detected if $(y_{WL} = 0)$ and

$$t \geq \frac{\log \frac{1}{2}}{\log(1 - p_{WL})} .$$

The same condition holds for the steam output measurement device.

Theorem 5.3. *Suppose that at a given cycle t exactly one component is broken. Furthermore, if the probability for the water level measuring device being OK in t if*

$$y^{(t)} \in \{y \in \Gamma : y_{P_1} = \ldots = y_{PM_4} = 0, y_{WL} \in \{1, \ldots, 4\}, y_{WL} + y_{WLDIF} \leq 4\}$$
$$\cup \{y \in \Gamma : y_{P_1} = \ldots = y_{SO} = 0, y_{WLDIF} \in \{0, \ldots, 5\}\}$$

is equal to or less than $\frac{1}{2}$, and the probability for the steam outcome measuring device for being OK in t if

$$y^{(t)} \in \{y \in \Gamma : y_{P_1} = \ldots = y_{SO} = 0, y_{WLDIF} \in \{0, \ldots, 5\}\}$$

is equal to or less than $\frac{1}{2}$, the failure of the component is detected by the failure manager.

In case that one more device fails, the failure detection in general behaves as follows: If the measurement devices do not fail, failures of more pump/pump monitor couples can be detected in all cases. If one of the measurement devices fails, the difference between the water level and steam outcome thresholds increase. Therefore, the detection of further failures becomes increasingly complex. The failure of the steam output measuring device can be detected independently from the failure of any other components and does only depend on system constants.

Reliability of the Failure Manager. Let $\alpha_K(t)$ be the probability that the failure manager has decided right about the device K in the cycles $1 \ldots t$. This probability can be calculated very easily: For example if the failure manager has decided in cycle t that the state of K is *OK*, then

$$\alpha_K(t) = \mathbb{P}_t(\mathcal{K}_t) \cdot \alpha_K(t-1) .$$

If the failure manager decided that K is *DEF* in t, then we set

$$\alpha_K(n) := 1 \quad \forall n > t$$

until K has been repaired. This seems to be reasonable as the failure manager does not deal with K as long as it is considered *DEF*. Thus we have for the probability $\alpha(t)$ that the failure manager decided correct in the cycles $1 \ldots t$

$$\alpha(t) = \prod_{K=P_1}^{SO} \alpha_K(t) .$$

As we have shown before, a failure of a pump/pump controller combination can always be detected. This is not the case with the measurement devices: It could be that a failure of a measurement device is not detected in the cycle the device becomes broken but in a later cycle. Thus we need to calculate the probability that the failure manager decides right about K (K a measurement device) in t the probability that it corrects a wrong decision that it made before. This is only possible if a failure has not been detected: If the failure manager decided that a device is *DEF* and this is wrong, the decision can not be corrected.

For K the water level or steam outcome measurement device let $\beta_K(t)$ be the probability that the failure manager corrects a wrong decision. Then we have for $\beta_K(t)$

$$\beta_K(t) = \sum_{n=1}^{t} \sum_{l=n+1}^{t} (1-p_K)^{n-1} p_K \prod_{j=n}^{l-1} \mathrm{prob}[\mathbb{P}_j(\overline{\mathcal{K}}_t) < \frac{1}{2}] \cdot \mathrm{prob}[\mathbb{P}_j(\overline{\mathcal{K}}_t) \geq \frac{1}{2}] \ ,$$

and it is not yet clear how to calculate both probabilities $\mathrm{prob}[\mathbb{P}_j(\overline{\mathcal{K}}_t) < \frac{1}{2}]$ and $\mathrm{prob}[\mathbb{P}_j(\overline{\mathcal{K}}_t) \geq \frac{1}{2}]$. But because we have for $\gamma(t)$, the probability that the failure manager decides right about the state of the system at cycle t

$$\gamma(t) = \alpha(t) + \beta_{WL}(t) + \beta_{SO}(t) \ ,$$

it is clear that $\alpha(t)$ is a lower bound for $\gamma(t)$: The probability, that the failure manager decides right about the system is at least $\alpha(t)$. Thus $\alpha(t)$ can be used to evaluate the reliability of the failure manager.

Computing the Water Level Adjusted Value. The final task of the Failure Manager is the computation of an approximation of the water level. If the water level measurement device is considered intact, we use the measured value. The important thing about it is that $\alpha_{WL}(t) + \beta_{WL}(t)$ gives us the probability that the water level measurement device is intact that is that the measured value is correct. If this probability becomes too low, adjusted values could be used and one could have the water level measurement device checked. If the water level measurement device is considered to be defect, we use the adjusted values described in the task description. Here we get the probability that the adjusted values are right by computing $\alpha_K(t)$ respectively $\alpha_K(t) + \beta_K(t)$ or all K used in the adjusted values. If this probability becomes too low, appropriate actions could be taken (maintenance or emergency stop for example).

6   Evaluation and Comparison

1. The solution starts with a system identification phase which results in an architecture specification and the subpartition of the task into three problems.
   Each of these problems is addressed with a different formalism, verification is applied where necessary and useful.

2. The solution does comprise an implementation of the control program which has been linked to the FZI simulator. Experimentation has been done using the failure files included in the FZI simulation as well as in a general way.
3. Cattel and Duval (see CD-ROM CD.) present a solution completely in Lustre. Their verification is lacking the stochastic discussion presented in our paper. Another comparable approach is the work of XiaoShan and JuAn (see CD-ROM JW.). Their paper deals with finding an optimal design, but contains no probabilistic reasoning in our sense as well. Therefore, our solution is unique in that the indeterminism of the failure detection is reflected in the indeterminism of our model. Additionally, our system identification phase is unique, but a number of contributions discuss the appropriateness of their model. All of them, though, give no constructive way to the choice of the formalism.
4. Focus Part: 1 Week. LUSTRE Part: 1 Week. Probability Part: 3 months. Implementation and Test: 2 Weeks.
5. For a good understanding of the solution, experience with the notation used is necessary. Basic knowledge of probability theory is needed for understanding our concept of the failure manager, but not necessary for an implementation.

References

Abr95a. J. R. Abrial. Additional information concerning the physical behaviour of the steam boiler. Technical report, 1995.
Abr95b. J. R. Abrial. Steam boiler control specification. Technical report, 1995.
Hal93.   N. Halbwachs. *Synchronous Programming of Reactive Systems*. Kluwer Academic Publishers, 1993.
KSW.   Frank Dederichs Ketil Stolen and R. Weber. Assumption/commitment rules for networks of asynchronously communicating agents. Technical report.
LL94.   Claus Lewerentz and Thomas Lindner, editors. *Formal Development of Reactive Systems*. LNCS 891. Springer-Verlag, 1994.
MBW.   C. Dendorfer M. Fuchs T. F. Gritzner M. Broy, F. Dederichs and R. Weber. The design of distributed systems – an introduction to focus. Technical Report SFB 342/2/92, Technische Universität München.
SD93.   R. Schlör and W. Damm. Specification and verification of system-level hardware designs using timing diagrams. In *The European Conference on Design Automation with the European Event in ASIC Design*, pages 518–524, 1993.

Refining Abstract Machine Specifications of the Steam Boiler Control to Well Documented Executable Code

Christoph Beierle, Egon Börger, Igor Đurđanović, Uwe Glässer, Elvinia Riccobene

[1] Fernuniversität-GH Hagen, Germany, christoph.beierle@fernuni-hagen.de
[2] Università di Pisa, Italy, boerger@di.unipi.it
[3] Universität-GH Paderborn, Germany, igor@uni-paderborn.de
[4] Universität-GH Paderborn, Germany, glaesser@uni-paderborn.de
[5] Università di Catania, Italy, riccobene@dipmat.unict.it

Abstract. We use the steam boiler control specification problem to illustrate how the evolving algebra approach to the specification and the verification of complex systems can be exploited for a reliable and well documented development of executable, but formally inspectable and systematically modifiable code. A hierarchy of stepwise refined abstract machine models is developed, the ground version of which can be checked for whether it faithfully reflects the informally given problem. The sequence of machine models yields various abstract views of the system, making the various design decisions transparent, and leads to a C++ program. This program has been demonstrated during the Dagstuhl-Meeting on Methods for Semantics and Specification, in June 1995, to control the Karlsruhe steam boiler simulator satisfactorily.

The abstract machines are evolving algebras and thereby have a rigorous semantical foundation, allowing us to formalize and prove, under precisely stated assumptions, some typical sample properties of the system. This provides insight into the structure of the system which supports easily maintainable extensions and modifications of both the abstract specification and the implementation.

1 Introduction

We solve the steam boiler problem to illustrate how the evolving algebra approach to design and verification of complex systems can be used for a well documented development of executable but nevertheless formally inspectable and systematically modifiable code. We go through a hierarchy of stepwise refined abstract machine models the ground version of which can be shown to faithfully reflect the informally given problem. The sequence of mathematical models provides various useful levels which reflect each a different design decision and starting from which the solution can be easily modified; it eventually leads to a C++ program which has been demonstrated during the Dagstuhl-Meeting on Methods for Semantics and Specification, in June 1995, to control the Karlsruhe Steam Boiler (see Chap. L. of this book) satisfactorily.

The models are evolving algebras and thereby have a rigorous semantical foundation [12]. They are related by stepwise refinements which reflect the systematic use of strongest information hiding and modularization techniques offered by the abstraction mechanism built into the notion of evolving algebra. The systematic use of successive refinements represents an important methodological software engineering principle, namely to avoid over-specification and to postpone premature design decisions as much as possible. The refinements also permit to state and prove interesting system properties at the appropriate level of abstraction; this is how the technique of building hierarchies of stepwise refined levels of abstraction has found its way into the evolving algebra methodology (see [2, 10]) where it has been used since then extensively (see for ex. [7, 6, 8, 15, 14, 13, 11, 9], see also [5] for an explanation why evolving algebras provide the framework par excellence for the most general realization of the refinement idea). We investigate some typical sample properties of the system which we formulate and prove, under precisely stated assumptions, in the abstract models. This provides insight into the structure of the system and yields useful directives for the definition of provably correct system components. Our proofs are traditional (not formalized) mathematical proofs and are viewed by us not in opposition to machine-checked proofs but as a possible guideline for constructing such detailed fully formalized deductions within (the implementation of) a specific proof system[6].

The most abstract model is a *ground model* in the sense of [3], i.e. the result of a formalization process of the informally given description which remains conceptually and notationally as close as possible to the informal problem statement and thereby can be inspected by the user for its adequacy. In order to illustrate how evolving algebras offer the greatest possible flexibility in adapting the formalization to the peculiarities of the given application domain, our ground model follows Abrial's text as closely as possible without committing to any particular implementation. As a result we obtain as starting point for the definition of the program a mathematical model—what usually is called a *formal requirement specification*—whose domains and functions directly reflect the basic objects and operations of the steam boiler system, avoiding any extraneous encoding or other formal overhead. Such a model provides a transparent and faithful link between the customer's world - where the application problem resides - and the system designer's and programmer's world - where the program has to be developed.[7] In particular the ground model allows one to "show" by

---

[6] For an illustration of this point see [1, 16] which report on machine verifications for some of the refinement steps introduced for the evolving algebra based correctness proof of a general compilation scheme of Prolog programs to WAM code in [10].

[7] Obviously this "link" holds only for those system parts or properties which are specified in the ground model. Stated otherwise, a ground model should contain all those parameters, actions and conditions which are relevant for the customer. An example in this paper is the treatment of error handling for equipment failures; we cannot discuss it appropriately unless we explicitly identify and describe the relevant features, as we do here in the refinement section 5.2. See [5] for further discussion of this point.

pointing to the model that it really reflects the informal description of the problem. (See [5] for a discussion of the role of these ground models for the foundation of applications of programming to the real world.)

We develop the model refinements up to a point where it becomes evident how executable C++-code can be obtained by translating—almost mechanically—the abstract machine instructions into C++-procedures. These procedures are executed in a context of basic routines which implement the semantics of our abstract machines. Via this translation the rules of the abstract machine models "show" the structure of the executable C++-code (which has been connected successfully to the Karlsruhe steam boiler simulator). In this way the successively refined abstract models constitute a documentation of the executable code, including the relevant information on the design process—each refinement step directly expresses some design decisions and can be used as reference point for possible modifications or extensions. The projection of the abstract machine models into the C++-program makes the C++-code inspectable by mathematical (formal) methods. We consider this possibility as a particularly challenging research direction and hope that further developments of the method will lead to useful techniques for the design of transparent, inspectable software.

In this paper we make no attempt to analyze or bridge the discrepancy between the few assumptions on the physical behaviour of the system which are contained in the informal problem description and the many additional assumptions which have been made by Anne Lötzbeyer for the design of the Karlsruhe steam boiler simulator. Along our way we list those assumptions which are needed to make the abstract models consistent. In the appendix on proofs for system properties some more assumptions are listed without which the proofs could not be carried through. In order to be able to link our executable C++-code successfully to the Karlsruhe steam boiler simulator, we had to take into account also the additional assumptions made for the design of the simulator; we do not list those assumptions here, they concern mainly the physical model of the steam boiler (dynamic.C). This is also the reason why we do not attempt to prove the "correctness" of the executable code with respect to the abstract evolving algebra models. Note however that in principle such a proof project could be carried through, using Wallace's [17] mathematical definition of the semantics of C++ as a reference model.

The sequence of successfully refined abstract machine models can be turned into a systematic modular architectural design. In this paper we abstain from doing this and focus our attention on the appropriateness of the formal requirement specification defined by the ground model and on how we can map refinements of this model into executable code.

As a technical consequence of the attempt to be faithful to Abrial's text we describe only the control part and not the physical behaviour of the steam boiler system. In particular we comply to the discrete control program view of it which avoids to have to consider any hybrid, real-time or distributed feature. This reduces the problem to cyclical reading of information coming from the physical components and reacting by triggering of corresponding actions (through sending out messages to those components). Our model is however abstract enough so that it could be refined to a distributed system which works in real-time, using the notions of distributed real-time evolving algebra runs developed in [12, 8, 14, 13].

As is to be expected from every seriously mathematical approach to system or program development, during the formalization process we have discovered numerous (probably deliberate) holes in the informal description which had to be filled in order to avoid inconsistencies or other unreasonable behaviour. Each time this happens we make the additional assumptions explicit and also give hints how the abstract machine model could easily be adapted to alternatives. These are typical examples of points where the evolving algebra approach allows us to easily formulate, in a language which is understandable to the customer, precise questions about further decisions to be taken.

The paper is organized as follows. Section 2 reviews some basic semantical concepts of abstract machines as far as they are required here. Section 3 addresses certain global aspects concerning the overall behaviour of the steam boiler control unit with respect to its embedding into the physical environment. The detailed behaviour of the control program depending on the given mode of operation is specified in Sect. 4. The resulting model is then refined in Sect. 5 by introducing a message passing interface, which allows us to deal also with error handling and detection of equipment failures. Section 6 explains the encoding of our most refined evolving algebra model into an executable C++-program (see CD-ROM Annex BBDGR.D). In CD-ROM Annex BBDGR.B we exemplify the formal verification process by proving a number of selected properties of our mathematical model. CD-ROM Annex BBDGR.C contains a Glossary summarizing the formal definitions; some of these definitions represent a possible refinement step.

2 The Concept of Abstract Machines

An *evolving algebra* $\mathcal{A}$ with *program P*—consisting of a finite number of *transition rules* of a form indicated below—and (a class of) *initial state*(s) S_0 models the operational semantics of a discrete dynamic system $\mathcal{S}$ by specifying its *observable behaviour* in terms of state transitions, where mathematical structures— i.e. collections of domains equipped with functions and predicates defined on them—serve as abstract representations for the *concrete states* of $\mathcal{S}$. W.r.t. the particular system class considered here (distributed control systems), a crucial system characteristic to be captured by the mathematical model is the *reactive* behaviour: the ongoing interaction between $\mathcal{S}$ and the *environment* $\mathcal{E}$ into which $\mathcal{S}$ is embedded.

State transitions of $\mathcal{A}$ may be effected in two possible ways: *internally*, through the rules of P, or *externally*, through actions in the environment $\mathcal{E}$. This offers a conceptual means to specify *concurrency* and *interdependency*. The dependency of $\mathcal{S}$ from $\mathcal{E}$ is reflected by the concept of *externally alterable* and of *oracle functions*[8]: these oracle functions refer to an *abstract interface* attaching the model to an external world (e.g. the environment $\mathcal{E}$). In contrast to a closed world assumption, where every relevant detail is included into the model, the approach taken here relies on an *open system view*.

A computation of $\mathcal{S}$ is modeled through a finite or infinite *run* ρ of $\mathcal{A}$ as a sequence of states $S_0 S_1 S_2 \ldots$ such that i) S_0 is an *initial state*; and ii) the internally controlled part of each state S_{i+1}, for $i = 1, 2, \ldots$, is obtained by *simultaneously* firing all those rules of P which are enabled on S_i. Each rule can be thought of as having the form ' `if` *Cond* `then` *Updates* ' where *Cond* is any first-order expression and *Updates* a set of function updates

$$f(t_1, \ldots, t_n) := t \ .$$

The semantical meaning of firing such a rule is that if in a given algebra *Cond* evaluates to true, then the value of f at the argument place $(t_1, \ldots, t_n)$ is set to t. For a more precise definition we refer the reader to CD-ROM Annex BBDGR.A.

In a distributed evolving algebra $\mathcal{A}$ *multiple* autonomous agents cooperatively model a *concurrent computation* of a system $\mathcal{S}$ in an asynchronous manner[9]; each agent a executes its own *single-agent program* $Prog(a)$ as specified by the *module* associated with a. More precisely, an agent a has a partial view $\text{View}(a, S)$ of a given global state S as defined by its subvocabulary (i.e. the function names occurring in $Prog(a)$) on which it fires the rules specified by $Prog(a)$. The underlying semantic model ensures that the order in which the agents of $\mathcal{A}$ perform their operations is always such that no conflicts between the update sets computed for distinct agents can arise. For further details we refer to [12].

The evolving algebra defined below models the behaviour of the steam boiler control program from the point of view of a single agent. A complete description of the entire control model—i.e. a distributed evolving algebra with additional agents specifying the behaviour of the various physical units—can be obtained as a straightforward extension of the model presented here.

3 Overall Operation of the Program

In this section we consider three global aspects concerning the embedding of the control unit into the given physical environment, namely: (1) the timing behaviour of the underlying message passing communication protocol; (2) the physical units to be distinguished by the control program with respect to error handling; (3) the detection of failures of control components.

---

[8] An *oracle function* of $\mathcal{A}$ may only be read but not be affected by (the transition rules of) $\mathcal{A}$, an externally alterable function can change due to an action of the environment (but it may also be internally updatable, i.e. due to firing of a transition rule of $\mathcal{A}$). See [5].

[9] The term 'distributed', as it is used here, actually refers to the distribution of control rather than to the distribution of data.

3.1 Modeling of Timing Behaviour

[*The program follows a cycle and a priori does not terminate. This cycle takes place each five seconds and consists of the following actions: reception of messages coming from the physical units, analysis of informations which have been received, transmission of messages to the physical units.*
To simplify matters, and in first approximation, all messages coming from (or going to) the physical units are supposed to be received (emitted) simultaneously by the program at each cycle.]

The timing behaviour of the program can be modeled by means of two nullary dynamic functions: *curr_time* is an oracle function used to represent a global clock; *last_time* is an internally updatable function used to indicate the beginning of the current cycle.

$$curr\_time, last\_time : NAT$$

As an integrity constraint on *curr_time* we require that the value of *curr_time* increases monotonically to the limit ∞ (**Cond I**). The condition $curr\_time - last\_time = 5$ triggers the start of a new cycle. Using the nullary function *curr_cycle* : *NAT* as an internally updatable cycle counter, we associate with each cycle a unique natural number. Each cycle consists of three consecutive phases, namely: *reading, executing,* and *writing*. The nullary function *phase* represents the current phase within a given cycle:

$$phase : \{reading, executing, writing\}.$$

Without loss of generality we assume that the above functions are initialized as follows (**Cond II**): $S_0(curr\_time) = S_0(last\_time) = S_0(curr\_cycle) = 0$ and $S_0(phase) = reading$.

The reading phase triggers the reception of incoming messages (and the reading of values for oracle functions). During the executing phase the program evaluates the incoming messages and the used oracle functions to compute the new state and the outgoing signals. The latter are sent during the writing phase. This timing behaviour is modeled by the following three timing rules:

$T1$: **if** $phase = reading \wedge curr\_time - last\_time = 5$
  then $ReadMessages$
    $phase := executing$
    $last\_time := curr\_time$
    $curr\_cycle := curr\_cycle + 1$

$T2$: **if** $phase = executing$ **then** $phase := writing$

$T3$: **if** $phase = writing$ **then** $SendMessages$
    $phase := reading$

Global Prerequisities. In the following we will restrict our attention to those non-final states S_i where the phase does change—i.e. such that $S_i(phase) \neq S_{i+1}(phase)$; at the level of analysis suggested by the informal specification they cover all the substantial information about the system behaviour. We further assume that the condition '*phase = executing*' specifies a global precondition extending the guards of all the rules in Sects. 4.1-4.5 and 5.2 below.

3.2 The Physical Environment

The physical environment of the steam boiler control unit consists of a number of physical units which interact with the control program via message-passing communication. These units are formally represented as elements of the following domains:

$PUMP = \{pump\text{-}1, \ldots, pump\text{-}4\}$
$PUMP\_CTRL = \{pump\_ctrl\text{-}1, \ldots, pump\_ctrl\text{-}4\}$
$UNIT = PUMP \cup PUMP\_CTRL \cup \{level\_measuring\_unit, steam\_measuring\_unit\}$

In addition to these physical units, the informal description identifies two more devices: a *valve* and an *operator desk*. However, at the given abstraction level these devices are never explicitly addressed nor are there any failures associated with them. Therefore they need not to be represented as objects in the formal model.

3.3 Failure Detection

A particularly important issue in the specification of the steam boiler control unit is a precise definition of the system reactions to failures of control components. The informal description distinguishes two basic classes of failures, namely: (1) failures of individual physical units (*physical unit failures*); (2) failures of the transmission system (*transmission failures*).

Physical Unit Failures Our ground model reflects the detection of physical unit failures by means of a unary predicate

$$Failure: \quad UNIT \to BOOL$$

indicating for each physical unit its status. In order to separate different concerns, the conditions depending on which a unit is considered as faulty are not considered here but will be defined later by further refinement steps (see Sect. 3.3).

For the sake of conciseness and uniformity of description, we define two further failure predicates as shorthands to refer to certain failure classes:

$$PumpFailure \equiv \exists p \in PUMP : Failure(p)$$
$$PumpCtrlFailure \equiv \exists c \in PUMP\_CTRL : Failure(c)$$

To distinguish the case that all physical units are assumed to operate correctly from those cases in which at least one of these units is assumed to have a failure, we will use the predicate *AllPhysicalUnitsOk* with the following meaning:

$$AllPhysicalUnitsOk \equiv \forall x \in UNIT : \neg Failure(x)$$

Transmission Failures The detection of a transmission failure is expressed in the ground model by means of a nullary predicate *TransmissionFailure* : *BOOL*. The meaning of this predicate will be defined through stepwise refinements (see Sect. 3.3 and the definitions in the Glossary).

4 Operation Modes of the Program

The observable behaviour of the control program depends on the current mode of operation:

[*The program operates in different modes, namely: initialization, normal, degraded, rescue, emergency stop.*]

In the ground model these operation modes are represented through a nullary dynamic function *mode* taking values in the following domain:

$$MODE = \{initialization, normal, degraded, rescue, emergency\_stop\}$$

For a succinct formulation of program modes and mode updates we will use abbreviations, such as:

$$InitMode \equiv\ mode = initialization$$
$$EnterNormalMode \equiv\ mode := normal$$

4.1 Global Requirements

Regardless of the mode in which the program is operating there are certain conditions forcing the system to immediately enter the emergency stop mode:

[*STOP: When the message has been received three times in a row by the program, the program must go into emergency stop.*]

[*A transmission failure puts the program into the mode emergency stop.*]

In the ground model these requirements are formalized using two predicates *ExternalStop* (indicating that the message STOP has been received by the program three times in a row) and *TransmissionFailure* which will be refined later on.

The informal description contains another emergency stop condition which may as well be considered as a global condition, namely:

[*If the water level is risking to reach one of the limit values M_1 or M_2 the program enters the mode emergency stop.*]

taking into account the following exception: as long as the system operates in initialization mode it never "is risking to reach one of the limit values M_1 or M_2" [10]. To model the required behaviour, we introduce a predicate *ReachingLimitLevel* with that intended interpretation. This implies in particular that (**Cond III**) for every state S_i ($i = 0, 1, \ldots$) of a regular run ρ of the steam boiler algebra the following condition is supposed to hold:

$$S_i \models \mathit{InitMode} \Rightarrow \neg \mathit{ReachingLimitLevel}$$

The informal description leaves open how the risk of reaching one of the limit values M_1 or M_2 is to be estimated. We thus define our model abstracting from such details and do not further address this aspect here[11]. Using the predicates introduced above, we are now able to express the specified behaviour by defining the following *emergency stop rule*:

$$G1: \quad \textbf{if } \mathit{EmergencyStop} \textbf{ then } \mathit{EnterEmergencyStopMode}$$

where the externally alterable predicate *EmergencyStop* is defined by

$$\mathit{EmergencyStop} \equiv \mathit{ExternalStop} \vee \mathit{ReachingLimitLevel} \vee \mathit{TransmissionFailure}$$

In order to avoid inconsistency of the model, the negation of *EmergencyStop* has to appear in the guards of all rules that may cause a change of mode other than changing it to emergency stop (see Sects. 4.3-4.6).

In addition to $G1$ another global rule $G2$ is used to specify the control of the water level depending on the current mode of operation. Although this is not explicitly stated in the informal description, one can reasonably argue that the operations of adjusting the water level to a default value or of maintaining its value within an admissible range are essentially the same for any $mode \in \{normal, degraded, rescue\}$:

> [*The normal mode is the standard operating mode in which the program tries to maintain the water level in the steam boiler between N_1 and N_2 with all physical units operating correctly. As soon as the water level is below N_1 or above N_2 the level can be adjusted by the program by switching the pumps on or off. The corresponding decision is taken on the basis of the information which has been received by the physical units.*]

> [*The degraded mode is the mode in which the program tries to maintain a satisfactory water level despite of the presence of failure of some physical unit.*]

---

[10] This particular interpretation reflects only one possible choice out of several reasonable alternatives.

[11] Note that the primary purpose of the predicate *ReachingLimitLevel*, as it is used here, is to identify and mark a 'loose end' in the specification such that its intended meaning is still to be fixed by further refinements.

[*The rescue mode is the mode in which the program tries to maintain a satisfactory water level despite of the failure of the water level measuring unit.*]

In initialization mode, however, the operational behaviour is different:

[*If the quantity of water in the steam boiler is above N_2 the program activates the valve of the steam boiler in order to empty it. If the quantity of water in the steam boiler is below N_1 then the program activates a pump to fill the steam boiler.*]

Despite of the distinctions to be made, the functionality required to control the water level can be expressed by a single rule using parameterized operations:

$G2$:   **if** *WaterLevelAdjusted* $\land \neg EmergencyStop$
      then *RetainWaterLevel(mode)*
      else *AdjustWaterLevel(mode)*

From the information given in the informal description it is not clear whether the predicate *WaterLevelAdjusted* should have different interpretations in different operation modes of the control program. So far, we can precisely specify the meaning of *WaterLevelAdjusted* only in initialization mode by stipulating that (**Cond IV**) for every state S_i ($i = 0, 1, \ldots$) in a regular run of the steam boiler algebra the following condition holds:

$$S_i \models InitMode \land N_1 \leq q \leq N_2 \Rightarrow WaterLevelAdjusted$$

Though one could indeed imagine that *WaterLevelAdjusted* has a fixed meaning irrespective of the current operation mode, there are also good reasons to anticipate more complex interpretations for modes other than initialization[12]. We do not address this aspect any further, but show sample refinements for *AdjustWaterLevel(m)* and *RetainWaterLevel(m)* in initialization mode.

In the definition of *AdjustWaterLevel(m)* it is necessary to include the condition *SteamBoilerWaiting* which triggers the effective start of the steam boiler initialization operation[13] (see Sect. 4.2):

---

[12] Taking the current state and the dynamics of the system into account as well, for instance, would allow us to reduce the tolerance limits in the physical layout of the system.

[13] Some authors argue that the informal problem description should have divided the initialization into two models in order to bring out explicitly the two different phases of the initialization process.

$AdjustWaterLevel\,(m)$
$\equiv$ **if** $SteamBoilerWaiting$
   thenif $WaterLevelBelowMin$
    then $RaiseWaterLevel\,(m)$
    else $ReduceWaterLevel\,(m)$

$ReduceWaterLevel\,(m)$
$\equiv$ **if** $m = initialization$
   then $StopPumps$
         $OpenValve$
   else $StopSomePumps$

$RaiseWaterLevel\,(m)$
$\equiv$ $ActivateSomePumps$
   if $m = initialization$
   then $CloseValve$

$RetainWaterLevel\,(initialization)$
$\equiv$ $StopPumps$
   $CloseValve$

ActivateSomePumps and *StopSomePumps* are used as abstract actions which leave space for non-deterministic choices. At the given abstraction level, we are not concerned with any operational details specifying how the exact number of pumps to be switched on or off is calculated depending on the dynamics of the system. The macros *ActivateSomePumps* and *StopSomePumps* are typical examples for how we suggest to systematically use 'well-defined holes' in the semantic definition of the steam boiler control. The missing details are filled in by specifying the particular model of the physical behaviour of the steam boiler which is to be used in conjunction with the control logic defined through our model. In this way, the control logic on the one hand and the physical model on the other hand can be separated explicitly and be treated independently from each other.

For the sake of simplicity, we assume that (**Cond V**) the operations which effectively activate or stop the pumps and open or close the valve do behave in a robust way; i.e., they will be realized such that they do not cause any effects on the state of the addressed device (a pump or the valve) whenever the current state of that device is already identical to the requested state. In the mathematical model this corresponds to a 'robustness' property of assignment.

4.2 Initialization Mode

Among the operation modes of the program the initialization mode takes a special role in that it deals with the inspection of the initial system state:

[*The initialization mode is the mode to start with.*]

The purpose of the initialization phase is to lead the system from some given initial state to a regular starting state ensuring that those conditions which are vital for a secure operation of the steam boiler hold. In case that this is not possible (due to intolerable malfunctioning of physical units or of the interconnecting communication system) the initialization attempt is to be aborted when detecting an emergency stop condition.

The informal description leaves certain details undefined which are required to fix the assumptions about initial states. To cope with that problem in our formal model, we add some reasonable requirements (not explicitly stated in the informal description) as *integrity constraints* on initial states; namely, we assume that every admissible initial state S_0 satisfies the following conditions: (1) the valve is initially closed; (2) the pumps are initially switched off. These requirements are formalized using a nullary predicate *ValveClosed* and a unary predicate *SwitchedOff* defined on pumps (**Cond VI**):

$$S_0 \models \textit{ValveClosed} \land (\forall x \in PUMP : \textit{SwitchedOff}(x))$$

In order to avoid logical inconsistencies in the specification, further assumptions about external conditions have to be made in conjunction with the informally stated requirements addressing the intended dynamic behaviour of the system. Those assumptions will be defined on the way.

The behaviour of the control program when operating in the initialization mode is specified by the *initialization rules $I1$ - $I3$* as defined below.

[*The program enters a state in which it waits for the message STEAM-BOILER_WAITING to come from the physical units. As soon as this message has been received the program checks whether the quantity of steam coming out of the steam boiler is really zero. If the unit for detection of the level of steam is defective—that is, when v is not equal to zero—the program enters the emergency stop mode.*]

To indicate that the message STEAM-BOILER_WAITING has been received by the program (either in the current cycle or in any of the previous cycles), we introduce a predicate *SteamBoilerWaiting* (to be refined at a later stage).

On the basis of the above definition (in conjunction with the reasonable assumption that the heating system of the steam boiler remains inactive during the entire initialization phase) we can now identify a concrete condition that leads to the recognition of a steam measuring unit failure: (**Cond VII**) for every state S_i ($i = 0, 1, \ldots$) in a regular run of the steam boiler algebra the following assertion holds:

$$S_i \models \textit{InitMode} \land (v > 0) \Rightarrow \textit{Failure}(\textit{steam\_measuring\_unit})$$

v is the 0-ary function (variable) which describes the quantity of steam coming out of the steam boiler. In a similar way all the variables of the informal problem description are represented in our evolving algebra models.

[*If the program realizes a failure of the water level detection unit it enters the emergency stop mode.*]

$I1$: **if** *InitMode* $\land$ *SteamBoilerWaiting*
$\land$ (*Failure*(*steam_measuring_unit*) $\lor$ *Failure*(*level_measuring_unit*))
then *EnterEmergencyStopMode*

[*As soon as a level of water between N_1 and N_2 has been reached the program can send continuously the signal PROGRAM_READY to the physical units until it receives the signal PHYSICAL_UNITS_READY which must necessarily be emitted by the physical units.*]

In the rule below the predicate *PhysicalUnitsReady* indicates whether the program has received the signal PHYSICAL_UNITS_READY (either in the current cycle or any of the previous cycles). The macro *IndicateProgramReady* is used as a shorthand to refer to the operation which sends the signal PROGRAM_READY to the physical units.

$I2:$ **if** *InitMode* $\wedge$ *SteamBoilerWaiting*
          $\wedge$ *WaterLevelAdjusted* $\wedge$ $\neg$*PhysicalUnitsReady*
     then *IndicateProgramReady*

Note that the control program repeats the sending of the PROGRAM_READY signal until it eventually receives the PHYSICAL_UNITS_READY signal, which has the following meaning:

[*As soon as this signal has been received, the program enters either the mode normal if all the physical units operate correctly or the mode degraded if any physical unit is defective.*]

In order to avoid a subtle error in the dynamics of the system, the system should behave as required above only if the water level is still between N_1 and N_2. Imagine that the water level becomes inadmissible (due to some mechanical defect of the steam boiler or because of a faulty pump which cannot be switched off) while the program is waiting for the signal PHYSICAL_UNITS_READY to be sent by the physical units. Now, the operation of adjusting the water level may still be in progress (and the water level outside the admissible range) when receiving the signal PHYSICAL_UNITS_READY. To switch to mode normal or degraded could therefore mean to effectively start the steam boiler in a state in which the water level is already outside the limiting values M_1, M_2.

It seems therefore reasonable to add the requirement that the system behaves as stated in the informal description only if the water level is adjusted and switches to mode emergency stop otherwise (**NB.**). At the same time, it must be ensured that rule $I3$ cannot switch to mode normal or degraded in case that rule $I1$ fires (recall that more than one rule may fire simultaneously):

$I3:$ **if** *InitMode* $\wedge$ *SteamBoilerWaiting* $\wedge$ *PhysicalUnitsReady*
     thenif *WaterLevelAdjusted* $\wedge$ $\neg$*Failure(level_measuring_unit)*
          $\wedge$ $\neg$*Failure(steam_measuring_unit)* $\wedge$ $\neg$*EmergencyStop*
       thenif *AllPhysicalUnitsOk*
         then *EnterNormalMode*
         else *EnterDegradedMode*
       else *EnterEmergencyStopMode*

4.3   Normal Mode

[*As soon as the program recognizes a failure of the water level measuring unit it goes into rescue mode.*]

[*Failure of any other physical unit puts the program into degraded mode.*]

$N1:$ **if** $NormalMode \land \neg EmergencyStop \land \neg AllPhysicalUnitsOk$
     thenif $Failure(level\_measuring\_unit)$
        then $EnterRescueMode$
        else $EnterDegradedMode$

Note that if a failure of the water level measuring unit and a failure of the steam measuring unit occur simultaneously it could be more effective to switch to emergency stop mode immediately rather than to switch to rescue mode and then to emergency stop (with one cycle delay). However, as this would also mean to change the required behaviour (which might have been defined in this way for other reasons), our model behaves in the prescribed way.

4.4 Degraded Mode

[*The degraded mode is the mode in which the program tries to maintain a satisfactory water level despite the presence of failure of some physical unit. It is assumed however that the water level measuring unit in the steam boiler is working correctly. The functionality is the same as in the preceding case.*]

[*As soon as the program sees that the water level measuring unit has a failure, the program goes into mode rescue.*]

[*Once all the units which were defective have been repaired, the program comes back to normal mode.*]

$D1:$ **if** $DegradedMode \land \neg EmergencyStop$
     thenif $AllPhysicalUnitsOk$
        then $EnterNormalMode$
        elif $Failure(level\_measuring\_unit)$
        then $EnterRescueMode$

4.5 Rescue Mode

[*The rescue mode is the mode in which the program tries to maintain a satisfactory water level despite of the failure of the water level measuring unit. The water level is then estimated by a computation which is done taking into account the maximum dynamics of the quantity of steam coming out of the steam boiler. For the sake of simplicity, this calculation can suppose that exactly n litres of water, supplied by the pumps, do account for exactly the same amount of boiler contents (no thermal expansion). This calculation can however be done only if the unit which measures the quantity of steam is itself working and if one can rely upon the information which comes from the units controlling the pumps.*]

[*As soon as the water measuring unit is repaired, the program returns into mode degraded or into mode normal.*]

[*The program goes into emergency stop mode if it realizes that one of the following cases hold: the unit which measures the outcome of steam has a failure, or the units which control the pumps have a failure, or the water level risks to reach one of the limiting values*[14] *.*]

$R1$: **if** *RescueMode*
    thenif *PumpCtrlFailure* ∨ *Failure(steam_measuring_unit)*
      then *EnterEmergencyStopMode*
      elif ¬*Failure(level_measuring_unit)* ∧ ¬*EmergencyStop*
    thenif *AllPhysicalUnitsOk*
      then *EnterNormalMode*
      else *EnterDegradedMode*

4.6 Emergency Stop Mode

[*The emergency stop mode is the mode into which the program has to go, as we have seen already, when either the vital units have a failure or when the water level risks to reach one of its two limit values.*]

This is ensured by the individual rules which define the program behaviour depending on the respective mode of operation.

[*This mode can also be reached after detection of an erroneous transmission between the program and the physical units. This mode can also be set directly from outside.*]

This is ensured by rule $G1$.

[*Once the program has reached the emergency stop mode, the physical environment is then responsible to take appropriate actions, and the program stops.*]

Notice that our rules do not care about actions which have been triggered when switching to emergency stop mode; in particular, this also means that such actions are not *canceled*. The emergency stop mode represents the final state within the ground model because there is no applicable rule by means of which the program could escape from emergency stop, once it has reached this mode. This is the reason why all our rules contain the negation of *EmergencyStopMode* in their guard. As the program stops, it cannot read any new input nor produce any further output nor update any function.

5 Message Passing Interface

The steam boiler control unit interacts with the physical environment through a message passing interface. In order to comply to the fairly abstract view suggested by the informal description, we model this message passing interface without specifying any operational details of how messages are sent or received. In particular, we do not address the exact timing behaviour—leaving open whether the communication model is synchronous or asynchronous—nor do we uniquely

---

[14] Remember that this third clause has been taken into account already by rule $G1$.

identify the physical units which are considered as sender or receiver of certain messages.

In our mathematical model messages are represented as abstract objects of a dynamic domain *MESSAGE*. The various message types specified in the informal description are introduced as elements of the domain *MSGTYPE*. Since the set of physical units is fixed and a priori known, it is convenient to encode the unit addresses directly into the message types—*MSGTYPE* thus contains objects such as *OPEN_PUMP_1, OPEN_PUMP_2,* ... etc. At the same time, we also refine *PUMP_STATE* and *PUMP_CONTROL_STATE* to *PUMP_OPEN* and *PUMP_CLOSED* resp. *PUMP_CONTROL_FLOW* and *PUMP_CONTROL_NO_FLOW*. Note that our definition of message types implies that messages are uniquely identified by their type among those messages which are sent or received within the same cyle.

To represent the actual message content, for instance as required for messages of type *MODE, LEVEL,* or *STEAM*, we assume to have the domain *MSGCONT*. For a straightforward formalization of messages having been sent or received prior to the current cycle, also the number of the cycle at which a message comes into life is attached to the message. We access this information by the three functions *type, cont, cycle* from *MESSAGE* to *MSGTYPE, MSGCONT, NAT* respectively.

5.1 Sending and Receiving

Think of the domain *MESSAGE* as being partioned into two (dynamically growing) subsets *IN* and *OUT* such that *IN* refers to the messages which have been received and *OUT* to those which have been sent by the control unit. The operation of creating a new message to be sent from the control unit to one or more of the physical units is explicitly modeled through the following macro:

$$CreateMssg(Type, Cont) \equiv \textbf{extend } MESSAGE \textbf{ with } x$$
$$type(x) := Type$$
$$cont(x) := Cont$$
$$cycle(x) := curr\_cycle$$
$$\textbf{endextend}$$

(Note that the actual send operation, as expressed by the *SendMessages* macro (cf. Sect. 3.1), becomes effective in the subsequent writing phase of the current cycle.) For the sake of brevity, we will use *CreateMssg(Type)* as a shorthand for *CreateMssg(Type, undef)* when dealing with messages for which the relevant information is just the message type.

Messages $m \in IN$ ($m \in OUT$) with $cycle(m) = curr\_cycle$ are considered as being received (sent) within the current cycle. Note that *IN*, in contrast to *OUT*, is not updated by the program but by the external environment (the physical units). To check whether a message of a certain type has been transmitted (sent or received) in the *current cycle*, it is convenient to use an extra predicate

$transmitted : MSGTYPE \to BOOL$ with the following meaning:

$transmitted(Type)$
$\equiv \exists m \in MESSAGE : type(m) = Type \land cycle(m) = curr\_cycle$

Similarly, we will also refer to messages which have been sent or received in the preceding or antepreceding (etc.) cycle using a special notation:

$transmitted(Type)^-$
$\equiv \exists m \in MESSAGE : type(m) = Type \land cycle(m) = curr\_cycle - 1$

$transmitted(...)^{--}$ is defined accordingly.

Although our message passing model reflects in direct manner the view of the informal description—a view which is not committed to any particular implementation—, it allows us to specify the transmission of messages with the necessary precision and detail as follows.

[*MODE(m): The program sends, at each cycle, its current mode of operation to the physical units.*]

Recall that the value of *mode* may be affected by the rules defined in Sects. 4.1-4.5. It is the value of mode, possibly updated at the end of the current cycle, that is to be sent to the physical units. A proper synchronization of the required operations can easily be achieved by refining rule $T2$ (introduced in Sect. 3.1) into two subrules, $T2.1$ and $T2.2$, effectively splitting the *executing* phase into two internal *microphases* as expressed below:

$T2.1:$ **if** $phase = executing$ $T2.2:$ **if** $phase = executing'$
  then $phase := executing'$ **then** $phase := writing$
                                         $CreateMssg(MODE, mode)$

[*STOP: When the message has been received three times in a row by the program, the program must go into emergency stop.*]

The abstract condition *ExternalStop* used in the definition of *EmergencyStop* in $G1$ can now be refined as follows:

$ExternalStop$
$\equiv transmitted(STOP) \land transmitted(STOP)^- \land transmitted(STOP)^{--}$

Most of the message types defined in the informal description are related to error handling. The corresponding error handling protocols are specified in Sect. 5.2, while the detection of equipment failures is considered in Sect. 3.3.

5.2 Error Handling Protocols

The error handling protocols dealing with failures of physical units require the availability of some status information about the units. We thus define a unary dynamic function $status : UNIT \rightarrow \{regular, defective, acknowledged\}$ specifying, for a given unit, one of three possible situations: the unit is considered as operating correctly (*regular*); a failure of this unit has occurred but the corresponding error message of the control program has not yet been acknowledged (*defective*); the error message has been acknowledged but so far no message has been received (*acknowledged*) from the unit telling that the latter has been repaired.

The reaction of the program to unit failures as identified by the predicate *Failure* (see Sect. 3.3) is specified by the following three error handling rules where, for the sake of definiteness, we assume (**Cond VIII**) that a failure detection message will be acknowledged before the environment sends a repaired message.[15]

$E1:$ **var** x **ranges over** $UNIT$
       if $status(x) = regular \wedge Failure(x)$
       then $status(x) := defective$
            $CreateMssg(FailureDetectionMssg(x))$

$E2:$ **var** x **ranges over** $UNIT$
       if $status(x) = defective$
       thenif $transmitted(FailureAcknowledgeMssg(x))$
          then $status(x) := acknowledged$
          else $CreateMssg(FailureDetectionMssg(x))$

$E3:$ **var** x **ranges over** $UNIT$
       if $status(x) = acknowledged \wedge transmitted(RepairedMssg(x))$
       then $status(x) := regular$
            $CreateMssg(RepairedAcknowledgeMssg(x))$

where $FailureDetectionMssg(x)$, $FailureAcknowledgeMssg(x)$, $RepairedMssg(x)$, and $RepairedAcknowledgeMssg(x)$ refer to the corresponding error handling messages depending on the device type and the device number of the particular unit.

5.3 Detection of Equipment Failures

In this section, we will define the meaning of the two up to now abstract failure predicates *Failure* and *TransmissionFailure*. The interpretation of these predicates is of vital importance for the overall behaviour of the entire model. In order to derive their meaning systematically by stepwise refinements, we

---

[15] Different solutions are possible of course if repairing takes less time than sending an acknowledgement of failure detection.

introduce a number of auxiliary predicates; for the definiton of these auxiliary (locally definable) predicates we refer to the Glossary, except for the predicate *Defective(x)* which is used as abbreviation for $status(x) \neq regular \land \neg transmitted(RepairedMssg(x))$.

[PUMP: *(1) Assume that the program has sent a start or stop message to a pump. The program detects that during the following transmission that pump does not indicate its having effectively been started or stopped. (2) The program detects that the pump changes its state spontaneously.*]

$$\text{for } p \in PUMP:$$
$$Failure(p) \Leftrightarrow NonReactingPump(p) \lor$$
$$IrregularPumpAction(p) \lor Defective(p)$$

[PUMP_CONTROLLER: *(1) Assume that the program has sent a start or stop message to a pump. The program detects that during the second transmission after the start or stop message the pump does not indicate that the water is flowing or is not flowing; this despite of the fact that the program knows from elsewhere that the pump is working correctly. (2) The program detects that the unit changes its state spontaneously.*]

$$\text{for } p \in PUMP\_CTRL:$$
$$Failure(p) \Leftrightarrow (NonReactingPumpCtrl(p) \land \neg Failure(Pump(p)))$$
$$\lor\ IrregularPumpCtrlEvent(p) \lor Defective(p)$$

[WATER_LEVEL_MEASURING_UNIT: *(1) The program detects that the unit indicates a value which is out of the valid static limits–i.e. between 0 and C. (2) The program detects that the unit indicates a value which is incompatible with the dynamics of the system.*]

$$Failure(level\_measuring\_unit) \Leftrightarrow$$
$$OutOfRangeWaterLevel \lor$$
$$IncompatibleWaterLevel \lor Defective(level\_measuring\_unit)$$

[STEAM_LEVEL_MEASURING_UNIT: *(1) The program detects that the unit indicates a value which is out of the valid static limits–i.e. between 0 and W. (2) The program detects that the unit indicates a value which is incompatible with the dynamics of the system.*]

$$Failure(steam\_measuring\_unit) \Leftrightarrow$$
$$OutOfRangeSteamValue \lor$$
$$IncompatibleSteamValue \lor Defective(steam\_measuring\_unit)$$

[TRANSMISSION: *(1) The program receives a message whose presence is aberrant. (2) The program does not receive a message whose presence is indispensable.*]

$$TransmissionFailure \Leftrightarrow$$
$$AberrantMessage \lor MissingMessage$$

(See notes 28 and 29 of the Glossary for the definition of the predicates *AberrantMessage* and *MissingMessage*.)

5.4 The Abstract Machine Program

Below we give a complete listing of the abstract machine program. Note that we assume the condition '$\neg EmergencyStop$' to be a global precondition extending the guards of all rules—except for the global rules $G1, G2$; we further assume that the condition '$phase = executing$' specifies an additional precondition for the following rules: the global rules ($G1, G2$), the mode rules ($I1$-$I3$, $N1$, $D1$, $R1$) and the error handling rules ($E1$-$E3$).

Timing Rules

$T1$: **if** $phase = reading$
$\quad\quad \land\ curr\_time - last\_time = 5$
$\quad$ **then** $ReadMessages$
$\quad\quad phase := executing$
$\quad\quad last\_time := curr\_time$
$\quad\quad curr\_cycle :=$
$\quad\quad\quad curr\_cycle + 1$

$T2.1$: **if** $phase = executing$
$\quad$ **then** $phase := executing'$

$T2.2$: **if** $phase = executing'$
$\quad\quad$ **then** $phase := writing$
$\quad\quad\quad CreateMssg(MODE, mode)$

$T3$: **if** $phase = writing$
$\quad\quad$ **then** $SendMessages$
$\quad\quad\quad phase := reading$

Global Rules

$G1$: **if** $EmergencyStop$
$\quad$ **then** $EnterEmergencyStopMode$

$G2$: **if** $WaterLevelAdjusted$
$\quad\quad \land\ \neg EmergencyStop$
$\quad$ **then** $RetainWaterLevel(mode)$
$\quad$ **else** $AdjustWaterLevel(mode)$

Initialization Mode

$I1$: **if** $InitMode$
$\quad\quad \land\ SteamBoilerWaiting$
$\quad\quad \land\ (Failure(steam\_measuring\_unit)$
$\quad\quad \lor\ Failure(level\_measuring\_unit))$
$\quad$ **then** $EnterEmergencyStopMode$

$I2$: **if** $InitMode$
$\quad\quad \land\ SteamBoilerWaiting$
$\quad\quad \land\ WaterLevelAdjusted$
$\quad\quad \land\ \neg PhysicalUnitsReady$
$\quad$ **then** $IndicateProgramReady$

$I3$: **if** $InitMode$
$\quad\quad \land\ SteamBoilerWaiting$
$\quad\quad \land\ PhysicalUnitsReady$
$\quad$ **thenif** $WaterLevelAdjusted$
$\quad\quad \land\ \neg Failure(level\_measuring\_unit)$
$\quad\quad \land\ \neg Failure(steam\_measuring\_unit)$
$\quad\quad \land\ \neg EmergencyStop$
$\quad$ **thenif** $AllPhysicalUnitsOk$
$\quad\quad$ **then** $EnterNormalMode$
$\quad\quad$ **else** $EnterDegradedMode$
$\quad$ **else** $EnterEmergencyStopMode$

Normal Mode Rule

$N1$: **if** *NormalMode*
　　$\wedge \neg EmergencyStop$
　　$\wedge \neg AllPhysicalUnitsOk$
　thenif *Failure(level_measuring_unit)*
　　then *EnterRescueMode*
　　else *EnterDegradedMode*

Rescue Mode Rule

$R1$: **if** *RescueMode*
　thenif *PumpCtrlFailure*
　　$\vee$ *Failure(steam_measuring_unit)*
　　then *EnterEmergencyStopMode*
　elif $\neg Failure(level\_measuring\_unit)$
　　$\wedge \neg EmergencyStop$
　thenif *AllPhysicalUnitsOk*
　　then *EnterNormalMode*
　　else *EnterDegradedMode*

Degraded Mode Rule

$D2$: **if** *DegradedMode*
　　$\wedge \neg EmergencyStop$
　thenif *AllPhysicalUnitsOk*
　　then *EnterNormalMode*
　　elif *Failure(level_measuring_unit)*
　　then *EnterRescueMode*

Error Handling Rules

$E1$: **var** x **ranges over** *UNIT*
　　if $status(x) = regular \wedge Failure(x)$
　　then $status(x) := defective$
　　　　CreateMssg (FailureDetectionMssg (x))

$E2$: **var** x **ranges over** *UNIT*
　　if $status(x) = defective$
　　thenif *transmitted(FailureAcknowledgeMssg (x))*
　　　then $status(x) := acknowledged$
　　　else *CreateMssg (FailureDetectionMssg (x))*

$E3$: **var** x **ranges over** *UNIT*
　　if $status(x) = acknowledged \wedge transmitted(RepairedMssg (x))$
　　then $status(x) := regular$
　　　　CreateMssg (RepairedAcknowledgeMssg (x))

6 Implementation

The evolving algebra specification of the steam boiler control program defined in the preceding sections can be implemented by a C++ program in such a way that the abstract specification represents the structure of the executable code. This makes the code easily inspectable by formal means and provides useful interfaces for possible modifications of the program. We believe that this approach to program documentation—i.e. providing a sequence of stepwise refined abstract models leading to executable code—contributes to the reliability of the produced software.

For the implementation of the evolving algebra model for the steam boiler control we have translated the rules, the signature (including initialization) and the abstract definitions (macros) of the model into C++ code. In order to make this work we had to program also the underlying semantics of evolving algebras (including the concurrency of the executions). For the connection to the Karlsruhe steam boiler simulator we also had to program the physical model. This physical model is realized through a collection of C++ functions (see files `dynamic.H` and `dynamic.C`) refining macros like `AdjustWaterLevel`, `OpenSomePumps`, `CloseSomePumps` and will not be addressed any further here.

In the following we focus on the embedding of the control model into C++, where we can identify three basic aspects, namely: *i*) the implementation of evolving algebra *core routines* (Sect. 6.1), *ii*) the mapping of program rules (Sect. 6.2) and *iii*) the communication with the simulator (Sect. 6.3). Section 6.4 presents some statistics on the code development. The complete code is available at: http://www.uni-paderborn.de/fachbereich/ AG/agklbue/staff/igor/ea/dag/c++/.

6.1 Evolving Algebra Core Routines

For the implementation of evolving algebra core routines we restrict here to those routines which are relevant for the controller specification effectively implementing a subclass of evolving algebras (whereas a complete model of executable evolving algebras can be found in [11]). The main aspect in the translation of evolving algebra states into C++ is the representation and handling of function values (see `eav.H`). A given function value is represented by the template class `EAV<T>` (*evolving algebra value*) in the form (*val,def,time*), where *val* is a C++ value of type `T`, *def* is a flag which masks the value *val* in case that it refers to the distinguished element *undef*, and *time* is a time stamp indicating the phase in which the value was assigned (resp. '0' for initially defined values). Time here is measured by an integer variable `ea_clock` which serves as a global phase counter.

An additional template class `EA<T>` extends the value representation scheme defined above by introducing a *history mechanism* such that *past values*— i.e. values that have already been updated—can be accessed (within a limited range[16]) in much the same way as *current values* by applying the postfix operator `[]`; if `f` is an r-ary function name and `t=` $t_1, \ldots, t_r$, where the t_i's are terms, then an expression of the form `f(t)[`$-k$`]`, $k = 1, 2, \ldots$, refers to the preceding, antepreceding, etc. value of `f(t)`. The interpretation of `f(t)` in phase k is defined by the tuple v in the history of `f(t)` (provided it exists) such that $v = (x, y, z)$, $z < k$ and there is no $v' = (x', y', z')$ in the history of `f(t)` with $z' < k$ and $z < z'$. Through the use of circular buffers the history mechanism is efficiently implemented such that the history length is a free parameter and can be chosen independently for each individual function.

---

[16] For the steam boiler control we need at most three values: the present value, the preceding value and the antepreceding value.

By combining the value representation scheme with the history mechanism the order in which the updates are computed within a given phase becomes irrelevant as implied by of the following two facts: *i*) all function values which are updated within the current phase get the same time stamp, viz. the value of `ea_clock`; *ii*) none of these updates can become effective prior to the next phase[17]. This in fact means that at the C++ level the parallel execution model of evolving algebra rules is transformed into a sequential execution model which is equivalent w.r.t. the resulting observable behaviour.

Encoding of Initial States. Each function used in our evolving algebra model must explicitly be declared as an instance of the template class `EA<T>`; for each function the desired history length must be specified if more than two values need to be stored[18]. The resulting collection of function declarations defines the *vocabulary* (or *signature*) of our evolving algebra model (see `vocab.H`); a proper initialization of these functions is defined in the file `vocab.C`. The only exceptions are the water level and steam value prediction functions, `level_comp_min`, `level_comp_max` resp. `steam_comp_min`, `steam_comp_max`, for which the very first messages `LEVEL(x)` and `STEAM(y)` sent by the simulation environment are taken as initial values.

Static universes are realized as C++ enumerations as exemplified by the encoding of the universe *UNIT*:

```
enum UNIT {
  LEVEL = 0, STEAM,
  PUMP_1, PUMP_2, PUMP_3, PUMP_4,
  PUMP_CTRL_1, PUMP_CTRL_2, PUMP_CTRL_3, PUMP_CTRL_4,
  UNIT_last
};
```

where *U_last* is a fictive generic serving as an end marker.

6.2 Mapping of Program Rules

In order to bridge syntactical differences between C++ and evolving algebras, the file `def.H` defines a number of C++ macros. The basic idea is illustrated by the following scheme:

   `IF a THEN b ELSE c ENDIF`   stands for: `if ( a ) { b } else { c }` .

As an example of the resulting mapping of our evolving algebra rules (see `rules.ea`[19]) consider the encoding of the error handling rule *E1*:

---

[17] Note that read operations in case of updated values do always refer to the values of the preceding phase (as explained above).

[18] The default history length of two is the minimum length required for the sequentialization of the execution model.

[19] Note that each rule in `rules.ea` is decorated by an additional `FIRE(R)` macro which is used for tracing purposes only.

$E1:$ **var** x **ranges over** $UNIT$
        if $status(x) = regular \land Failure(x)$
        then $status(x) := defective$
              $SendMssg\ FailureDetectionMssg\,(x))$

```
// Rule E1:
    VAR_RANGES_OVER( x, UNIT )
       IF status(x) == regular AND Failure(x)
       THEN
          status(x) = defective;
          CreateMssg(FailureDetectionMssg(x));
       ENDIF
    ENDVAR
```

Here the macro `VAR_RANGES_OVER(x,U)` stands for traversing the universe `U`, i.e. a C++–loop statement starting at the first and ending with the last element of `U`. The evolving algebra execution model is realized as an infinite loop (see `main.C`) within which the included rules (`#include "rules.ea"`) are executed and the phase counter `ea_clock` is incremented.

Macros. Macros used in the specification are either realized as C++ macros (see `macros.H`), in case of simple macros, or as C++ functions (see `macros.C`), in case of complex macros. Some macros which concern the physical model are further refined to reflect the implementation. As an example of the encoding of macros consider the following:

NoFlowIndication(pump_ctrl-i)
   $\equiv transmitted(OPEN\_PUMP\_i)^{--} \land$
   $\neg transmitted(CLOSE\_PUMP\_i)^{-} \land$
   $transmitted(PUMP\_CONTROL\_i\_NO\_FLOW)$

```
#define NoFlowIndication( i ) \
      ( transmitted__( open_PUMP_1 + i) AND \
   NOT transmitted_ (close_PUMP_1 + i) AND \
      transmitted(PUMP_CTRL_1_no_flow + i) )
```

6.3   Communication with the Simulator

Communication between the control program and the Karlsruhe simulator is based on message passing through pipes. On top of this communication model, which offers the usual low-level communication primitives, a more abstract communication model is realized (see `comm.H`, `comm.C`); this abstract model directly reflects the view of the formal specification. The dynamic universe *MESSAGE* is implicitly modeled through an array `message` of boolean instances of the class `EA<BOOL>`:   `EA<BOOL>  message[ MSGTYPE_last ];` (where the array represents the function $message : MSGTYPE \to BOOL$). The creation of a message x is expressed by the following macro:

```
#define CreateMssg(x)    message[x] = true .
```

With each entry of the array **message** we associate a current value and history. The history length is limited, i.e. old messages may be discarded; however it is long enough to ensure a proper functioning of the program. At the C++ level the **SendMessages** operation initiated by rule $T1$ (see Sect. 3.1) has the following meaning: the program checks the presence of messages for each individual message type (as indicated by the array **message**) and sends for each logically present message a physical message (a string representation) to the simulator. Similarly, the **ReadMessages** operation indicates the presence of messages sent by the simulator by updating the corresponding entries of **message**. Finally, **ResetInports** and **ResetOutports** delete all present input/output messages by resetting the entries of **message** to **false**.

6.4 Statistics and Experiments

The first implementation, the one which was presented at the Dagstuhl workshop, took about two weeks of work for one person. After the workshop a new version which also reflects improvements in the specification was produced within a week. The size of the resulting program is 1720 lines of source code (about 38 KB) and 93 KB executable (compiled with SUN's 4.1 C++ compiler on a SUN workstation with Solaris 2.5).

The major cause for the problems we encountered during the tests of the first version of our program was the lack of a rigorous description of the physical model and of the communication and timing behavior which have been used for running the Karlsruhe simulator. This incompleteness of the informal problem description is also the reason why certain errors can not be detected without additional explicit design decisions; an example is the simultaneous breakdown of pumps and pump controllers and similar cases which have been analyzed furthermore in some other contributions to this work. On the basis of a formalization of the physical model for the steam boiler, it would be possible to provide a correctness proof for the translation of evolving algebra rules into C++ code using the precise semantic model for C++ in [17]—but such an endeavor is out of the scope of this case study.

7   Evaluation and Comparison

1. We provide a formal specification of the control program as an abstract machine which is very close to the informal description and therefore can easily be compared to it for checking its appropriateness from the application point of view. This specification is then refined to more detailed abstract machines and eventually translated into executable C++ code. Our abstract machine models incorporate an architectural design. Some characteristic examples of properties possessed by the abstract models are formulated and proved mathematically.

2. In the last refinement step we translate our abstract machine specification into an executable C++ program. It is available at http://www.uni-paderborn.de/fachbereich/AG/agklbue/staff/igor/ea/dag/c++/ and on the CD included in this volume. This program was linked successfully to the Karlsruhe steam boiler simulator. Several experiments were done with the control program and the steam boiler simulator. The control program was tested (and its second and final version passed) with the provided example scenarios. Our tests led to various changes in the design of the Karlsruhe simulator; they showed the incompleteness of the informal description and the need for a complete formalization of the physical model (which we provide in our solution not as part of the abstract specification but only through the C++ program, which reflects the decisions made for the physical system behaviour for the implementation of the Karlsruhe simulator). Such a complete description allows one to detect e.g. a simultaneous breakdown of pumps and pump controllers.

3. The specification of the static parts of our abstract machines is similar to and complemented by what one finds in the contributions using algebraic specification or state based temporal logics, see BCPR, GDK, CW1, LM. Our notion of abstract machines (i.e. evolving algebras) is similar in spirit to the corresponding notion underlying Abrial's method B. In our abstract models we deliberately did not make any assumptions on the physical behaviour of the system (in order to remain faithful to Abrial's original problem description), so that all the solutions which focus on a detailed description and mathematical analysis of the physical behaviour complement our work. Due to the lack of a sufficiently complete informal description of the physical system behaviour we did not push our mathematical analysis of system properties which is complemented by all the solutions in this book which focus on a more detailed analysis and proofs of system properties.

4. For the specification of the control program approximately three weeks were used for its first version and another two weeks for polishing it. Implementation time for the C++ program was two weeks for its first version, and one week for the second (and final) version.

An "average programmer" (who is supposed to have some basic knowledge of traditional mathematical reasoning) can learn to use evolving algebras in about half a week of training. We believe that the ease with which an experienced programmer can learn the use of evolving algebras is a distinguishing feature of the suggested approach to systematic formally supported programming.

5. Our experience with abstract machine models shows that an average programmer can understand the given specification without any previous knowledge of evolving algebras; it suffices to read the rules as abstract pseudo-code. A more detailed understanding of the proofs of system properties, in particular where details of the underlying precise semantics of evolving algebras are needed, may require one or two days of familiarizing oneself with the used language of transition systems.

Acknowledgement We thank two anonymous referees for their valuable criticism.

References

1. Wolfgang Ahrendt. Von Prolog zur WAM. Verifikation der Prozedurübersetzung mit KIV. Diploma thesis, University of Karlsruhe, Dec. 1995.
2. Egon Börger. A logical operational semantics for full Prolog. Part I: selection core and control. In E. Börger, H. Kleine Büning, M.M. Richter, editors, *CSL'89. 3rd Workshop on Computer Science Logic*, Springer LNCS, vol. 440, 1990, pages 36-64.
3. Egon Börger. Logic programming: the evolving algebra approach. In B. Pehrson and I. Simon (Eds.) *IFIP 13th World Computer Congress 1994*, Volume I: *Technology/Foundations*, Elsevier, Amsterdam, 391-395.
4. Egon Börger. Annotated bibliography on evolving algebras. In E. Börger, editor, *Specification and Validation Methods*, Oxford University Press, 1995, pages 37-51.
5. Egon Börger. Why use evolving algebras for hardware and software engineering. In *Proc. of SOFSEM'95 (Nov. 25 - Dec. 2, 1995, Bratislava, Czech Republic)*, LNCS 1012, Springer-Verlag, 1995, pages 236-271.
6. Egon Börger and Igor Đurđanović. Correctness of compiling Occam to Transputer code. Computer Journal, 1996, vol. 39, pages 52-92.
7. E. Börger, I. Durdanović, and D. Rosenzweig. Occam: Specification and compiler correctness. Part I: Simple mathematical interpreters. In E.-R. Olderog (Ed.), *Proc. PROCOMET'94 (IFIP Working Conference on Programming Concepts, Methods and Calculi)*, pages 489-508, North-Holland, 1994
8. E. Börger, Y. Gurevich, and D. Rosenzweig. The bakery algorithm: yet another specification and verification. In E. Börger, editor, *Specification and Validation Methods*. Oxford University Press, 1995, pages 231-243.
9. Egon Börger and Silvia Mazzanti. A correctness proof for pipelining in RISC architectures. In DIMACS TR 96-22, July 1996, pages 1-60.
10. E. Börger and D. Rosenzweig. The WAM – definition and compiler correctness. In L. C. Beierle and L. Plümer, editors, *Logic Programming: Formal Methods and Practical Applications*, Series in Computer Science and Artificial Intelligence. Elsevier Science B.V./North–Holland, 1995, pages 20-90 (Chapter 2).
11. G. Del Castillo, I. Đurđanović and U. Glässer. An evolving algebra abstract machine. In H. Kleine Büning, editor, *Computer Sience Logic (Proc. of CSL'95)*, LNCS, Springer-Verlag, 1996, pages 191-214.
12. Yuri Gurevich. Evolving Algebra 1993: Lipari Guide. In E. Börger, editor, *Specification and Validation Methods*, Oxford University Press, 1995, pages 9-36.
13. Y. Gurevich and J. Huggins. The railroad crossing problem: an experiment with instantaneous actions and immediate reactions. In H. Kleine Büning, editor, *Proc. of Computer Sience Logic – CSL'95*, LNCS, Springer-Verlag, 1996, pages 266-290.
14. Y. Gurevich and R. Mani. Group membership protocol: specification and verification. In E. Börger, editor, *Specification and Validation Methods*, Oxford University Press, 1995, pages 295-328.
15. J. Huggins. Kermit: specification and verification. In E. Börger, editor, *Specification and Validation Methods*, Oxford University Press, 1995, pages 247-293.
16. Cornelia Pusch. Verification of compiler correctness for the WAM. In *Proc. TPHOLs '96*, LNCS, Springer-Verlag (to appear).
17. C. Wallace. The semantics of the C++ programming language. In E. Börger, editor, *Specification and Validation Methods*. Oxford University Press, 1995, pages 131-164.

An Algebraic Specification of the Steam-Boiler Control System

Michel Bidoit [1], Claude Chevenier [2], Christine Pellen [2]
and Jérôme Ryckbosch [2]

[1] LIENS, C.N.R.S. U.R.A. 1327 & Ecole Normale Supérieure
45 Rue d'Ulm, F–75230 Paris Cedex 05, France

[2] Direction des Etudes et Recherches, Electricité de France
1 Avenue du Général de Gaulle, F–92141 Clamart, France

Abstract. We describe how to derive an algebraic specification of the Steam-Boiler Control System starting from the informal requirements provided to the participants of the Dagstuhl Meeting *Methods for Semantics and Specification*, organized jointly by Jean-Raymond Abrial, Egon Börger and Hans Langmaack in June 1995. The aim of this formalization process is to analyze the informal requirements, to detect inconsistencies and loose ends, and to translate the requirements into a formal, algebraic, specification. During this process we have to provide interpretations for the unclear or missing parts. We explain how we can keep track of these additional interpretations by localizing very precisely in the formal specification where they lead to specific axioms. Hence we take care of the traceability issues. We also explain how the formal specification is obtained in a stepwise way by successive refinements. Emphasis is put on how to specify the detection of the steam-boiler failures. Finally we discuss validation and verification issues. For this case study we use the PLUSS algebraic specification language and the LARCH PROVER.

1 Introduction

Our aim is to explain how one can solve the "Steam-boiler control specification problem" described in Chapter AS using classical algebraic specification techniques. First, let us stress that we understand the given problem in a rather strict way, i.e. we consider that we must provide a formal specification of the **program** which serves to control the level of water in the steam-boiler (and later on we must demonstrate that our specification can be refined to an implementation, i.e. a program, cf. Chapter AS, Section 1), but we do not consider that we must specify the whole system, i.e. the control program and its physical environment as described in Chapter AS, Section 2. In the following this control program will be referred to as the "Steam-Boiler Control System".

Our work plan is depicted in Figure 1 and can be described as follows:

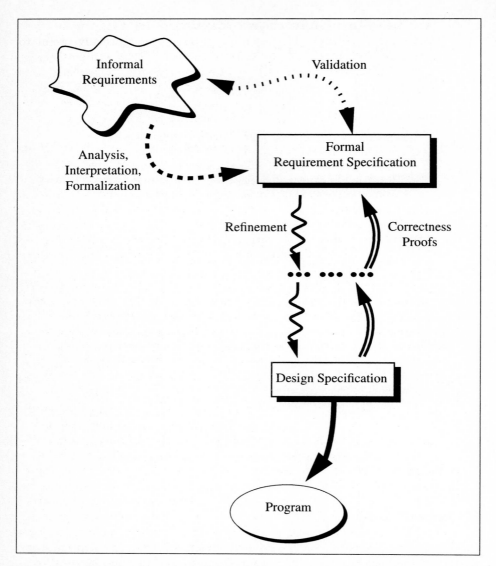

Figure 1. Specification of the Steam-Boiler Control System: the global work plan

1. The main task is to derive from the informal requirements a formal, algebraic, requirement specification of the Steam-Boiler Control System. In particular, this task involves the following activities:
    (a) We must proceed to an in-depth analysis of the informal requirements. Obviously, this is necessary to gain a sufficient understanding of the problem to be specified, and this preliminary task may not seem worth

```
spec MESSAGES-SENT is
  use BASICS;
sort S_Message = {MODE(Modes), PROGRAM_READY, VALVE,
                  OPEN_PUMP(PumpNumber), CLOSE_PUMP(PumpNumber),
                  FAILURE_DETECTION(PhysicalUnit),
                  REPAIRED_ACKNOWLEDGEMENT(PhysicalUnit)};
end MESSAGES-SENT.
```

```
spec MESSAGES-RECEIVED is
  use BASICS, VALUES;
sort R_Message =
       {STOP, STEAM_BOILER_WAITING, PHYSICAL_UNITS_READY,
        PUMP_STATE(PumpNumber,PumpState),
        PUMP_CONTROL_STATE(PumpNumber,PumpControllerState),
        LEVEL(Value), STEAM(Value),
        REPAIRED(PhysicalUnit),
        FAILURE_ACKNOWLEDGEMENT(PhysicalUnit),
        junk};
end MESSAGES-RECEIVED.
```

Note that for the received messages, in addition to the messages specified in Chapter AS, Section 6, we add an extra constant message **junk**. This message will represent any received message which does not belong to the class of recognized messages. We do not add a similar message to the messages sent, since we may assume that the Steam-Boiler Control System will only send proper messages. Obviously, the reception of a **junk** message will lead to the detection of a failure of the message transmission system.

In the **STEAM-BOILER** specification we describe the various constants that characterize the Steam-Boiler.

We are now ready to start to specify the Steam-Boiler Control System. We use the following method to describe an automaton as an abstract data type. We introduce a sort for the actions (**SBCS_Actions**), and one for the states (**SBCS_States**). In our case we have only one action (**Tick**), which corresponds to the activation of the Steam-Boiler Control System every five seconds (we do not associate an explicit action to message sending, as explained below). The **Tick** action takes as argument a (possibly empty) set of received messages. This means that we implicitly assume that our automaton will be synchronized with a clock delivering a **Tick** signal every five seconds. Note that the evolution of states is not triggered by the arrival of messages, but by this implicitly assumed clock, and when a clock signal (i.e. a **Tick**) arrives, we just collect the messages received since the last **Tick**. Values of sort **SBCS_States** are obtained from an initial state **Init** by successive activations of the **Tick**

```
par ITEM is
  sort Item;
end ITEM.

generic spec SET-OF[ITEM] is
  use NATURAL;
sort Set[Item];
generated by:
  empty : -> Set[Item];
  __ + __ : Set[Item], Item -> Set[Item];
operations:
  __ + __ : Set[Item], Set[Item] -> Set[Item];    % Set Union
  __ - __ : Set[Item], Set[Item] -> Set[Item];    % Set Difference
  __ ∈ __ : Item, Set[Item] -> Bool;              % Membership
  __ ⊆ __ : Set[Item], Set[Item] -> Bool;         % Set Inclusion
  card : Set[Item] -> Nat;                        % Cardinal
axioms:
  sort Set[Item] partitioned by ∈;
    % Note: This implies that the addition of an element
    % is a commutative and idempotent operation.ᵃ
  ¬(x ∈ empty);
  x ∈ (S + y) <=> (x = y ∨ x ∈ S);
  x ∈ (S1 + S2) <=> (x ∈ S1 ∨ x ∈ S2);
  x ∈ (S1 - S2) <=> (x ∈ S1 ∧ ¬(x ∈ S2));
  S1 ⊆ S2 <=> (∀x. x ∈ S1 => x ∈ S2);
  card(empty) = 0;
  x ∈ S => card(S + x) = card(S);
  ¬(x ∈ S) => card(S + x) = card(S) + 1;
where: x,y:Item; S,S1,S2:Set[Item];
end SET-OF[ITEM].
```

[a] This axiom, using the *partitioned by* construction, is logically equivalent to:
(S1 = S2) <=> (∀x. x ∈ S1 <=> x ∈ S2).

```
spec SET-OF-MESSAGES is
  use SET-OF[ITEM => MESSAGES-RECEIVED];
  % Induces a renaming of Set[Item] into Set[R_Message].
  use SET-OF[ITEM => MESSAGES-SENT];
  % Induces a renaming of Set[Item] into Set[S_Message].
  % In the following we will need also to reason about sets of pumps...
  use SET-OF[ITEM => PUMPNUMBER];
  % Induces a renaming of Set[Item] into Set[PumpNumber].
end SET-OF-MESSAGES.
```

```
spec STEAM-BOILER is
  use VALUES;
constants: C, M1, M2, N1, N2, W, U1, U2, P : Value;
constants: dt : Value;  % Time duration between two cycles (5 sec.)
axioms:
  % These constants must verify some obvious properties:
  0 < M1; M1 < N1; N1 < N2; N2 < M2; M2 < C;
  0 < W; 0 < U1; 0 < U2; 0 < P;
end STEAM-BOILER.
```

action. Each activation is reflected by the $\{\_,\_\}$ operation. Hence all states of the Steam-Boiler Control System automaton are denoted by terms of the form $\{\{\ldots\{\{\text{Init},\text{Tick}(\text{msgs1})\},\text{Tick}(\text{msgs2})\},\ldots\},\text{Tick}(\text{msgsN})\}$ which reflect the history of events leading to the corresponding states (cf. Figure 2). Thus in our approach states are dealt with explicitly, and they will occur as explicit arguments of the operations we will specify. Note however that we do not specify states as being some tuple of "state variables", and we still adopt a rather abstract view of what the states of the Steam-Boiler Control System are. Indeed, at this stage it would be difficult to determine which state variables are needed, and this decision should not be part of the requirements formalization process, but rather of a later stage of design specification. We nevertheless need to retrieve some informations concerning the current state of the Steam-Boiler

```
draft STEAM-BOILER-CONTROL-SYSTEM is
  use SET-OF-MESSAGES;
sort SBCS_Actions;
generated by:
  % Each five seconds, Tick triggers a new cycle:
  Tick : Set[R_Message] -> SBCS_Actions;
sort SBCS_States;
generated by:
  Init : -> SBCS_States;
  % Evolution of states triggered by actions:
  { __ , __ } : SBCS_States, SBCS_Actions -> SBCS_States;
operations:
  % Current mode of operation:
  __ .Mode : SBCS_States -> Modes;
  % Set of messages sent by the Steam-Boiler Control System:
  MessagesSent : SBCS_States, Set[R_Message] -> Set[S_Message];
  % Set of pumps ordered to start or to stop:
  ChosenPumps:
    SBCS_States, Set[R_Message], PumpState -> Set[PumpNumber];
end STEAM-BOILER-CONTROL-SYSTEM.
```

Control System, such as the current mode of operation. This is done by means of "observers", in this case the operation .Mode. Similarly, we define an operation **MessagesSent** which, given a state and a set of received messages, returns the set of messages sent by the Steam-Boiler Control System (after having performed the analysis of the received messages). We introduce also a **ChosenPumps** operation which will return the set of pumps to which a start or stop order is sent (depending on its third argument). Note that it is still too early to specify these operations, and therefore the STEAM-BOILER-CONTROL-SYSTEM draft specification does not include any axiom.

Since we have decided not to specify the initialization phase, the initial state **Init** will represent the state immediately following the reception of the **PHYSICAL_UNITS_READY** message (cf. Chapter AS, Section 4.1).

The next step is the specification of the various operating modes in which the Steam-Boiler Control System operates. (As explained above we do not take into account the **Initialization** mode in this specification.) According to Chapter AS, Section 4, the operating mode of the Steam-Boiler Control System depends on which failures have been detected (cf. e.g. "all physical units [are] operating correctly", "a failure of the water level measuring unit"). We will therefore first introduce, in the SBCS-PREDICATES draft specification, a number of predicates which should reflect the failures detected by the Steam-Boiler Control System. It is important to make a subtle distinction between the actual failures about which we basically know nothing, and the failures detected by the Steam-Boiler Control System. In our specification, the behaviour of the Steam-Boiler Control System is induced by the failures detected, whatever

---

draft SBCS-PREDICATES is
  enrich STEAM-BOILER-CONTROL-SYSTEM;
  % Given a known state and newly received messages, the following
  % predicates give an up-to-date view of the physical environment.
operations:
  % true iff we rely on the message transmission system:
  TransmissionOk : SBCS_States, Set[R_Message] -> Bool;
  % true iff we rely on the Pn pump:
  PumpOk : SBCS_States, Set[R_Message], PumpNumber -> Bool;
  % true iff we rely on the Pn pump controller:
  PumpControllerOk : SBCS_States, Set[R_Message], PumpNumber -> Bool;
  % true iff we rely on the output of steam measuring device:
  OutputOfSteamOk : SBCS_States, Set[R_Message] -> Bool;
  % true iff we rely on the water level measuring device:
  WaterLevelOk : SBCS_States, Set[R_Message] -> Bool;

Continued on next page

```
                    Draft SBCS-PREDICATES continued
  % redundant but useful:
  PhysicalUnitOk : SBCS_States, Set[R_Message], PhysicalUnit -> Bool;
  % true iff we believe that all physical units operate
  % correctly (including the message transmission system):
  AllPhysicalUnitsOk,
  % true iff we have received the STOP message three times in a row:
  AskedToStop,
  % true iff we estimate that the water level risks
  % reaching the min (M1) or max (M2) limits:
  DangerousWaterLevel,
  % true iff we rely on the output of steam measuring device
  % and at least on some pump and its controller:
  SystemStillControllable,
  % true iff we know there is some good reason to enter the
  % emergency stop mode:
  EmergencyStop       : SBCS_States, Set[R_Message] -> Bool;
axioms:
  AllPhysicalUnitsOk(s,msgs)   <=>
    ( TransmissionOk(s,msgs) ∧
      (∀PhysUnit. PhysicalUnitOk(s,msgs,PhysUnit)) );
  SystemStillControllable(s,msgs)   <=>
    ( OutputOfSteamOk(s,msgs) ∧
      (∃Pn. PumpOk(s,msgs,Pn) ∧ PumpControllerOk(s,msgs,Pn)) );

  PumpOk(s,msgs,Pn) = PhysicalUnitOk(s,msgs,Pump(Pn));
  PumpControllerOk(s,msgs,Pn) =
                         PhysicalUnitOk(s,msgs,PumpController(Pn));
  OutputOfSteamOk(s,msgs) = PhysicalUnitOk(s,msgs,OutputOfSteam);
  WaterLevelOk(s,msgs) = PhysicalUnitOk(s,msgs,WaterLevel);
where:
  s:SBCS_States; msgs:Set[R_Message];
  Pn:PumpNumber; PhysUnit:PhysicalUnit;
end SBCS-PREDICATES.
```

the actual failures are. Most of the predicates introduced in SBCS-PREDICATES have self-explanatory names and do not deserve any further comment. Note, however, that for the sake of clarity we also introduce some additional predicates (PhysicalUnitOk, AllPhysicalUnitsOk, SystemStillControllable) that are redundant (in the sense they are defined by means of the other predicates) but will prove useful for increasing the readability of the specification.

The aim of the predicate SystemStillControllable is to characterize the conditions under which the Steam-Boiler Control System will operate in Rescue mode. Let us point out that the corresponding part of the informal requirements (see Chapter AS, Section 4.4) is not totally clear, in particular the exact meaning of the sentence "if one can rely upon the information which comes from the units for controlling the pumps". There is a double ambiguity here: on the one hand it is unclear whether "the pumps" means "all pumps" or "at least one pump"; on the other hand there are two ways of "controlling" each pump (the information sent by the pump and the information sent by the pump controller), and it is unclear whether "controlling" refers to both of them or only to the pump controller. Our interpretation will be as follows: we consider it is enough that at least one pump is "correctly working", and for us correctly working will mean we rely on both the pump and the associated pump controller. As for all interpretations made during the formalization process, in principle we should interact with the designers of the informal requirements in order to clarify what was the exact intended meaning and to check that our interpretation is adequate. The important point is that our interpretation is entirely localized in the axiomatization of **SystemStillControllable**, and it will therefore be fairly easy to update our specification in case of misinterpretation.

Once these predicates are introduced, we can specify the mode in which the Steam-Boiler Control System operates. At first glance the informal requirements (cf. Chapter AS, Section 4) look quite complicated, mainly because they explain, for each operating mode, under which conditions the Steam-Boiler Control System should stay in the same operating mode or switch to another one. Here again we have to interpret the informal requirements: in Chapter AS, Section 4.2, it is said that the Steam-Boiler Control System should switch from **Normal** mode to **Rescue** mode as soon as a failure of the water level measuring unit is detected. However, in Chapter AS, Section 4.4, it is explained that the Steam-Boiler Control System can only operate in **Rescue** mode if some additional conditions hold (represented by our predicate **SystemStillControllable**). We decide therefore that when in **Normal** mode, if a failure of the water level measuring unit is detected, the Steam-Boiler Control System will switch to **Rescue** mode only if **SystemStillControllable** holds, otherwise it will switch (directly) to **EmergencyStop** mode.[4] Once this interpretation is made, a careful analysis of the requirements shows that, except for the **EmergencyStop** mode, we can determine the new operating mode (after reception of some messages) without taking into account the previous one, thanks to our predicates. The resulting axiomatization, introduced in the **SBCS-MODES** draft specification, is therefore both simple and clear.

---

[4] If our interpretation is incorrect, then in some cases we may have replaced a sequence **Normal** -> **Rescue** -> **EmergencyStop** by a sequence **Normal** -> **EmergencyStop**. Note that a sequence **Normal** -> **Rescue** -> **Normal** or **Degraded** is not possible since several cycles are necessary between a failure detection and the decision that the corresponding unit is again fully operational, cf. Section 3, i.e. we must have a sequence of the form **Normal** -> **Rescue** -> ... -> **Rescue** -> **Normal** or **Degraded** in such cases.

```
draft SBCS-Modes is
  enrich SBCS-Predicates;
axioms:
  Init.Mode = Normal;

  % Emergency stop mode
  EmergencyStop(s,msgs)  <=>
    ( s.Mode = EmergencyStop  ∨
      AskedToStop(s,msgs)     ∨
      ¬TransmissionOk(s,msgs) ∨
      DangerousWaterLevel(s,msgs) ∨
      (¬WaterLevelOk(s,msgs) ∧
       ¬SystemStillControllable(s,msgs)) );
  EmergencyStop(s,msgs) =>  {s,Tick(msgs)}.Mode = EmergencyStop;

  % Normal mode
  ¬EmergencyStop(s,msgs)    ∧
  AllPhysicalUnitsOk(s,msgs)    =>  {s,Tick(msgs)}.Mode = Normal;

  % Degraded mode
  ¬EmergencyStop(s,msgs)  ∧
  TransmissionOk(s,msgs)  ∧
  WaterLevelOk(s,msgs)    ∧
  ¬AllPhysicalUnitsOk(s,msgs) => {s,Tick(msgs)}.Mode = Degraded;

  % Rescue mode
  ¬EmergencyStop(s,msgs)  ∧
  TransmissionOk(s,msgs)  ∧
  ¬WaterLevelOk(s,msgs)   ∧
  SystemStillControllable(s,msgs) => {s,Tick(msgs)}.Mode=Rescue;

where: s:SBCS_States; msgs:Set[R_Message];
end SBCS-Modes.
```

Remember we do not take into account the initialization phase in our specification. For the sake of simplicity we will assume that the Steam-Boiler Control System starts in the **Normal** mode, which explains the axiom **Init.Mode = Normal**.

As a side remark, let us point out that the **SBCS-Modes** draft specification illustrates a specification technique that will be used systematically in the sequel. Note that we have, on the one hand, some operations having as (main) arguments the current state of the Steam-Boiler Control System and the newly received messages (cf. e.g. the predicates introduced in the **SBCS-Predicates**

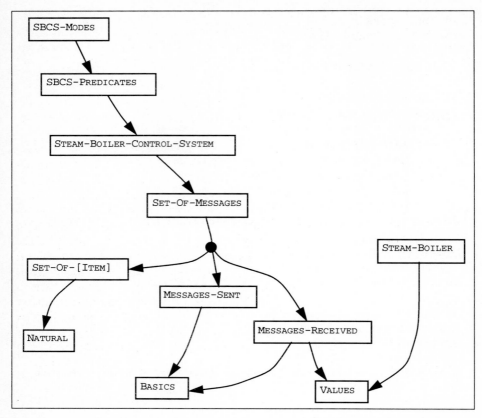

Figure 3. Structure of the specification of the Steam-Boiler Control System (part I)

draft specification). On the other hand, we have operations having the current state of the Steam-Boiler Control System as (main) argument (cf. e.g. the **Mode** operation). Then the axioms make explicit the mutual recursive relationships between all these operations. For instance, **Mode** is defined on a new state {s,Tick(msgs)} on the basis of the values of the predicates **EmergencyStop**, **TransmissionOk**, etc. (applied to the previous state s and the newly received messages **msgs**), while the predicates themselves (here only **EmergencyStop**) are defined (for a given state s and received messages **msgs**) using the value of **Mode** in the state s.[5]

We are now done with this first draft specification of the Steam-Boiler Control System and the overall structure of this specification is displayed in Figure 3. In

---

[5] This point is further illustrated by e.g. **FailureStatus** and **FailureDetected** in the FAILURE-STATUS-EVOLUTION draft specification, see Section 3.2.2.

the next steps of our formalization process, we will specify the predicates introduced in the SBCS-PREDICATES draft specification, which amounts to specifying the detection of equipment failures, and then we will be ready for specifying the operations MessagesSent and ChosenPumps. This will be the topic of the next two sections.

3 Specifying the Detection of Equipment Failures

The detection of equipment failures is described in Chapter AS, Section 7. It is quite clear that this detection is the most difficult part to formalize, mainly because both our intuition and the requirements (cf. e.g. "knows from elsewhere", "incompatible with the dynamics") suggest that we should take into account some inter-dependencies when detecting the various possible failures. For instance, if we ask a pump to close, and if in the next cycle the pump state still indicates that the pump is open, we may in principle infer either a failure of the message transmission system (e.g. the closing order was not properly sent or was not received, or the message indicating the pump state has been corrupted) or a failure of the pump (which was not able to execute the closing order or which sends incorrect state messages). Our understanding of the requirements is that in such a case we must conclude there has been a failure of the pump, not of the message transmission system. Let us stress again that it is important to distinguish between the actual failures of the various pieces of equipment, and the diagnosis we will make. Only the latter is relevant in our specification.

3.1 Understanding the Detection of Equipment Failures

Before starting to specify the detection of equipment failures, we must proceed to a careful analysis of Chapter AS, Section 7, in order to clarify the inter-dependencies mentioned above. Only then we will be able to understand how to structure our specification of this crucial part of the problem.

A first rough analysis of the part of Chapter AS, Section 7, devoted to the description of potential failures of the physical units (i.e. of the pumps, the pump controllers and the two measuring devices) shows that these failures are detected on the basis of the information contained in the received messages: we must check that the received values are in accordance with some *expected* values (according to the history of the system, i.e. according to the "dynamics of the system" and to the messages previously sent by the Steam-Boiler Control System). In particular, the detection of failures of the physical units relies on the fact that we have effectively received the necessary messages. If we have not received these messages, then we should conclude there has been a failure of the message transmission system (see below), and in these cases (cf. the SBCS-MODES draft specification), the Steam-Boiler Control System switches to the EmergencyStop mode. The further detection of failures of the physical units (in addition to the already detected failure of the message transmission system) is therefore irrelevant in such cases.

Let us now consider the message transmission system. The part of Chapter AS, Section 7, devoted to the description of potential failures of the message transmission system is quite short. Basically, it tells us that we should check that the Steam-Boiler Control System has received all the messages it was expecting, and that none of the messages received is aberrant. However, it is important to note that the involved analysis of the received messages combines two aspects: on the one hand, there is some "static" analysis of the received messages in order to check that all messages that must be present in each transmission are effectively present (cf. Chapter AS, Section 6). These messages are exactly the messages required to proceed to the detection of the failures of the physical units (cf. above). On the other hand, the Steam-Boiler Control System expects to receive (or, on the contrary, not to receive) some specific messages according to the history of the system (for instance, the Steam-Boiler Control System expects to receive a "failure acknowledgement" from a physical unit once it has detected a corresponding failure and sent a "failure" message to this unit, but not before), and here some "dynamic" analysis is required. Obviously, the "static" analysis of the messages can be made on the basis of the received messages only, while the "dynamic" analysis must take into account, in addition to the received messages, the history of the system, and more precisely the history of the failures detected so far and of the "failure acknowledgement" and "repaired" messages received so far.

From this first analysis we draw the following conclusions:

1. The detection of the message transmission system failures (hence the specification of `TransmissionOk`) should be split into two parts: a "static" analysis and a "dynamic" analysis of the received messages. The "dynamic" analysis interacts with the detection of the failures of the physical units.
2. For the detection of the failures of the physical units, we can freely assume that the "static" analysis of the received messages has been successful.

Let us now focus on the "dynamic" part of the analysis of the received messages. To perform this "dynamic" analysis, we must check that we receive "failure acknowledgement" and "repaired" messages when appropriate. In order to do this, we must keep track of the failures detected and of the "failure acknowledgement" and "repaired" messages received. Since the same reasoning applies for all physical units, we can do the analysis in a generic way. For each physical unit, we will keep track of its status, which can be either `Ok`, `FailureWithoutAck` or `FailureWithAck`. The status of a physical unit will be updated accordingly with respect to the detection of failures, and reception of "failure acknowledgement" and "repaired" messages. This analysis leads to the diagram given in Figure 4. Intuitively, the predicate `FailureDetected` (applied to the previous state, newly received messages and some physical unit) should report whether a failure of the corresponding physical unit has been detected. Note that for performing the "dynamic" analysis of the received messages, it is not necessary to have already specified this predicate **FailureDetected**. It is sufficient to assume that this predicate will be specified later with the expected meaning. Indeed the specification of **FailureDetected** amounts to the specification of the detection of the failures of the physical units, and this should therefore be done when specifying the detection of these kinds of failures. The important point here is that the introduction of the predicate **FailureDetected** allows us to specify the (dynamic part of the) detection of the message transmission system failures, even if we have not (yet) specified the detection of the failures of the physical units.

Let us now consider the inter-dependencies to be taken care of when detecting the failures of the physical units:

1. First we consider the detection of the failures of the pumps and of the pump controllers. The inter-dependency between both kinds of failures is obvious. If we ask for a pump closing and observe later on that the pump status is **Closed** while the corresponding pump controller status is **Flow**, obviously something is wrong, but it may be either the pump or the pump controller. However, following the informal requirements, we will consider failures of the pumps first, and then we will describe failures of the pump controllers under the assumption that the corresponding pumps are working correctly. (Hence in our example we will infer a failure of the pump controller, not of the pump.) This seems correct since in the informal requirements the description of pump failures is "self-contained" while the description of pump controller failures contains the (somewhat cryptic) "knows from elsewhere" statement.
2. The detection of failures of the steam output measuring device is the simpler one, since here there is no interaction with the detection of other kinds of

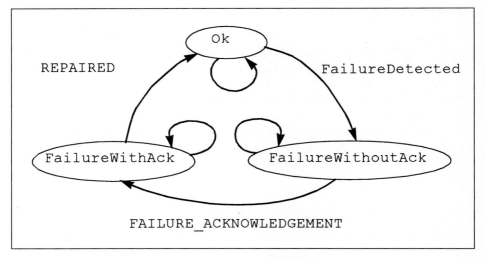

Figure 4. Physical Unit Status Evolution

failures. This is due to the fact that for the steam output measuring device, "incompatible with the dynamics of the system" only refers to the maximum gradients of increase/decrease of the output of steam.
3. The detection of the failures of the water level measuring device is more complex, since in this case, to assess whether the water level measured is "compatible with the dynamics of the system", we must take into account the quantity of water poured into the boiler (and to estimate this quantity, we have to take care of possibly "broken" pumps) as well as the quantity of water that has exited the boiler in the form of steam (and to estimate this quantity, we must know whether we rely or not on the steam output measuring device).

From the above analysis, we conclude that we will specify the detection of equipment failures as follows:

1. In a first step we specify the detection of the message transmission system failures (hence `TransmissionOk`). This will leads to a **MESSAGE-TRANSMISSION-SYSTEM-FAILURE** piece of specification. As explained above, this specification task will be split into two parts:
   (a) First we specify the "static" analysis of the received messages.
   (b) Then we specify the "dynamic" analysis of the received messages. To do this, we will have first to specify how to keep track of the "status" of the physical units (assuming that the predicate `FailureDetected` will be properly specified later on).
2. Then we specify the detection of the failures of the physical units in the following order:
   (a) Failures of the pumps, leading to a **PUMP-FAILURE** piece of specification.
   (b) Failures of the pump controllers, leading to a **PUMP-CONTROLLER-FAILURE** piece of specification.
   (c) Failures of the steam output measuring device, leading to a **OUTPUT-OF-STEAM-FAILURE** piece of specification.
   (d) Failures of the water level measuring device, leading to a **WATER-LEVEL-FAILURE** piece of specification.

These various pieces of specification are described in the next subsections.

3.2 Detection of the Message Transmission System Failures

As explained above, we first specify the "static" analysis of the received messages, and then we specify the "dynamic" analysis of these messages.

Static Analysis of the Received Messages. To specify the "static" analysis of messages, it is necessary to introduce additional predicates and operations on sets of received messages. This is done in the **MESSAGES-PREDICATES** and **MESSAGES-OPERATIONS** specifications.

To check whether there is a transmission failure, we must in particular check that all "indispensable" messages are present. In addition, a set of received messages is "acceptable" if there are no "duplicated" messages in this set. Since we have specified the collection of received messages as a set, we cannot have several occurrences of exactly the same message in this set. (Note that this means that our choice of using "sets" instead of "bags", for instance, is therefore not totally innocent: either we assume that the reception of several occurrences of exactly the same message will never happen, and this is an assumption on the environment, or we assume that this case should not lead to the detection of a failure of the message transmission system, and this is an assumption on the requirements.) However, specifying the collection of received messages as a set does not imply that a set of received messages cannot contain several **LEVEL(v)** messages, with distinct values (for instance). Hence, we must check this explicitly, and this is the aim of the predicates **Contains1LevelMessage**, **Contains1SteamMessage**, **Contains1PumpStateMessageFor** and **Contains1PumpControllerStateMessageFor**.

Remember that the reception of "unknown" messages (i.e. messages that do not belong to the list of messages as specified in Chapter AS, Section 6) is taken into account via the extra constant **junk** message (cf. the specification

```
spec MESSAGES-PREDICATES is
  use SET-OF-MESSAGES;
operations:
  __ Contains1LevelMessage,
  __ Contains1SteamMessage :  Set[R_Message] -> Bool;
  __ Contains1PumpStateMessageFor __,
  __ Contains1PumpControllerStateMessageFor __ :
                         Set[R_Message], PumpNumber ->Bool;
  __ ContainsAFailureAckFor    __,
  __ ContainsARepairedMessageFor __,
  __ ContainsAFailureAckAndARepairedMessageFor __ :
                         Set[R_Message], PhysicalUnit -> Bool;
  AllIndispensableMessagesArePresentOnce: Set[R_Message] -> Bool;
axioms:
  msgs Contains1LevelMessage  <=>
    ((∃v. LEVEL(v) ∈ msgs) ∧
     (∀v1.∀v2. LEVEL(v1) ∈ msgs ∧ LEVEL(v2) ∈ msgs => v1 = v2));

  msgs Contains1SteamMessage  <=>
    ((∃v. STEAM(v) ∈ msgs) ∧
     (∀v1.∀v2. STEAM(v1) ∈ msgs ∧ STEAM(v2) ∈ msgs => v1 = v2));

                   Continued on next page
```

Spec **Messages-Predicates** continued

```
msgs Contains1PumpStateMessageFor Pn   <=>
  ((∃PState. PUMP_STATE(Pn,PState) ∈ msgs) ∧
   (∀PState1.∀PState2.
    PUMP_STATE(Pn,PState1) ∈ msgs ∧
    PUMP_STATE(Pn,PState2) ∈ msgs  => PState1 = PState2) );

msgs Contains1PumpControllerStateMessageFor Pn   <=>
  ((∃PCState. PUMP_CONTROL_STATE(Pn,PCState) ∈ msgs) ∧
   (∀PCState1.∀PCState2.
    PUMP_CONTROL_STATE(Pn,PCState1) ∈ msgs ∧
    PUMP_CONTROL_STATE(Pn,PCState2) ∈ msgs
       => PCState1 = PCState2) );

msgs ContainsAFailureAckFor PhysUnit   <=>
  FAILURE_ACKNOWLEDGEMENT(PhysUnit) ∈ msgs;

msgs ContainsARepairedMessageFor PhysUnit   <=>
  REPAIRED(PhysUnit) ∈ msgs;

msgs ContainsAFailureAckAndARepairedMessageFor PhysUnit   <=>
  (msgs ContainsAFailureAckFor PhysUnit ∧
   msgs ContainsARepairedMessageFor PhysUnit );

AllIndispensableMessagesArePresentOnce(msgs)   <=>
  (msgs Contains1LevelMessage ∧
   msgs Contains1SteamMessage ∧
   (∀Pn. msgs Contains1PumpStateMessageFor Pn) ∧
   (∀Pn. msgs Contains1PumpControllerStateMessageFor Pn) );

where: msgs:Set[R_Message]; PState,Pstate1,Pstate2:PumpState;
  PCState,PCState1,PCState2:PumpControllerState; v,v1,v2:Value;
  Pn:PumpNumber; PhysUnit:PhysicalUnit;
end MESSAGES-PREDICATES.
```

Messages-Received). We believe also that we cannot simultaneously receive a failure acknowledgement and a repaired message for the same physical unit, i.e. that at least one cycle is needed between acknowledging the failure and repairing the unit. We will check this using the **ContainsAFailureAckAndARepairedMessageFor** predicate.[6]

---

[6] We must confess that this belief is induced by our intuition about the behaviour of the system. Indeed nothing in the requirements allows us to make either this interpretation or the opposite one. Although not essential, this assumption will simplify the axiomatization.

```
spec MESSAGES-OPERATIONS is
  use MESSAGES-PREDICATES;
operations:
  % Useful to retrieve the water level and output of steam values, etc.
  % These operations need to be specified only when the set of received
  % messages is "acceptable".
  Level, Steam : Set[R_Message] -> Value;
  PumpState : PumpNumber, Set[R_Message] -> PumpState;
  PumpControllerState:
    PumpNumber, Set[R_Message] -> PumpControllerState;
axioms:
  Level(msgs + LEVEL(v)) = v;
  % Note: This axiom is enough since we can prove that:
  %   msgs Contains1LevelMessage =>
             ∃msgs1.∃!v. msgs = msgs1 + LEVEL(v)
  Steam(msgs + STEAM(v)) = v;
  PumpState(Pn, msgs + PUMP_STATE(Pn,PState)) = PState;
  PumpControllerState(Pn, msgs + PUMP_CONTROL_STATE(Pn,PCState))
                                            = PCState;
where: msgs:Set[R_Message]; v:Value; Pn:PumpNumber;
  PState:PumpState; PCState:PumpControllerState;
end MESSAGES-OPERATIONS.
```

The "static" analysis of received messages can now be expressed as follows:

% Static analysis:
¬(junk ∈ msgs) ∧
(∀PhysUnit.
  ¬(msgs ContainsAFailureAckAndARepairedMessageFor PhysUnit)) ∧
AllIndispensableMessagesArePresentOnce(msgs)

Dynamic Analysis of the Received Messages: Keeping Track of the Physical Unit Status. We focus now on the "dynamic" part of the analysis of the received messages. As explained in Section 3.1, to perform this "dynamic" analysis, we must check that we receive "failure acknowledgement" and "repaired" messages when appropriate, and to do this, we must keep track of the status (Ok, FailureWithoutAck or FailureWithAck) of each physical unit. The status of a physical unit will be updated accordingly with respect to the detection of failures, and reception of "failure acknowledgement" and "repaired" messages, as shown by the diagram given in Figure 4 (cf. Section 3.1). This diagram, which depicts some kind of automaton, is formalized by the FAILURE-STATUS-EVOLUTION draft specification. In this draft specification, we introduce a sort Status and

```
draft FAILURE-STATUS-EVOLUTION is
  enrich SBCS-PREDICATES;
  use MESSAGES-OPERATIONS;
sort Status = {Ok, FailureWithoutAck, FailureWithAck};
operations:
  FailureStatus : SBCS_States, PhysicalUnit -> Status;
  FailureDetected :
    SBCS_States, Set[R_Message], PhysicalUnit -> Bool;
axioms:
  FailureStatus(Init,PhysUnit) = Ok;

  FailureStatus(s,PhysUnit) = Ok ∧
  FailureDetected(s,msgs,PhysUnit)
    => FailureStatus({s,Tick(msgs)},PhysUnit) = FailureWithoutAck
       ∧ ¬PhysicalUnitOk(s,msgs,PhysUnit);

  FailureStatus(s,PhysUnit) = Ok ∧
  ¬FailureDetected(s,msgs,PhysUnit)
    => FailureStatus({s,Tick(msgs)},PhysUnit) = Ok
       ∧ PhysicalUnitOk(s,msgs,PhysUnit);

  FailureStatus(s,PhysUnit) = FailureWithoutAck ∧
  (msgs ContainsAFailureAckFor PhysUnit)
    => FailureStatus({s,Tick(msgs)},PhysUnit) = FailureWithAck
       ∧ ¬PhysicalUnitOk(s,msgs,PhysUnit);

  FailureStatus(s,PhysUnit) = FailureWithoutAck ∧
  ¬(msgs ContainsAFailureAckFor PhysUnit)
    => FailureStatus({s,Tick(msgs)},PhysUnit) = FailureWithoutAck
       ∧ ¬PhysicalUnitOk(s,msgs,PhysUnit);

  FailureStatus(s,PhysUnit) = FailureWithAck ∧
  (msgs ContainsARepairedMessageFor PhysUnit)
    => FailureStatus({s,Tick(msgs)},PhysUnit) = Ok
       ∧ PhysicalUnitOk(s,msgs,PhysUnit);

  FailureStatus(s,PhysUnit) = FailureWithAck ∧
  ¬(msgs ContainsARepairedMessageFor PhysUnit)
    => FailureStatus({s,Tick(msgs)},PhysUnit) = FailureWithAck
       ∧ ¬PhysicalUnitOk(s,msgs,PhysUnit);

where: s:SBCS_States; msgs:Set[R_Message]; PhysUnit:PhysicalUnit;
end FAILURE-STATUS-EVOLUTION.
```

two operations: `FailureStatus`, which allows us to observe the status of a physical unit in a given state, and `FailureDetected`, which reports whether a failure of the corresponding physical unit has been detected (given the previous state of the Steam-Boiler Control System and the newly received messages). Remember that we do not specify the operation `FailureDetected` for the moment, we just assume that this operation will be properly specified later on with the expected meaning. At this point, to perform the dynamic analysis of the received messages, we only need to specify the behaviour of the `FailureStatus` operation. Note also that, since we have assumed that the Steam-Boiler Control System starts in the `Normal` mode, we specify that the initial status of each physical unit is `Ok`.[7]

As a last remark, we would like to point out that the three operations `FailureStatus`, `FailureDetected` and `PhysicalUnitOk` have obvious relationships (some of them are explicitly given by the axioms of FAILURE-STATUS-EVOLUTION, e.g. we specify the predicate `PhysicalUnitOk` by means of `FailureDetected`). Let us stress again that during this formalization process there is no need for minimality, and on the contrary we will see in Section 5 how to exploit such redundancies during the validation of the formal specification.

Detection of the Message Transmission System Failures. We now have all the ingredients required to specify the `TransmissionOk` predicate, taking into account both static and dynamic aspects. This is done in the MESSAGE-TRANSMISSION-SYSTEM-FAILURE draft specification.

We introduce a new predicate, `ThereIsAnAberrantMessage`, which reflects the dynamic part of the analysis. The axiomatization of this predicate is now straightforward, provided the following remarks:

1. We understand that for each failure signaled by the Steam-Boiler Control System, the corresponding physical unit will send just one failure acknowledgement.
2. We will specify the Steam-Boiler Control System in such a way that when it receives a "repaired" message, the Steam-Boiler Control System acknowledges it immediately. Hence, if there is no problem with the message transmission system, and due to the fact that transmission time can be neglected, the Steam-Boiler Control System must in principle receive only one repaired message for a given failure. Note that this is not contradictory with the "until..." part of the sentences describing the "repaired" messages in the informal requirements (cf. Chapter AS, Section 6).

---

[7] The reader may detect that the FAILURE-STATUS-EVOLUTION draft specification is not totally correct. However, we prefer to give here the text of the specification as it was originally written, and we will explain in Section 5 how we detect, when validating the specification of the Steam-Boiler Control System, that something is not correct, and how the problem can be fixed.

```
draft MESSAGE-TRANSMISSION-SYSTEM-FAILURE is
  enrich FAILURE-STATUS-EVOLUTION;
operations:
  ThereIsAnAberrantMessage : SBCS_States, Set[R_Message] -> Bool;
axioms:
  ThereIsAnAberrantMessage(s,msgs)  <=>
    (% Messages which are unexpected when no more in initialization mode:
    (¬(s.Mode = Initialization) ∧
      (STEAM_BOILER_WAITING ∈ msgs ∨ PHYSICAL_UNITS_READY ∈ msgs))
    ∨
    % The program receives for some physical unit a failure acknowled-
    % gement without having previously sent the corresponding failure
    % detection message, or receives redundant failure acknowledgements:
    (∃PhysUnit.
        msgs ContainsAFailureAckFor PhysUnit ∧
          (FailureStatus(s,PhysUnit) = Ok ∨
           FailureStatus(s,PhysUnit) = FailureWithAck))
    ∨
    % The program receives for some physical unit a repaired message
    % but the unit is ok or its failure is not yet acknowledged:
    (∃PhysUnit.
        msgs ContainsARepairedMessageFor PhysUnit ∧
          (FailureStatus(s,PhysUnit) = Ok ∨
           FailureStatus(s,PhysUnit) = FailureWithoutAck))  );

  % It is now quite easy to specify TransmissionOk:
  TransmissionOk(s,msgs)  <=>
    % Static analysis:
    ¬(junk ∈ msgs) ∧
    (∀PhysUnit.
      ¬(msgs ContainsAFailureAckAndARepairedMessageFor PhysUnit))
    ∧ AllIndispensableMessagesArePresentOnce(msgs) ∧
    % Dynamic analysis:
    ¬ThereIsAnAberrantMessage(s,msgs) ;

where: s:SBCS_States; msgs:Set[R_Message]; PhysUnit:PhysicalUnit;
end MESSAGE-TRANSMISSION-SYSTEM-FAILURE.
```

3.3 Detection of the Pumps and Pump Controllers Failures

The specification of the detection of the failures of the pumps and of the pump controllers follows in a straightforward way the informal requirements and is given in the appendix (see CD-ROM Annex BCPR.A.1).

3.4 Detection of the Steam Output Measuring Device Failures

The specification of the detection of the failures of the steam output measuring device is given in the appendix (see CD-ROM Annex BCPR.A.2).

3.5 Detection of the Water Level Measuring Device Failures

The specification of the detection of the failures of the water level measuring device is given in the appendix (see CD-ROM Annex BCPR.A.3).

This ends the specification of the detection of equipment failures. To complete our formalization process, we now have to specify the **MessagesSent** and **ChosenPumps** operations.

4 Specification of MessagesSent and of ChosenPumps

The specification of the **MessagesSent** and **ChosenPumps** operations is given in the appendix (see CD-ROM Annex BCPR.B).

The formalization of the Steam-Boiler Control System requirements is now completed (except for the initialization phase). The last step is to put together all the draft specification pieces. This leads to the **SBCS-WITHOUT-INITIALIZATION** specification (see CD-ROM Annex BCPR.B), where we also specify the **AskedToStop** predicate (this trivial point has not yet been considered).

The various pieces of specification introduced in Sections 3 and 4 and their relationships are displayed in Figure 5.

5 Validation of the Formal Specification

Once the formalization of the informal requirements is completed, we must now face the following question: is our formal specification adequate? Answering this question is a difficult issue since there is no formal way to establish the adequacy of a formal specification w.r.t. informal requirements, i.e. we cannot *prove* this adequacy. However, we can try to *test* it, by performing various "experiments". If these experiments are successful, we only increase our confidence in the formal specification. If some experiment fails, then we can examine the specification and try to understand the causes of the failure, possibly detecting some flaw in the specification. Let us stress that all the experiments we may think of are necessarily biased, since during the formalization phase we have tried to reflect the informal requirements, and the aim of the experiments will be to check that we have indeed reflected them in a satisfactorily way. Hence, if the informal requirements contained misleading informations that were not (cannot be) detected during the formalization phase, it is very likely that such

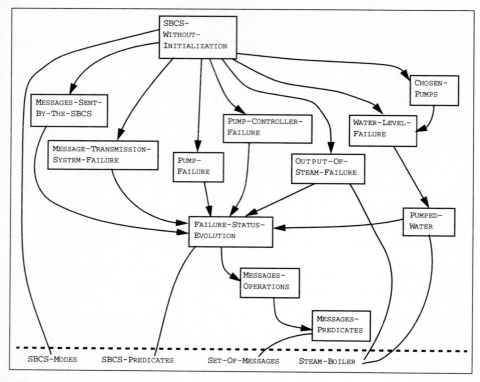

Figure 5. Structure of the specification of the Steam-Boiler Control System (part II)

misleading information will not be detected during the validation phase either.[8] This means that the aim of the validation process is to check the adequacy of the formal specification w.r.t. the informal requirements, but in no way to check the adequacy of the formal specification (or of the informal requirements) per se.

In our case, we will base our validation process on theorem proving, i.e. we will check that some formulas are logical consequences of our axioms. For this purpose we use the LARCH PROVER [5]. During this validation process we can

---

[8] To make this point clear, assume that the informal requirements describe the factorial function, and assume moreover that we know nothing about factorial. Then, if the informal requirements state that the factorial function, when applied to zero, returns three, we will of course very easily reflect this in our formal specification, by the axiom fact(0) = 3. Then, whatever the experiments we perform, we will conclude that our formal specification is adequate w.r.t. the informal requirements, which is in this case a trivial issue. But detecting that we should have specified fact(0) = 1 (hence that the informal requirements are incorrect) requires extra knowledge about the context, here about factorial.

consider two kinds of proof obligations:

1. We can check that some expected logical consequences of our specification hold. This can be considered as some kind of "internal validation" of the formal specification. In principle we must check that our formal specification is consistent (i.e. has at least one model), but in our framework this property is undecidable in general.
2. We can check that some expected properties inferred from the informal requirements are logical consequences of our specification ("external validation"). To do this, we must first reanalyze the informal specification, state some expected properties, translate them into formulas, and then attempt to prove that these formulas are logical consequences of our specification. This task is not easy, since in general one has the feeling that all expected properties were already detected and included in the axioms during the formalization process.

The application of these principles to the specification of the Steam-Boiler Control System leads to various proofs (see CD-ROM Annex BCPR.C). Below we give as an example some proofs related to the detection of equipment failures.

Remember that there are a lot of redundancies in our specification. We will take profit of these redundancies to check that our axiomatization of the detection of equipment failures is adequate. More precisely, we investigate the relationships between **PhysicalUnitOk** and **FailureStatus**:

1. If a physical unit is "**Ok**", then no failure for this unit was detected during the analysis of the last received messages, and its status is therefore "**Ok**" as well:

```
prove   PhysicalUnitOk(s,msgs,PhysUnit)  <=>
    FailureStatus({s,Tick(msgs)},PhysUnit) = Ok                    □
```

2. If a physical unit is not "**Ok**", then the status of this unit is "**Failure**" (either with or without acknowledgement):

```
prove   ¬PhysicalUnitOk(s,msgs,PhysUnit)  <=>
    FailureStatus({s,Tick(msgs)},PhysUnit) = FailureWithAck   ∨
    FailureStatus({s,Tick(msgs)},PhysUnit) = FailureWithoutAck     □
```

3. If we detect a failure of some physical unit, then it is not true that all physical units are "**Ok**":

```
prove   FailureDetected(s,msgs,PhysUnit) =>
                        ¬AllPhysicalUnitsOk(s,msgs)
```

But here **the proof fails**! A careful analysis of the proof attempt shows that the proof fails since it could be the case that, simultaneously with the reception of a repaired message, we nevertheless detect another failure of the same unit.[9] From this analysis we conclude that the axiom (cf. the FAILURE-STATUS-EVOLUTION draft specification):

```
FailureStatus(s,PhysUnit) = FailureWithAck  ∧
(msgs ContainsARepairedMessageFor PhysUnit)
   =>  FailureStatus({s,Tick(msgs)},PhysUnit) = Ok
       ∧ PhysicalUnitOk(s,msgs,PhysUnit);
```

is not correct. Indeed, the problem can be outlined by the following proof attempt:

```
prove  ( FailureStatus(s,PhysUnit) = FailureWithAck  ∧
         msgs ContainsARepairedMessageFor PhysUnit  ∧
         FailureDetected(s,msgs,PhysUnit) )  =>
                         PhysicalUnitOk(s,msgs,PhysUnit)
```

which succeeds while it should not! Hence we must correct the FAILURE-STATUS-EVOLUTION draft specification and replace the above axiom by the two following ones:

```
FailureStatus(s,PhysUnit) = FailureWithAck  ∧
(msgs ContainsARepairedMessageFor PhysUnit) ∧
FailureDetected(s,msgs,PhysUnit)
   =>  FailureStatus({s,Tick(msgs)},PhysUnit) = FailureWithoutAck
       ∧ ¬PhysicalUnitOk(s,msgs,PhysUnit);

FailureStatus(s,PhysUnit) = FailureWithAck  ∧
(msgs ContainsARepairedMessageFor PhysUnit) ∧
¬FailureDetected(s,msgs,PhysUnit)
   =>  FailureStatus({s,Tick(msgs)},PhysUnit) = Ok
       ∧ PhysicalUnitOk(s,msgs,PhysUnit);
```

Once the axiomatization is corrected, we can check that the above proof (which succeeded while it should not have) now fails as expected, and moreover that the expected property holds:

```
prove   FailureDetected(s,msgs,PhysUnit) =>
                         ¬AllPhysicalUnitsOk(s,msgs)                □
```

To be on the safe side, we check that all properties proved so far still hold with the new axiomatization (by replaying the corresponding proof scripts).

---

[9] One of the main advantages of the LARCH PROVER is that "it has been designed to assist in reasoning by carrying out routine (and possibly lengthy) steps in a proof automatically and by providing useful information about why proofs fail, if and when they do". Moreover, the LARCH PROVER has been designed to make proofs (of flawed conjectures) fail "quickly".

6 Evaluation and Comparison

In this paper we have shown how to derive an algebraic specification of the Steam-Boiler Control System starting from the informal requirements. In particular, we have shown how this formalization process was useful to detect inconsistencies and loose ends in the informal requirements, and how to keep track in the formal specification of the interpretations we had to make in order to overcome the unclear or missing parts of the requirements. An interesting aspect of this case study is the demonstration of a method for deriving the formal specification in a stepwise way by successive refinements. We hope we have convinced the reader that the resulting specification is both clear and rather easy to understand. To achieve these aims we have used the PLUSS algebraic specification language which proves useful not only to express specifications in a structured and modular way, but also to make explicit the development process of the whole specification. An important part (although unfortunately too often neglected) of the specification task is the **validation** of the formal specification with respect to the informal requirements. One of the main advantages of the (classical) algebraic framework is that it allows us to benefit from the most advanced existing automated tools, and in particular we have shown how to use the LARCH PROVER to validate our algebraic specification. The importance of this validation task (and the suitability of the LARCH PROVER) was clearly demonstrated in our case study, and in particular we have shown that thanks to the proofs performed, we have detected a flaw in our specification (that was easily corrected).

Hence in this paper we have provided a formal requirements specification of the control program. Moreover, various safety properties are studied and proven formally. We have not specified the initialization phase of the Steam-Boiler Control System due to lack of time. However, it is obvious that the method used in the paper will apply to the specification of the initialization phase as well, and even if it is clear that the resulting specification would be a bit more complex (due to the intricacies of the initialization phase), there is absolutely no problem extending our specification by the description of the initialization phase. According to the work plan described in the introduction, we should in principle have derived from our requirements specification a design specification and finally a program. Again, this was not done due to lack of time, but we would like to stress that this second part of the work plan, although technical, is not really difficult. Our requirements specification is very detailed and contains a precise specification of the control program including specifications for the detection of equipment failures. Hence much of what would go into a functional design in other approaches is already contained in our requirements specification. Similarly, we have not provided a separate architectural design, but our algebraic requirements specification is structured in such a way that part of the architecture of the program will be derived from it quite easily.

The solution proposed by M.-C. Gaudel, P. Dauchy and C. Khoury (see

Chapter GDK) is very similar to ours, the main difference being that they use a slightly different variant of algebraic specifications with implicit states.

We believe algebraic specifications provide just the right trade-off between expressive power and (relative) simplicity to specify systems like the Steam-Boiler Control System. Clearly, some expertise and training is required to *write* the algebraic specification (we believe height weeks training should be enough for an average programmer familiar with the basic concepts of logic), but we would like to emphasize the fact that almost no prior knowledge of algebraic specifications is required to *understand* it (only some basic background in logic is required). We estimate that the global time spent to write and validate the specification is one man-month (this does not include the time required to write the paper). Our claim that our algebraic specification of the Steam-Boiler Control System can be understood by people not trained in algebraic specifications is supported by concrete experience with some Electricité de France engineers.

References

1. M. Bidoit, M.-C. Gaudel, and A. Mauboussin. How to make algebraic specifications more understandable? An experiment with the PLUSS specification language. *Science of Computer Programming*, 12(1):1–38, June 1989.
2. M. Bidoit and R. Hennicker. Modular correctness proofs of behavioural implementations. Available by WWW: http://www.pst.informatik.uni-muenchen.de/~hennicke/, 1995. A short version appeared as: Proving the correctness of behavioural implementations, in *Proc. of AMAST'95*, Springer-Verlag L.N.C.S. 936, pages 152–168, 1995.
3. Michel Bidoit. PLUSS, *un langage pour le développement de spécifications algébriques modulaires*. Thèse d'Etat, Université Paris-Sud, Orsay, France, May 1989.
4. Michel Bidoit. Development of modular specifications by stepwise refinements using the PLUSS specification language. In *Proc. of the IMA Unified Computation Laboratory Conference (Stirling, Scotland, July 1990)*, pages 171–192. Oxford University Press, 1992.
5. S. Garland and J. Guttag. An overview of LP, the Larch Prover. In *Proc. of the Third International Conference on Rewriting Techniques and Applications*, pages 137–151. Springer-Verlag L.N.C.S. 355, 1989. See also on WWW: http://larch.lcs.mit.edu:8001/larch/LP/overview.html.
6. John V. Guttag and James J. Horning. *Larch: Languages and Tools for Formal Specification*. Springer-Verlag, 1993.
7. F. Orejas, M. Navarro, and A. Sànches. Implementation and behavioural equivalence: A survey. In *Recent Trends in Data Type Specification*, pages 93–125. Springer-Verlag L.N.C.S. 655, 1993.
8. D. Sannella and A. Tarlecki. Model-theoretic foundations for program development: basic concepts and motivation. Available by WWW: http://www.dcs.ed.ac.uk/staff/dts/pub/mtf.ps, 1995.
9. M. Wirsing. *Algebraic specification*. Handbook of Theoretical Computer Science. Elsevier Science Publishers B. V., 1990.

A Steam-Boiler Control Specification with Statecharts and Z

Robert Büssow, Matthias Weber

Technische Universität Berlin[*]

1 Introduction

This report presents a solution to the steam-boiler control problem (see Chapter AS, this book). The main idea is to integrate a mathematical specification technique with a well-known engineering technique for the specification of safety-critical control systems. Our starting point is the technology of statecharts, which is currently being adopted in industry for the specification of embedded systems. To cope with the growing complexity and the safety requirements of these systems, we propose a combination of the specification language Z and statecharts, Z being used to model the data structures and data transformations within the system [9].

The next section sketches some key ideas relating to this combination. In the subsequent sections, we present key elements of a solution to the steam-boiler control specification problem. Throughout the presentation, we attempt to adhere closely to the original specification of the problem (see Chapter AS, this book), especially with respect to the physical interface of the control software. The full specification can be found in an Appendix (see CD-ROM Annex BW).

2 Specification Methodology

A widely used technique in modern software engineering is to model a system by a combination of different – but semantically compatible – "views" of that system. The primary benefit of such an approach is to keep very complex systems manageable and to detect misconceptions or inconsistencies at an early stage. In the approach presented here, we divide the modeling into three views: the architectural model of the system, the reactive model of the system, and the functional model of the system (Figure 1).

The *architectural model* of a system describes the relationships between the classes of components used in the system as well as the actual configuration of the system components themselves. For the description of this model, we adopt the object-oriented modeling paradigm [2, e.g.]: We view an embedded control system as a hierarchically structured collection of objects that change their state and interact with each other throughout their lifetime. The relationships between object classes are described using well-known elements of class diagrams, i.e. diagrams depicting classes and their structural relationships, such as aggregation and inheritance.

---

[*] Forschungsgruppe Softwaretechnik, Sekretariat FR5-6, Franklinstr. 28/29, D-10587 Berlin, Germany (email: {buessow,we}@cs.tu-berlin.de). This work was partly carried out within the Espress project (http://www.first.gmd.de/~espress), supported by the German Federal Ministry of Education, Science and Technology

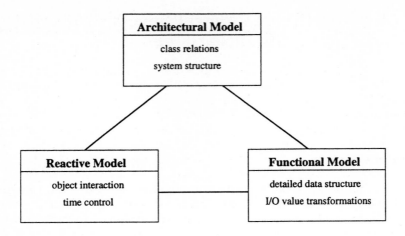

Fig. 1. The Three Modeling Views of an Embedded System

The two other views are primarily concerned with the specification of the behavior of single components of the embedded control system. We make a fundamental distinction with respect to the behavior of system components. The *functional model* of a component comprises data definitions, data invariants, and data transformation relations. In particular, for anyone component, its encompasses its local state and the input/output relation of its operations. Constraints, e.g. relating to safety properties, on the components states can be derived from these descriptions. The *reactive model* comprises the life-cycle of components, i.e. interactions with other components and the control of time during these interactions. Reactive behavior is modeled by specifying how and under which timing constraints, operations from external objects are requested or supplied (or both) in the state changes of objects.

We specify reactive behavior using an appropriate variant of timed hierarchical state transition diagrams, i.e. with a variant of statecharts [3]. There are two reasons for this choice: firstly, statecharts have proved to be sufficiently expressive for modeling complex component interactions and time control, and secondly, the use of statecharts, or close variants of statecharts, is currently spreading in industry. This also enables us to use existing analysis and simulation tools for this notation.

Note: In this report, we use two kinds of transition labels in our statecharts:

<operation> [condition] / <external activities>

after *time limit*: <internal event> / <external activities>

The first label is used for transitions that are triggered externally from the environment by calling an operation. The second label is used for *timeout transitions*, which are internally triggered events: after remaining in the source state for a certain time the transition is triggered internally. The first kind of transition may be additionally guarded by a condition. Both kinds of transition may subsequently trigger a number of external activities, e.g. by requesting services from other system objects.

A variety of formal semantics for statecharts have been developed [8]. The present paper is more in line with current work on embedding statecharts into an object-oriented setting [6, 4]. We therefore wish to make two remarks about the basic semantic concepts of the statechart notation as used in this report:

- The basic communication mechanism is point-to-point communication rather than broadcasting. Requesting an operation from an object can be interpreted as sending a message to an object; providing an operation to an object can be interpreted as receiving a message from an object. As pointed out in the architectural view, communications can be synchronous or asynchronous. Following the approach in [6], operation transitions are thus based on the concepts of request and provision of operations rather than on the concept of event.
- The execution of a transition is not timeless, and external messages may arrive at any time. Consequently, the system may not be able to react immediately to a message. Thus, incoming messages must be queued and then worked off individually. By convention, if there is no transition for a particular message, the system does not change its state.

Further experience gained by conducting case studies should guide the development of the notations and the semantics on which our approach is based.

Often, functional behavior in state-based systems is specified by textual or formal descriptions of pre- and postconditions and of data invariants. In our approach, we specify the functional behavior of objects using the state-based formal specification language Z [7]. There are two main reasons for using Z: firstly, in our view, Z has proved to be particularly useful for modeling complex functional data transformations; and secondly, both in academia and industry, Z has become one of the most widely used formal specification notations. Since we aim at a practical approach when modeling functionality, we try to stick to a constructive subset of Z, i.e. a subset that can be compiled into efficient code, whenever this is reasonable in a particular application. The use of a mathematical notation for modeling functional behavior enables us to prove abstract safety properties about the control system, such as provisions that the system may never enter certain hazardous states. Safety conditions imposed on data structures and data relationships should, of course, be specified using the full expressive power of the Z language.

The discussion so far should have made clear that we are *not* advocating a monolithically formal approach. Rather, our goal is to systematically embed mathematical elements into industrially used engineering techniques. As will be seen, this leads to an approach some parts of which are "hard", i.e. completely precise, while others remain "softer", i.e. allow for a certain range of interpretations. In our view, such an approach may still serve to prove interesting safety properties, while at the same time providing the flexibility needed for adaptation to the concrete situation in a specific industrial application context.

3 Informal Specification

Assuming that the reader is acquainted with the informal specification of the steam-boiler, we merely recapitulate here on its physical properties.

The steam-boiler is characterized by its overall capacity C (liters), its maximal steam output W (liter per second), the maximal increase and decrease gradients

U_1 and U_2 (liter per second^2), the number of pumps NP, the pump throughput P (liter per second), and the time a pump needs to start *pump_delay* (seconds). The water level must be between M_1 and M_2 and should be between N_1 and N_2 (liters). The control program communicates with the steam-boiler at intervals of T (seconds) length.

$C : \mathbb{N}$	[capacity of the tank (l)]
$W : \mathbb{N}$	[maximal steam output (l/s)]
$U_1 : \mathbb{N}$	[max. decrease of the steam output (l/s^2)]
$U_2 : \mathbb{N}$	[max. increase of the steam output (l/s^2)]
$NP : \mathbb{N}$	[number of pumps]
$P : \mathbb{N}$	[pump throughput (l/s)]
pump_delay $: \mathbb{N}$	[pump starting time (s)]
$T : \mathbb{N}$	[sampling rate (s)]
$M_1, M_2, N_1, N_2 : \mathbb{N}$	[absolut and normal water-level limits (l)]

$0 \leq M_1 \leq N_1 \leq N_2 \leq M_2 \leq C$
$NP = 4;\ T = \text{pump\_delay} = 5$
$T * NP * P < N_2 - N_1 \wedge T * W < N_2 - N_1$ \hfill [*]

We add an additional constraint (*) ensuring that the water level changes from low to high (or vice versa) during one interval T.

Finally, we wish to mention that in contrast to the original specification (see Chapter AS, this book), here a defective pump does not affect the degraded or rescue modes as long as the pump controller is still working. This leads to a more orthogonal specification; the overall behavior remains the same.

4 Architectural Model

In the previous section, we presented the informal requirements of the steam-boiler control problem. Since this is a very small example, the analysis and architectural design is straightforward. The results are summarized in the class and instance diagrams presented in this section. We use notations inspired by OMT [5] and Booch [2]. However, choice of notations is by no means crucial, and it should not be difficult for the experienced readers to translate the information content of the following diagrams into their favorite notation.

Structurally, the steam-boiler is an object that controls several physical units (see Figure 2). The physical units have already been mentioned above. The introduction of a separate unit manager represents an important architectural decision: after some analysis of the problem, we found it very useful for reasons of conceptual transparency and logical simplicity of the control system, to distinguish between the non-periodic processing of signals and their periodic transmission as achieved by the message-transmission system. These two tasks, which appear to be rather confused in the informal specification, are kept conceptually separate throughout this presentation.

On the structural level, this concern is reflected by the *unit manager*. This component encapsulates two things: the sampling of sensor data from the sensors and status information from the actors, and the periodic transmission of this data to

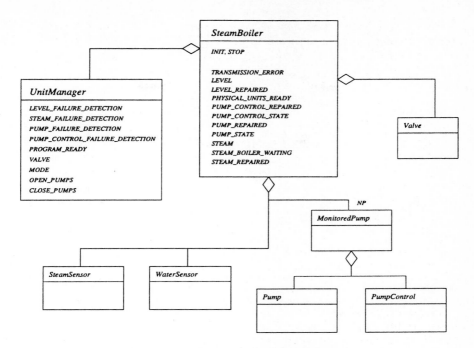

Fig. 2. Main Components and Their Services

the main control system. Furthermore, it encapsulates the processing of all control messages sent from the central steam-boiler control system to the physical units. For convenience, we allow to explicitly specify super- and subobject names, within angle brackets, along aggregation links. Figure 3 illustrates the message flow between components within the steam-boiler system.

The unit manager periodically requests the sensors to deliver their sensors' data, and the actors, i.e. the pumps, to deliver their acting state. The unit manager then transmits this information by sending appropriate messages to the steam-boiler control system. Furthermore, it periodically receives data messages about the current operating mode of the control system and transmits them to each physical unit. If anything goes wrong during these transmissions the unit manager sends a special data transmission error message to the main control system.

The unit manager also receives control messages (such as failure detections from the main control system, repair messages from physical units, etc.) from the main control system or the physical units. It likewise sends control messages (such as the reception of a failure detection message by a physical unit or the acknowledgment of a repair by a physical unit. If any aberrant control message arrives, i.e. a garbled acknowledgment or a transmission timeout, a transmission error message is sent to the main control system. Finally, the system controls a water valve, which is of minor importance here as it is needed only during initialization of the steam-boiler. All messaging below the unit manager is considered as given, e.g. in the form of physical links. The software specification is only concerned with the message flow between the unit manager and the steam-boiler.

Note the distinction between public and protected services of the steam-boiler

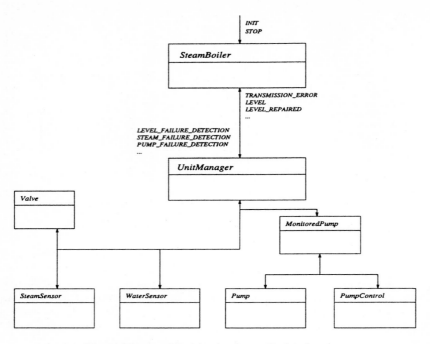

Fig. 3. Message Flow in the Steam-Boiler System

control system. The only public operations are initialization and stop; the other operations may be accessed only from subcomponents. Note also that, we do not indicate message parameters here, this forms part of the functional model. Nor do we not model in detail the interface between the unit manager and the physical units, since this is considered to be beyond the scope of the case study. The two software components in this system are the unit manager and the control of the steam-boiler object. This structural diagram could be enhanced by more specific information about the interfaces to the hardware components, e.g. specific program or store addresses, etc. Of course, many alternatives designs are possible, and one is briefly discussed at the end of this report. However, the question as to which design is best is not our concern here, because our primary aim is to explain our approach in general terms. Notations for the structural model, as well as notations for the dynamic and functional models, are discussed at length in [10].

5 Reactive Model

We focus here on the reactive behavior of the steam-boiler control. As is often the case with statecharts, we present the reactive model as a sequence of top-down refinements.

5.1 Top-Level Behavior

On this level, we have a very simple view of the steam-boiler: it either works normally or it is in state of emergency. This is shown in Figure 4.

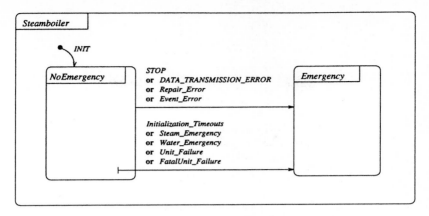

Fig. 4. Top-Level Reactive Behavior of the Steam-Boiler

There are several[2] events that may cause the system to enter emergency status, regardless of its working state. First of all, a stop event may be sent from the outside. Other reasons may be an error with respect to data transmission, e.g. due to a broken link. Also, there may be a problem because of the arrival of certain control signals in unexpected situations (*Event_Error*[3]). Finally, an emergency may be caused because of repair messages about units for which no failure has been detected (*Repair_Error*).

Besides these fatal errors, there are several events that cause the system to enter emergency mode from certain substates of its working state *NoEmergency*. These events may be timeouts during initialization, dangerous water or steam indications, or the failure of some set of units. These transitions are described in detail in the following subsection in which the state *NoEmergency* is further refined.

5.2 Normal Behavior

Normal system behavior is further refined in Figure 5.

The state *NoEmergency* is divided into an initialization state, followed by the state in which the system is currently running. Note that we have refined the detection of unit failures into the following four transitions:

$$Unit\_Failure \equiv SteamSensor\_Failure$$
$$\text{or } WaterSensor\_Failure$$
$$\text{or } Pump\_Failure$$
$$\text{or } PumpControl\_Failure$$

In the initialization state, an error might occur because outgoing steam is measured, although the steam-boiler is not yet supposed to be running (*Steam_Emergency*). Similarly, if the steam-boiler is running, an error might occur because a dangerous (low or high) water level is measured:

$$Water\_Emergency \equiv LEVEL[WaterDanger']$$

---

[2] As a shorthand form for writing multiple arrows, disjunctive events are separated by **or**.
[3] All definitions can be found in the full version of the specification in the Appendix.

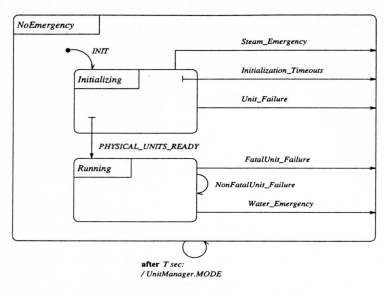

Fig. 5. Normal Behavior of the Steam-Boiler

Guards, shown in square brackets, can be associated with transitions. Primed conjuncts inside guards act like what might be called *post conditionals*, i.e. the transition takes place only if its effect would make these conjuncts true. In the example, these particular transitions describe situations in which the transmission of certain data to the system is unacceptable and therefore results in a transmission error.

The following unit failures are considered fatal in the running state:

$$FatalUnit\_Failure \equiv WaterSensor\_Failure[Degraded]$$
$$\text{or} SteamSensor\_Failure[Rescue]$$
$$\text{or} PumpControl\_Failure[Rescue]$$

The conditions *Degraded* and *Rescue* describe the two failures operating modes (see Chapter AS, this book). They are defined in the functional model.

The detection of a water sensor failure can be defined precisely as follows:

$$WaterSensor\_Failure \equiv LEVEL[WaterSensorBroken']$$
$$/UnitManager.LEVEL\_FAILURE\_DETECTION$$

This completes the discussion of all possible error transitions. Finally as shown, in Figure 5, the system periodically transmits its operating mode to the physical units. More precisely, the transmission is repeated at a sampling interval of T seconds.

5.3 Control of the Running Steam-Boiler

Figure 6 shows the main operation cycle for the steam-boiler operates its pumps. New incoming water-level messages may cause to open or close pumps. Note that the situation is not completely symmetric, since opening the pumps gives rise to an intermediate state in which the pumps balance the pressure of the steam-boiler.

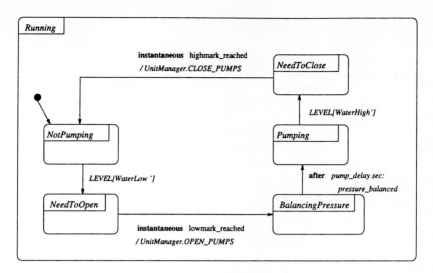

Fig. 6. Pump Control

5.4 Reactive Behavior of the Unit Manager

The unit manager is responsible for communication between the physical units and the main control system. The unit manager periodically reads the sensors and transmits their readings to the main control unit. Similarly, it regularly transmits the outgoing system messages to all the units. More precisely, every T seconds it performs the following tasks:

- It collects all incoming messages from the physical units.
- The incoming messages are analyzed. If a data-transmission message is missing or a repair-acknowledgment message is missing or erroneous, a transmission error is signaled in the main control system.
- If no transmission error has been detected, the incoming messages are translated one-to-one into requests for operation of the main control system. The order of the requests does not matter, except that incoming data messages have to be processed in the following order:

    LEVEL, STEAM, PUMP_CONTROL_STATE, PUMP_STATE.

- When processing these operations, the main control system will in turn request the unit manager to perform several operations, in order to send messages to the physical units. These messages are not transmitted individually to the physical units, but rather collected and then transmitted together.

The unit manager periodically reads the sensors and transmits their readings to the main control unit. Similarly, it regularly transmits the system's mode to all the units. Using appropriate statecharts, one can model these and many additional behavioral aspects such as constraints on transmission time etc. Furthermore, the unit manager transmits events between the main control and the physical units. For example, it collects ready or waiting signals from the units, and keeps track of failure and repair acknowledgments. Depending on the concrete circumstances, these components could also be specified by some other means, or even directly coded.

6 Functional Model

The functional model of the steam-boiler control system is presented by defining the state space and state invariants of the steam-boiler control system and then the data transformations on this state space effected by the various operations and events. The internal state space is largely made up of appropriate *internal models* of the physical components. These models contain all the information necessary for the control system to decide which action to take. In order to avoid a confusion of terms, we introduce a systematic naming convention: the internal model of a physical unit U is named *UModel*.

6.1 States of Physical Units

The state space of the steam-boiler control system contains appropriate information about each of the physical units. Part of this information is about the current state of each unit. The various units may each be in a subset of basic states. For example, the sensor units are either working or broken.

$UnitStates ::= working \mid broken \mid closed \mid opening \mid open \mid flow \mid noflow$
$SensorStates == \{working, broken\}$

6.2 Water Sensor Model

The water-level sensor is modeled by its state and lower and upper approximations of the water-level value. These approximations are used if nothing specific is known about the sensor value. As will be seen, during normal operation of the steam-boiler, the two approximations are equal, representing the unique water-level value. Initially, these approximations are set to the physical bounds of the system.

┌─ *WaterSensorModel* ─────────────
│ $qst : SensorStates$
│ $qa_1, qa_2 : \mathbb{N}$
├──────────────────────────────
│ $0 \leq qa_1 \leq qa_2 \leq C$

┌─ *WaterSensorInit* ──────────────
│ *WaterSensorModel'*
├──────────────────────────────
│ $qst' = working$
│ $qa_1' = 0$
│ $qa_2' = C$

Depending on the value of the approximations, various different water level states can be defined (Figure 7).

The second disjunct of the predicate for *WaterDanger* corresponds to a situation in which the level is considered defective and the two approximations have diverged beyond the normal range. In a sense, the water is both low *and* high, and it is therefore not clear what to do. We have therefore classified it as dangerous. This characterization is complete., i.e. no cases have been omitted.

$WaterSensorModel \vdash WaterLow \lor WaterHigh \lor WaterNormal \lor WaterDanger$

$$WaterLow == [WaterSensorModel \mid M_1 \leq qa_1 < N_1 \wedge qa_2 \leq N_2]$$
$$WaterHigh == [WaterSensorModel \mid N_1 \leq qa_1 \wedge N_2 < qa_2 \leq M_2]$$
$$WaterNormal == [WaterSensorModel \mid (N_1 \leq qa_1 \wedge qa_2 \leq N_2)]$$
$$WaterTolerable == [WaterSensorModel \mid M_1 \leq qa_1 \wedge qa_2 \leq M_2]$$
$$WaterDanger == [WaterSensorModel \mid qa_1 < M_1 \vee M_2 < qa_2$$
$$\vee (M_1 \leq qa_1 < N_1 \wedge N_2 < qa_2 \leq M_2)]$$
$$WaterSensorWorking == [WaterSensorModel \mid qst = working]$$
$$WaterSensorBroken == [WaterSensorModel \mid qst = broken]$$

Fig. 7. Predicates Involving the Water Level

6.3 Steam Sensor Model

Like the water sensor, the steam sensor is modeled by its state and the lower and upper approximations of the steam value. Initially, the steam output is assumed to be zero.

$\underline{\quad SteamSensorModel \quad\quad\quad\quad\quad}$
$vst : SensorStates$
$va_1, va_2 : \mathbb{N}$
$\overline{0 \leq va_1 \leq va_2 \leq W}$

$\underline{\quad SteamSensorInit \quad\quad\quad\quad\quad}$
$SteamSensorModel'$
$\overline{vst' = working}$
$va_1' = 0 = va_2'$

As in the case of the water sensor, a number of useful predicates can be defined in obvious terms (see CD-ROM Annex BW).

$SteamZero, SteamSensorWorking, SteamSensorBroken$

6.4 Monitored Pumps Model

Each pump may be closed, opening, open, or broken.

$$PumpStates == \{closed, opening, open, broken\}$$

Like the sensors, an individual pump is modeled by its state and the approximations of its capacity.

$\underline{\quad PumpModel \quad\quad\quad\quad\quad\quad\quad\quad\quad}$
$pst : PumpStates$
$pa_1, pa_2 : \mathbb{N}$
$\overline{pa_1 \leq pa_2}$
$pst \in \{closed, opening\} \Rightarrow pa_1 = 0 = pa_2$
$pst = open \Rightarrow pa_1 = P = pa_2$

Two constraints define the capacity approximations for a non-defective pump. Each pump is controlled by a monitor, which may indicate that the water is flowing or not flowing. Additionally, the monitor itself may be broken.

$$MonitorStates == \{flow, noflow, broken\}$$

A monitored pump is a pump together with a monitor. The main purpose of the monitor is to define the pump's capacity approximations in case it is broken. Other constraints exclude certain combinations of monitor and pump states.

―― $MonitoredPumpModel$ ――――――――――――――――――――
$PumpModel$
$mst : MonitorStates$
――――――
$pst = broken \Rightarrow$
$\qquad (mst = flow \Rightarrow pa_1 = P = pa_2) \land$
$\qquad (mst = noflow \Rightarrow pa_1 = 0 = pa_2) \land$
$\qquad (mst = broken \Rightarrow pa_1 = 0 \land pa_2 = P)$
$pst \in \{closed, opening\} \Rightarrow mst \in \{noflow, broken\}$
$pst = open \Rightarrow mst \in \{flow, broken\}$

The definition of a monitored pump is lifted to a sequence of monitored pumps as follows:

―― $MonitoredPumpsModel$ ――――――――――――――――――――
$Ps : \text{seq}_1 \, MonitoredPumpModel$
$pa_1, pa_2 : \mathbb{Z}$
――――――
$\#Ps = NP$
$pa_1 = P * \#\{i : 1 .. NP \mid (Ps\ i).pa_1 = P\}$
$pa_2 = P * \#\{i : 1 .. NP \mid (Ps\ i).pa_2 = P\}$

A number of simple conditions involving monitored pumps can be defined (see CD-ROM Annex BW).

$PumpsWorking, PumpControlsWorking, PumpControlBroken, ...$

6.5 Valve Model

The required information about the water valve is very simple, and the Z specification is straightforward.

$$ValveStates == \{open, closed\}$$

―― $ValveModel$ ――
$vlv : ValveStates$

6.6 Physical Steam-Boiler Model

Given all these component models, it is sufficient to model very general modes of the physical steam-boiler:

$PhysModes ::= waiting \mid adjusting \mid ready \mid running \mid stopped$
$Alarm ::= ON \mid OFF$

―― Modes ――――――――――――――
$st : PhysModes$
$alarm : Alarm$
――――――――――
$st = stopped$
$\quad \Rightarrow alarm = ON$

―― RunningOk ――――――――――――――
$Modes$
――――――――――
$st = running$
$alarm = OFF$

6.7 Steam-Boiler State

A main feature of the steam-boiler is its ability to continue running safely even when failures of the physical units occur. We need to distinguish here between two classes of failures in the steam-boiler system: in the first class, which is relevant for the initialization phase of the system, at least one physical unit is broken, in the second class, relevant for the running system, the water level sensor together with either the steam-output sensor or one of the pump control sensors is broken.

$NoDefects == WaterSensorWorking \land SteamSensorWorking$
$\qquad \land PumpsWorking \land PumpControlsWorking$
$TolerableDefects == WaterSensorWorking$
$\qquad \lor (SteamSensorWorking \land PumpControlsWorking)$

The state space of the steam-boiler control system can now be defined as follows:

―― SteamBoiler ――――――――――――――
$WaterSensorModel$
$SteamSensorModel$
$MonitoredPumpsModel$
$ValveModel$
$Modes$
――――――――――
$st \in \{waiting, adjusting, ready\} \Rightarrow$
$\qquad (alarm = OFF \Leftrightarrow NoDefects \land SteamZero)$
$st = running \Rightarrow$
$\qquad (alarm = OFF \Leftrightarrow WaterTolerable \land TolerableDefects)$
$(st = running \lor PumpsOpen) \Rightarrow ValveClosed$

There are two constraints defining precisely which condition can be guaranteed in the various working modes. These constraints can be seen as a formal specification of the informal modes (see Chapter AS, this book). Finally, a third constraint states that the pumps should never operate while the valve is open.

We can now define two important conditions used in the reactive model.

$Degraded == SteamBoiler \wedge RunningOk \wedge WaterSensorWorking$
$\qquad \wedge (SteamSensorBroken \vee (PumpControlBroken \setminus (i?)))$
$Rescue == SteamBoiler \wedge RunningOk \wedge WaterSensorBroken$
$\qquad \wedge SteamSensorWorking \wedge PumpControlsWorking$

Having defined the state space of the steam-boiler, we now define the functionality of its various services and internal events.

6.8 Initialization

The steam-boiler is initialized by initializing all its parts. To begin with, all pumps are assumed to be closed, and all monitors are assumed to indicate no flow of water. Furthermore, the valve is closed, the steam-boiler is in waiting mode, and the alarm is off.

$SteamBoilerInit == WaterSensorInit \wedge SteamSensorInit \wedge \ldots$

6.9 Data Transmission Services

Data are transmitted in four steps: the water-level data, the steam data, the pump-control data, and the pump data.

The level transmission recognizes a level device failure by checking whether the fresh level-sensor values remain within the anticipated lower and upper approximations. The approximations are calculated as stated in the informal specification (if the valve is open, the level may drop at an unknown rate, hence the lower approximation is set to zero).

―― *CalculatedLevelBounds* ――――――――――――――――――
$SteamBoiler$

$qc_1, qc_2 : \mathbb{N}$
――――――――――――――――――
$qc_1 =$ **if** $vlv = open$ **then** 0
$\qquad$ **else** $max\{0, qa_1 - (va_2 + U_1 \text{ div } 2 * T) * T + pa_1\}$
$qc_2 = min\{C, qa_2 - (va_1 - U_2 \text{ div } 2 * T) * T + pa_2\}$
――――――――――――――――――

Depending on the comparison of the fresh value and the anticipated bounds, it then decides whether the sensors' state is considered defective and updates the sensor-value approximations. Note that on the functional model the input parameter of the level-transmission message becomes explicit.

$\boxed{\begin{array}{l}\underline{\mathit{LEVEL}}\\ \Delta\mathit{SteamBoiler}\\ \Xi\mathit{WaterSensorModel};\ \Xi\mathit{MonitoredPumpsModel};\ \Xi\mathit{ValveModel}\\ q?:\mathbb{N}\\ \hline alarm = OFF;\ st' = st\\ \exists\, qc_1, qc_2 : \mathbb{N}\ |\ \mathit{CalculatedLevelBounds}\\ \bullet\ qst' = \textbf{if}\ qc_1 \leq q? \leq qc_2\ \textbf{then}\ qst\ \textbf{else}\ broken\\ \quad \wedge\ (qa_1', qa_2') = \textbf{if}\ qst' = working\ \textbf{then}\ (q?, q?)\ \textbf{else}\ (qc_1, qc_2)\end{array}}$

The transmission of the steam value, the pump value, and the pump-monitor value is treated analogously.

6.10 Control Messages

The actual control of the water level is handled by several internal events. The operations *lowmark_reached* and *highmark_reached* effect the opening and closing of the pumps, respectively. After the opening of the pumps has been completed, the internal event *pressure_balanced* occurs. To illustrate this, let us specify *lowmark_reached*. The remaining control messages can be specified analogously.

$\boxed{\begin{array}{l}\underline{\mathit{lowmark\_reached}}\\ \Delta\mathit{SteamBoiler}\\ \Xi\mathit{WaterSensorModel};\ \Xi\mathit{SteamSensorModel};\ \Xi\mathit{ValveModel}\\ \hline \mathit{RunningOk}\\ \forall\, i : 1\mathrel{..} NP\ \bullet\\ ((Ps'\ i).pst = (\textbf{if}\ (Ps\ i).pst = closed\ \textbf{then}\ opening\ \textbf{else}\ (Ps\ i).pst)\\ \quad \wedge\ (Ps'\ i).mst = (Ps\ i).mst)\\ (pa_1', pa_2') = (pa_1, pa_2);\ st' = st\end{array}}$

6.11 Repair Messages

If a level-repair message arrives, the level's state is changed accordingly. Note that the approximate values do not have to altered.

$\boxed{\begin{array}{l}\underline{\mathit{LEVEL\_REPAIRED}}\\ \Delta\mathit{SteamBoiler}\\ \Xi\mathit{SteamSensorModel};\ \Xi\mathit{MonitoredPumpModel};\ \Xi\mathit{ValveModel}\\ \hline alarm = OFF\\ st' = (\textbf{if}\ qst = broken\ \textbf{then}\ st\ \textbf{else}\ stopped)\\ qst' = working\\ (qa_1', qa_2') = (qa_1, qa_2)\end{array}}$

6.12 Analysis of the Functional Model

The preconditions of these steam-boiler operations and events we have been considering can be calculated as follows:

SteamBoiler ⊢

pre *LEVEL* ⇔ *alarm* = *OFF*

pre *LEVEL_REPAIRED* ⇔ pre *lowmark_reached* ⇔ *RunningOk*

7 Alternative Design with Fine-Grained Objects

From a methodological point of view, one might argue that in our solution the steam-boiler control system is too large, i.e. not sufficiently decomposed, thus offering only a small degree of reusability. This criticism is, in our view, partly justified, in it being feasible to identify fine-grain subobjects encapsulating those of the control variables that represent the various sensors' models, and to define appropriate attributes and constraints of these subobjects.

This restructuring would introduce more modularity, but it would not affect overall system behavior. Furthermore, it would then appear feasible that these new subobjects be reused in similar control applications. We have chosen here not to follow this approach, mainly because, in our view, the cohesion between the attributes of the steam boiler control object seemed to be to high to justify the overhead of separating out subobjects. This illustrates the general problem of how to forecast future reuse. Ultimately, the decision will probably depend on experience gained in solving a variety of control problems in this particular application area.

8 Consistency between the Reactive and the Functional Views

The reactive and functional views of an embedded system can be checked against each other in many interesting ways. The basic idea is to systematically and consistently relate the state hierarchy and the transitions introduced in the statecharts to the state spaces and operations as defined by the Z schemas.

8.1 Relating States

A straightforward way to relate states between the two different views is to map every statechart state S on an appropriate Z schema S_z, and then to formulate various proof obligations for this mapping to be adequate.

Assuming such a mapping to be given for a particular component, the consistency conditions can be presented in three steps. For an arbitrary state S from the reactive model of this component, we distinguish between the following three cases:

- S is an elementary state, i.e. there is no decomposition of S in the reactive model. In this case, one has to verify that the associated Z state S_z is nonempty, i.e.

Consistency: $\vdash \exists S_z$.

- S is an OR-composed state, i.e. in the reactive model S is decomposed into exclusive substates $S_1, S_2, \cdots, S_n$ $(n > 0)$ with associated Z schemas $S_z, S_{1z}, S_{2z}, \cdots, S_{nz}$. In this case, one has to check sufficiency, necessity, and disjointness of the decomposition.

   Sufficiency: $S_{1z} \vee S_{2z} \vee \cdots \vee S_{nz} \vdash S_z$.
   Necessity: $S_z \vdash S_{1z} \vee S_{2z} \vee \cdots \vee S_{nz}$
   Disjointness: $S_z \vdash \neg (S_{iz} \wedge S_{jz})$ for all $i, j \in \{1, \cdots, n\}$, where $i \neq j$.

Of course, the top-level statechart of a component must be related to the Z schema defining the overall state space of the component.

8.2 Relating Steam-Boiler States

Following our general methodology for ensuring consistency, for the state boxes introduced in Figure 4, we define the direct substates as follows.

$Emergency == [SteamBoiler \mid alarm = ON]$

$NoEmergency == [SteamBoiler \mid alarm = OFF]$

In order to ensure consistency, we have to prove that these two state classes are disjoint and that their disjunctive composition yields the overall state space:

$SteamBoiler \vdash \neg (Emergency \wedge NoEmergency)$

$Emergency \vee NoEmergency \vdash SteamBoiler$

$SteamBoiler \vdash Emergency \vee NoEmergency$

Furthermore, since the state class *Emergency* is considered primitive, i.e. not further refined in the dynamic model, we have to show that it is non-empty, i.e. consistent.

$\vdash \exists Emergency$

Note that the non-emptiness of primitive states does, of course, ensure non-emptiness of all composed states. According to the statechart structure, similar proofs have to be carried out with respect to the remaining state definitions.

8.3 Relating Operations

In the functional view, we have defined a Z schema for each service, internal event, or guard in the statechart. Based on the association of a Z schema with each statechart box, one can verify conformity between the statechart transitions and the Z definitions.

The idea is to consider an arbitrary state and an arbitrary operation and then check for consistency w.r.t. the transitions leaving that state. More precisely, given an arbitrary operation Op and state S, we have to prove that each transition leaving S and labeled Op (and possibly some guard) behaves as expected, i.e. results in the desired state. We also have to prove that, if the operation or event Op occurs and neither one of the guards of those transitions is true, the application of Op preserves this state.

First, we distinguish the case that no transitions labeled Op are leaving S. In such a case, we have to show that application of S preserves this state.

Preservation: $S_z \land Op_z \land \neg P_{Op,S} \vdash S'_z$.

S_z and Op_z are the Z schemata associated with S and Op.

Here, $P_{Op,S}$ denotes the *priority condition* of Op in S. The priority condition $P_{Op,S}$ of an operation Op with respect to a state S is defined as the disjunction $C_1 \lor \cdots \lor C_n$ of all guards of transitions labeled Op and leaving from any superstate of S. This condition is needed since we assume these transitions to fire with priority over the transitions leaving S.

We still have to deal with the case that $n > 0$. Assume that transitions $t_1, \cdots, t_n$ ($n > 0$) are the transitions labeled Op and guards $C_1, \cdots, C_n$ (the default guard is *true*) are leaving from S to states $S_1, \cdots, S_n$. We check for consistency of these transitions as follows:

Applicability: $S_z \vdash \text{pre } Op_z$.
Explicit Correctness: $S_z \land Op_z \land C_{iz} \land \neg P_{Op,S} \vdash S'_{iz}$, for $1 \leq i \leq n$.
Implicit Correctness: $S_z \land Op_z \land \neg (C_{1z} \lor \cdots \lor C_{nz}) \land \neg P_{Op,S} \vdash S'_z$, if S_z is primitive.

C_{iz} and S_{iz} are the Z schemata associated with C_i and S_i. Note that the applicability check, i.e. any state from which a transition labeled with Op is leaving must imply the precondition of Op. Note also that implicit correctness has only to be checked for primitive states, as this induces implicit correctness for composed states.

Note that implicit correctness is trivial in cases where the disjunction of the guards is complete, e.g. in the large number of cases where $n = 1$ and $C_1 \Leftrightarrow true$.

This defines a large number of proof obligations for ensuring consistency. Fortunately, one can often significantly reduce the number of necessary proofs by using two properties related to the applicability of operations. The first property is the fact that, if an operation is applicable to an OR-superstate of S, it is also applicable to S. This property usually significantly reduces the number of applicability checks.

The second property is related to inapplicability of an operation Op in a state S. Op is inapplicable in S if we can prove that

Inapplicability: $S_z \vdash \neg \text{pre } Op_z$.

The property then states that, if an operation is inapplicable to an OR-superstate of S, it is also inapplicable to S. The use of this property often allows us to omit the checks of explicit and implicit correctness.

8.4 Relating Steam-Boiler Operations

Again, we will systematically discuss the proof obligations, based on the hierarchical structuring of the dynamic view.

In the top-level statechart (Figure 4), we notice first of all that no operation is applicable in the *Emergency* state. This implies all applicability obligations about this state. There are eight transitions leaving the emergency state directly, giving rise to the same number proof obligations for explicit correctness. For example, one of the transitions subsumed under *Repair_Event* (Figure 4) is labeled with *LEVEL_REPAIRED*; it transforms from the state *NoEmergency* to the state *Emergency* (Figure 8).

Fig. 8. A transition from Figure 4

We have to prove that this is indeed consistent with the definition of the repair message for the water level sensor, i.e. we must prove the following correctness property about the transition:

NoEmergency ∧ *LEVEL_REPAIRED*
∧ *WaterSensorWorking* ⊢ *Emergency'*

Since *NoEmergency* is a composed state, we do not have to prove any implicit correctness obligation. The remain proof obligations for transitions labeled with *LEVEL* can be generated in analogously.

9 Evaluation and Comparison

Our solution contains a functional and architectural design of the system's components, its safety, and basic aspects of its liveness. Performance (in the sense of response time) is not taken into consideration. It is assumed that the system is always fast enough. Note that our operational modeling approach avoids a harsh distinction between requirements and design specification, but follows the general philosophy of using a phase-independent set of notations. Our objective was to provide a specification technique with a varying degree of formality. All data structures and data transformation are formally specified in Z. These specifications are type-checked mechanically with the ESZ type-checker[4]. Since we do not decide upon the formal semantics for all aspects of the statecharts, the specification of reactive system behavior remains semi-formal.

As regards the verification of the functional/architectural design, we derived formal proof obligations, which validate the consistency between overlapping aspects of the different models, and (mentally) examined some of these formulas for validity.

The solution has not been implemented and hence has, not been linked to the simulator. In the future, tools for translating such specifications into code should be developed or adapted. For statecharts, such tools are commonly available. Concerning Z specifications, we would argue to stick to an operational modeling style, allowing the generation of efficient code. The degree to which such a style can be reasonably adopted probably depends largely on the particular application area in hand. For example, the Z specification used in this example can be directly mapped into efficient code. Within the Espress project, there is encouraging progress towards automatic code generation.

The use of the Z notation makes our solution similar to VDM-based solutions (see Chapters LP and S, this book). The main difference is that our solution uses Z for

---

[4] available through ftp://ftp.cs.tu-berlin.de/pub/local/uebb/

modeling the static functional behavior only, the reactive behavior being delegated to statecharts. Concerning the reactive behavior itself, it is interesting to compare our semi-formal use of statecharts with solutions using formal models of state machines (e.g. see Chapter LL, this book). Finally, an interesting extension to our solution might also model the continuous components of the system using an appropriate extension of state machines (see Chapter HW, this book).

The development of the specification was a highly iterative process, taking about two weeks. Writing up the solution took about another two weeks. Polishing this paper took perhaps another week.

The state diagrams and the "mathematics-like" Z notation should already provide a rough idea of what is meant, without any special knowledge being required. One or two weeks suffice to understand the basic elements of Z and semi-formal statecharts. However, operational knowledge of Z and statecharts is necessary in order to be able to produce a solution to such a problem within our framework. To not only understand but also know how to effectively apply these notations to nontrivial problems requires at least one to two months experience.

In summary, the proposed combination of statecharts and Z for modeling embedded control systems proved to be both semantically and pragmatically interesting. In the Espress project, important experiments with the aim of identifying useful recommendations, guidelines, and heuristics for the process of developing such combined specification are conducted. Meanwhile, the development of the combination is continued.

Acknowledgements

Our thanks go to all colleagues from the Espress project. Dr. Maritta Heisel provided many comments on an early version. Special thanks to Phil Bacon for polishing up the English.

References

1. Jean-Raymond Abrial. Steam-boiler control specification problem. Material for the Dagstuhl meeting *Methods for Semantics and Specification*, June 4-9, 1995., August 1994.
2. G. Booch. *Object-Oriented Analysis and Design with Applications*. Benjamin Cummings, second edition, 1994.
3. D. Harel. Statecharts: A visual formalism for complex systems. *Science of Computer Programming*, 8(3):231–274, 1987.
4. D. Harel and E. Gery. Executable Object-Modeling with Statecharts. In *Proc. ICSE 18*, 1996.
5. J. Rumbaugh et al. *Object-Oriented Modeling and Design*. Prentice-Hall, 1991.
6. B. Selic, G. Gullekson, and P. T. Ward. *Real-Time Object-Oriented Modeling*. John Wiley & Sons, 1994.
7. M. Spivey. *The Z Notation, A Reference Manual*. Prentice Hall, 2nd edition, 1992.
8. M. von der Beeck. A comparison of statecharts variants. In *Symposium on Fault-Tolerant Computing*, LNCS. Springer, 1994.
9. M. Weber. Combining Statecharts and Z for the Design of Safety-Critical Control Systems. In Marie-Claude Gaudel and James Woodcock, editors, *FME'96: Industrial Benefits and Advances in Formal Methods*, volume 1051 of *LNCS*, pages 307–326. Springer Verlag, 1996.
10. M. Weber. Integrating Mathematical Techniques in the Development of Embedded Control Systems. internal report; TU-Berlin, 1996.

An Action System Approach to the Steam Boiler Problem

Michael Butler[1], Emil Sekerinski[2], Kaisa Sere[3]

[1] Dept. of Electronics and Computer Science, University of Southampton,
Southampton, United Kingdom, M.J.Butler@ecs.soton.ac.uk.
[2] Dept. of Computer Science, Åbo Akademi University, Turku, Finland,
Emil.Sekerinski@abo.fi.
[3] Dept. of Computer Science and Applied Mathematics, University of Kuopio,
Kuopio, Finland, Kaisa.Sere@uku.fi.

Abstract. This paper presents an approach to the specification of control programs based on action systems and refinement. The system to be specified and its physical environment are first modelled as one initial action system. This allows us to abstract away from the communication mechanism between the two entities. It also allows us to state and use clearly the assumptions that we make about how the environment behaves. In subsequent steps the specifications of control program and the environment are further elaborated by refinement and are separated. We use the refinement calculus to structure and reason about the specification. The operators in this calculus allow us to achieve a high degree of modularity in the development.

1 Introduction

The action system formalism, introduced by Back and Kurki-Suonio [4], is a state based approach to distributed computing. A set of guarded actions share some state variables and may act on those variables. The two main development techniques we use on action systems in this case study are *refinement* and *parallel decomposition*. Refinement allows us to replace abstract state variables with more concrete representations such that the behaviour of the refined action system satisfies the behaviour of the abstract action system. Parallel decomposition allows us to split an action system into parallel sub-systems by partitioning state variables and actions.

An important aim of this case study has been to produce an action system specification of the Steam Boiler problem (see Chapter AS, this book) that is easy to understand, and thus easier to validate, and then derive a controller from this specification that is close to the desired implementation. We achieve a simplified specification in two main ways. Firstly, we model both the controller and its physical environment; this allows us to simplify the description of the interaction between the controller and it's environment. Secondly, we use abstraction to describe a very general view of the required behaviour of the system and then elaborate this view using refinement. A controller specification is then derived using refinement and decomposition.

Overview. We start by briefly describing the action system framework and the refinement calculus to the required extent. Section 3 provides more detail on our approach to this particular case study. Section 4 presents the abstract specification of the controller and its environment. Section 5 refines the ways in which the system may fail and (as a byproduct) refines the modes as well. CD-ROM Annex BSS.1 gives the refinement rules to the extent needed. CD-ROM Annex BSS.2 describes further refinement steps towards a distributed implementation in an imperative language. CD-ROM Annex BSS.3 contains the implementation in Pascal.

2 Action Systems

The action systems formalism combined with the refinement calculus has proved to be very suited to the design of parallel and distributed systems [7, 6, 5]. Action systems are similar to the UNITY programs of Chandy and Misra [11] which have an associated temporal logic. The design and reasoning about action systems is carried out within the refinement calculus that is based on the use of predicate transformers. The refinement calculus for sequential programs has been studied by several researchers [2, 13, 14]. The main refinement technique used in our specification is data refinement [8] that is related to e.g. the refinement mapping technique of Abadi and Lamport [1].

Actions. An action is any statement in an extended version of Dijkstra's guarded command language [12]. This language includes assignment, sequential composition, conditional choice and iteration, and is defined using *weakest precondition* predicate transformers. We remove Dijkstra's law of "Excluded Miracle" which says that no statement is miraculous (i.e., can establish any postcondition), and take the view that an action is only *enabled* in those initial states in which it behaves non-miraculously. The *guard* of an action A is the condition $g(A)$, defined by

$$g(A) = \neg wp(A, false).$$

The action A is said to be enabled when the guard is true. The action A is said to be *always enabled*, if $wp(A, false) = false$ (i.e., $g(A) = true$).

We also use the following constructs:

- *Abort:* The action **abort** behaves arbitrarily, making changes to variables or not terminating at all. It represents undesired behaviour.
- *Guarding:* The action $P \rightarrow A$, where P is a predicate and A is an action, is disabled when P is false, otherwise it behaves as A.
- *Nondeterministic assignment:* The action $x := x'.P$, where P is a predicate relating x and x', assigns a value x' satisfying P to state variable x. The arbitrary nondeterministic assignment $x := ?$ is a shorthand for $x := x'.true$.
- *Choice:* The action $A_1 \; [] \; A_2$ tries to choose an enabled action from A_1 and A_2, the choice being nondeterministic when both are enabled.

- *Sequencing:* The action $A_1 \,; A_2$ first behaves as A_1 if this is enabled, then as A_2 if this is enabled at the termination of A_1, otherwise the whole sequence $A_1 \,; A_2$ is not enabled.
- *Always enabled:* The action $\overline{A}$ behaves as A when A is enabled, otherwise it behaves as skip, thus $\overline{A}$ is always enabled.
- *Simultaneous execution:* For actions A_1 and A_2 that are non-aborting, i.e. terminating when executed in an enabled initial state, $A_1 \parallel A_2$ is the simultaneous execution of A_1 and A_2, e.g.,

$$(P \to x := E) \parallel (Q \to y := F) \;=\; P \wedge Q \to x, y := E, F,$$

and when the actions are always enabled we have e.g.,

$$x := E \parallel y := F \;=\; x, y := E, F.$$

Action Systems. An *action system* has the form:

$$\mathcal{A} \;=\; |[\, \textbf{var } x.I \,;\, \textbf{do } A_1 \,[]\, \ldots \,[]\, A_m \, \textbf{od} \,]|\,:\, z.$$

The action system $\mathcal{A}$ is initialised by action I. Then, repeatedly, an enabled action from $A_1 \ldots A_m$ is nondeterministically selected and executed. The action system terminates when no action is enabled, and aborts when some action aborts.

The *local* variables of $\mathcal{A}$ are the variables x and the *global* variables of $\mathcal{A}$ are the variables z. The local and global variables are assumed to be distinct. Each variable is associated with an explicit type. The *state variables* of $\mathcal{A}$ consist of the local variables and the global variables. The actions are allowed to refer to all the state variables of an action system. In the sequel, we use the keywords **global** and **var** to distinguish global and local variables.

Parallel Composition. Consider two action systems $\mathcal{A}$ and $\mathcal{B}$

$$\mathcal{A} \;=\; |[\, \textbf{var } x.I \,;\, \textbf{do } A_1 \,[]\, \ldots \,[]\, A_m \, \textbf{od} \,]|\,:\, z$$
$$\mathcal{B} \;=\; |[\, \textbf{var } y.J \,;\, \textbf{do } B_1 \,[]\, \ldots \,[]\, B_n \, \textbf{od} \,]|\,:\, v$$

where $x \cap y = \emptyset$. We define the *parallel composition* $\mathcal{A} \parallel \mathcal{B}$ of $\mathcal{A}$ and $\mathcal{B}$ to be the action system

$$\mathcal{C} \;=\; |[\, \textbf{var } x.I \,;\, y.J \,;\, \textbf{do } A_1 \,[]\, \ldots \,[]\, A_m \,[]\, B_1 \,[]\, \ldots \,[]\, B_n \, \textbf{od} \,]|\,:\, z \cup v.$$

Thus, parallel composition will combine the state spaces of the two constituent action systems, merging the global variables and keeping the local variables distinct.

The behaviour of a parallel composition of action systems is dependent on how the individual action systems, the *reactive components*, interact with each other via the global variables that are referenced in both components. We have for instance that a reactive component does not terminate by itself: termination is a global property of the composed action system. More on these topics can be found in [3].

Refinement. Action systems are intended to be developed in a stepwise manner within the refinement calculus. In the steam boiler example, data refinement is used as a main tool. Here we briefly describe these techniques. Data refinement of action systems is studied in detail in [3].

The refinement calculus is based on the following definition. Let A, A' be actions. The action A is *refined* by action A', denoted $A \leq A'$, if

$$\forall Q.(wp(A, Q) \Rightarrow wp(A', Q)).$$

This usual refinement relation is reflexive and transitive. It is also monotonic with respect to most of the action constructors used here, e.g. guarding, choice, sequencing and simultaneous execution, see [8]. (Refinement between actions does not necessarily imply refinement between action systems.)

Let now A be an action referring to the variables x, z, denoted $A : x, z$, and A' an action referring to the variables x', z. Then statement A is *data refined* by statement A' using *abstraction relation* $R(x, x', z)$, denoted $A \leq_R A'$, if

$$\forall Q.(R \land wp(A, Q) \Rightarrow wp(A', \exists x.R \land Q)).$$

Note that $\exists x.R \land Q$ is a predicate on the variables x', z.

Data Refinement of Action Systems. Let $\mathcal{A}$ and $\mathcal{A}'$ be the two action systems

$$\mathcal{A} = |[\textbf{ var } x.I\,; \textbf{do } A \textbf{ od }]| : z$$
$$\mathcal{A}' = |[\textbf{ var } x'.I'\,; \textbf{do } A' \textbf{ od }]| : z$$

Let $R(x, x', z)$ be an abstraction relation on the local variables x, x', and global variables z. Assume I, I' do not access but only assign to x, x', respectively. The action system $\mathcal{A}$ is data refined by $\mathcal{A}'$ using R, denoted $\mathcal{A} \leq_\mathcal{R} \mathcal{A}'$ if:

(i) *Initialisation:* $I \leq_R I'$,
(ii) *Main actions:* $A \leq_R A'$,
(iii) *Exit condition:* $R \land gA \Rightarrow gA'$.

If $\mathcal{A} \leq_\mathcal{R} \mathcal{A}'$, then the behaviour of $\mathcal{A}'$ satisfies the behaviour of $\mathcal{A}$ in the sense that all possible state traces of $\mathcal{A}'$ are possible state traces of $\mathcal{A}$. This is described in detail in [9].

3 Approach

In this section, we discuss some features of our approach to the Steam Boiler problem using the action system formalism.

Single-Language Framework. We use the same formalism (action systems) to describe specifications and designs. Thus, the initial formal description of the behaviour we require of the system is given as an abstract action system rather than as a set of properties in some variant of temporal logic. By a series of data-refinement steps, this abstract action system is transformed into a concrete action system more closely resembling the eventual implementation. Refinement is the main form of proof we use.

Elaboration by Refinement. Rather than embody all the requirements in the initial specification, we have chosen instead to introduce some of the requirements in successive refinement steps. This is achieved by using data abstraction to generalise the requirements. For example, instead of modelling all the different equipment failures in the initial specification, we just have one general notion of failure, which is elaborated into the different forms of failure in subsequent refinement steps.

Usually, refinement is used as a way of verifying the correctness of an implementation w.r.t. a specification. But here we also use refinement as a way of structuring the requirements such that they are easier to validate. One consequence of this approach is that the abstraction relations used in refinement steps really form part of the formal description of the requirements; our abstract action system model is intended to represent the essence of the required behaviour of the system, and the abstraction relations show how this essence relates to the extra requirements being introduced in a refinement (elaboration) step.

Environment and Controller as One System. Our initial action system is intended to model the behaviour of the overall system, that is, the physical environment and the controller together. After some refinement steps, we use parallel decomposition to separate the controller and the physical environment into two interacting action systems, thus arriving at a specification of the controller itself.

Modelling the environment and the controller as a single action system allows us to abstract away from the communication mechanism between them. For example, all sensors are modelled as state variables which are updated by the environment actions and may be read directly by the controller actions. Only in later refinement steps do we introduce an explicit mechanism for passing the values of sensors from the environment to the controller. Similarly, device actuators are modelled initially as state variables that are updated by the controller and read by the environment and these are refined later.

Another reason for modelling the environment is that it allows us to state and use assumptions that we make about how the environment behaves. For example, when we introduce a mechanism in the controller for estimating the water level in the steam boiler, we need to model the way in which the water level may change in the environment.

Timing and Discreteness The requirements state that the environment sends messages to the controller once every five seconds giving updated sensor values, and that the controller then responds to these by sending out new values for the actuator states. We model this as a simple alternation between an environment action and a controller action. We do not use any explicit model of time, rather we simply assume that the environment action occurs once every five seconds, and that the controller action is fast enough to respond within that five seconds.

This discrete model of the environment is not a true reflection of the behaviour of the physical environment which is a continuous system. However it is sufficient for us to be able to model our assumptions about the environment.

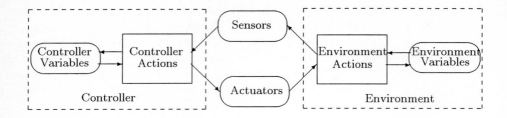

Fig. 1. Structure of the system specification

Modularity. As well as using refinement to structure the specification as mentioned above, we also use the refinement calculus composition operators, such as choice and simultaneous assignment, to structure the descriptions of actions. An important feature of these operators is that they are compositional with respect to refinement allowing us to achieve a high degree of modularity in the development.

4 Abstract Specification

The steam boiler system is specified by actions which represent the evolution of the physical environment and the reactions of the controller. Values that the control program needs to measure are modelled as variables which are read by the controller actions and modified by the environment actions. Devices that control the physical environment are modelled as variables which are modified by the controller actions and read by the environment actions (see Fig. 1).

Abstraction in the initial specification is achieved by:

1. unifying the different ways of system failure into one notion of failure,
2. reducing the number of modes by unifying the *Normal*, *Degraded*, and *Rescue* modes to a single *Operating* mode,
3. modelling all actuators, sensors, and controller variables as part of a state space, thus abstracting from their distribution and the message passing protocol.

Variables. The abstract view of the system state consists of the following variables. The current measure for the water level (in litres) in the boiler is given by q:

global q : Num

The current measure for the steam output (in litres/sec) is given by v:

global v : Num

The current measures for the water input through the pumps during an interval of $T = 5$ seconds are given (in litres/T secs)[4] by $p_1, \ldots, p_4$:

global $p_1, \ldots, p_4$: Num

The current measure for the water output through the evacuation valve during an interval of $T = 5$ seconds is given (in litres/T secs) by e:

global e : Num

The water pumps and evacuation valve are controlled by the following variables:

global *pumps* : *on* | *off*

global *valve* : *open* | *closed*

During normal operation, the water level should be between N_1 and N_2, and it is unsafe for it to go below M_1 or above M_2:

const N_1, N_2, M_1, M_2 : Num **where** $M_1 < N_1 < N_2 < M_2$

The variable *reliable* is an abstraction for whether the information available to the controller is reliable or not. If it is *true*, then we can rely on the measures, otherwise we should go to emergency mode:

var *reliable* : Bool

The boiler control has three basic modes, the initialisation mode, the normal operating mode, and the emergency mode:

global *mode* : *Init* | *Operating* | *Emergency*

System Structure. The steam boiler system is specified by a repeated alternation between the physical environment and the controller, with the possibility of a failure making the measures unreliable. This alternation assumes that the actions of the controller take a negligible amount of time giving the controller the chance to react to changes of the environment:

$System \ \widehat{=} \ |[\ \textbf{var}\ reliable : \text{Bool}.I\ ;\ \textbf{do}\ Environment\ ;\ Controller\ \textbf{od}\ ]|$

The controller consists of sets of actions which control the water pumps, the evacuation valve and the mode and are executed in parallel. The strict alternation is used here as a modelling feature. It will be removed later when the environment and the controller are separated into systems of their own.

---

[4] In the problem statement (Chapter AS, this book), $p_1, \ldots, p_4$ and e are in litres/sec.

Safety Condition. The system should remain in operating mode, only if the water level is safe. More precisely, after reacting to the environment's messages, if the controller remains in operating mode, then the water level should be safe:

$$safety \; \widehat{=} \; mode = Operating \Rightarrow M_1 \leq q \leq M_2$$

This is checked by ensuring that $wp(I, safety)$ holds, i.e. the initialisation establishes *safety*, and $safety \Rightarrow wp(Environment; Controller, safety)$, i.e. the body of *System* preserves *safety*.

Initialisation. Initially, we start in initialisation mode and assume that the pumps are off and the evacuation valve is closed. We further assume that we are in a reliable state. We make no assumptions about the current measures $(q, p_1, \ldots, p_4, v, e)$.

$$I \; \widehat{=} \; pumps := \mathit{off} \,;\, valve := closed \,;\, mode := Init \,;\, reliable := true$$

4.1 Physical Environment Specification

In the abstract specification, we allow the environment to make arbitrary assignments to the water level, input and output levels, and the reliable flag:

$$Environment \; \widehat{=} \; q, p_1, p_2, p_3, p_4, v, e, reliable := ?,?,?,?,?,?,?,?$$

4.2 Controller Specification

Init Mode. Initialisation of the boiler involves bringing the water level to between N_1 and N_2, as long as no failure is detected. If the water level is above N_2, then the following action opens the evacuation valve:

$$Valve_1 \; \widehat{=} \; \begin{array}{l} mode = Init \\ reliable = true \\ q > N_2 \end{array} \rightarrow valve := open$$

(When describing actions we use the syntax above. Here the conjunction of the three predicates on the left hand side of the arrow constitutes the guard, which in this case is $mode = Init \wedge reliable = true \wedge q > N_2$. The lines on the right hand side form the action body, here $valve := open$.)

Once the water level is at or below N_2, the evacuation valve is closed:

$$Valve_2 \; \widehat{=} \; \begin{array}{l} mode = Init \\ reliable = true \\ q \leq N_2 \end{array} \rightarrow valve := closed$$

If the water level is below N_1, then the following action switches on the pumps:

$$Pumps_1 \; \widehat{=} \; \begin{array}{l} mode = Init \\ reliable = true \\ q < N_1 \end{array} \rightarrow pumps := on$$

The pumps are switched off if the water level is above N_2:

$$Pumps_2 \;\widehat{=}\; \begin{array}{l} mode = Init \\ reliable = true \\ q > N_2 \end{array} \;\rightarrow\; pumps := \mathit{off}$$

Once the level is between N_1 and N_2, the system enters operating mode, provided the evacuation valve is closed and no failure of the gauges is detected:

$$Mode_1 \;\widehat{=}\; \begin{array}{l} mode = Init \\ reliable = true \\ valve = closed \\ N_1 \leq q \leq N_2 \end{array} \;\rightarrow\; mode := Operating$$

Observe that the requirement of the valve being closed is not explicitly mentioned in the informal requirements specification. It is though a reasonable restriction and hence, included here.

In case of failure of the gauges, the system goes to emergency mode:

$$Mode_2 \;\widehat{=}\; \begin{array}{l} mode = Init \\ reliable = false \end{array} \;\rightarrow\; mode := Emergency$$

Operating Mode. Operating mode simply involves switching the pumps on and off as appropriate, in order to maintain the proper water level. If the water level is below N_1, the pumps are switched on. If the water level is above N_2, the pumps are switched off. If the water level is in between, we leave it open whether the pumps are on or off. Thus, if it is below N_2, the pumps may be switched on:

$$Pumps_3 \;\widehat{=}\; \begin{array}{l} mode = Operating \\ reliable = true \\ q < N_2 \end{array} \;\rightarrow\; pumps := on$$

If the water level is above N_1, the pumps may be switched off. Note that the guards of the actions are overlapping; if q is between N_1 and N_2 the pumps can be either switched on or off. (This nondeterminism is reduced in the next refinement step.)

$$Pumps_4 \;\widehat{=}\; \begin{array}{l} mode = Operating \\ reliable = true \\ q > N_1 \end{array} \;\rightarrow\; pumps := \mathit{off}$$

If a failure is detected or the water level is unsafe, the system goes into emergency mode:

$$Mode_3 \;\widehat{=}\; \begin{array}{l} mode = Operating \\ (reliable = false \;\vee \\ q < M_1 \;\vee\; q > M_2) \end{array} \;\rightarrow\; mode := Emergency$$

Emergency Mode. We make no assumptions about what happens during emergency mode, so the system may abort:

$$Fail \;\widehat{=}\; mode = Emergency \;\rightarrow\; \mathbf{abort}$$

Controller Action. The control of the water pumps, the evacuation valve and the mode is given by the actions *Pumps*, *Valve*, and *Mode*, respectively:

$Pumps \;\widehat{=}\; Pumps_1 \;[]\; \ldots \;[]\; Pumps_4$

$Valve \;\widehat{=}\; Valve_1 \;[]\; Valve_2$

$Mode \;\widehat{=}\; Mode_1 \;[]\; Mode_2 \;[]\; Mode_3$

The controller executes all these actions, as well as the *Fail* action, in parallel, if they are enabled:

$Controller \;\widehat{=}\; \overline{Pumps} \;\|\; \overline{Valve} \;\|\; \overline{Mode} \;\|\; \overline{Fail}$

5 Refining Failures

In this step, we distinguish possible failures and specify more precisely the required behaviour for failures. We also make the behaviour of the environment more deterministic.

Variables. The following variables are introduced. The status of the water level gauge, the four water pump gauges and the steam output gauge are given by q_gauge, p_1_gauge, ..., p_4_gauge and v_gauge, respectively:

var $q\_gauge, p_1\_gauge, \ldots, p_4\_gauge, v\_gauge : ok \;|\; failed$

If a gauge fails, then the value transmitted from the gauge to the controller does not necessarily correspond to the real measure. The transmitted values of the water level gauge, the four water pump gauges and the steam output gauge are given by $trans\_q, trans\_p_1, \ldots, trans\_p_4$ and $trans\_v$, respectively:

var $trans\_q, trans\_p_1, \ldots, trans\_p_4, trans\_v : \text{Num}$

In case the water level gauge fails, an estimate of the minimal and maximal water level is maintained by the controller. These adjusted values are given by qa_1 and qa_2, respectively:

var $qa_1, qa_2 : \text{Num}$

Abstraction Relation. The variable *reliable* is refined by the above variables under the following abstraction relation R, which is made up of several parts.

For each gauge, if it is working properly (and its measure is transmitted correctly), then the transmitted value corresponds to the real measure:

$$R\_trans \;\widehat{=}\; \begin{array}{l} (q\_gauge = ok \Rightarrow trans\_q = q) \;\wedge \\ (p_1\_gauge = ok \Rightarrow trans\_p_1 = p_1) \;\wedge \\ \ldots \;\wedge \\ (p_4\_gauge = ok \Rightarrow trans\_p_4 = p_4) \;\wedge \\ (v\_gauge = ok \Rightarrow trans\_v = v) \end{array}$$

Here we identify transmission errors with gauge failures.

At initialisation, at least the water level gauge is working:

$$R\_init \;\widehat{=}\; \begin{array}{l}(mode = Init \Rightarrow \\ (reliable = true \Leftrightarrow q\_gauge = ok))\end{array}$$

We distinguish four conditions of the gauges, denoted by *NormalCond*, *DegradedCond*, *RescueCond* and *EmergencyCond*, respectively:

$$NormalCond \;\widehat{=}\; \begin{array}{l} q\_gauge = ok \wedge \\ p_1\_gauge = ok \wedge \ldots \wedge p_4\_gauge = ok \wedge \\ v\_gauge = ok \end{array}$$

$$DegradedCond \;\widehat{=}\; \begin{array}{l} q\_gauge = ok \wedge \\ (p_1\_gauge = failed \vee \ldots \vee p_4\_gauge = failed \vee \\ v\_gauge = failed) \end{array}$$

$$RescueCond \;\widehat{=}\; \begin{array}{l} q\_gauge = failed \wedge \\ p_1\_gauge = ok \wedge \ldots \wedge p_4\_gauge = ok \wedge \\ v\_gauge = ok \end{array}$$

$$EmergencyCond \;\widehat{=}\; \begin{array}{l} q\_gauge = failed \wedge \\ (p_1\_gauge = failed \vee \ldots \vee p_4\_gauge = failed \vee \\ v\_gauge = failed) \end{array}$$

If the pump gauges are ok and the steam output gauge is ok, then the real water level is between its lower and upper estimate:

$$R\_est \;\widehat{=}\; \begin{array}{l} mode = Operating \Rightarrow \\ (NormalCond \vee DegradedCond \Rightarrow qa_1 = q = qa_2) \wedge \\ (RescueCond \Rightarrow qa_1 \leq q \leq qa_2) \end{array}$$

For the relation of the abstract variable *reliable* to the concrete variables q_gauge, p_1_gauge, ..., p_4_gauge, v_gauge we note that not every failure of a gauge of the refined specification corresponds to a failure in the abstract specification. In initialisation mode, only the water level gauge has to work properly. In operating mode, the water level gauge has to work properly or otherwise all other gauges have to work properly and the water level must be safe based on the estimated values:

$$R\_reliable \;\widehat{=}\; \begin{array}{l} (mode = Operating \Rightarrow \\ (reliable = true \Leftrightarrow \\ (NormalCond \vee DegradedCond \vee RescueCond) \wedge \\ (M_1 \leq qa_1 \wedge qa_2 \leq M_2) \wedge (N_1 \leq qa_1 \vee qa_2 \leq N_2))) \end{array}$$

In the refinement, we assume that the evacuation valve is closed in operating mode:

$$R\_valve \;\widehat{=}\; mode = Operating \Rightarrow e = 0$$

The abstraction relation for the operating mode is

$$R\_oper \ \widehat{=} \ R\_est \land R\_reliable \land R\_valve$$

The complete abstraction relation is the conjunction of $R\_trans, R\_init,$ and R_oper:

$$R \ \widehat{=} \ R\_trans \land R\_init \land R\_oper$$

System Structure. The steam boiler system is refined by a repeated sequential composition of the environment (changing the "real" measures), an action keeping track of the estimated water level, and the controller:

$$System' \ \widehat{=} \ |[\ \textbf{var}\ q\_gauge, p_1\_gauge, \ldots, p_4\_gauge, v\_gauge,\\ trans\_q, trans\_p_1, \ldots, trans\_p_4, trans\_v, qa_1, qa_2.\\ I'\ ;\ \textbf{do}\ Environment'\ ;\ Estimate\ ;\ Controller'\ \textbf{od}\ ]|$$

For verifying the refinement $System \leq_R System'$, we have to establish following conditions:

$$I \ \leq_R \ I' \quad (1)$$

$$Environment\ ;\ Controller \ \leq_R \ Environment'\ ;\ Estimate\ ;\ Controller' \quad (2)$$

$$R \land g(Environment\ ;\ Controller) \ \Rightarrow \\ g(Environment'\ ;\ Estimate\ ;\ Controller') \quad (3)$$

For the purpose of implementation, we will consider *Estimate* to be part of the controller. For the purpose of verifying the refinement, we will consider it to be part of the environment; this is allowed by the associativity of sequential composition. Hence (2) is established by (see CD-ROM Annex BSS.1 for the rule):

$$Environment \ \leq_R \ Environment'\ ;\ Estimate \quad (4)$$

$$Controller \ \leq_R \ Controller' \quad (5)$$

Furthermore, we will design *Environment'*, *Estimate* and *Controller'* such that they are always enabled. Hence (3) is established by (see the CD-ROM Annex BSS.1 for the rules):

$$g(Environment') \ = \ true \quad (6)$$

$$g(Estimate) \ = \ true \quad (7)$$

$$g(Controller') \ = \ true \quad (8)$$

In summary, this leads to the proof obligations (1), (4) - (8).

Initialisation. Initially, we assume that the gauges are working properly but make no assumptions about the transmitted values $trans\_q$, $trans\_p_1$, ..., $trans\_p_4$, $trans\_v$, and the water level estimates qa_1, qa_2:

$$I' \;\widehat{=}\; \begin{array}{l} pumps := \mathit{off}\,;\, valve := closed\,;\, mode := Init\,; \\ q\_gauge, p_1\_gauge, \ldots, p_4\_gauge, v\_gauge := ok, ok \ldots, ok, ok \end{array}$$

In order to prove (1), using the rule for partwise data refinement (see CD-ROM Annex BSS.1), it is sufficient to establish

$$\begin{array}{l} reliable := true \;\leq_R \\ q\_gauge, p_1\_gauge, \ldots, p_4\_gauge, v\_gauge := ok, ok \ldots, ok, ok \end{array}$$

This amounts to proving

$$R \;\Rightarrow\; R[reliable, q\_gauge, p_1\_gauge, \ldots, p_4\_gauge, v\_gauge := \\ true, ok, ok, \ldots, ok, ok]$$

which holds by the laws of predicate calculus.

5.1 Physical Environment Refinement

The refined environment is specified by the (physically) possible changes of the measures during the interval $T = 5\;sec$. This is expressed by the following assignments, explained below:

$$Env_1 \;\widehat{=}\; \begin{array}{l} p_1 := p_1'.(p_1' \in \{0..P*T\})\,;\,\ldots\,;\, p_4 := p_4'.(p_4' \in \{0..P*T\})\,; \\ e := \mathbf{if}\; valve = open\; \mathbf{then}\; E*T\; \mathbf{else}\; 0\,; \\ v := v'.v\_variation(v, v')\,; \\ q := q'.q\_variation(q, v, p_1 + p_2 + p_3 + p_4, e, q') \end{array}$$

where

$$q\_variation(q, v, p, e, q') \;\widehat{=}\; \begin{array}{l} q\_min(q, v, p, e) \leq q' \leq q\_max(q, v, p, e) \;\wedge \\ 0 \leq q' \leq C \end{array}$$

$$q\_min(q, v, p, e) \;\widehat{=}\; q - v*T - (1/2)*U_1*T^2 + p - e$$

$$q\_max(q, v, p, e) \;\widehat{=}\; q - v*T + (1/2)*U_2*T^2 + p - e$$

$$v\_variation(v, v') \;\widehat{=}\; \begin{array}{l} v - U_2*T \leq v' \leq v + U_1*T \;\wedge \\ 0 \leq v' \leq W \end{array}$$

If the evacuation valve is opened, the water output through the valve is E litres per second, otherwise 0; it is assumed that the evacuation valve never fails.

Each of the four water pumps might be switched on or off or not be working properly. Hence the throughput of each individual pump is between 0 and P per second, not necessarily depending on the value of *pumps*.

The steam output will decrease by at most $U_2 * T$ within T seconds and increase by at most $U_1 * T$. In any case, the steam output is between 0 and W.

The water level increases by the amount p of water flowing through the pumps and decreases by the amount e of water flowing through the evacuation valve. Furthermore, within T seconds, the water level may decrease by at most $v * T - (1/2) * U_1 * T^2$ and increase by at most $v * T + (1/2) * U_2 * T^2$. The water level is always between 0 and the maximal capacity C of the steam boiler.

All of the gauges may fail independently at any time. In case they are ok, the transmitted values correspond to the real measures.

$$Env_2 \;\widehat{=}\; \begin{array}{l} q\_gauge := \;?\;;\, p_1\_gauge := \;?\;;\, \ldots\;;\, p_4\_gauge := \;?\;;\, v\_gauge := \;?\;; \\ trans\_q := trans\_q'.(q\_gauge = ok \Rightarrow trans\_q' = q)\;; \\ trans\_p_1 := trans\_p_1'.(p_1\_gauge = ok \Rightarrow trans\_p_1' = p_1)\;; \\ \ldots\;; \\ trans\_p_4 := trans\_p_4'.(p_4\_gauge = ok \Rightarrow trans\_p_4' = p_4)\;; \\ trans\_v := trans\_v'.(v\_gauge = ok \Rightarrow trans\_v' = v) \end{array}$$

The refined environment action is given by:

$$Environment' \;\widehat{=}\; Env_1\;;\, Env_2$$

The controller needs to adjust the water level estimates. The minimal and maximal estimates of the water level correspond to the real measure in case the water level gauge is ok. Otherwise the estimate is based on the transmitted values of the pump input and the steam output. In case one of the pump gauges or the steam output gauge fails, no estimates can be made, i. e. the estimates will be assigned arbitrary values.

$$Estimate \;\widehat{=}\; \begin{array}{l} qa_1 := \textit{if } q\_gauge = ok \textit{ then } trans\_q \textit{ else} \\ \quad q\_min(qa_1, trans\_v, trans\_p_1 + \ldots + trans\_p_4, 0)\;; \\ qa_2 := \textit{if } q\_gauge = ok \textit{ then } trans\_q \textit{ else} \\ \quad q\_max(qa_2, trans\_v, trans\_p_1 + \ldots + trans\_p_4, 0) \end{array}$$

The proof obligation (4) amounts to:

$$q, p_1, p_2, p_3, p_4, v, e, reliable := \;?,?,?,?,?,?,?,?\quad \leq_R \quad Env_1\;;\, Env_2\;;\, Estimate$$

Using the rules for the partwise data refinement and for merging assignments, this is implied by:

$$\begin{array}{l} p_1, p_2, p_3, p_4 := \;?,?,?,?\quad \leq_R \\ p_1 := p_1'.(p_1' \in \{0..P*T\})\;;\; \ldots\;;\, p_4 := p_4'.(p_4' \in \{0..P*T\}) \end{array} \tag{9}$$

$$\begin{array}{l} e, q, v, reliable := \;?,?,?,?\quad \leq_R \\ \quad e, q, v, p_1\_gauge, \ldots, v\_gauge, trans\_p_1, \ldots, trans\_v, qa_1, qa_2 := \\ \quad e', q', v', p_1\_gauge', \ldots, v\_gauge', trans\_p_1', \ldots, trans\_v', qa_1', qa_2'.Q \end{array} \tag{10}$$

where

$$Q \;\widehat{=}\; \begin{array}{l}(valve = open \Rightarrow e' = E * T) \wedge (valve = closed \Rightarrow e' = 0) \wedge \\ q\_variation(q, v, p_1 + p_2 + p_3 + p_4, e', q') \wedge v\_variation(v, v') \wedge \\ (p_1\_gauge' = ok \Rightarrow trans\_p_1' = p_1') \wedge \\ \ldots \wedge \\ (v\_gauge' = ok \Rightarrow trans\_v' = v') \wedge \\ (q\_gauge' = ok \Rightarrow qa_1' = trans\_q' \wedge qa_1' = trans\_q') \wedge \\ (q\_gauge' = failed \Rightarrow \\ \quad qa_1' = q\_min(qa_1, trans\_v', trans\_p_1' + \ldots + trans\_p_4', 0) \wedge \\ \quad qa_2' = q\_max(qa_2, trans\_v', trans\_p_1' + \ldots + trans\_p_4', 0))\end{array}$$

Data refinement (9) reduces to a simple refinement of a nondeterministic assignment, as variables $p_1, \ldots, p_4$ are not refined by R. Data refinement (10) holds according to the data refinement rule for nondeterministic assignments if:

$$Q \wedge R \;\Rightarrow\; (\exists reliable'.R')$$

where

$$R' \;\widehat{=}\; \begin{array}{l} R[reliable, e, q, v, p_1\_gauge, \ldots, v\_gauge, trans\_p_1, \ldots, trans\_v, \\ qa_1, qa_2 := reliable', e', \ldots, qa_2'] \end{array}$$

As $R = R\_trans \wedge R\_init \wedge R\_oper$, the proof can be carried out considering the three phases separately. The full proof is omitted for brevity.

Proof obligations (6) and (7) immediately follow from the rules for calculating guards (see CD-ROM Annex BSS.1).

5.2 Controller Refinement

Init Mode. In initialisation mode, only proper functioning of the water level gauge is required. If the water level gauge is ok, the transmitted water level $trans\_q$ corresponds to the real measure q, and the appropriate decisions to open or close the evacuation valve can be made:

$$Valve_1' \;\widehat{=}\; \begin{array}{l} mode = Init \\ q\_gauge = ok \\ trans\_q > N_2 \end{array} \to\; valve := open$$

$$Valve_2' \;\widehat{=}\; \begin{array}{l} mode = Init \\ q\_gauge = ok \\ trans\_q \leq N_2 \end{array} \to\; valve := closed$$

Similarly, if the water level gauge is ok, appropriate decisions to open or close the water pumps can be made:

$$Pumps_1' \;\widehat{=}\; \begin{array}{l} mode = Init \\ q\_gauge = ok \\ trans\_q < N_1 \end{array} \to\; pumps := on$$

$$Pumps'_2 \,\widehat{=}\, \begin{array}{l} mode = Init \\ q\_gauge = ok \\ trans\_q > N_2 \end{array} \to pumps := \mathit{off}$$

Once the water level is between N_1 and N_2, the system enters operating mode, otherwise it remains in initialisation mode, provided no failure of the water level gauge is detected:

$$Mode'_1 \,\widehat{=}\, \begin{array}{l} mode = Init \\ q\_gauge = ok \\ valve = closed \\ N_1 \le trans\_q \le N_2 \end{array} \to mode := Operating$$

In case the water level gauge fails, the system goes to emergency mode:

$$Mode'_2 \,\widehat{=}\, \begin{array}{l} mode = Init \\ q\_gauge = failed \end{array} \to mode := Emergency$$

Operating Mode. The actions of the operating mode are refined in two ways, depending on the status of the water level gauge: If the water level gauge is working properly, safe decisions can be made for switching the pumps on and off. This holds in normal and degraded operating mode:

$$Pumps'_3 \,\widehat{=}\, \begin{array}{l} mode = Operating \\ NormalCond \lor DegradedCond \\ trans\_q < N_1 \end{array} \to pumps := on$$

$$Pumps'_4 \,\widehat{=}\, \begin{array}{l} mode = Operating \\ NormalCond \lor DegradedCond \\ trans\_q > N_2 \end{array} \to pumps := \mathit{off}$$

If the water level is between N_1 and N_2, we decide to leave the pumps as they are, either on or off, in order to minimise the number of on/off switches.

If the water level is unsafe, the system goes into emergency mode:

$$Mode'_3 \,\widehat{=}\, \begin{array}{l} mode = Operating \\ (NormalCond \lor DegradedCond) \\ (trans\_q < M_1 \lor trans\_q > M_2) \end{array} \to mode := Emergency$$

If the water level gauge has failed, decisions about whether to switch pumps on and off have to be based on the minimal estimate qa_1 and maximal estimate qa_2. If the water level is between M_1 and N_2, the pumps are switched on (this corresponds to cases 1 and 2 as described in the problem statement – Chapter AS, this book):

$$Pumps'_5 \,\widehat{=}\, \begin{array}{l} mode = Operating \\ RescueCond \\ M_1 \le qa_1 \land qa_2 \le N_2 \end{array} \to pumps := on$$

If the water level is between N_1 and M_2, the pumps are switched off (cases 5,6):

$$Pumps_6' \;\widehat{=}\; \begin{array}{l} mode = Operating \\ RescueCond \\ N_1 \leq qa_1 \wedge qa_2 \leq M_2 \end{array} \rightarrow pumps := \mathit{off}$$

If the water level is between N_1 and N_2, we decide to leave the pumps as they are, either on or off (case 4). If the lower estimate is below N_1 and the upper estimate is above N_2, this is considered a failure, and the pumps remain as they were (case 3).

If the water level is considered unsafe based on the estimates or if the estimate is so vague that is does not allow sensible operation (case 3), or if additionally one of the other gauges fails, the system goes into emergency mode.

$$Mode_4' \;\widehat{=}\; \begin{array}{l} mode = Operating \\ (RescueCond \wedge \\ \quad (M_1 > qa_1 \vee qa_2 > M_2 \vee \\ \quad (qa_1 < N_1 \wedge N_2 > qa_2)) \vee \\ EmergencyCond) \end{array} \rightarrow mode := Emergency$$

Controller Actions. The control of the water pumps, the evacuation valve and the mode is given by the actions *Pumps*, *Valve*, and *Mode*, respectively:

$$Pumps' \;\widehat{=}\; Pumps_1' \;[]\; \ldots \;[]\; Pumps_4'$$

$$Valve' \;\widehat{=}\; Valve_1' \;[]\; Valve_2'$$

$$Mode' \;\widehat{=}\; Mode_1' \;[]\; \ldots \;[]\; Mode_4'$$

The refined controller consists of different parts, which correspond to those of the previous specification:

$$Controller' \;\widehat{=}\; (\overline{Pumps'} \;\|\; \overline{Valve'} \;\|\; \overline{Mode'} \;\|\; \overline{Fail})$$

The verification of the refinement (5) can be carried out for the pumps, valve, mode actions separately, leading to:

$$\overline{Pumps} \leq_R \overline{Pumps'} \tag{11}$$

$$\overline{Valve} \leq_R \overline{Valve'} \tag{12}$$

$$\overline{Mode} \leq_R \overline{Mode'} \tag{13}$$

For the pump actions we note that the four abstract actions are replaced by six concrete actions. The only requirement is that each concrete action refines some abstract action; hence (11) is implied by (see CD-ROM Annex BSS.1):

$$Pumps_1 \leq_R Pumps_1'$$
$$Pumps_2 \leq_R Pumps_2'$$
$$Pumps_3 \leq_R Pumps_3' \;[]\; Pumps_5'$$
$$Pumps_4 \leq_R Pumps_4' \;[]\; Pumps_6'$$
$$R \wedge (g(Pumps_1') \vee \ldots \vee g(Pumps_6')) \Rightarrow$$
$$g(Pumps_1) \vee \ldots \vee g(Pumps_4)$$

Because all actions above assign only to variables which are not refined, there is only a proof obligation for the guards, i.e. the four refinements above are equivalent to:

$$R \wedge g(Pumps_1) \Rightarrow g(Pumps'_1)$$
$$R \wedge g(Pumps_2) \Rightarrow g(Pumps'_2)$$
$$R \wedge g(Pumps_3) \Rightarrow g(Pumps'_3) \vee g(Pumps'_5)$$
$$R \wedge g(Pumps_4) \Rightarrow g(Pumps'_4) \vee g(Pumps'_6)$$

The refinement of *Valve* and *Mode* leads to similar proof obligations. They can be discharged with the rules of predicate calculus.

Finally, proof obligation (6) follows immediately from the enabledness of the constituents of *Controller'*. This completes the proof of this refinement step.

Further refinement steps, which lead towards a distributed implementation, are described in CD-ROM Annex BSS.2.

6 Evaluation

In this section, we answer the evaluation questions posed by the editors.

1. The whole system, the control program and the steam boiler plant, is specified. The steam boiler specification includes that of the water level, the steam sensor, the pump actuator, the pumps, the drain, as well as the transmission system, but only to the extent required for the development of the control program. The full specification is constructed from an initial abstract specification, with a simple view of failures and no consideration of the distribution, in two refinement steps. The first adds failure treatment and the second adds communication between controller and steam boiler. All steps are specified formally, but are not checked mechanically.
2. A Pascal implementation has been derived from the final refinement step. It is very similar to the final action system specifications but implements the simultaneous composition $A_1 \parallel A_2$ of actions by an appropriate sequential composition and guarded choice by an if statement. The implementation is around 170 lines long and written in SunOS Pascal.
   The implementation has been linked to the FZI simulator. The I/O conventions and the system constants have been adapted to the FZI simulator.
   Experimentation has been done with the control program and did not reveal any errors, after solving the technical problems with linking. However, as the simulation transmits incorrect values in case of gauge failure (in fact it transmits the old values), this is not detected by the control program (see the conclusions section).
3. Abrial's solution (see Chapter A, this book) using B AMN is closely related in that the refinement calculus notation and B AMN have a similar semantic basis. Also, Abrial uses refinement as a way of structuring requirements as in our approach. However, Abrial doesn't model the environment only the controller.

The Z specifications of the controller produced by other groups (see Chapter BW, this book, and [10]) resemble our most detailed refinement of the controller. While they concentrated on accurately representing all the details of the controller, we placed more emphasis on using abstraction to make validation easier.

4. About 4 person months were spent in producing the solution.

   In order to produce a solution to such a problem, familiarity with the specification notation and a practical understanding of data-refinement are required. It is also necessary to understand proof techniques used. This would take about 2 weeks training.

5. For a good understanding of the solution, familiarity with a Pascal like programming language, the additional specification notation and a practical understanding of data-refinement are required. It is not necessary to understand proof techniques or the semantics of actions and action systems.

   An average programmer should be able understand the solution.

   In order to be able to understand the individual steps of the solution, 1 hour will be necessary for a programmer to learn what is needed.

7 Conclusions

The development presented describes both the controller *and* the physical environment. Specifying the environment was used for deriving the updates of the water level estimates of the controller in case of water level gauge failure.

The environment is specified by an action which describes the possible evolution during a period of 5 seconds. It does not completely determine the behaviour of a concrete environment, but only its view by the controller every 5 seconds. In particular, there might be peaks of the water level below M_1 and above M_2 in between. Hence the safety requirement should be interpreted such that it holds only every 5 seconds. (This is all what is required by the informal specification: the system is in danger if the water level is below M_1 or above M_2 for *more* than 5 seconds.)

The refined controller guarantees safe functioning despite gauge failure as long as the information about gauge failure is reliable. If this is unreliable, e.g., the water level gauge pretends to function properly but does not, no safe decisions are possible at all (except going to emergency mode). The informal specification suggests that to cope with this, the controller should check whether the measure are "compatible with the dynamics of the system". However, this is problematic. Besides introducing nondeterminism (which gauge is to blame?) neither does it guarantee reliable operation (we could blame the wrong gauge or not detect a gauge failure for a long time). This strategy can only be used for making the operation more reliable with a certain *probability*. Although this would be in accordance with engineering practices, probabilistic specification of gauge failures and reasoning about the probabilistic reliability of a controller is outside our approach. It is suggested for further research.

Acknowledgements

The work reported here is carried out within the projects Irene and Formet. These projects are supported by the Academy of Finland and the Technology Development Centre of Finland (Tekes).

References

1. M. Abadi and L. Lamport. The existence of refinement mappings. In *Proc. of the 3rd Annual IEEE Symp. on Logic In Computer Science*, Edinburgh, pp. 165–175, 1988.
2. R. J. R. Back. *On the Correctness of Refinement Steps in Program Development.* PhD thesis, Department of Computer Science, University of Helsinki, Helsinki, Finland, 1978. Report A–1978–4.
3. R. J. R. Back. Refinement calculus, part II: Parallel and reactive programs. In J. W. de Bakker, W.–P. de Roever, and G. Rozenberg, editors, *Stepwise Refinement of Distributed Systems: Models, Formalisms, Correctness. Proceedings. 1989*, volume 430 of *Lecture Notes in Computer Science*. Springer–Verlag, 1990.
4. R. J. R. Back and R. Kurki-Suonio. Decentralization of process nets with centralized control. In *Proc. of the 2nd ACM SIGACT-SIGOPS Symp. on Principles of Distributed Computing*, pages 131–142, 1983.
5. R. J. R. Back, A. J. Martin, and K. Sere. Specifying the Caltech asynchronous microprocessor. *Science of Computer Programming*, North-Holland. Accepted for publication.
6. R.J.R. Back and K. Sere. Stepwise refinement of action systems. *Structured Programming*, 12:17-30, 1991.
7. R. J. R. Back and K. Sere. From modular systems to action systems. Proc. of *Formal Methods Europe'94*, Spain, October 1994. *Lecture Notes in Computer Science*. Springer–Verlag, 1994.
8. R. J. R. Back and J. von Wright. Refinement calculus, part I: Sequential nondeterministic programs. In J. W. de Bakker, W.–P. de Roever, and G. Rozenberg, editors, *Stepwise Refinement of Distributed Systems: Models, Formalisms, Correctness. Proceedings. 1989*, volume 430 of *Lecture Notes in Computer Science*, pages 42–66. Springer–Verlag, 1990.
9. R. J. R. Back and J. von Wright. Trace Refinement of Action Systems In B. Jonsson, J. Parrow, editors, *CONCUR '94: Concurrency Theory. Proceedings. 1994*, volume 836 of *Lecture Notes in Computer Science*, pages 367–384. Springer–Verlag, 1994.
10. P. Bernard A Z specification of the boiler. Presented at seminar on *Methods for Semantics and Specification*, Schloss Dagstuhl, June 1995.
11. K. Chandy and J. Misra. *Parallel Program Design: A Foundation.* Addison–Wesley, 1988.
12. E. W. Dijkstra. *A Discipline of Programming.* Prentice–Hall International, 1976.
13. C. C. Morgan. The specification statement. *ACM Transactions on Programming Languages and Systems*, 10(3):403–419, July 1988.
14. J. M. Morris. A theoretical basis for stepwise refinement and the programming calculus. *Science of Computer Programming*, 9:287–306, 1987.

The Steam Boiler Problem in Lustre

Thierry CATTEL, Gregory DUVAL
Laboratoire de Téléinformatique, Ecole Polytechnique Fédérale
CH-1015 Lausanne, Switzerland, {cattel,duval}@di.epfl.ch

Abstract - This paper reports the results of specifying, verifying and implementing the Steam Boiler problem with Lustre. The model is detailed and is able to drive the system and takes device failures (pumps, pump controllers, water, steam and transmission) and emergency stop into account. Safety properties have been checked on the model with Lesar, the Lustre model-checker. An implementation of the system have been made using the C code produced by Lustre from the model and linked with the TCL/TK simulation. This application shows Lustre´s suitability for developing safe control process problems from specifications.

1 Introduction

Among the french synchronous reactive languages Esterel, Signal and Lustre[3], Lustre has shown to be well adapted for the specification of hardware[3] and process control systems [1]. It is provided with Lesar, a model-checker that allows the verification of safety properties. Compositional verification is also possible. Assumptions about the program environment may be expressed. Finally C code is generated automatically from the Lustre description.
Our objective was the design of a complete controller for the steamboiler problem (see Chapter AS, this book), the verification of some properties and then the derivation of a straightforward implementation from the Lustre specifications.
In section 2, we briefly introduce the Lustre framework. In section 3, we describe the architecture of our solution. In section 4, we describe our detailed models. In section 5, we explained the verifications performed. At the end of the paper, an attempt is made to evaluate the criteria proposed in Chapter. AS, this book. We do not present the Steam Boiler problem, this may be found in this volume (see Chapter. AS, this book).
Although we tried to follow the informal description as best as we could, we still had to make some design decisions due to incompleteness and ambiguities of the original text. These decisions concern mainly the following items:
- We did not understand from the problem description how to distinguish between failures of the pumps and failures of the pump controllers. We have therefore identified these two error conditions according to the following table (table 1.). Notice that table 1. takes into account the asymmetric behaviour of pumps after a *close* or *open* command. After a *close* command, in the normal case, the supposed status of the pump (*Pump Status*) is *off* and it must be confirmed by the real *Pump State* and *Pump Controller State*. On the contrary, after an *open* command, in the normal

case, *Pump Status* is *switched_on*, *Pump State* must be *on* (the pump has taken the order into account) but the flow is no yet effective (*Controller State* is *off*). In all other situations, either the pump or the controller or both are in a failed state.
- The problem description gives no indication of how to decide how many pumps should be opened or closed in one cycle. A solution for this was chosen.
- In Emergency mode we assume that the system should not do anything but terminate its execution.

TABLE 1. Pump's and Controller's Failure

Pump status	Pump State	Controller State	Flow	Failure
off	off	off	NO	none
off	off	on	NO	Controller
off	on	off	NO	Pump
off	on	on	YES	Pump
on	off	off	NO	Pump
on	off	on	YES	Pump
on	on	off	YES	Controller
on	on	on	YES	none
switched_on	off	off	NO	Pump
switched_on	off	on	NO	Pump Controller
switched_on	on	off	NO	none
switched_on	on	on	NO	Controller

2 Lustre

Lustre[2] is a synchronous data-flow language. Any Lustre variable or expression represents the sequence of values it takes during the whole execution of the program. Lustre operators operate globally over these sequences. The synchronous nature of the language consists of assuming that all the variables and expressions in a program take the n-th value of their respective sequences at the same time. A program is intended to have a cyclic behaviour where an execution cycle consists of computing the n-th value of each variable or expression. A program is considered as being a set of equations and an equation $V=E$, where V is a variable and E an expression, means that the sequences of values associated with V and E are identical.

The real-time capabilities of the language are derived form this synchronous interpretation. The real physical time is considered as an external event, which has no privileged nature. This is the multiform time point of view where time may be counted in seconds as well as in meters. In Lustre, an event is modelled by a boolean variable whose value is true whenever the event occurs. The synchronous interpretation is an abstract point of view which considers the program reaction time to be negligible with respect to the reaction time of its environment. This only assumes that the reaction time is short enough to distinguish and order the incoming events. In practice, it can be checked by measuring the maximum reaction time of the program.

A Lustre program is structured into nodes: a node is a subprogram specifying a relation between its inputs and its outputs. The following Lustre node computes the rising edges of a boolean signal:
1. node REDGE(S:bool) returns(REDGE:bool);
2. let REDGE= S -> S and not(pre(S)); tel;

The -> operator is useful for initializing a sequence. The expression *E1->E2* specifies that the first value of the expression is *E1* and subsequent values are defined by *E2*. The *pre* operator allows the previous value of a sequence to be denoted. The expression *pre(E)* denotes the undefined *nil* value in the first execution step and the previous value of *E* in the other executions steps.

The data types available in Lustre are *bool*, *int* and *real*. It is also possible to define arrays and express recursive nodes. A subarray of an array *A* is denoted as *A[inf..sup]*, where *inf* and *sup* are positive integer indexes. The first element of an array is numbered 0. The recursion is specified with the *with* statement and needs to be based on an integer constant. The following node computes the sum of the elements of an integer array:
1. node sum(const n:int;a:int^n) returns(sum:int);
2. let sum=with n=1 then a[0] else a[n-1]+sum(n-1,a[0..n-2]);tel;

It is also possible to construct arrays (*[]*) and concatenate them (|). All the Lustre operators are potentially vectorial.

The Lustre tool is composed of a graphical simulator, and a model-checker. The model-checker Lesar performs either a standard model-checking by exhaustive enumeration of the states of the program or a symbolic model-checking based on BDDs. Safety property may be expressed as the invariance of some boolean Lustre expression[5]. Properties may also be modularly verified [4], (see CD-ROM, Annex CD.1). The general method for verifying a *Property* for a *Circuit* is shown in Fig.1. *Property* is a boolean variable which is checked to be always true (at each instant).

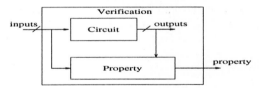

FIGURE 1. Verification

As an example of verification we propose the following example where we specify a node for computing the falling edges of a signal in two different ways and prove that both are equivalent. The node *Verification* calls the nodes to be checked (*REDGE*, *FEDGE1* and *FEDGE2*) and computes a single boolean signal (*property_ok*) that the model-checker verifies to be always *true*. This also illustrates the possibility of expressing assumptions about the program environment (*assert* for *property2*), which is useful for open systems.

```
1. node FEDGE1(S:bool) returns(FEDGE1:bool);
2. let FEDGE1= not(S) -> not(S) and pre(S); tel;
3.
4. node FEDGE2(S:bool) returns(FEDGE2:bool);
5. let FEDGE2= REDGE(not(S)); tel;
6.
7. node Verification(S1,S2:bool) returns(property_ok:bool);
8. var property1,property2:bool;
9. let property1=FEDGE1(S1)=FEDGE2(S1);
10.     assert(S2=not(S1));
11.     property2=FEDGE1(S1)=REDGE(S2);
12.     property_ok=property1 and property2;
13.tel;
```

Lesar is well adapted to Lustre programs with boolean variables. Numeric variables are more difficult to manage, but Lesar knows how to solve static linear constraints such as $(a > 0) \Leftrightarrow \neg(a \leq 0)$. For instance the following is accepted to be always *true*:

```
1. node Verification(i:int) returns(property_ok:bool);
2. let property_ok=(if i>0 then i else 1)>0; tel;
```

However the general problem of solving dynamic linear constraints is undecidable and the property verified by the following node, though trivially *true*, is rejected as *false*:

```
1. node Verification() returns(property_ok:bool);
2. var f:int;
3. let f=1->pre(f);
4.     property_ok=(f>0);
5. tel;
```

This unfortunately has an important consequence on Lustre usability. This means that, for the moment, no Lustre program involving numeric variables defined with expressions including *pre* operator may be verified with Lesar. Some extensions are currently being studied [8].

3 Design & Architecture

The paradigm commonly adopted for reactive systems (Fig.2) clearly separates the plant from its controller. Some contributions (see Chapter LP, this book) advocate designing the controller by first specifying the plant and all its possible states, including the undesired ones. Then the controller is derived by adding constraints preventing the plant to go in undesired states. Although Lustre allows both closed and opened systems to be expressed, we did not specify the plant itself but only the controller.

Our solution is mainly based on the TLA contribution regarding the general architecture (see Chapter LM, this book). We performed some adaptations and corrected a few omissions and inconsistencies. In particular we did not follow the dynamics described in the requirement document (see Chapter AS, this book). Since the steam and water devices never fail together without leading the system into emergency mode, it is always possible to evaluate the failure level, taking into account the correct level and the pump´s outflow. Furthermore, at implementation time, it appeared that the available

TCL/TK simulation had slightly different dynamics. We first followed the dynamics described in Chapter AS, this book, but the related implementation failed to drive the simulation. We had to modify the model to adapt its dynamics to the simulation. The solution to this problem is presented here.

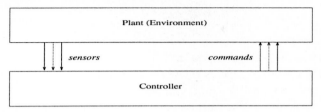

FIGURE 2. Reactive systems

The general method adopted here consists of drawing hierarchical dataflow diagrams: the application architecture, from which the Lustre specification will be derived. The next paragraph will show the declarative power of Lustre and hopefully convince that a Lustre specification is natural. Nevertheless, such a specification may become quite complex, depending on the nature of the problem and some kind of consistency verification is necessary. This is achieved by stating safety properties in Linear Temporal Logic[7], translating them into Lustre and submitting them for verification as explained in paragraph 2. The eventual C implementation is then obtained by applying the Lustre compiler to the Lustre specification. The correction of this implementation relies on the compiler´s implementation.

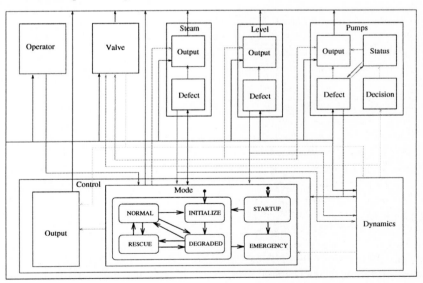

FIGURE 3. Boiler Architecture

The Steam boiler problem is decomposed into several subprocesses as is shown in Fig.3. The system reads all values coming from the physical units and broadcasts them to its processes which compute intermediary results or send commands to the plant. The general role of all the processes is now described.

- *Operator*: It counts the stop events coming for the operator desk of the plant and sends a stop request event to the *ControlMode* process when *NB_stop* stop events have occurred in a row.
- *Valve*: It opens the valve in *initialization* mode if the water level is too high and closes it when the water level is below this same level.
- *Steam*: It is composed of two processes. *SteamDefect* is for the detection of failure. It takes the steam level, the steam failure acknowledgement and the steam reparation indication as input and computes the status of the steam device. *SteamOuput* monitors the emission of messages to the physical unit. Depending on the operation mode of the controller, the status of the steam device computed by *SteamDefect* and the steam reparation indication, *SteamOuput* actually sends the steam failure detection and the steam reparation acknowledgement to the plant.

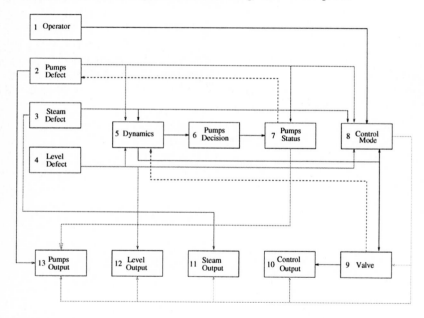

FIGURE 4. Dataflow diagram

- *Level*: Composed of two processes, *LevelDefect* for the detection of failure and *LevelOutput* for the emission of messages to the physical unit. Similar to *Steam*.
- *Pumps*: Composed of four processes, *PumpsDefect* detects the failures on pumps or Controllers and decides if water flows from pumps. It directly implements table 1 presented above. *PumpsDecision* determines how many pumps should be provid-

ing water to the boiler. This depends on the current values of steam outcome, water level and nominal power of each pump. *PumpsStatus* calculates the new status of each pump and pump controller, taking into account the previous status and the states of the pump and pump controller. *PumpsOutput* sends messages to the pumps and controllers including open or close commands, failure detection or reparation acknowledgement messages.
- *Dynamics*: Estimates the water and steam levels, taking into account the possible water and steam failures and the values coming from the physical units.
- *Control*: Composed of two processes, *ControlMode* manages the operation mode of the system by taking into account errors detected and the levels calculated and *ControlOutput* manages the initial protocol with the plant.

Fig.4 shows the data dependencies between the modules. The dotted lines refer to dependencies related to previously calculated values. If we forget these lines and unfold the bottom part of the diagram to the left, the numbers indicated suggest a possible ordering of calculation.

4 Modelling

We present some of the modules of the controller for showing how simple the Lustre solution is. The complete program is shown elsewhere (see CD-ROM, Annex CD.2). We need to consider the constants that characterize the plant and the controller. *N_pump* is the number of pumps of the boiler, *C* to *P* (line 3-7 below) define the dynamics of the boiler. Our model is also parametrized by *NB_stop*, the number of stop events needed in a row to stop the plant, and *Dt* the length of a sampling cycle expressed in seconds. Other constants define the operation mode of the boiler (lines 10-12), the state of the valve (line 13), the state of the pump and/or pump controllers (lines 14-15) and the status of errors for all the different devices (lines 16-17). The *none* constant is used to indicate the absence of a value. This will be used for detecting transmission failures.

```
1.  const
2.  -- Boiler characteristics
3.     N_pump       = 4;            C            = 1000;
4.     M1           = 150;          N1           = 400;
5.     N2           = 600;          M2           = 850;
6.     V            = 10;           W            = 25;
7.     P            = 15;
8.  -- Controller characteristics
9.     NB_stop      = 3;            Dt           = 5;
10.    startup      = 1;            initialize   = 2;
11.    normal       = 3;            degraded     = 4;
12.    rescue       = 5;            emergency    = 6;
13.    open         = 1;            closed       = 0;
14.    off          = 0;            on           = 1;
15.    switched_on  = 2;            none         = 3;
16.    ok           = 0;            signalled    = 1;
17.    acked        = 2;
```

4.1 Operator

The *stop* events coming for the operator desk of the plant are count and *stop_request* is set to *true* when *NB_stop* stop events have occurred in a row. The module *ControlMode* will take this information into account in order to set the operation mode to *emergency*.

```
1. node Operator(stop:bool) returns (stop_request:bool);
2. var nb_stops:int;
3. let nb_stops    = (if stop then 1 else 0) ->
4.                   if stop then pre(nb_stops)+1 else 0;
5.     stop_request = (nb_stops>=NB_stop);
6. tel;
```

4.2 Defect

Failure detection uses a generic module that manages the state of the devices. Given the previous state, the failure condition and the presence of error acknowledgement message or reparation message it sets the new state of the device to *signalled*, *acked* or *ok*. When the state is *signalled* that means the error was freshly detected and the modules in charge of outputs (*PumpsOutput*, *LevelOutput* or *SteamOutput*) will have to signal the error to the plant by sending the appropriate message.

```
1. node Defect(statein:int; fail_cond,ack_chan,repair_chan:bool)
2. returns(stateout:int);
3. let   stateout = if (statein=ok) then
4.                    if fail_cond then signalled else ok
5.                  else
6.                    if (statein=signalled) then
7.                       if ack_chan then acked else signalled
8.                    else -- statein=acked
9.                       if repair_chan then ok else acked;
10.tel;
```

4.3 SteamDefect

We now see in more detail how a particular failure is detected. The module *SteamDefect* takes as input the messages related to the steam device coming from the plant plus the steam level and calculates the state of the steam device using the *Defect* module seen previously. We suppose that there is now failure at the very start of the system (line 4). The failure condition is calculated by module *steam_failure_detect*, and it is *true* whenever the steam level is out of *[0..W]*.

```
1. node SteamDefect(steam_failure_acknowledgement,
2.                  steam_repaired:bool;steam:int;)
3. returns( SteamDefect:int);
4. let   SteamDefect = ok -> Defect(pre(SteamDefect),
5.                                  steam_failure_detect(steam),
6.                                  steam_failure_acknowledgement,
7.                                  steam_repaired);
8. tel;
9. node steam_failure_detect(steam:int)
10.returns(steam_failure_detect:bool);
11.let   steam_failure_detect = ((steam < 0) or (steam > W)); tel;
```

4.4 Dynamics

The module *Dynamics* takes into account the steam and water levels and their defect states, whether the valve is open or closed, as well as if the water is flowing out of the pumps. It calculates a new water level in litres (q) and a new steam outcome in litres/sec (v). These will be the same values coming from the plant in the absence of failure. In the event of failure, the values will be calculated as explained later. It also computes the quantity of water leaving the pumps in litre/sec (p).

When a steam failure is detected, the steam coming out of the boiler is evaluated using the difference between the previous water level, the actual water level (*level*) and the quantity of water provided by the pumps to the boiler. We then have the following estimation:

$$v = \frac{pre(level) - level}{Dt} + \sum_{i=1}^{N\_pump} p[i]$$

When a water failure is detected, the level of water in the boiler is evaluated at each cycle of the system using the previously calculated values of water, the amount of water provided by the pumps (p), the quantity of steam which comes out of the boiler (*steam*) and if the valve is open or closed (*valve_state*):

$$q = pre(q) - steam \times Dt + \sum_{i=1}^{N\_pump} p[i] \times Dt \qquad \text{if valve\_state=closed}$$

$$q = pre(q) - steam \times Dt + \sum_{i=1}^{N\_pump} p[i] \times Dt - V \times Dt \qquad \text{if valve\_state=open}$$

$p[i] = P$ if flow[i] else 0

The module *sum* sums of all the elements of an integer array as presented in section 2.

```
1. node Dynamics(
2.          valve_state,level,steam,level_defect,steam_defect:int;
3.          flow:bool^N_pump;)
4. returns (q,v:int; p:int^N_pump);
5. let q = level->
6.          if level_failure(level_defect) then
7.              pre(q) - steam*Dt + sum(N_pump,p)*Dt -
8.              (if valve_state=open then V*Dt else 0)
9.          else level;
10.     v = steam->
11.         if steam_failure(steam_defect) then
12.             (pre(q) -q)/Dt + sum(N_pump,p)*Dt
13.         else steam;
14.     p[0..N_pump-1] = 0^N_pump->
15.                     if (not(flow[0..N_pump-1])) then
16.                         0^N_pump
17.                     else P^N_pump;
18.tel;
```

4.5 ControlMode

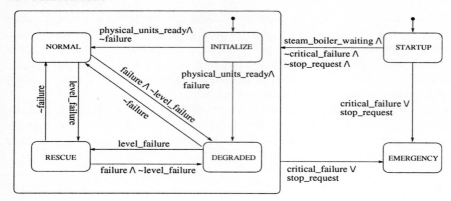

FIGURE 5. Operation modes

ControlMode manages the operation mode of the system by taking into account errors detected, the steam level coming from the plant, at initialization, and the adjusted water level. As in Chapter LM, this book, we added an extra mode *startup* for the very beginning of the program. The program remains in this mode while it has not received the *steam_boiler_waiting* message from the plant and there is no critical failure. Whenever the system detects a critical failure or a *stop_request* it goes into mode *emergency* and will stay there forever. The other evolutions correspond to Fig.5.

```
1. node ControlMode(
2.    steam_boiler_waiting,physical_units_ready,stop_request:bool;
3.    steam,level_defect,steam_defect:int;
4.    pump_defect,pump_control_defect:int^N_pump;
5.    q:int; pump_state:int^N_pump)
6. returns(op_mode :int);
7. let op_mode= startup->
8.              if (critical_failure(pre(op_mode),
9.                                   steam,
10.                                  level_defect,
11.                                  steam_defect,
12.                                  pump_defect,
13.                                  q,
14.                                  pump_state) or
15.                 stop_request or pre(op_mode)=emergency)
16.             then emergency
17.             else
18.                if (pre(op_mode)=startup) then
19.                   if steam_boiler_waiting then initialize
20.                   else startup
21.                else
22.                   ...
23.tel;
```

4.6 Complete System

To specify the whole system *BoilerController*, it suffices to assemble all the components together following the dataflow diagram of Fig.4.

```
1. node BoilerController(
2.      stop,steam_boiler_waiting,physical_units_ready : bool;
3.      level, steam :int; pump_state : int^N_pump;
4.      pump_control_state,pump_repaired,
5.      pump_control_repaired : bool^N_pump;
6.      level_repaired,steam_repaired : bool;
7.      pump_failure_acknowledgement,
8.      pump_control_failure_acknowledgement : bool^N_pump;
9.      level_failure_acknowledgement,
10.     steam_failure_acknowledgement:bool)
11.returns(
12.     program_ready : bool; mode : int; valve : bool;
13.     open_pump,close_pump,pump_failure_detection,
14.     pump_control_failure_detection : bool^N_pump;
15.     level_failure_detection,
16.     steam_outcome_failure_detection : bool;
17.     pump_repaired_acknowledgement,
18.     pump_control_repaired_acknowledgement : bool^N_pump;
19.     level_repaired_acknowledgement,
20.     steam_outcome_repaired_acknowledgement: bool);
21.var stop_request : bool; op_mode : int;
22.     q,v : int; p : int^N_pump; steam_defect : int; ...
23.let stop_request  = Operator(stop);
24.     steam_defect  = ok->
25.                     SteamDefect(
26.                       steam_failure_acknowledgement,
27.                       steam_repaired,
28.                       steam);
29.     ...
30.     (q,v,p)       = (level,steam,0^N_pump)->
31.                     Dynamics(
32.                       pre(valve_state),
33.                       level,steam,
34.                       level_defect,steam_defect,
35.                       flow);
36.     op_mode       = startup->
37.                     ControlMode(
38.                       steam_boiler_waiting,
39.                       physical_units_ready,
40.                       stop_request,steam,
41.                       level_defect,steam_defect,
42.                       pump_defect,pump_control_defect,
43.                       q,pump_state);
44.     ...
45.tel;
```

5 Verification

Once the Lustre specification is complete, it is interesting to check its consistency, before generating the implementation. A clean way of doing this, is to enumerate the requirements the Lustre specification should hold, first in natural language, then in a suitable temporal logic using the specification variables. It then remains to code the requirements in Lustre and submit to verification using Lesar as explained in paragraph 2. When Lesar checks a property P, represented as an output boolean variable, it checks it is *true* at each instant. This means that the considered property is in fact $\Box P$. The history of the program is reachable thanks to variable values and the *pre* Lustre operator. This means that only the past linear temporal logic operators will be implementable in Lustre. If some temporal property encompassing some future operators have to be checked, they will have first to be transformed into an equivalent with past operators only, except the box (always) operator at the outermost level. We first present a general way of coding the past operators of [7], but formulae of the form $\Box(a \to g)$ are generally sufficient. The logical implication is straightforwardly coded:

1. `node implies(p,q:bool) returns(implies : bool);`
2. `let implies = if p then q else true; tel;`

The strong and weak previous operators (-) and (~) are easily coded with the *pre* operator:

1. `node sprevious(p:bool) returns(sprevious:bool);`
2. `let sprevious = false -> pre(p); tel;`
3. `node wprevious(p:bool) returns(wprevious:bool);`
4. `let wprevious = true -> pre(p); tel;`

The *once* operator <-> is coded using the expansion formula $<->p \Leftrightarrow p \vee (-)<->p$:

1. `node once(p:bool) returns(once:bool);`
2. `let once = p or sprevious(once); tel;`

The *sofar* operator [-] is coded using the expansion formula $[-]p \Leftrightarrow p \wedge (\sim)[-]p$:

1. `node sofar(p:bool) returns(sofar:bool);`
2. `let sofar = p and wprevious(sofar); tel;`

The *since* operator S is coded using the expansion formula $pSq \Leftrightarrow q \vee (p \wedge (-)pSq)$:

1. `node since(p,q:bool) returns(since : bool);`
2. `let since = q or (p and sprevious(since)); tel;`

The backto operator B is coded using the formula $pBq \Leftrightarrow pSq \vee [-]p$:

1. `node backto(p,q:bool) returns(backto : bool);`
2. `let backto = since(p,q) or sofar(p); tel;`

We found more than 30 different properties of the Steam Boiler in all of the contributions. They may be grouped according to several classes. Properties stating the operational mode conforming to Fig.5, for instance:

- **P1**: The mode is in {startup, initialize, normal, degraded, rescue, emergency}
- **P2**: Once the mode is emergency it is forever.
- **P3**: In normal mode no device is signalled to be in failure
- **P4**: In normal mode the water level is maintained in [M1..M2]

Properties related to failure protocol, for instance:
- **P5**: If a (water/steam/pump/controller) failure is detected, the failure is signalled until the related failure acknowledgement is received.

Properties related to valve correct state, for instance:
- **P6**: Valve commands are issued only in initialize mode.
- **P7**: In initialize mode if water level is greater than N2 the valve is open.

Properties related to legal pump usage:
- **P8**: We never try to use a pump which is in a failure state unless it is repaired
- **P9**: The open and close pump commands are exclusive

In general, all the properties may be checked on the whole specification (*BoilerController* in paragraph 4.6) which is parametrized by the number of pumps *N_pump*. For *N_pump* equal to 1 or 2, the verification generally completes but after a long computation time and high memory consumption. For greater values of *N_pump* the combinatory explosion precludes this possibility and it is necessary to apply modular verification techniques. A comprehensive example of modular verification is provided in CD-ROM, Annex CD.1, but in the case of the SteamBoiler, it is even easier to apply the technique since the property checked is always related to inputs and ouputs of a single module. Since only one module may produce a given output, it is possible to isolate and test this particular module separately. For instance, property *P1* may be verified on the whole *BoilerController* or only on the *ControlMode* module. Table 2 below shows the resource usage on a 50 Mhz Sun Sparc 20 workstation with 128 Mbytes of memory:

TABLE 2. P1 verification

Resource	Cpu time (s)				Memory (Mb)			
Pumps	1	2	3	4	1	2	3	4
BoilerController	9.16	∞	∞	∞	5	5	5	5
ControlMode	1.72	2.22	2.95	3.83	4	4	4	4

This confirms that modular verifications are mandatory. For being more precise about *P1*, $\Box(op\_mode \in \{startup, initialize, normal, rescue, degraded, emergency\})$ is the related LTL expression that is easily coded in Lustre as:
1. p1 = (op_mode=startup or ... or op_mode=emergency);

P2 may be is expressed as: $\Box((op\_mode = emergency) \rightarrow \Box(op\_mode = emergency))$ but it should only use past operators. It may be shown that a stronger formula may be: $\Box((\text{-})(op\_mode = emergency) \rightarrow (op\_mode = emergency))$ which is easily coded in Lustre as:
1. p3 = implies(sprevious(op_mode=emergency),op_mode=emergency);

Which may be proved equivalent to:
1. p3 = true->implies(pre(op_mode=emergency),op_mode=emergency);

Surprisingly this does not hold for our model, though the following does:
1. p3 = true->implies(pre(op_mode)=emergency,op_mode=emergency);

This is due to the fact that if v is of a numeric type, $(pre(v)=N)=(pre(v=N))$ cannot be proved, though trivial, since it is a dynamic numeric property. Property P3 may be expressed as $\Box((op\_mode = normal) \rightarrow nofailure)$, where *nofailure* means no water level, no steam, no pump and controller or transmission failure. It is coded as follows, where AND is a node that checks that all the element of a boolean array are true:

```
1. p3 = implies(
2.           op_mode=normal ,
3.           level_defect=ok and
4.           steam_defect=ok and
5.           AND(N_pump,pump_defect=ok) and
6.           AND(N_pump,pump_control_defect=ok) and
7.           not transmission_failure(pump_state));
```

$P4=\Box((op\_mode = normal) \rightarrow (M1 \leq q \wedge q \leq M2))$ is the most important safety property. Finally the Lustre code for verifying *P1*, *P2*, *P3* and *P4* is:

```
1. node Verification(
2.           steam_boiler_waiting:bool;...;pump_state:int^N_pump)
1. returns(properties_ok:bool);
1. var op_mode :int;
1.     p1,p2,p3,p4:bool;
1. let op_mode= ControMode(steam_boiler_waiting,...,pump_state);
1.     p1 = ...;p2 = ...;p3=...;p4= ...;
1.     properties_ok = p1 and p2 and p3 and p4;
1. tel;
```

Property *P5* could be checked for the steam device, as $\Box(D(steam) \rightarrow (S \text{ W } A))$ where: $D(steam) = steam\_failure\_detect(steam)$, $S = steam\_failure\_detection$, $A=steam\_failure\_acknowledgement$, or equivalently $\Box(\neg S \rightarrow (\neg D(steam) \text{ B } A))$, but the usage of *pre* on the numeric output result precludes the verification. *P6* is tested on module *Valve*: $\Box((mode \neq initialize) \rightarrow \neg valve)$ and *P9* on module *PumpsOutput* as $\forall i \in [0,N\_pump - 1] \bullet \Box(\neg(openpump[i] \wedge closepump[i]))$. These last properties where successfully checked in a few seconds.

6 Evaluation and Conclusions

We attempt to answer the evaluation questions included in Chapter AS, this book, and then make out a conclusion.

1. The solution comprises a requirement specification for the system components as well as for the safety of the system. Liveness is out of the scope of the verification possibilities of the Lesar model-checker. No requirement is given for the performance of the system, but it might be measured for the implementation on a given machine, since Lustre generate automata with linear time execution.

1´. The requirement specification of the solution is presented in linear temporal logic, from which the Lustre verification code may be systematically derived.

2. The solution gives a comprehensive specification of each part of the controller. The plant behaviour is not specified (e.g. input signals such as Operator *stop*, or *repair* and *ack* messages coming from physical devices), only possible constraints (e.g. non simultaneity of two input signals occurrence) are expressed. Thus the specification is open in the sense that the system is not closed by its environment specification.

2´. The functional design of the solution has been formally verified against the requirement specification using the automatic model-checker Lesar.

3. The solution gives an architectural design in terms of dataflow graphical representation.

3´. The architectural design served for the rigorous derivation of the Lustre skeletons. This was done manually, but could be automatized by using available tools.

4. The solution comprises a C implementation of the control program automatically generated by the tool Lustre from the complete functional specification. The controller has been linked to the FZI simulator and interactive experiments (simulated failures, repairing,...) have been done as well as failure files usage. This confirmed that the controller works correctly w.r.t safety requirements.

5. The architecture of the solution was inspired by the TLA contribution (see Chapter LM, this book). TLA allows specification of open as well as closed systems. When specifying open systems, it is possible, as in Lustre, to state constraints on the environment behaviour. The dynamics of the TLA solution were slightly adapted to fit the FZI simulator dynamics that differs from the requirement document (see Chapter AS, this book), although the original dynamics was also taken into account in Lustre as an alternative and verified against the safety requirements. Neither Signal nor Esterel contribution exists, ours is thus the only one based on synchronous programming.

6. The SPIN solution (see Chapter DC, this book) was developed beforehand by the authors of the Lustre solution. It relies on an automatic model-checker and is inspired by the same dynamics. However, this has a different architecture and takes liveness requirements into account.

The other contributions all complement the Lustre solution to some extent. Regarding the system specification style, some contributions use frameworks particularly adapted for specifying complex data type such as algebraic specifications (see Chapters BBDGR, BCPR, GM, GDK, OKW, this book), model-oriented specification (see Chapters A, BW, LP, S, this book). Regarding the type of requirements expressed and verified, almost all contributions address qualitative properties (safety and liveness), but some also address qualitative properties (performance, response time,...) (see Chapters AL, HW, LW, RS, this book). The way verifications are formally performed (apart from testing) belong roughly to two classes. Few contributions use automatic model-checkers (see Chapters DC,CD,WS, this book), the other contributions rely on formal deductive proofs supported by tools or not. And last, the style of verification may or not involve refinements. Among those using refinements, let us quote those of Chapters A, AL, CW1, VH, this book. Using refinements leads automatically to performing compositional verification, but compositional verification may be used without refinements, this is in particular possible with Lustre.

7. In terms of effort, one person spent about three to four weeks on the solution but the results of the Spin contribution (see Chapter DC, this book) where partially reused, in particular, the dynamics and parts of the controller implementation. We believe two weeks (self-)training should be sufficient to become able to produce a solution to a **problem similar to the steamboiler´s.**

8. A detailed knowledge of the used formalism is not needed for understanding the proposed solution. An overview of the language and tools usage through a tutorial is enough for an average programmer to understand the solution and would need 3 to 4 days.

As a general conclusion we can say that Lustre is an adequate framework for developing process control applications. First, it fits nicely into a simple lifecycle, where the architecture is drawn as dataflow-graphs. The skeletons of the functional model may be systematically derived from the architecture. In fact, commercial tools exist on top of Lustre for this purpose. The declarative nature of Lustre and the possibility to structure the functional specification hierarchically makes it very natural and readable. The controller implementation is obtained automatically by C code generation and needs only limited adaptations related to communication with the plant. We succeeded in providing a complete model for the problem, and this was confirmed by the ability to drive the provided simulation. We were able to evaluate Lesar´s power and become aware that verifications are generally possible only if they are modularly achieved. Models involving numeric dynamic properties are for the moment problematic, but could be addressed thanks to the combined usage of Lustre and some theorem prover[9].

For more details on our solution (code, demos) see *http://ltiwww.epfl.ch/~cattel/steamboiler.html*.

7 References

1. C. Lewerentz, T. Lindner (Eds.). Formal Development of Reactive Systems. Case Study Production Cell. Springer-Verlag, 1993.
2. Halbwachs N., A Tutorial of Lustre, IMAG, Grenoble, 1993.
3. Halbwachs N., Synchronous programming of reactive systems, Kluwer Academic, 1993
4. Halbwachs N., Lagnier F., Ratel C., Programming and verifying real-time systems by means of the synchronous data-flow language LUSTRE, IIE Trans. on Soft. Engin., Special issue on The Specification and Analusis of Real-Time Systems, September, 1992.
5. Bouajjani A., Fernadez J.-C., Halbwachs N., On the verification of safety properties. TR Spectre L12, IMAG, Grenoble, 1990.
6. Holzmann G.J., Design and Validation of Computer Protocols, 512 pgs, ISBN 0-13-539925-4, Publ. Prentice Hall, (c) 1991 AT&T Bell Laboratories.
7. Manna Z., Pnueli A., Temporal Verification of Reactive Systems: Safety, Springer-Verlag, 1995.
8. R. Alur, C. Courcoubetis, N. Halbwachs, T. Henzinger, P. Ho, X. Nicollin, A. Olivero, J. Sifakis, S. Yovine. The Algorithmic Analysis of Hybrid Systems. Theoretical Computer Science B, Vol. 138, pp.3-34. January 1995.
9. S. Bensalem, P. Caspi, C. Parent, Handling data-flow programs in PVS, http://www.imag.fr/VERIMAG/PEOPLE/Catherine.Parent, 1996.

The Steam-Boiler Problem — A TLT Solution

Jorge Cuéllar*, and Isolde Wildgruber

Siemens R&D, ZFE T SE 1, D-81730 Munich, Germany

Abstract. This paper presents the TLT specification of the steam-boiler control-program described in Chapter AS. The text of the TLT specification of the control program is short and easily understandable. Due to the chosen abstraction level, the proofs that it satisfies the specification of Chapter AS are very simple. TLT has the advantage that the algorithm may be directly described as performing macro-steps. A macro step is specified *not* as a sequence of micro-steps but rather as a set of *constraints* (which may be formulated in first-order logic). These constraints relate the current state of the controller (i.e. the *information* that the controller has about the environment), the current input and the corresponding reaction (and change of state) of the controller. (Of course, the macro-step is *implemented* as a sequence of micro-steps). Thus, to argue about the program we may rely more heavily on propositional or first-order logic rather than on temporal logic.

1 Introduction

Temporal Language of Transitions (TLT) is a specification language for the compositional specification and verification of distributed programs [CWB94, CH95, BC95]. A subset of the TLT language is a programming language, for which a compiler [CBH96a] exists. TLT modules can be translated to TLA [Lam94] or interpreted as infinite boolean ω-automata, similar to [Kur94, CGS91]. The composition of TLT modules is defined logically as conjunction (or product automata or intersection of ω-languages) and the refinement of TLT modules is defined logically as implication (or language inclusion).

Abstractly, a TLT program may be defined as follows: consider a signature for a set of variables $\mathcal{V}$ and a set of *actions* $\mathcal{A}$. Given a fixed interpretation of the data types, a *program* in $(\mathcal{V}, \mathcal{A})$ is a first-order predicate describing a transition relation, that is, a subset of $\Sigma \times \mathbb{A} \times \Sigma$, where Σ is the set of all states (all possible valuations of the variables in $\mathcal{V}$) and $\mathbb{A}$ is the set of all possible actions (all possible valuations of each action in $\mathcal{A}$). A valuation of a variable $v \in \mathcal{V}$ is a value in the domain of v, denoted by $\mathcal{D}(v)$. A valuation of an action $A \in \mathcal{A}$ is a value in the extended domain of A, denoted by $\mathcal{D}_\perp(A)$ and defined as $\mathcal{D}(A) \dot\cup \{\perp\}$. If in a transition A evaluates to $\perp$, we say that A "does not happen", if it evaluates to $val \in \mathcal{D}(A)$, we say that it happens and that it passes the value val.

The purpose of TLT is to have a notation to present specifications in several abstraction levels which have simple mathematical models. The proofs of

---

* Correspondence to: Jorge.Cuellar@zfe.siemens.de

properties and of refinement relations is a separate issue and it can be carried out in temporal logic (eg., in TLA) or be done semantically. This provides flexibility, we can use model checking when possible, theorem proving when necessary, and mathematical arguments when the generality permits (to reduce complexity). Our theorem prover is based on a shallow embedding of TLA in the Lambda tool [Bus95]). We also make extensive use of the model-checking tool SVE [FSS+94].

In this paper we apply the TLT framework to the steam-boiler specification-problem of Chapter AS. With respect to the sub-language of TLT used in this paper, we will present a short introduction to essential aspects below. We however refer to the references for more information regarding TLT.

One interesting aspect of the problem statement in Chapter AS is, that, "on a first approximation, one may wish to assume that all inputs happen at the same time and that the corresponding outputs happen in zero-time delay." What does "approximation" mean? It could mean that for the real-time analysis, we first let $t_{input} = t_{output}$ (where t_E is the time where E happens) and then let $t_{input} \sim t_{output}$. But what does that mean for the *discrete* controller? (which may be seen as an automata or as a transition system). If we let input and output happen simultaneously, will then the implementation be a *refinement* (say, in TLA sense) of this initial *"approximation"*? We think it should. But in a real-time implementation the inputs will happen *before* the outputs. The solution to this is almost standard. The specification describes a relation between *observable* states and next-states (called steps). The intermediate states between two observable states are of no interest at all. In one step (or "macro-step") a sequence of "mini-steps" is performed, including an input and an output, for instance. The specification program may well specify that inputs and outputs happen both in the same macro-step. But if we do this, we may run into a problem: control programs on realistic real-time operating-systems have different cycles for different tasks, have interrupts, set internal clocks, etc. It is rather easy to write a specification which only "runs" under the unrealistic assumption that *all* inputs are processed at the same time. The challenge is to write a specification, that can be implemented under this assumption *as well as* under more realistic real-time assumptions. For example, it should not be requested that the inputs from the physical units are synchronized with "stop" or "repaired" messages from the console.

Now, we briefly describe the contents of the paper. The Appendix (see CD-ROM Annex CW1) gives a brief overview of TLT, syntax and semantics. Section 2 contains the notation for the steam-boiler control-program, i.e. data types, messages, variables, constants, events and actuators used in the control program. In Section 3, the TLT specification of the control program is presented after discussing the chosen abstraction level on the failure management module.

2 Notation

In this section we set our notation for the steam-boiler control program, presented in the next section. First, we list the data types or, more properly, domains that will be used in the sequel. In the next subsections we describe the "messages", in TLT called observable actions, of the program. We need more constants than in the original specification (particularly for time control). At last we list the internal events and actions of the program.

2.1 Data Types

The types that we need are:

Real		the real numbers
Nat		the natural numbers
Bool	:=	$\{0, 1\}$
Intervals	:=	$\{[x, y] \mid x, y \in$ Real$, x \leq y\}$ (closed intervals)
Physical_Units	:=	Pumps $\uplus$ Pump_Controllers $\uplus$ $\{l, s\}$
Pumps	$\subseteq$	Physical_Units
Pump_Controllers	$\subseteq$	Physical_Units with $|$Pumps$| = |$Pump_Controllers$|$
Mode_Type	:=	$\{init, normal, rescue, degraded, emstop\}$

where l and s denote the physical units associated with the level and steam, respectively. $v \notin$ Physical_Units denotes the valve.

There is a (constant) function *Pump_Controller*: Pumps $\rightarrow$ Pump_Controllers. The elements of Pumps are denoted informally by pump$_k$, $k \in \{1, \ldots, |$Pumps$|\}$, and the corresponding Pump_Controllers pump_controller$_k$. Let the functions *min, max*: Intervals $\rightarrow$ Real (defined by: if $I = [x, y]$ then $x = min\ I$ and $y = max\ I$) be also denoted by the subindices 1 and 2, respectively. That is, $I = [I_1, I_2]$.

2.2 Messages from Sensors

In the problem specification, the messages from the sensors or measurement devices are: PUMP_STATE(n,b), PUMP_CONTROL_STATE(n,b), LEVEL(q), STEAM(v). For the sake of parametrization we denote all four messages by READ(X,val), where X is a physical unit and val is a real value[2]. It is assumed that for each physical unit X only one value of val may be transmitted. This is simply achieved by defining the type of the action READ as $\mathcal{F}$(Physical_Units,Real), the set of (non trivial)[3] partial functions of Physical_Units to Real. That means, each time that READ "happens", this action assigns to one or more physical units a corresponding value[4]. If we had taken the type of READ as Real$^{\text{PHYSICAL\_UNITS}}$, the set of

---

[2] That is, READ(X,val) = PUMP_STATE(n,val) if X is the n-th pump, or READ(X,val) = PUMP_CONTROL_STATE(n,val) if X is the n-th pump controller, etc.

[3] That is, excluding the function nowhere defined.

[4] Nevertheless, we never write READ(X) = val, but READ(X,val).

total functions from Physical_Units to Real, then we would have assumed that each time that READ "happens", *all* physical units are assigned a value. (Of course this *should* happen, but it is not an assumption: this should be checked by the control program).

Formally, the declarations are

Actions
   In   READ: $\mathcal{F}$(Physical_Units, Real)

with $[(X \in \text{Pumps} \uplus \text{Pump_Controllers}) \wedge \text{READ}(X, val)] \Rightarrow val \in \{0,1\} = $ Bool.

2.3 Messages for Actuators

The messages received by the physical units are the commands to open or close the pumps and to activate the valve. We found no reason to open or close only *some* pumps.[5] In a more realistic situation, the decision of how many pumps should be opened and which ones will depend on many other factors not discussed here. We will just open or close *all* pumps with the actions OPEN_PUMPS and CLOSE_PUMPS. For the valve, we learned from the simulation that the VALVE action opens *and* closes the valve, toggling its state.

Actions
   Out   OPEN_PUMPS, CLOSE_PUMPS: ()
   Out   VALVE: ()

2.4 Messages for Failure Management

We have adopted here a different notation from the one used in the problem statement of Chapter AS. Instead of writing PUMP_REPAIRED(n, b), or LEVEL_REPAIRED, we write REP(X), where $X \in$ Physical_Units. The same is done for all failure-management messages. If needed, an extra small TLT module could translate from our notation to the one used in Chapter AS. For instance: ▮ REP(l) ⇒ LEVEL_REPAIRED would be one line in such a module.

The formal declaration of the TLT actions is:

Actions
   In   FAIL_ACK, REP: $\mathcal{F}$(Physical_Units, ())
   Out FAIL_DET, REP_ACK: $\mathcal{F}$(Physical_Units, ())

Note that the type is non-trivial partial functions from Physical_Units to the unit type. This domain is equivalent to the set of non-empty subsets of Physical_Units.

---

[5] Actually, for the purpose of detecting failures on the water-level sensoring-unit it is probably better to open or close *all* pumps, in order to ensure that the water level is indeed changing all the time. See section 5.4 of Chapter CW2.

2.5 Other Messages from/to the Console

In	STOP: ()
In	STEAM_BOILER_WAITING: ()
In	PHYSICAL_UNITS_READY: ()
Out	PROGRAM_READY: ()
Out	MODE: {*init, normal, rescue, degraded, emstop*}

as described in Chapter AS.

2.6 Variables and Constants

The internal (local) variables of the program will be explained later, as they are used. *N* and *M* are the intervals of normal and maximal water level, as in Chapter AS. *P* (the pump throughput), *C* (maximal water capacity), *W* (maximal steam output) are also defined in Chapter AS. All other constructs are not contained in Chapter AS. They are introduced here for several reasons. One is to increase the reusability by not coding the constants into the code. (For instance, the specification says, that the pump should react in one cycle. If you use a higher sampling velocity, you will have to change the code). *No_stops* (in Chapter AS defined as 3) is the number of stops that should be sent "in a row" (here: with a separation of at most δ^{STOP} units of time). T ($= 5$ in Chapter AS) is the cycle time. ϵ is a small quantity which plays a secondary role. The same variable ϵ will be used for many different small constants. In more realistic specifications, those variables ϵ would be distinguished. δ^{PUMP} is the time that the pump needs to react, from the perspective of the program.[6] δ^{VALVE} is the corresponding value for the valve. Some events from the environment are expected to happen after the program has issued a corresponding command. δ^{READY} and $\delta^{\text{FAIL\_ACK}}$ are the maximal time in which the events PHYSICAL_UNITS_READY and FAIL_ACK should follow PROGRAM_READY and FAIL_DET, respectively. The constant *Vv_max* is the maximal quantity of water exiting the steamboiler through the valve. The functions *F* and *G*, defined in Chapter CW2, Section 3.5, should have the property that if the real (physical) value of the steam at a point in time T_i is in an interval I, then after one cycle (that is at time $T_i + T$), this value is in $G(I, T)$ and the total amount of exited steam in this interval is contained in $F(I, T)$.

Local	t:	Real INIT 0
Local	pRd, physURd:	Bool INIT 0
Local	steam_boiler_waiting:	Bool INIT 0
Local	emergency:	Bool INIT 0
Local	read:	Physical_Units $\rightarrow$ Real
Local	fail:	Physical_Units $\rightarrow$ Bool INIT 0
Local	pCmd, vCmd:	Bool INIT 0

---

[6] Let us call Δ^{PUMP} the maximal real reaction time of the pump. Due to the sampling in cycles, the program may "see" somewhat larger reaction times. We choose δ^{PUMP} to be the least multiple of T (the cycle time) larger than to Δ^{PUMP}.

Local	past:	$\{\mathbf{v}, \mathbf{s}\} \cup$ Pumps $\to$ Intervals INIT [0,0]	
Local	calculated, adjusted:	$\{\mathbf{l}, \mathbf{s}\} \to$ Intervals INIT [0,0]	
Local	pastPumps:	Intervals INIT [0,0]	
Local	count:	Nat INIT 1	
Local	state:	$\{0, 1, 2\}$ INIT 0	
Constant	N, M:	Intervals	
Constant	P, Vv_max, C, W:	Real	
Constant	No_stops:	Nat	
Constant	T, ϵ:	Real	
Constant	δ^{READY}:	Real	
Constant	$\delta^{\text{PUMP}}, \delta^{\text{VALVE}}$:	Real	
Constant	$\delta^{\text{FAIL\_ACK}}$:	Real	
Constant	δ^{STOP}:	Real	
Constant	F:	Intervals $\times \mathbb{R}^+ \to$ Intervals	
Constant	G:	Intervals $\times \mathbb{R}^+ \to$ Intervals	

2.7 Internal Events and Actuators

These will be explained later.

Actions
    Local   TRANSMISSION_ERROR_READ: ()
    Local   TRANSMISSION_ERROR_CONTROL: ()
    Local   TRANSMISSION_ERROR_FAIL: ()
    Local   FAIL: $\mathcal{F}$(Physical_Units, ())
    Local   TICK: ()

2.8 Further Notation

If x is a boolean valued variable, we use x and x' to represent the predicates $x = 1$ and $x' = 1$, respectively (when no confusion is possible). $\neg x$ and $\neg x'$ represent $x = 0$ and $x' = 0$, respectively.

t_{EVENT} is a history variable which contains the value of the variable t, the last time that the event EVENT happened. (That is, it is a time stamp). Initially it has value $-\infty$ and it is updated to the value t' each time that t_{EVENT} happens.

If x is a program variable, $x\!\uparrow$ is an abbreviation for $x' > x$, $x\!\downarrow$ for $x' < x$ and $x\!\updownarrow$ for $x' \neq x$.

3 The TLT Specification of the Control Program

In this section we describe the control program. We preferred to choose a high abstraction level, which is more compact and easier to verify. But, as consequence, several actions that are intuitively thought as happening in a sequence may happen in one "step". Therefore, first we discuss the abstraction level by means of the failure management module. Then, we present and comment the rest of the program.

3.1 Choosing an Abstraction Level and the Failure Management

As already mentioned in the introduction we believe that it is important to write a specification at a high abstraction level (performing for example input and output on one step) but allowing implementations in different real-time embeddings. The design should be relatively independent of the real-time operating system under which it runs. The difficulty is clear when we focus on the failure management. We want the messages FAIL_DET, FAIL_ACK, REP, FAIL_ACK[7] to happen at (almost) any point in time. It is unreasonable that those messages exclude each other, and then for instance the failure can only be repaired after two read cycles having been detected, or that the FAIL_ACK and the REP are not allowed to happen almost at the same time, both in *one* cycle. Let us compare three levels of abstraction: one with a very fine granularity, where the events happen only in a strict order, a second one where first the inputs are received[8] (READ and/or FAIL_ACK and/or REP) and then the corresponding outputs (FAIL_DET and/or REP_ACK) happen, and a third more abstract level where the inputs and the outputs triggered by them happen in a single step.

Let us consider the first low-level granularity. At the beginning of a cycle the controller reads the measured value by the action READ and checks instantaneously whether this value is acceptable. If not, an error is detected and the message FAIL_DET is sent. Then, one waits until the acknowledge FAIL_ACK is received. When REP is received, the values of the physical unit are read and analyzed again[9]. If they correspond to the expected ones, after REP_ACK the process returns to the initial state and starts again. If not, also REP_ACK is sent but then, the control program is in a state where the following action is FAIL_DET.

In Figure 1 this process is expressed as an automaton.

Figures 2 and 3 show this automaton for abstraction levels 2 and 3, respectively. These automata can be generated automatically from the first one.

The advantage of automata as being graphically represented vanishes at certain levels of complexity. Still, automata may be represented compactly by tables or, for example, by a TLT specification. Let us translate (and complete) the automaton of Figure 3 to TLT. (For the sake of readability, the automata of Fig-

---

[7] We drop the parameter X. In this way, we will construct a "generic" module FM (Failure Management) which may be replicated for each X in Physical_Units producing instances, FM(X). FM(X) is the same program text as FM, but with all actions and variables renamed, from, say v and A, to $v(X)$ and A(X), respectively. Then, FM is defined as the conjunction of FM(X). For the details of the syntax and semantics of parametrization in TLT see [CBH96b].

[8] One or more inputs at a time.

[9] Another choice of sequencing is also possible, but this is irrelevant for our purposes.

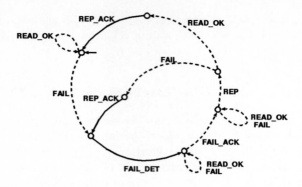

Fig. 1. The Automaton of the Failure Management in Abstraction Level 1. The initial state is left when a FAIL action happens.

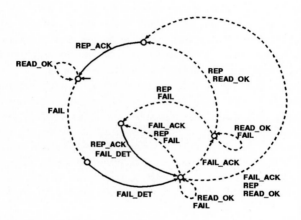

Fig. 2. The Automaton of the Failure Management in Abstraction Level 2

ures 1-3 are incomplete. For instance, the situations in which aberrant messages happen, or when an expected acknowledge don't arrive, are not shown).

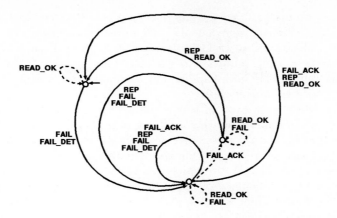

Fig. 3. The Automaton of the Failure Management in Abstraction Level 3

The Failure Management.

Always *(* Failure Management *)*
∥ REP ⇒ REP_ACK
∥ state = 0 ∧ FAIL ⇒ FAIL_DET ∧ state′ = 1 ∧ fail′
∥ state = 1 ∧ FAIL_ACK ∧ ¬REP ⇒ state′ = 2
∥ state = 1 ∧ FAIL_ACK ∧ REP ∧ ¬FAIL ⇒ state′ = 0 ∧ ¬fail′
∥ state = 1 ∧ FAIL_ACK ∧ REP ∧ FAIL
          ⇒ state′ = 1 ∧ FAIL_DET
∥ state = 2 ∧ REP ∧ ¬FAIL ⇒ state′ = 0 ∧ ¬fail′
∥ state = 2 ∧ REP ∧ FAIL ⇒ state′ = 1 ∧ FAIL_DET
∥ state = 0 ∧ (FAIL_ACK ∨ REP) ⇒ TRANSMISSION_ERROR
∥ state = 1 ∧ (REP ∧ ¬FAIL_ACK)
          ⇒ TRANSMISSION_ERROR
∥ state = 2 ∧ FAIL_ACK ⇒ TRANSMISSION_ERROR
∥ state = 1 ∧ t↑ ∧ ¬FAIL_ACK ∧ (t′ − $t_{\text{FAIL\_DET}} > \delta^{\text{FAIL\_ACK}}$)
          ⇒ TRANSMISSION_ERROR

The automaton has three internal states, which we will denote with a variable state with values 0, 1, 2. First of all, notice that in all transitions where REP happens, also REP_ACK happens and vice versa. This means that REP_ACK is triggered (only) by REP. For the rest of the module we may ignore REP_ACK. Let us consider *state* = 0. If a FAIL event happens (FAIL is the result of a detection of a FAIL condition in another module which is responsible for the monitoring of the physical unit) a FAIL_DET message is created and the system moves to *state* = 1. Further, a *fail* flag is set to 1. (The fail flag is not identical to the fail condition that generates FAIL: the flag may only be reset to 0 if a REP message arrives. The fail condition may (should) disappear before that). Three cases are distinguished when $state = 1$. They correspond to the 3 (non-stutter) arrows starting at $state = 1$ in Figure 3. For instance, the new state is 0 if in a single step, FAIL_ACK and REP both happen, but the FAIL event was not reported. In this case the fail flag is reset to 0.

Similarly, when $state = 2$: in this situation the system waits until a REP message arrives. Depending on what FAIL happens or not, the next state is 1 or 0, respectively. This completes the picture as shown in Figure 3. Now, the table may be augmented to include the unexpected cases of TRANSMISSION_ERROR. It is easy to see in which states which events are aberrant. If in state i, A is aberrant (A may be a boolean combination of messages, excluding stutter) all we have to do is to include the line $state = i \wedge A \Rightarrow T_E$. The last line in the specification states that T_E also happens if we are in $state = 1$, the clock is updated ($t\uparrow$ is an abbreviation for $t' > t$) and FAIL_ACK does not occurr in this step, but *it should* because we have been waiting for it for a long time. This time was measured using the history variable $t_{\text{FAIL\_DET}}$ that records the last time that FAIL_DET happened (and therefore the moment starting from which we are waiting for FAIL_ACK).

3.2 The Other Modules of the TLT Specification

Now, we describe the other modules of the TLT specification, see Figure 4, where each box corresponds to a module which will be explained below.

The Module Reader. The system uses an internal event, TICK, to start a new cycle. The system only starts to TICK when STEAM_BOILER_WAITING has already happened (that is, when *steam_boiler_waiting* is true). This event TICK updates the internal clock, $t' = t + T$, and triggers other events in other modules. At the beginning of each cycle (see Chapter CW2) the information from the physical units is read in. The data received is recorded in the internal variable *read*, which will also be used in later cycles. A TRANSMISSION_ERROR is found, if STEAM_BOILER_WAITING happens twice or if at the beginning of a cycle some value of a physical unit is missing.

Always *(* Reader *)*
 ∥ STEAM_BOILER_WAITING $\Rightarrow$ steam_boiler_waiting$'$
 ∥ READ $\wedge$ steam_boiler_waiting$'$ $\Rightarrow$ TICK
 ∥ TICK $\Rightarrow$ $t' = t + T$
 ∥$_{\text{X:Physical\_Units,val:Real}}$ TICK $\wedge$ READ(X, val) $\Rightarrow$ read$'$(X) = val
 ∥ STEAM_BOILER_WAITING $\wedge$ steam_boiler_waiting
   $\Rightarrow$ TRANSMISSION_ERROR_READ
 ∥ TICK $\wedge$ $\exists_X$ ¬READ(X) $\Rightarrow$ TRANSMISSION_ERROR_READ

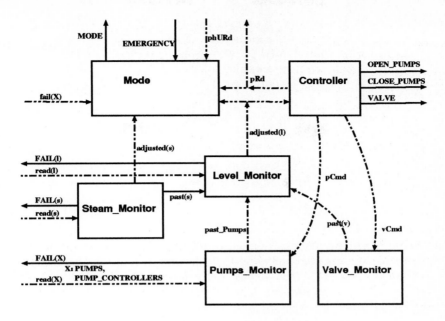

Fig. 4. The System without Failure Management and Reader

The Controller. The controller has to control the water level of the steamboiler by opening and closing the valve and the pumps. The value of the water level that it uses is *adjusted*(l) which represents the interval where the controller believes that the real value of water level is. If its second value is less than N_1, i.e. $adjusted'(l) \prec N$, then the pumps are opened. When the pumps are opened, the valve has to be closed.

The initial phase ends when PHYSICAL_UNITS_READY is received. If the program is in initial mode and the water level is within the interval $N := [N_1, N_2]$, then PROGRAM_READY is sent by the control program. The message PHYSICAL_UNITS_READY must be received after PROGRAM_READY has been sent. If it is received unexpectedly ($physURd \vee \neg pRd$) or if it is not received after a while ($t' - t_{\text{PROGRAM\_READY}} > \delta^{\text{READY}}$) then the local action TRANSMISSION_ERROR_CONTROL is performed which forces the system into emergency stop (see the module Mode). The variable t_{COMMAND} contains the timestamp of the last COMMAND; if there has been no COMMAND event yet, the value is $-\infty$.

Always *(* Controller *)*
  ‖ $pRd\uparrow \wedge \neg emergency' \Rightarrow$ PROGRAM_READY
  ‖ PHYSICAL_UNITS_READY $\Rightarrow physURd'$

‖ PHYSICAL_UNITS_READY ∧ (physURd ∨ ¬pRd)
    ⇛ TRANSMISSION_ERROR_CONTROL
‖ t↑ ∧ (pRd ∧ ¬physURd') ∧ (t' − $t_{\text{PROGRAM\_READY}}$ > δ^{READY})
    ⇛ TRANSMISSION_ERROR_CONTROL
‖ adjusted(1)↕ ∧ ¬pRd ∧ (adjusted'(1) ≻ N) ⇛ vCmd' ∧ ¬pCmd'
‖ adjusted(1)↕ ∧ ¬pRd ∧ (adjusted'(1) ⊆ N)
    ⇛ ¬vCmd' ∧ ¬pCmd' ∧ pRd'
‖ adjusted(1)↕ ∧ (adjusted'(1) ≺ N) ⇛ pCmd' ∧ ¬vCmd'
‖ adjusted(1)↕ ∧ pRd ∧ (adjusted'(1) ≻ N) ⇛ ¬pCmd'
‖ pCmd↑ ⇛ OPEN_PUMPS
‖ pCmd↓ ⇛ CLOSE_PUMPS
‖ vCmd↕ ⇛ VALVE

The Valve Monitor. The valve monitor calculates the interval describing the minimum/maximum amount of water that has been emitted in the last cycle T. If the valve is open ($vCmd$) or it has been closed during the last cycle (($t' - t_{\text{VALVE}}) \leq \delta^{\text{VALVE}}$) then $past'(v) = [0, Vv\_max] \times T$. $past'(v)$ is used in the level monitor to calculate the expected water level.

Always *(* Valve Monitor *)*
‖ TICK ∧ vCmd ∨ (t' − t_{VALVE}) ≤ δ^{VALVE} ⇛ $past'(v) = [0, Vv\_max] \times T$
‖ TICK ∧ ¬vCmd ∧ (t' − t_{VALVE}) > δ^{VALVE} ⇛ $past'(v) = [0,0]$

The Pumps Monitor. A failure FAIL(X) is detected by the pumps monitor if the read value (for a pump or a pump controller) has not been expected. Moreover, this monitor calculates how much water has flown through all pumps into the steamboiler during the last cycle T (similar to $past'(v)$ in the valve monitor). This interval is used in the level monitor to calculate the expected water level.

Always *(* Pumps Monitor *)*
  ‖$_{\text{X:Pumps}}$ TICK ∧ (read'(X) ≠ pCmd) ⇛ FAIL(X)
  ‖$_{\text{X:Pump\_Controllers}}$ TICK ∧ [(read'(X,1) ∧ ¬pCmd)
        ∨ (read'(X,0) ∧ pCmd ∧ (t' − $t_{\text{OPEN\_PUMPS}}$ > δ^{PUMP})]
      ⇛ FAIL(X)
  ‖$_{\text{X:Pumps}}$ TICK ∧ (read'(X) ≠ read(X) ∨ fail(X) ∨ fail'(X)
        ∨ fail(Pump_Controller(X)) ∨ fail'(Pump_Controller(X)))
      ⇛ past'(X) = [0,T]
  ‖$_{\text{X:Pumps}}$ TICK ∧ (read'(X) = read(X) ∧ ¬fail(X) ∧ ¬fail'(X)
        ∧ ¬fail(Pump_Controller(X)) ∧ ¬fail'(Pump_Controller(X)))
      ⇛ past'(X) = [read(X), read(X)] ×T
‖ TICK ⇛ pastPumps' = P× $\sum_{\text{X:Pumps}}$ past'(X)

The Steam Monitor. The dynamics of the steam imply that if the value of steam, v, is in an interval I, then after a time T it is in the interval $G(I,T)$, and the amount of exited steam is included in $F(I,T)$. If the interval *adjusted*(s) contains v on a certain activation, then at the next one (after time T) v will be in $G(adjusted(s),T)$. This is the calculated value for the steam, $calculated'(s)$. During the same cycle, the amount of exiting steam was contained in $past'(s) = F(adjusted(s), T)$. If the read value of steam is not in $calculated'(s)$, then an error must have occurred. The program will attribute this error to the steam sensor: FAIL(s). If steam has no error, the read value is taken as the next adjusted value (plus or minus some accuracy level ϵ), else, the value of $calculated'(s)$ is used.

Always *(* Steam Monitor *)*
 ∥ TICK ⇒ $past'(s) = F(adjusted(s), T)$
 ∥ TICK ∧ pRd ⇒ $calculated'(s) = G(adjusted(s), T)$
 ∥ TICK ∧ $read'(s) \notin calculated'(s)$ ∧ ⇒ FAIL(s)
 ∥ TICK ∧ $fail'(s)$ ⇒ $adjusted'(s) = calculated'(s)$
 ∥ TICK ∧ $\neg fail'(s)$ ⇒ $adjusted'(s) = (read'(s) \oplus [-\epsilon, \epsilon]) \cap [0, W]$

The Level Monitor. The level monitor calculates the water level by adding the quantity of water pumped into the steamboiler and subtracting the exited amount. If the new read value is not in $calculated'(l)$ then a level failure is detected.

The new adjusted value depends on whether a level failure has occurred or not: in the first case $adjusted'(l) = calculated'(l)$, else $adjusted'(l) = read'(level)$ (plus/minus ϵ).

Always *(* Level Monitor *)*
 ∥ TICK ⇒ $calculated'(l) = [0, C] \cap$
     $(adjusted(l) \ominus past'(s) \oplus pastPumps' \ominus past'(v))$
 ∥ TICK ∧ $read'(l) \notin calculated'(l)$ ⇒ FAIL(l)
 ∥ TICK ∧ $fail'(l)$ ⇒ $adjusted'(l) = calculated'(l)$
 ∥ TICK ∧ $\neg fail'(l)$ ⇒ $adjusted'(l) = (read'(l) \oplus [-\epsilon, \epsilon]) \cap [0, C]$

The Module Emergency Detection. An emergency state is reached in several cases: if there has been detected a transmission error, if there occurred a level or a steam failure during the inital phase, if the water level risks to leave the interval M, if there has been already detected a level failure and a failure of another physical unit occurred, or if the operator issued STOP commands at the console.

If the STOP messages arrives No_stops consecutive times (consecutivity is expressed by $t' - t_{STOP} \leq \delta^{STOP}$), then a counter is incremented to No_stops and the program enters emergency stop.

Always *(* Emergency Detection *)*
- ⫾ Transmission_Error_Control ∨ Transmission_Error_Read
  ∨ Transmission_Error(X) ⇒ emergency$'$
- ⫾ Tick ∧ ¬physURd$'$ ∧ *(*fail$'$*(1)* ∨ fail$'$*(s))* ⇒ emergency$'$
- ⫾ Tick ∧ pRd ∧ *(*adjusted*(1)* ⊖ F*(*adjusted*(s)*, T*)* ⊕ $P \times [0,T] \not\subset M$*)*
  ⇒ emergency$'$
- ⫾ Tick ∧ fail$'$*(1)* ∧ *(*∃$_{X:Physical\_Units}$ $(X{\neq}1$ ∧ fail$'(X))$ ⇒ emergency$'$
- ⫾ Stop ∧ *(*t$'$ − t$_{STOP}$ ≤ δ^{STOP}*)* ⇒ count$'$ = count + *1*
- ⫾ Stop ∧ *(*t$'$ − t$_{STOP}$ > δ^{STOP}*)* ⇒ count$'$ = *1*
- ⫾ count↑ ∧ *(*count$'$ ≥ *No_stops)* ⇒ emergency$'$

The Module Mode. The module "Mode" checks if there is an emergency condition or if there are any errors detected and sets the value accordingly to the specification text. This value is sent via Mode to the console.

Always *(* Mode *)*
- ⫾ Tick ∧ ¬physURd$'$ ∧ ¬emergency$'$ ⇒ Mode*(init)*
- ⫾ Tick ∧ ∀$_{X:Physical\_Units}$ ¬fail$'(X)$ ∧ physURd$'$ ∧ ¬emergency$'$
  ⇒ Mode*(normal)*
- ⫾ Tick ∧ physURd$'$ ∧ ¬emergency$'$ ∧ fail$'$*(1)*
  ∧ ∀$_{X:Physical\_Units}$ $(X{\neq} 1$ ⇒ ¬fail$'(X))$ ⇒ Mode*(rescue)*
- ⫾ Tick ∧ physURd$'$ ∧ ¬emergency$'$ ∧ ¬fail$'$*(1)*
  ∧ ∃$_{X:Physical\_Units}$ $(X{\neq} 1$ ∧ fail$'(X))$ ⇒ Mode*(degraded)*
- ⫾ Tick ∧ emergency$'$ ⇒ Mode*(emstop)*

3.3 Loose Ends

Some parts of the specification need some clarification. We found several loose-ends in the text:

1. The specification assumes that it is possible to safely distinguish pump failures from their corresponding pump-controller failures. In the description of the rescue mode it states: "This ... can however be done only ... *if one can rely upon the information [from the pump-controllers]*." We show in Chapter CW2 that it is impossible to safely distinguish pump and pump-controller failures. Therefore, the quoted statement simply means: "... can however be done only if there is no pump or pump-controller failure."
2. We do not assume that the program may receive messages from the console only at the beginning of each cycle. In the specification problem text it says that "when the message Stop has been received three times in a row ...". We assumed that *in a row* means that the time between the Stop events was (each time) smaller than a constant δ^{stop}.

3. We suppose that the failure management should work like this: if at time T_1 the event FAIL_DET(X) happens (for some X: Physical_Units) then it is expected that before time $T_1 + \delta^{\text{FAIL\_ACK}}$ the FAIL_ACK(X) happens. Else ("transmission error") the program goes into emergency stop mode.

   On the other hand, while one expects that then the event REP(X) happens *sometime*, if this event is missing, there is not necessarily a reaction: the program may work forever in degraded/rescue mode if this event never happens.

4. It is not clear whether we can safely assume or not that if the event REP(X) happens, the physical unit X is working correctly in reality (at least in this moment). That is, $\forall_X$ REP(X) $\Rightarrow \neg real\_failure(X)$. This would simplify things.

5. In the problem statement it says that after PROGRAM_READY the message PHYSICAL_UNITS_READY must be received. This may mean "... in order to continue", or "... else, the program goes to emergency_stop". We assume the second interpretation. δ^{READ} is the maximal time in which this reaction of the environment is expected.

6. In a certain condition it is not clear if a pump-controller failure should be detected or not: assume that after an OPEN_PUMPS the pump does not react (and a pump failure is detected). What value should the pump controller give on the next cycle? (Or more precisely, after $\delta^{\text{OPEN\_PUMPS}}$). The reason why this is unclear is that the pump X may have failed to turn on (thus the value reported by READ(X,val), val $= 0$, is *correct*) or the pump X may be on but the value reported (READ(X,val), val $= 0$) is *incorrect*. In the first case, a correctly working pump_controller will report (after $\delta^{\text{OPEN\_PUMP}}$) that no water is flowing, in the second one, that water is flowing.

7. Should the pumps be closed when STEAM_BOILER_WAITING happens? We assume, yes.

8. When the pump and the pump controller, respectively, have been repaired, what value should they give? For example, assume that at time T_1 the pumps were opened but pump X did not react. At time T_2 (say, $T_2 > T_1 + \delta^{\text{OPEN\_PUMP}}$) pump X and the corresponding pump controller $\tilde{X}$ were repaired. Should now the pump be on or off? If the pump is now off, is that a new failure? We assume, *yes*. When the pumps are repaired, they should be in the state which the last command OPEN_PUMPS or CLOSE_PUMPS defines.

9. There should be some relations between the (designed) constants |Pumps| N, T, and the physical constants W, P, and $\delta^{\text{Pump\_Controller}}$.

10. When PROGRAM_READY is sent, the water level is within the interval N. What should be done before receiving PHYSICAL_UNITS_READY: should still the water level be in N when PHYSICAL_UNITS_READY arrives? (We assume: not necessarily).

11. When does heat (and steam) start? After PROGRAM_READY, PHYSICAL_UNITS_READY or after the mode is set to a value different from *init*? We assume the last possibility. Moreover, we understood the statement

"[in the initialization mode]... as soon as [STEAM_BOILER_WAITING] has been received, the program checks wheter the quantity of steam coming out of the steam boiler is really zero" as stating that in the initialization mode steam different from 0 is considered as steam error.

3.4 Discrete Properties of the Program

The specification text of Chapter AS is an informal textual description of certain discrete properties that the program should have. This textual description can be translated into temporal logic, automata and/or, under some assumptions, a first-order logic formula which represents the 1-step transition-relation between certain states.

The specification text has almost no proper temporal properties (say, infinitely often, or eventually always). *Most* of the statements of the specification text can be formulated in the form "if A happens then B happens" and they may be understood as stating that A $\Rightarrow$ B holds on all transition steps, since the reaction B is supposed to happen in the same cycle as the triggering action A. For example, "[in init mode]... if the quantity of water is above N_2, the program activates the valve...", can be translated as a simple implication.

This is of course one important reason for choosing an abstraction level in which a step includes both input and output, as we did (see Section 3.1). If one chooses a lower abstraction level or if the reaction B is not assumed to happen necessarily in the same cycle as A, then this properties would be written in temporal logic as $\Box(A \Rightarrow \Diamond B)$ or $A \mapsto B$ (A leads-to B), instead of the much easier assertion that A $\Rightarrow$ B holds on all transition steps. Our program was actually *constructed* starting with such implications.

But, before doing this, the specification text was carefully reformulated to distinguish the real physical values, the sensor values and the program variables (which encode the information or knowledge that the controller has). More precisely, most of the specification text may be formulated as "if the controller is in a certain state (i.e. has a certain *knowledge* about the environment) and a certain input happens, then the controller has a new state (another knowledge about the environment) and has to produce a certain output." This may be written as

$$state \land Input \Rightarrow state' \land Output$$

The problem with the specification text is that (as do most specifications in this area) it does not properly separate between 1) *physical variables*, 2) the *knowledge* (or *supposition*) of the program about the physical variables and 3) the inputs, i.e. sensor values, (which may be incorrect) that the program receives. For instance, "... if the quantity of water is above N_2" should be interpreted as meaning 2) or 3). Other parts of the specification text (like: "the program *tries* to maintain a satisfactory water level despite of the presence of failure ...") cannot be directly formally translated. But this informal specification implies that the program tries to model the physical variables and keep track of how much water could have entered through the pumps, etc. etc. This means, more concretely, that in certain *scenarios* (that have to be found!), the program variables *approximate* the real physical variables *and* that the physical variables are indeed controlled as intended. Therefore, the motivation for some parts of the design of the program (including the additional information concerning the physical behavior of the steam boiler of Chapter AS) can only be fully understood on the basis of an environment model (see Chapter CW2).

For the purpose of a formal requirement specification, the notion of transmission failure must be made precise. Some messages at some states are aberrant. Other states (for with the specification states: "... is supposed to happen") must happen in a certain time. If an aberrant message happens or an expected message does not arrive on time, we say that a "real transmission failure" happens (as opposed to a TRANSMISSION_ERROR is detected). One has to show that if a real transmission failure happens, then it is detected, and vice versa, if it is detected then indeed a real transmission failure happened. Let us look more closely at the formalization of the real transmission failure, denoted by t_f, which may be defined as the disjunction of $t\_f\_read \vee \bigvee_{X:Ph\_Units} t\_f\_failure\_management(X) \vee t\_f\_control$. t_f_read is set to 1 in a transition if STEAM_BOILER_WAITING happens for a second time (or more) *or*, after STEAM_BOILER_WAITING some sensor value is missing in a cycle.

Formally, the history variable t_f_read is defined by the "equations" (in TLT):

INIT $\quad t\_f\_read = 0$
Always $\quad \|$ STEAM_BOILER_WAITING $\wedge$ steam_boiler_waiting $\Rightarrow t\_f\_read'$
$\qquad \|$ STEAM_BOILER_WAITING $\wedge$ READ
$\qquad \wedge\ \exists_X(\neg\exists_{val} \text{READ}(X, val)) \Rightarrow t\_f\_read'$

Observe that t_f_read is not a program variable, but an auxiliary "environment" variable. It describes whether a given sequence of observable messages from the environment is normal or not.

Now, the proof that any t_f_read will be detected is trivial: in any transition where TICK happens it is true that $(t\_f\_read' \Leftrightarrow \text{TRANSMISSION\_ERROR\_READ})$ (by simple predicate logic). From the program text it also follows, for the same transition, that TRANSMISSION_ERROR_READ $\Rightarrow emergency'$ and $emergency' \Rightarrow$ MODE(*emstop*).

Similarly, $t\_f\_control$ and $t\_f\_failure\_management(X)$ are defined and corresponding properties are proved. It follows that any real transmission failure will be detected and vice versa, any detected transmission failure indeed happened.

4 Evaluation and Comparison

In this section, we answer the evaluation questions posed by the editors.

1. As explained in Section 3.4 the informal specification text has been translated (quite trivially) to a formal requirement specification. (The text of this requirement specification as such is not presented here). Moreover, a functional/an architectural design is given in Sections 2 and 3 which may be seen

as both. This functional/architectural design of the solution has been verified against the requirements specification. Proofs were performed by hand in a formal setting.

2.a. Yes, a C implementation has been derived from the TLT specification.

2.b. Yes.

2.c. Yes, our experimentation showed that some assumptions on the environment were incorrect. We also found some errors in the simulator. For instance, the simulator did not handle properly detected failures that, according to his perspective, had not happened. These detected but not really existing errors occur as we have seen because under some situations some errors may be confused, or after a REP we expected the pumps to be in a state different from the one by the simulation.

3. As a specification language and as a methodology, TLT is closest to TLA (see Chapter LM) and to Evolving Algebras (see Chapter BBDGR). One basic difference, from our point of view, is that a TLT module may contain additional information (eg., an action is either an input or an output action). TLT modules may be regarded as transducers (infinite Mealy automata) representing (nondeterministic) *functions* of inputs and outputs, and not only as automata accepting traces in inputs and outputs. At least for implementation purposes (eg., for a compiler), this information is quite relevant.

During the Dagstuhl Seminar two similarities to other solutions presented became apparent: the solution by Abrial (see Chapter A) uses at a certain point a data-flow argument to structure the order in which the different modules are scheduled. This is also exactly the case in our solution. Secondly, many of the consistency conditions that were found in the paper of Chapter BW were identical to the ones that we found.

Our solution is in two other respects more realistic than most other solutions and than the suggestions in Chapter AS. The program has a *weak* notion of *cycle*: not all input messages are expected to be received at the beginning of the cycle. For instance, the STOP messages of the console or the ones related to the failure management are not being delayed until a sensor event happens. Also, the events on the physical units are time-supervised: instead of counting how many cycles have passed since the last command, say to a pump, the controller calculates *how much time* has elapsed. In this way the specification is much more easily modifyable for new cycle times or different physical properties of the technical equipment.

4.a. The first version of the program (which is slightly different from the one presented here) required 2-3 person months, including the C program and the experimentation.

4.b. Assuming familiarity with first order logic and basic automata theory, about 1 week.

5.a. Familiarity with first order logic.

5.b. Yes.

5.c. 1 hour.

References

[BC95]   Dieter Barnard and Simon Crosby. The Specification and Verification of an ATM Signalling Protocol. In *Proc. of 15th IFIP PSTV'95*, Warsaw, June 1995.

[Bus95] H. Busch. A Practical Method for Reasoning About Distributed Systems in a Theorem Prover. In *Higher Order Logic Theorem Proving and its Applications - 8th International Workshop, Aspen Grove, UT, USA, Proceedings*, pages 106–121. Springer-Verlag, LNCS 971, September 1995.

[CBH96a] Jorge Cuéllar, Dieter Barnard, and Martin Huber. A Solution relying on the Model Checking of Boolean Transition Systems. In *The RPC-Memory Specification Problem*, to appear in LNCS, 1996.

[CBH96b] Jorge Cuéllar, Dieter Barnard, and Martin Huber. Rapid Protyping for an Assertional Specification Language. *TACAS'96, LNCS 1055*, March 1996.

[CGS91] C. Courcoubetis, S. Graf, and J. Sifakis. An Algebra of Boolean Processes. In *Proc. of CAV'91*, pages 454–465, 1991.

[CH95]   Jorge Cuéllar and Martin Huber. The FZI Production Cell Case Study: A distributed solution using TLT. In *Formal Development of Reactive Systems: Case Study Production Cell*, volume 891 of *LNCS*. Springer-Verlag, 1995.

[CWB94] J. R. Cuéllar, I. Wildgruber, and D. Barnard. Combining the Design of Industrial Systems with Effective Verification Techniques. In M. Naftalin, T. Denvir, and M. Betran, editors, *Proc. of FME'94*, volume 873 of *LNCS*, pages 639–658, Barcelona, Spain, October 1994. Springer-Verlag.

[FSS$^+$94] T. Filkorn, H.A. Schneider, A. Scholz, A. Strasser, and P. Warkentin. SVE User's Guide. Technical report, Siemens AG, ZFE T SE 1, D-81730 München, Germany, 1994.

[Kur94] R.P. Kurshan. *Computer Aided Verification of Coordinating Processes*. Princeton University Press, 1994.

[Lam94] L. Lamport. The Temporal Logic of Actions. *ACM Transactions on Programming Languages and Systems*, 16(3):872–923, May 1994.

The Real-Time Behavior of the Steam-Boiler

Jorge R. Cuéllar*, and Isolde Wildgruber

Siemens R&D, ZFE T SE 1, D-81730 Munich, Germany

Abstract. This paper presents a rigorous model of the dynamics of the physical process of the steam-boiler and of the embedding of the discrete program given in Chapter CW1 in the continuous environment. Program constants and variables are clearly distinguished from physical quantities, and it is shown, that under some conditions the program variables approximate the physical quantities. This relation between physical and program variables is, technically speaking, a *coupling invariant*. Hybrid and real-time aspects of the system are also considered, including several positive and negative results regarding the expected properties of the system. Some of them are *impossibility* results, in the sense that no (reasonable) controller program can solve some aspects of the problem properly.

This paper illustrates a simple method for combining different abstraction levels in the verification of industrial control programs. Programs are specified by their next-state relation (in first order logic) without real-time (but, if needed, using program clocks). The behavior of the environment is described by the physical laws that govern the dynamics of the process. Those two levels are bridged together by the specification of the real-time deadlines that the SW- and HW-implementation has to guarantee.

1 Introduction

The analysis of real-time industrial computing requires studying at least three different aspects: the discrete control-automata, the real-time constraints in the scheduling of the SW-system, and the mathematical control of the physical process. Automata-theoretical methods and/or temporal logics are used quite efficiently to formalize and verify discrete systems. One of the reasons for their efficiency is that they abstract away from time. Only the relative order of events is described in the formalism. On the other hand, real-time control programs have a well defined structure, and several methods for analyzing their timing behavior have been established, particularly, rate-monotonic analysis [KLR90]. Needless to say, the mathematics of control theory, see for instance [Son90], is quite independent of both automata theory and real-time operating-systems analysis.

---

* Correspondence to: Jorge.Cuellar@zfe.siemens.de. Supported in part by a BMBF grant from the German Federal Ministry of Education, Science, Research and Technology under contract number 01IS519A (project KORSYS).

One way of dealing with real-time constraints is to consider a new variable (denoting *time*) that never decreases and eventually increases unboundedly. The advantage is that one treats the whole hybrid system, from the discrete automata over the real-time deadline considerations up to real-time behavior of the physical process, with one uniform formalism at *one abstraction level*. This method works quite well in certain, rather simple, examples see [Lam93], [CBH96]. The main drawbacks are: a) such an approach increases quite substantially the complexity in the theorem-proving or model-checking procedures for the automata, b) the notions of "approximation" in analysis and "abstraction/refinement" in discrete systems do not agree in a common semantical setting, and c) the tasks of algorithm validation and real-time analysis are done usually by different people with different methods and tools.

In this paper we use a simple method for integrating these three levels, without introducing real-time aspects into automata, or "formal methods" into control theory. In order to understand, specify and verify control programs it is better to use a high-level of abstraction: in one step the program samples sensor values and reacts to the gathered data. This macro-step (read–react) is indeed performed as a sequence of micro-steps, which, strictly speaking, happen at different points in time, but the details about the real-time behavior of the micro-steps are irrelevant, at this point. All one needs to know is that the implementation indeed meets the specified deadlines and that all outputs are produced before the next cycle is due to start.

What we advocate, is to use at least *two different* abstraction levels to describe the system: one, at a high level of abstraction, sees only the (macro-) steps of the controller program, the other one sees the continuous flow of the physical values. The high level of abstraction simplifies the reasoning about the program itself, since much of the argumentation may be done in propositional or predicate logic, and not in temporal logic.

The program variable representing the water level (*read*, *adjusted* or *calculated*) is logically and possibly numerically different from the real physical variable water level. What does it mean that a control program is correct? A possible answer is that the physical variables are indeed controlled in a certain way. Then program specification and verification call for describing *both* the control program and the physical system and interrelating them. The difficulty lies in the fact that one has two abstraction (granularity) levels. The physical variables change continuously (all time points are visible, steps are "infinitesimal") while the program steps are (realistically) *time-consuming* discrete steps, where the internal intermediate states of the program are invisible.

Section 2 introduces a simple common semantical basis for continuous and discrete systems and presents the formulas that relate the program level and the physical environment. Sections 3 discusses the dynamics of the steam-boiler. Section 4 present the list of assumptions needed for the rest of the paper. Section 5 contains positive and negative results concerning expected properties of the system.

2 Real-Time Embeddings of Discrete Specifications

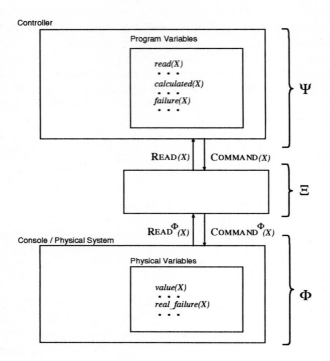

Fig. 1. The three levels of Design and Analysis. The formula Ψ of the "upper" level describes the program behavior in terms of the visible program events and the internal program variable. In particular, it expresses the next-step ("macro-step") relation. At the "lower" level (the physical one) the formula Φ describes the *trajectories* of the system. To bridge the gap between those two levels, we also need a formula Ξ that relates the visible program actions A to their corresponding physical events A^{Φ}.

As indicated in Figure 1, we use three types of formulas to specify the system. If a program trace π satisfies program (automaton) Ψ, will write

$$\pi \models \Psi.$$

We will denote by Ψ_0 the control program of Chapter CW1. We also refer to that chapter for the use of notation.

At the physical level Φ describes the *trajectories* **f** of the system. **f** is a vector of functions defined on $\mathbb{R}^+$ (= the time domain). The components of **f** are the sensor and actuator events and the value of the physical variables of the system. The sensor and actuator components of **f** are the "physical counterpart" to the observable actions of the control program. If A is any visible (input or output) event of the program, (for instance, READ, VALVE, OPEN_PUMPS or CLOSE_PUMPS) we will denote by A^Φ the corresponding physical event.[2] If A^Φ is a sensor or actor event, passing a value val in $\mathcal{D}$ then we consider A^Φ as a function

$$A^\Phi : \mathcal{D} \to Time \to \{0,1\}$$

with discrete support, that is, the intersection of the set $\{t \mid \exists_{val} A^\Phi(val)(t) \neq 0\}$ with any bounded time interval is finite. A function A^Φ with discrete support will be called in the sequel a *)physical) event*. In other words, we consider the visible actions of the program (i.e., the events $READ^\Phi$, $VALVE^\Phi$, etc.) as functions of time. The actions of type unit are treated as boolean-valued functions. $READ^\Phi$ is a vector of boolean-valued functions of time, $READ^\Phi(X, val)$: $\mathbb{R}^+ \to \{0,1\}$. We also use the symbol $READ^\Phi$ for the function $READ^\Phi(t) = max_{X, val} READ^\Phi(X, val)(t)$. This corresponds to

$$READ^\Phi \Leftrightarrow \exists_{X, val} READ^\Phi(X, val).$$

If $A^\Phi(val)(t) = 1$ we say that event A^Φ *happens* at time t and passes value val. We also write $A^\Phi(val)(t)$ or $A^\Phi(t) = val$, instead of $A^\Phi(val)(t) = 1$. The sensor values are communicated to the program via an action $READ^\Phi$. If X is a physical unit then $READ^\Phi(X, val)(t)$ states that at time t the sensor value for X is val.

Besides these sensor and actuator events **f** also contains the physical variables, and in particular, the "true" values of the physical units (different from the estimated values of the control program). For these functions we will use the following convention. If x is a parameterized (or vector valued) function with parameter X in $\mathcal{X}$ ($\mathcal{X}$ will usually be a set of physical units) and values in $\mathcal{D}$, we write $x : \mathcal{X} \to \mathcal{D}$ but, actually, x depends also on $t \in \mathbb{R}^+$ (the time). Therefore, x is a function in $\mathcal{X} \to \mathbb{R}^+ \to \mathcal{D}$.

We will assume, without loss of generality, that the first thing that happens, at time $t = 0$, is $STEAM\_BOILER\_WAITING^\Phi$ and that the first READ happens (if at all)[3] at some time $t > 0$. Further let us assume, that if MODE(emergency) happens, it is the last event that happens. Else, the system continues forever. Besides the visible actions A^Φ, the trajectory **f** also contains physical variables: the values of all physical units (plus the valve) and their internal state (coding the fact that they are failing or not). Basically, the formula Φ that describes **f** is a differential equation (or a piecewise differential equation with boundary conditions). When a trajectory **f** satisfies Φ we write

$$\mathbf{f} \models \Phi.$$

---

[2] It is necessary to distinguish between the program events A and the physical events A^Φ. This is due to the *asynchronous* relation between the program and the physical process, or from another viewpoint, due to the two different abstraction levels.

[3] The first READ may perhaps never happen, if, for instance, a transmission failure is detected before it could happen.

In Section 3, we present the formula Φ_0 which describes the physics of the system.

To bridge the gap between those two levels, we also need a formula Ξ that relates the "time" when the program sees a sensor or actor event with the real time when these events take place in the physical system. More precisely, it relates the visible program actions READ, OPEN_PUMPS, etc. to the physical events READ$^\Phi$, OPEN_PUMPS$^\Phi$, etc. Then, if Ξ relates π to $\mathbf{f}$ we write

$$(\pi, \mathbf{f}) \models \Xi$$

For the rest of this section we discuss the formula Ξ_0 that formalizes the assumptions on the real-time hardware- and software-environment.

The controller of Chapter CW1 performs only one step reading the sensors and producing the outputs. In real-time, these actions do not happen "in one step". *First*, the program reads the sensor values and *then*, after some internal calculations, in some *non-specified order* the outputs are delivered to the environment. All we know to relate these actions to the physical trajectory is that the sensor values are appropriately accurate (if they are not failing), they are read at the beginning of each cycle and that the outputs are produced before the next cycle starts.

But it is not the case that our TLT program in Chapter CW1 is so naive as to assume (or enforce) that all program traces are of the form

$$\sigma = (\gamma_0, \gamma_1, \gamma_2, \ldots, \gamma_{N-1}, \gamma_N, \ldots)$$

where at γ_0 STEAM_BOILER_WAITING happens, at each γ_i, $0 < i < N$, a READ happens and for $i \geq N$ (if $N < \infty$) γ_N is a stutter step where no action happens. Between two READ events, more interaction with the environment is allowed. For instance, STOP messages of the console are not being delayed until a READ event happens. Also, a second STEAM_BOILER_WAITING would be possible, but would lead to a transmission failure and thus to MODE(emergency). Also recall that the actions of failure management are not being synchronized somehow to the READ events.

Nevertheless, there is a notion of *cycle* in the program. A cycle starts each time that an event READ happens. This definition does not depend on the fact that the program performs "macro-steps". Even if the program performs many steps between two READ events, it is perfectly clear how to assign a cycle number to each event. If a program is refined by introducing extra "micro-steps" to perform one step, this *cycle* assignment should be preserved. There is *another* notion of cycle in the physical environment. There, a *cycle* starts when the physical event READ$^\Phi$ happens. The basic assumption is that those two notions of cycle coincide, in a sense to be made more precise. An event A happens in the program cycle i iff its corresponding physical event A$^\Phi$ happen in the physical cycle i.

Moreover, the real-time operating system gives us more information about the physical times where the cycles start: The SW-task that controls the steam boiler is a periodic task. The sensors are read at times $t = T_1, T_2, \ldots, T_i, \ldots$ with $T_{i+1} - T_i = T$ (the cycle time). (Also we could weaken the last assumption to: $T_{i+1} - T_i \approx T$. That means that $| T_{i+1} - T_i - T | < \epsilon$ for a known and convenient $\epsilon > 0$).

Now, let us formalize the notions of cycle for a trace π and a trajectory **f** and let us relate them. For this purpose we will use "auxiliary variables". Let us consider first a trace π of our program Ψ_0. Let *count* be the history (auxiliary) variable defined initially as 0 and incremented by 1 each time that READ happens. Formally, *count* is defined by the temporal formula

$$\Psi_{count} :\Leftrightarrow (count = 0) \land \Box(\text{READ} \Leftrightarrow count' = count + 1).$$

The rationale behind the introduction of an auxiliary variable is that it does not restrict the class of traces of the program. In other words, if π is a trace satisfying the program Ψ_0, then it is possible to augment π on each step with values of *count*, such that the new trace $\tilde{\pi}$ satisfies $\Psi_0 \land \Psi_{count}$. Thus the introduction of an auxiliary variable corresponds to the introduction of an existential quantifier, see [Lam94].[4]

If A is a visible event of the program, we will let A_i denote the temporal formula

$$A_i :\Leftrightarrow \Diamond(count' = i \land A).$$

Thus, the trace π satisfies A_i iff A happens in the i-th cycle of the program. For $i = 0$, $\pi \models A_0$ iff A happens before the first cycle (that is, before the first READ). The transition where the first READ happens satisfies the transition formula $count = 0 \land count' = 1 \land$ READ. Therefore, $\pi \models \text{READ}_1$, iff eventually a first READ happens, or, in other words, if a cycle starts. Similarly, $\pi \models \text{READ}_n$ iff there is an n-th cycle. Also, instead of writing, for instance, $(\text{READ}(X, val))_n$ we write $\text{READ}_n(X, val)$. Define N as the number of cycles (N may be $+\infty$). That is, $N := sup\{n \mid \text{READ}_n\}$. If $0 \leq i \leq N$, and v is a program variable, define v_i as the value that v has at the end of cycle number $i - 1$. (For $i > N$ the value v_i is arbitrary). In other words, if $i \leq N$, a formula $f(v_i)$ containing v_i is an abbreviation:

$$f(v_i) :\Leftrightarrow \Diamond \exists_{val}(count = i - 1 \land count' = i \land v = val \land f(val)).$$

As in the discrete case, it is useful to construct auxiliary variables (functions). Given a function $x(t)$, there are many examples of auxiliary functions that can be defined depending on the domain of x: integrals, derivatives, supremum, composition, convolution, etc., or even more general operators.

Given an function $A(t)$ that represents the occurrence of a physical event, (that is, $A(t)$ has discrete support), define

$$Count^-(A)(t) := | \{t_1 \mid 0 \leq t_1 < t \land A(t_1) = 1\} |$$
$$Count^+(A)(t) := | \{t_1 \mid 0 \leq t_1 \leq t \land A(t_1) = 1\} |.$$

---

[4] Notice that the formula Ψ_{count} is stutter-invariant.

Both functions count the number of times that A already happened. $Count^+(A)$ and $Count^-(A)$ only differ at the points where A happens and at those points $Count^-(A) + 1 = Count^+(A)$. For $i \geq 1$, the point T_i^A defined by

$$T_i^A := \sup\{t \in \mathbb{R}^+ \mid Count(A)(t) < i\}$$

has the properties that $A(T_i^A) = 1$, $Count^+(A)(T_i^A) = i$ and $Count^-(A)(T_i^A) = i-1$. If $T_1^A = 0$, then we call $\{T_i^A \mid i \geq 1\}$ the *observation points* of A. If $T_1^A > 0$ then let also $T_0^A := 0$ be included in the set of observation points. We will be interested in the particular case where $A = \text{READ}^\Phi$. Therefore, we *define the observation points of the system* by $T_i := T_i^{\text{READ}^\Phi}$.

To gap the passage from real-time to discrete systems we need discrete auxiliary variables. Given a function $x : \mathbb{R}^+ \to \mathcal{D}(x)$ and the observation points $0 = T_0 < T_1 < T_2 < \ldots < T_n \ldots$ in $\mathbb{R}^+$, a discrete auxiliary variable on $x(t)$ is given by an equation of the form

$$y_i = \text{a function of } x \mid_{[T_i^-, T_{i+1})}, \text{ or}$$
$$B_i = \text{a predicate (boolean function) of } x \mid_{[T_i^-, T_{i+1})}.$$

(We let y_i and B_i also depend on $x(T_i^-) = \lim_{t \to T_i^-} x(t)$). In particular, we will let y be a supremum, infimum or integral, and B_i be of the form:

$$\exists_t [T_i \leq t < T_{i+1} \wedge B(x(t))].$$

Now, we can complete the formulation of the formulas Ξ_0 which express the basic assumptions about the relation of the traces of the program to the trajectories of the system. For the READ events, we assume that

$$\text{READ}_i(X, val) \Leftrightarrow \text{READ}^\Phi(X, val)(T_i).$$

Therefore, a read event (with parameters X, val) happens in the i-th cycle of the program iff it happens at time T_i in the physical system. For all other observable events we assume a weaker property:

$$E_i(param) \Leftrightarrow \exists_t T_i \leq t < T_{i+1} \wedge E^\Phi(param)(t).$$

This is the precise formalization of the idea that the two notions of cycle coincide.

3 Modeling the Environment

In Chapter CW1 the formula Ψ_0 that describes the set of possible program traces was given. In the last section we presented the formula Ξ_0 that relates the program to the physical system. Now, in this section we present the formula Φ_0 that describes the physical evolution of the system.

3.1 The Physical Variables - Basic Dynamics

The physical variables that we will use are listed now. In this paper we will use the notation $\tilde{X}$ to denote $Pump\_Controller(X)$, $X \in$ Pumps.

- Vv: $\mathbb{R}$, the valve throughput
- $value$: Physical_Units $\uplus \{\mathbf{v}\} \to \mathbb{R}$, the real state of the physical units and the valve: for the water level, $value(\mathbf{l})$ is the real value of the water level in the steam boiler; similarly, $value(\mathbf{s})$ is the current exit of steam. Let X be a pump. If $\tilde{X}$ is the pump controller for the pump, then $value(\tilde{X})$ is 1 or 0 depending on whether water is flowing or not, respectively.

    We assume no *valve error* (including no transmission failure for valve commands to the environment). Therefore, the valve is open at time t iff the number of VALVE$^\Phi$ commands in the interval $[0, t)$ is odd. So, it is natural to define $value(\mathbf{v})(t)$ as the number of VALVE$^\Phi$ events in the interval $[0,t)$ modulo 2.

    For a pump X, it is possible to define $value(X)$ in several ways. Consider the function $open(X)$ defined to be 1 iff the motor is on (else 0). One option is to set $value(X) = open(X)$. There are two reasons for not doing this. The function $open(X)$ (which may be seen as part of **f**) is rather uninteresting, since it does not contain much more information than $value(\tilde{X})$. The other reason is that a different choice will make the definition of the physical variable *real-failure* more natural. We found convenient to define $value(X)$ as the state in which X *should be*, that is, equal to 1 *iff* the last command was OPEN_PUMPS$^\Phi$. Since we know from the program that OPEN_PUMPS$^\Phi$ and CLOSE_PUMPS$^\Phi$ alternate, $value(X)(t)$ can simply be *defined* as the number of (OPEN_PUMPS$^\Phi$ $\vee$ CLOSE_PUMPS$^\Phi$) events in the interval $[0,t)$.[5] In other words, the program variable $pCmd$ models the physical variable $value(X)$.
- $real\_failure$: Physical_Units $\to$ Bool. Part of the task is to detect failures of physical units. If a failure has been *detected*, this detection is recorded by a program variable. Correspondingly, on the physical side, there should be something like a "failure has *happened*" (whether detected or not) and we need a corresponding boolean variable (or propositional predicate) to express this. The informal statement "the physical unit X has a failure" is expressed by $real\_failure(X) = 1$. $real\_failure$ is a somewhat mysterious physical variable. All others are in a natural sense *visible*. You may measure the water-level, steam or valve output. But how do you exactly define $real\_failure$? It is not completely clear in the problem statement what, for instance, a "steam failure" is (in physical or mechanical terms). Is it supposed to be only a *sensor* problem? If steam *is* flowing when it should not, is that a steam failue? If the sensor is working correctly but the message gets corrupted over the transmission channel, is that a "steam failure"? We adopt a pragmatical approach: the meaning of the statement "there is no

---

[5] The OPEN_PUMPS$^\Phi$ $\vee$ CLOSE_PUMPS$^\Phi$ event is visible for the physical environmet. Therefore, $value(X)$ is a *visible* auxiliary (history) physical variable.

steam failure" will be given by a formula that relates the event READ(X, val) and the continuous variable $value(X)$. This will be given in Section 3.3.

3.2 Assumptions on Water Level and Valve

Let v, q and p denote $value(\mathbf{s})$, $value(\mathbf{1})$, and $P \times \sum_{\tilde{X}:\mathsf{Pump\_Controller}} value(\tilde{X})$, respectively. We assume that v and q are continuous functions and that p is piecewise constant. ($value(\tilde{X})$, for $\tilde{X}$: Pump_Controllers is piecemeal constant with values 0 or 1).

Recall that $Vv(t)$ is the valve throughput. We assume that

$$q(t) = q(t_0) + \int_{t_0}^{t} p(t)dt - \int_{t_0}^{t} v(t)dt - \int_{t_0}^{t} Vv(t)dt.$$

This means that water in the boiler only is provided by the pumps and evacuated by the valve or steam outlet, and that the units of q, p, v and Vv are normed to each other.

We assume that the valve opens and closes exactly at the time points when the command VALVE$^{\Phi}$ happens. Therefore, $Vv(t) > 0$ iff $value(\mathbf{v})(t) = 1$. We assume that if the valve is open and no water is flowing in, then eventually the water level will effectively decrease. Formally, we assume that there is a constant $\epsilon = \epsilon_{valve}$, such that $Vv(t) > 0$ is equivalent to $Vv(t) \geq \epsilon$. (This assumption can be weakened to : if $Vv(t) > 0$ for all $t \in [T_i, T_{i+1})$, then $\int_{T_i}^{T_i+1} Vv(t)dt \geq \epsilon$).

3.3 Assumptions on Failures

Failures on physical units could be very strange boolean function on time, but in order to understand them and prove properties of the system we will assume that they are "well-behaved". (Anyway, what does it mean that the failures are not measurable?) The meaning of $real\_failure(X)$ depends on what type of physical unit X is: *a pure sensor* (the steam-measuring unit, the water-level measuring-unit and the pump controllers, which measure whether water is flowing through the pump or not) or an *actuator/sensor unit* (the pumps).

A failure in a pure sensor is easy to understand: If the sensor is working properly, the value transmitted by the READ$^{\Phi}$ event is close to the physical value of the variable. More precisely, if X is a pure sensor, we assume that if $real\_failure(X)(t) = 0$ and READ$^{\Phi}(X,val)(t) = 1$, then there are two cases: if $value(X)$ is boolean valued, then $value(X)(t) = val$ and if $value(X)$ is real valued then $|\ value(X)(t) - val\ | \leq \epsilon$, where $\epsilon > 0$ is some accuracy level.

A naive interpretation of the failures of a pump is that the pump is only an actuator (with OPEN_PUMPS$^{\Phi}$ and CLOSE_PUMPS$^{\Phi}$ as commands) with no "sensor" function. But this is not correct: the pump is also giving a value (PUMP_STATE$^{\Phi}$, or, in our notation READ$^{\Phi}$). If there is a failure in the pump the failure may be that the "actuator part" is not working (the motor is not on or off when it should), or the "sensor part" is transmitting an incorrect value (the motor is working as it should, but the value PUMP_STATE$^{\Phi}$ transmitted to the control program is incorrect). The interpretation that the pump is only an actuator corresponds to the (unrealistic) assumption that this second type of failure never happens.

Let us now consider a pump X (actuator/sensor). If the pump is working correctly, when an OPEN_PUMPS$^\Phi$ event happens, the motor of the pump should be turned on immediately and the pump should respond with a READ$^\Phi$(X, 1). When the pump is turned off, the next sensor value should be READ$^\Phi$(X, 0). Therefore, if $real\_failure(X)(t) = 0$ and READ$^\Phi(X, val)(t) = 1$ then $value(X)(t) = val$. This turns out to be the same condition as for the pure sensors, but for a different reason: observe that $value(X)$ is *not* defined to be 1 iff the motor of the pump is on. Thus, the pump-sensor failure can be defined as READ$^\Phi(X, val)(T_i) = 1 \land val \neq value(X)$.

But the pump may also fail in a different sense: perhaps water is indeed flowing when it should not or vice versa. Recall that $value(X)$ records if OPEN_PUMPS$^\Phi$ was the last event, while $val(\tilde{X}) = 1$ iff water is flowing. For the pump controller $\tilde{X}$, the function $value(\tilde{X})$ *should* be 1 if $value(X) = 1$ and the time elapsed from the last OPEN_PUMPS$^\Phi$ is greater than a certain fixed time, that we denote by Δ^{Pump}. Δ^{Pump} is the maximal reaction time of the pump. $value(\tilde{X})$ *should* be 0 if $value(X) = 0$. In other words, a pump-actuator failure happens if $value(\tilde{X})(t) = 1$ when $value(X)(t) = 0$ or if $value(\tilde{X})(t) = 0$ when $value(X)(t_1) = 1$ for all t_1 in $[t - \Delta^{Pump}, t]$.

Summarizing, let us define *real-failure(X)(t)* by

$$real\_failure(X)(t) = 1 \Leftrightarrow (\text{READ}^\Phi(X, val)(t) = 1 \land val \neq value(X)) \lor$$
$$(value(X)(t) = 0 \land value(\tilde{X})(t) = 1) \lor$$
$$(\forall_{t-\Delta^{Pump} \leq t_1 \leq t} value(X)(t_1) = 1 \land value(\tilde{X})(t) = 0)$$

If a failure on the pump controller is present, the transmitted physical value READ$^\Phi(\tilde{X}, val)$ is not equal to the real value $value(\tilde{X})$. Note, that the program only receives the transmitted value, and it is therefore impossible to discriminate pump from pump-controller failures in the control program. For example, if after an OPEN_PUMPS$^\Phi$ command a pump answers that the pump motor is on, but (after δ^{PUMP} time) the pump controller indicates that no water is flowing, it is not clear whether water is not flowing and the pump has a problem or water is flowing but the pump controller has failed.

For a pump X, if $real\_failure(X) = 1$ for short intervals which do not contain a READ$^\Phi$ event, then the flow of water into the steam boiler may be very irregular, unpredictable and unrelated to the readings of the pumps and pump controllers. Except for those properties that do not depend on the behaviour of the pumps at all, we found no interesting property of the system that holds if the values of $real\_failure(X)(t)$ for $T_i < t < T_{i+1}$ are independent (in probabilistic sense) of the values at the extreme points of the interval. A probabilistic scenario would lead us too far; we therefore assume that $real\_failure(X)$(for $X \in$ Pumps) is a *persistent* function in the following sense: if $real\_failure(X)(t) = 1$ is true for some t then $real\_failure(X)(T_i) = 1$ or $real\_failure(T_{i+1}) = 1$, where $T_i \leq t < T_{i+1}$. In the simulation this assumption holds. (If failure happens, this failure persists until it is repaired, which in turn happens only after it is detected. To detect a failure, a READ$^\Phi$ is necessary).

Clearly, the value of *real_failure(X)(t)* for $t \neq T_i$ (at times where READ$^\Phi$ does not happen) is irrelevant if X is a pure sensor. (It should not concern us, whether the sensor is faulty if its value is not used). Therefore, without loss of generality, we may assume that *real_failure(X)* is also persistent for any pure sensor X.

3.4 Behavior of the Pumps

Observe that the control program has no exact information about the exact physical time where the last OPEN_PUMPS$^\Phi$ command happened. Therefore, the value Δ^{Pump} has to be modified to a value δ^{Pump} which represents the maximal reaction time of the pump to open from the point of view of the program. δ^{Pump} also depends on the cycle time, the definition is: $\delta^{\mathsf{Pump}} = T \times (\lfloor \Delta^{\mathsf{Pump}}/T \rfloor + 1)$, we can conclude from the assumptions of 3.3:

- If $value(X)(T_j) = 1$ (that is, before time T_j the last was OPEN_PUMPS$^\Phi$), and *real_failure(X)*(T_{j-1}) = *real_failure(X)*$(T_j) = 0$ (that is, no real failure has happened in the (closed!) interval $[T_{j-1}, T_j]$) then READ$^\Phi(X,1)(T_j)$, and if the last OPEN_PUMPS$^\Phi$ event happened in the interval $[T_i, T_{i+1})$ with $T_i + \delta^{\mathsf{Pump}} \leq T_j$ and *real_failure*$(\tilde{X})(T_{j-1})$ = *real_failure*$(\tilde{X})(T_j) = 0$, then $\int_{T_{j-1}}^{T_j} value(\tilde{X})(t)dt = T_j - T_{j-1} = T$.
- If $value(X)(T_j) = 0$ and *real_failure(X)*(T_{j-1}) = *real_failure(X)*$(T_j) = 0$ then READ$^\Phi(X,0)(T_j)$. If, additionally, the last CLOSE_PUMPS$^\Phi$ event happened in $[T_i, T_{i+1})$ with $i < j$ and *real_failure*$(\tilde{X})(T_{j-1}) = 0$ and *real_failure*$(\tilde{X})(T_j) = 0$ then $\int_{T_{j-1}}^{T_j} value(\tilde{X})(t)dt = 0$.

Notice, by the way, that from the point of view of the program the maximal reaction time for closing the pump is T. As a consequence, if a pump was closed in an interval $[T_i, T_{i+1})$, it may not be concluded that $\int_{T_i}^{T_{i+1}} value(\tilde{X})(t)dt = 0$. This is implicitly suggested in the problem specification: if a CLOSE_PUMPS$^\Phi$ event happens in $[T_i, T_{i+1})$ and we assume that there is no time delay in reading sensors and sending commands, the CLOSE_PUMPS$^\Phi$ should have happened at time $t = T_i$. Since the pump reacts immediately, then no water was flowing in the interval. This does not hold for instance in the simulation.

3.5 Assumptions on Steam

Recall that $-U_2$ and U_1 are the maximal gradients for the steam. That is, the steam $v(t)$ is assumed to be a continuous function satisfying

$$-U_2(t_2 - t_1) \leq v(t_2) - v(t_1) \leq U_1(t_2 - t_1),$$

and $0 \leq v(t) \leq W$. Let G: Intervals $\times \mathbb{R}^+ \to$ Intervals be the following function:

$$G_1(I, t) = max\{0, I_1 - U_2 t\}$$

$$G_2(I,t) = min\{W, I_2 + U_1 t\}.$$

Then, the conditions above simply state that v has the property, that if $v(t) \in I$, then $v(t+t_1) \in G(I, t_1)$, for all t, t_1. Since G is monotonic, this implies in particular: if $v(T_i) \in I$ then $v(t) \in G(I, T)$ for all $T_i \leq t \leq T_{i+1}$. Let us now consider F: Intervals $\times \mathbb{R}^+ \to$ Intervals defined as the integral of G:

$$F(I,t) = \int_0^t G(I, t_1) dt_1.$$

It follows that

- if $v(T_i) \in I$ then $v(T_{i+1}) \in G(I,T)$
- if $v(T_i) \in I$ then $\int_{T_i}^{T_{i+1}} v(t) dt \in F(I,T)$.

A further assumption on $v(t)$ is needed: before PROGRAM_READY$^\Phi$ there is no steam. Formally, $v(t) = 0$ *unless* PROGRAM_READY$^\Phi$, or in defined terms, $Count^-(\text{PROGRAM\_READY}^\Phi)(t) \Rightarrow v(t) = 0$.

4 Summary of Basic Functions, Definitions, Assumptions and Scenarios

Now, we give a complete list and formalize and summarize the definitions, assumptions and scenarios that will be used.

4.1 The Program

The program is the conjunction (or product) of the automata described in Chapter CW1, Section 4. They constrain a set of visible actions. We have called this specification Ψ_0. Now, we introduce, without loss of generality, some auxiliary program variables. (In other words, they do not constrain the sets of traces and trajectories more that Ψ_0 does. See also the remarks to auxiliary variables in Section 2.)

Definition of Auxiliary Program Variables.

(Progr Count)	$count = 0 \land \Box(\text{READ} \Leftrightarrow count' = count + 1)$
(Progr Event)	$A_i :\Leftrightarrow \Diamond(count' = i \land A)$
	(for all program events A)
(Progr Var)	$f(v_i) := \Diamond \exists_{val}(count = i - 1 \land count' = i$
	$\land v = val \land f(val))$
(Number Cycles)	$N := sup\{n \mid \text{READ}_n\}$

4.2 The Physical Environment

Now, we introduce the notation and intended meaning for the physical variables, the definition of auxiliary physical variables and the set of assumptions on them. Together, they may be seen as the formula Φ_0 describing the set of visible event sequences from the point of view of the physical environment.

Notation for the Physical Variables.

(X: Pumps*)* $\tilde{X} := Pump\_Controller(X)$: Pump_Controllers
($\tilde{X}$: Pump_Controllers*)* $value(\tilde{X})(t) = 1$ *iff water is flowing (else 0)* : Bool
(steam) $v(t) = value(\mathbf{s})(t) = steam\ output$: $\mathbb{R}$
(water-level) $q(t) = value(\mathbf{l})(t) = water\ level$: $\mathbb{R}$
(valve) $Vv(t) = valve\ output$: $\mathbb{R}$
(X: Physical_Units*)* $real\_failure(X)(t)$: Bool

Definition of Auxiliary Physical Variables.

(Phys Count) $Count^-(A^\Phi)(t) := |\{t_1 \mid 0 \leq t_1 < t \wedge A^\Phi(t_1) = 1\}|$
         (for all physical events $A^{\overline{\Phi}}$)
(Observation Points) $T_i := sup\{t \mid Count^-(\text{READ}^\Phi)(t) < i\}$
(Value of Pump X) $value(X) :=$
         $Count^-(\text{OPEN\_PUMPS}^\Phi \vee \text{CLOSE\_PUMPS}^\Phi)\ mod\ 2$
(Value of Valve) $value(v) := Count^-(\text{VALVE}^\Phi)\ mod\ 2$
(Past values) $past\_pump(X)_i := \int_{T_{i-1}}^{T_i} value(\tilde{X})(t)dt$
         $past\_pumps_i := \sum_{X:Pumps} past\_pumps(X)_i$
         $past\_valve_i := \int_{T_{i-1}}^{T_i} Vv(t)dt$
         $past\_steam_i := \int_{T_{i-1}}^{T_i} v(t)dt$

Assumptions on the Physical Environment.

(Ass_Fail_1)   (Failure and sensor. X: **l**, **s** or Pump_Controllers)
         $real\_failure(X)(T_i) = 1$
         $\Leftrightarrow \text{READ}^\Phi(X, val)(T_i) = 1 \wedge |value(X)(T_i) - val| \geq \epsilon$
(Ass_Fail_2)   (Failure and actuator. X: Pumps)
         $real\_failure(X)(t) = 1$
         $\Leftrightarrow (\text{READ}^\Phi(X, val)(t) = 1 \wedge val \neq value(X)) \vee$
         $(value(X)(t) = 0 \wedge value(\tilde{X})(t) = 1) \vee$
         $(\forall_{t-\Delta^{Pump} \leq t_1 \leq t} value(X)(t_1) = 1 \wedge value(\tilde{X})(t) = 0)$
(Ass_Fail_3)   (Persistensy)
         $real\_failure(X)(T_i) = real\_failure(X)(T_{i+1}) = 0$
         $\Rightarrow \forall_{t:T_i \leq t \leq T_{i+1}} real\_failure(X)(t) = 0$
(Ass_Level)   (Dynamics of the water level)
         $q(t) = q(t_0) + \int_{t_0}^{t} p(t)dt - \int_{t_0}^{t} v(t)dt - \int_{t_0}^{t} Vv(t)dt$
(Ass_Steam)   (Dynamics of the steam)
         $Count^-(A)(\text{PROGRAM\_READY}^\Phi) \Rightarrow v(t) = 0.$
         $v(t) \in I \Rightarrow v(t + t_1) \in G(I, t_1)$
         $v(t) \in I \Rightarrow \int_{t}^{t_1} v \in F(I, t_1 - t)$
(Ass_Valve)   (Dynamics of the valve)
         $(Vv(t) = 0 \Leftrightarrow value(\mathbf{v})(t) = 0) \wedge$
         $(Vv(t) \geq \epsilon \Leftrightarrow value(\mathbf{v})(t) = 1)$

4.3 The Real-Time Operational Environment

The assumptions Ξ_0 on the real-time environment are very simple: the sensors are read at times $T_i = i \cdot T$ and the notions of cycle are the same for the program and for the environment actions. (For the definition of *(No-trans-failure)*, see Subsection 4.5.

Assumptions on the Real-Time SW- and HW- Environment.

(Ass_Cycle_1) $\forall_i \, (T_i - T_{i-1} = T)$
(Ass_Cycle_2) *(No-trans-failure)* $\Rightarrow$ READ$_i(X, val) \Leftrightarrow$ READ$^\Phi(X, val)(T_i)$
(Ass_Cycle_3) *(No-trans-failure)* $\Rightarrow$ E$_i \Leftrightarrow \exists_t \, T_i \leq t < T_{i+1} \wedge E^\Phi(t)$
(for all visible program events E = E(param)
and their corresponding physical event E^Φ)

4.4 The Design Constants

The following assumptions on the design constants are reasonable:

(Constants_1) $T \times |\mathsf{Pumps}| \times P < l/2$
(Constants_2) $|\mathsf{Pumps}| \times P > W$

where $l := \min\{N_2 - N_1, M_2 - N_2, N_1 - M_1\}$. (Instead of *(Constants_1)*, we actually only need a weaker version).

4.5 The Real-Time Scenarios

The scenarios are further assumptions on the physical environment that are needed to prove certain hybrid properties. They are not implied by $\Phi_0 \wedge \Psi_0 \wedge \Xi_0$.

Automata may also be used to express properties of *trajectories*. A trajectory satisfies the specification given by the automaton if the order of events of the trajectory is accepted by the automaton. We use this method to describe the scenario *No transmission failure rest*. An automaton $\mathfrak{A}$ on actions $A_1^\Phi, A_2^\Phi, \ldots$ describes a set of trajectories with discrete support $\mathbf{f}(t) = (A_i^\Phi(t))$ as follows:

$$\mathbf{f} \models \Phi(\mathcal{A}) \text{ iff } order(\mathbf{f}) \models \mathfrak{A},$$

where $order(\mathbf{f}) = (\gamma_0, \gamma_1, \gamma_2, \ldots)$ is the trace defined (inductively) by the order of events in $\mathbf{f}$. More precisely, γ_0 is the first set of events that happen in $\mathbf{f}$, and if γ_k and T_k are already defined then

$$T_{k+1} = \inf\{t > T_k \mid \exists_i A_i^\Phi(t) = 1\}$$

and γ_k is the set of events A_i^Φ such that $A_i^\Phi(T_{k+1}) = 1$. (The variable i ranges over $\{1, 2, 3, \ldots, N\}$, where N is the number of READ events.)

The automaton *No-trans-failure-order* describes the correct *order* of the messages from the physical units. From the assumption *No-trans-failure-read* follows that they happen at the correct *time*.

Scenarios

(No pump failures) $\quad \forall_{X:Pumps \uplus Pump\_Contr} \forall_t real\_failure(X)(t) = 0$
(No steam failure) $\quad \forall_t real\_failure(\mathbf{s})(t) = 0$
(No water-level failure) $\quad \forall_t real\_failure(\mathbf{l})(t) = 0$
(Single pump failure) $\quad \forall_{X:Pumps} \forall_i (\forall_{T_{i-1} \leq t \leq T_i} \neg real\_failure(X)(t)$
$\qquad\qquad\qquad\qquad \vee \forall_{T_{i-1} \leq t \leq T_i} \neg real\_failure(\tilde{X})(t))$
(No-trans-failure-read) $\quad \forall_i \forall_X \exists!_{val} \text{READ}^\Phi(X, val)(T_i) = 1$
(No-trans-failure-order) $\quad$ (defined by an automaton).
(No-trans-failure) $\quad$ (*No-trans-failure-read* $\wedge$ *No-trans-failure*).

5 What Hybrid Properties Hold?

Given three sets of formulas Φ, Ψ and Ξ as we have (with or without scenarios), there are several ways of reasoning about traces and trajectories. One possibility is to "lift" the formulas Φ to discrete formulas over the traces using Ξ. This is possible, and in fact, we have done it here. This corresponds to describing the possible "macro" transition steps of the physical system. But sometimes it is convenient to "push" down the formulas Ψ to continuous formulas and argue with explicit time. In some cases we found this second method much easier, particularly since we want to show properties of the trajectories.

5.1 Positive Results

The first properties state that the program variables indeed approximate the physical quantities. In the case of no failures they are the basic induction step to show a coupling invariant between program and environment.

- Assuming single pump failure, it holds: $past\_pump(X)_i \in past(X)_i$, where $past(X)_i$ is the interval calculated by the program and $past\_pump(X)_i$ (Section 4.2) is the real amount of water that flew into the steam boiler through pump X.
- $v(T_i) \in adjusted(\mathbf{s})_i \Rightarrow v(T_{i+1}) \in G(adjusted(\mathbf{s})_i, T) = calculated(\mathbf{s})_{i+1}$
- $q(T_i) \in adjusted(\mathbf{l})_i \Rightarrow q(T_{i+1}) \in calculated(\mathbf{l})_{i+1}$

As defined here, not all transmission failures may be detected. (The program may not make sure that the cycles do start at $T_i = i \cdot T$. Even if the controller used a clock, a *clock* failure could be confused with a transmission failure). But

- if the cycles do start at the times $T_i = i \cdot T$, then any transmission failure will be detected and vice-versa, any detected failure really happened.

The next property states that the *first* failure will be detected, but a second failure on one pair $pump_i/pump\_controller_i$ may remain unrecognized.

- If no emergency condition happens and assuming single pump failure, then a failure on $\text{pump}_k/\text{pump\_controller}_k$ will be detected. And vice versa, if a failure in $\text{pump}_k/\text{pump\_controller}_k$ is detected, then a real-failure in the pair happened.

The next two properties state that the program does not "detect" failures that had not happened.

- Assuming single pump failure, if a steam/water-level failure is detected, then indeed, a steam or water-level failure has happened.
- If a *real-failure* never happens (including no transmission failure) after STEAM_BOILER_WAITING$^\Phi$, the initial mode is eventually left, the next state is normal, and this state is never left again.

We sketch the proof of the last property: since any detected failure indeed witnesses a *real-failure*, and we assume no failures, we have no detected failures. The crucial part of the proof is to show that if STEAM_BOILER_WAITING$^\Phi$ has been received, then the read values for the water level will be eventually in the interval $[N_1, N_2]$. When this happens, and since in this transition $adjusted'(1) = \text{READ}^{\Phi\prime}(1) \in [N_1, N_2]$ then the valve and pumps will be closed and a PROGRAM_READY$^\Phi$ will be emitted. Since a PHYSICAL_UNITS_READY$^\Phi$ is expected, either it happens or an emergency situation is found. To prove the claim first suppose that the water level value is less than N_1. First observe that the water level may never increase by more than $T \times |\text{Pumps}| \times P$ on one cycle (of duration T). Now, if all pumps are open for at least a time δ^{Pump}, if there is no steam and the valve is closed, then the real water level value increases by $T \times |\text{Pumps}| \times P$. If this number is less than $N_2 - N_1$ (this design decision is assumed) then, sometime a value between N_1 and N_2 will be reached. A similar argument shows that if the initial water-level value is greater than N_2, then READ$^\Phi(1)$ will be less or equal to N_2, i.e., in $[N_1, N_2]$ or less than N_1, returning to the case already discussed.

5.2 Negative Results

One problem detecting failures via plausibility checks is that if the sensor readings of several physical units are incompatible with each other, with the history so far, and/or with the dynamics of the system, it is still not clear which sensor value is incorrect. A typical solution is to arrange the physical units in groups in a certain pre-order, for example: pumps $\prec$ steam $\prec$ water level. A failure of one unit (or group) may be identified independently of the sensor values of the units which follow in that order. This is partly the case here: a failure of the pumps/pump controllers may (and must) be identified without using the values of the steam or water-level units.

Ideally, then the values for the steam are analyzed (and compared say to temperature and pressure), and then the values for the water level may be estimated and compared to the sensor values. We think that some extra sensor values (for example pressure and temperature, or an independent measure of the steam) are missing in this system. As a consequence, this steam-boiler example is not very realistic since failures in the steam and in the water-level units can not be properly discriminated (and this does not make much sense in view of the specification text).

- Pump and pump-controller failures may be confused (This is not too disturbing: still, a water-level interval may be safely calculated.) If we assume a "single-failure-scenario", it is possible to identify any failure on a pump/pump controller, although it does not seem feasible to clearly distinguish a pump failure from a pump-controller failure.
- Even very frequent steam failures may pass unadverted or are not properly identified. (Only in some special cases steam and water-level failures may be distinguished, thus, the failure management for the steam and water-level measuring-units in this specification is clearly insufficient). For example, if the steam sensor after reading a correct value v_0 starts sending the constant value v_0 (independently of the correct values of steam), this failure will remain unnoticed, since the sequence of values (v_0, v_0, ...) is always compatible with the dynamics of the steam. If in this situation a failure is identified (at all), it cannot be distinguished from a failure on the water level. (Proof: construct two identical sequences of sensor values, one with the assumption of a real failure only on steam unit, and the other with a real failure only on the water-level unit).
- In general (if the scenario *single pump failure* does not hold) pump and pump-control failures may remain unnoticed.
- A pump/pump-controller failure may hide a later level failure. Even without a pump/pump-controller failure, the uncertainty whether a pump is open or not may hide a level failure, at least for some time. It is therefore important to assume, that the times for pump-response are small, compared to times to fill from say N_1 to N_2.
- After one steam error has happened (and has been detected), a water-level error may remain totally unnoticed, even if the water level reaches its limiting values. Assume for example that a steam failure has been (correctly) detected. After each cycle, there is less and less information about the correct values of the steam, therefore, after some time *any* value of steam (between 0 and W) is possible. Now, assume that the water level unit fails but starts sending a constant value q_0. The water level values are then compatible with the dynamics of the system (independently of the state of the pumps or pump controllers: the amount of steam may be compensating exactly what is flowing in). Therefore, the water level may be reaching one of its two limit values, but this (realistic) failure will never be noticed.
- In general, it is possible that a water-level failure remains unnoticed, even if all other physical units are working correctly, and even if the physical values of the water level reaches its limiting values M_1 or M_2 (How realistic such an example is depends on the values of the static parameters W, P, U and T). One example is the following: assume that the steam and the water level

units are sending constant values W and q_0 respectively, and that all pumps are open. Let us assume that the pumps were designed to exactly compensate the maximal amount of steam. That is, if all pumps are open and the steam is constant at its maximal value, the water level remains constant. (Many other scenarios are easy to construct. This assumption makes the example easier to follow). Then, the values (W, W, ...) and (q_0, q_0, ...) will not indicate any failure. But, perhaps, in reality, the water level is increasing constantly; the steam values are correct, but the water level values are not. This can be happen, for instance, if the steam continuously oscillates and reaches a maximal value of W each time it is measured.

5.3 Improving the design

One remaining question: is it possible to design the system such that in many *realistic* scenarios any long-lasting real-failure will be noticed (with a high probability) *before* the water level reaches a critical value? As we have seen before, it is not always possible to distinguish pump from pump-controller failures or steam from level failures. Also, it is not always possible to identify *second* steam/level failures (that is, a level failure that occurred before a level failure has been successfully repaired, or vice versa). Nevertheless, under some assumptions the answer is yes. Let us assume that the values of the static parameters are such that if all pumps are open for a sufficiently long time, the water level has to be increasing, and if all pumps are closed for a sufficiently long time, the water level *has* to decrease (or else the controller program obtains from elsewhere the information that the steam boiler is *off*). Assume further that any "*realistic*" real failure in the water-level unit either sends a value outside of the possible range $[0, Q]$, or sends continuously an (almost) constant value q_0, or sends monotonic (increasing or decreasing) values, or sends (seemingly) random values in $[0, Q]$ (with any fixed distribution). Further, assume that when a REP$^\Phi$ message is received then, at that moment, the corresponding value of the physical unit is correct. Under those assumptions it is possible to design a program with the desired property. The strategy includes: if any failure is not repaired within a fixed time, go the emergency mode. The static values of the parameters must be chosen such that the calculated intervals for water level are *much smaller* than the possible values (of $[0, C]$).

Some other remarks to improve the design of the system are: first, extra sensors in a hierarchical pattern are missing, as described in Section 5.2. Second, some other conditions should lead to emergency: when many pumps are defect, then the interval *adjusted*(l) is too wide (say, if $|$ *adjusted*(l) $| \geq (N_2 - N_1) - T \times No\_Pumps \times P)$. With these assumptions one may prove that after STEAM_BOILER_WAITING$^\Phi$, the initial mode will be eventually left (independently of failures).

6 Conclusion

Many methods have been proposed to study hybrid systems. Some of them integrate two different levels of abstraction, one for the transitions of the program and one for the real-time evolution of the physical system. (See for instance the TLT-Duration Calculus proposal in [RCM+95]). This seems quite reasonable: it is much easier to reason about abstract programs, which perform "macro-steps" expressed as a set of constraints and postpone the details of the analysis of the real-time behavior, which can be carried out at another level.

References

[CBH96] Jorge Cuéllar, Dieter Barnard, and Martin Huber. A Solution relying on the Model Checking of Boolean Transition Systems. In *The RPC-Memory Specification Problem*, to appear in LNCS, 1996.

[KLR90] Mark H. Klein, John P. Lehoczky, and Ragunathan Rajkumar. Rate-Monotonic Analysis for Real-Time Industrial Computing. *Computer*, pages 24 – 33, January 1990.

[Lam93] Leslie Lamport. Hybrid Systems in TLA+. *Eds. Grossman, Nerode, Ravn, Rischel*, 736:77–102, 1993.

[Lam94] L. Lamport. The Temporal Logic of Actions. *ACM Transactions on Programming Languages and Systems*, 16(3):872–923, May 1994.

[RCM+95] Hans Rischel, Jorge Cuéllar, Simon Mørk, Anders P. Ravn, and Isolde Wildgruber. Development of Safety-Critical Real-Time Systems. In *Proceedings of the SOFSEM '95, M.Bartosek, J.Staudek, J.Wiedermann (Eds)*, volume 1012 of *LNCS*, November 1995.

[Son90] E. D. Sontag. *Mathematical Control Theory*. Springer, 1990.

Specifying and Verifying the Steam Boiler Problem with SPIN

Gregory DUVAL, Thierry CATTEL
Laboratoire de Téléinformatique, Ecole Polytechnique Fédérale
CH-1015 Lausanne, Switzerland, {cattel,duval}@di.epfl.ch

Abstract - This paper reports the results of specifying and verifying the Steam Boiler problem with Promela/SPIN. Several models of the system have been produced with different degrees of completeness. Each model represents an abstract level for capturing the original problem requirements. The last model is very detailed and gives a first solution to the steam boiler problem. The model is able to drive the system and takes device failures (pumps, pump controllers, steam and water) into account. Liveness and safety properties have been successfully checked on the models to insure that the system behaviour is correct. An implementation of the system has been made using Synchronous C++, a concurrent extension of C++, and linked with the TCL/TK simulation. A presentation of future evolutions of the system is also described. This application shows that SPIN is quite appropriate for developing control process problems from specifications.

Keywords - application, process control, LTL properties verification, concurrent programming.

Introduction

Though SPIN[3,4,5] was specifically designed for checking protocols, it appeared that it was also quite suitable for addressing other problems such as distributed algorithms and multiprocessor operating systems [2]. Process control systems may be seen as particular protocols and be verified as such. Furthermore, SPIN is well adapted for checking liveness and safety properties. Our objective was to design a system on which to check some properties and then to derive a straightforward implementation of the detailed Promela specifications into Synchronous C++ [1] a concurrent extension of C++ developed in our Labs, that will soon be integrated in Gnu distribution.

1. The Steam Boiler problem

The problem (see Chapter AS, this book) is to control the level of water in a steam boiler by using different kinds of devices. The water level has to be maintained between two limits (N1 and N2) and can not pass over two limits (M1 and M2) for more than five seconds. The system has to work correctly because the quantity of water present has to be neither too low nor too high. Otherwise the steam-boiler can be seriously damaged. The physical system comprises the following devises :

- The steam-boiler
- A device to measure the quantity of water in the steam-boiler
- A device to measure the quantity of steam which comes out of the steam-boiler
- Four pumps to provide the steam-boiler with water.
- Four pump-controllers to insure that all the pumps are working correctly
- A valve for evacuation of water from the steam-boiler. It is only used in the initialization phase.
- An operator desk to manually stop the system

The system has to take failures of the devices into account. All Pumps, Pump controllers, Steam level sensors and Water level sensors can fail and so the system has to adapt its behaviour by taking these failures into account. Because the valve is only used in the initialization phase we consider that it can not fail.

Although we tried to follow the informal description as best as we could, we still had to make some design decisions due to incompleteness and ambiguities of the original text. These decisions concern mainly the following items :

* We did not understand from the problem description how to distinguish between failures of the pumps and failures of the pump controllers. We have therefore identified these two error conditions by assuming that a pump does not need time to start pumping water into the boiler.

Pump Status	Pump State	Controller State	Flow	Failure Code
off	off	off	NO	OK
off	off	on	NO	Controller
off	on	off	NO	Pump
off	on	on	YES	Pump
on	off	off	NO	Pump
on	off	on	YES	Pump
on	on	off	YES	Controller
on	on	on	YES	OK

* The problem description gives no idea of how to decide how many pumps shall be opened or closed in one cycle. We give a solution for this.

* In Emergency mode we assume that the system should not do anything but terminate its execution.

The system communicates with the physical units through messages transmitted via communication channels. Time for message transmission is supposed to be zero. The program follows a cycle which consists of the following actions :

* Reception of messages coming from the units
* Analysis of information which has been received
* Transmission of messages to the physical units

2. Promela/SPIN

Promela/SPIN [3,4,5] is a generally distributed automated verification system that is slowly evolving into an academic and industrial standard for on-the-fly LTL model checking. The two main advantages of the tool are that it is firmly founded on formal automata theory, and it can handle applications of full-scale industrial size.

The description formalism Promela allows expression of a concurrent or distributed system as a set of independent processes (*proctype*) communicating through synchronous (rendez-vous) and asynchronous channels (*chan*). The data types available are restricted to simple types such as integers and boolean and may serve for building richer types with structure and array constructors. Promela's syntax is very close to that of C and of CSP. There are also extra statements for defining *atomic* action sequences. It is possible to specify logical assertions (*assert*) inside the models and also more general linear temporal requirements applied to state sequences. Some labels are provided for marking the system states in order to track deadlocks (*end*) or cycles without progress (*accept, progress*).

The SPIN tool is composed of a simulator, a model-checker and a graphical debugger. More interesting is the model-checker that walks through systems of several millions states. If an error is found, a trace is stored that guides the simulator for a diagnosis session. SPIN is provided with two optimization algorithms that insures its great efficiency : one for random-walk searches and another one for partial-order reduction techniques. The tool also provides the facility to express LTL[6] properties.

3. Design & Architecture

The final model of the system was produced in several steps. First, we considered a failure free system and then we have progressively added the failure management (pump failures, steam and water failures).

3.1 Architecture representation

In process control applications, we may separate processes into two different types. Each kind of process has a particular role as follows :
- The Environment objects are used to model the physical units of the system (pumps, controllers, ...).
- The Controller uses input data (levels, pumps states, ...) to calculate any orders which have to be sent to the pumps

To design the architecture of the Steam Boiler system, we have used a tool called Software Architect's Assistant. Using the following formalism, we progressively built the architecture of the system.

This formalism is called Darwin and allows the design of concurrent systems using processes (components) and communication points (read or write). For example, a system composed of two processes communicating may be represented as follow :

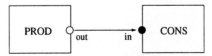

FIGURE 1. Producer - Consumer architecture

A direct mapping between a Darwin component and a Promela process may be done. Therefore, an easy translation can be realized from a graphical Darwin architecture to a Promela model. For example, the corresponding Promela code for the Producer-Consumer may be written as follow :

```
chan PC = [0] of {bit};
proctype PROD (chan out) {
bit v;
do
:: produce(v);
   out!v;
od; }
proctype CONS (chan in) {
bit v;
do
:: out?v;
   consume(v);
od; }
init {
run PROD(PC);
run CONS(PC);
}
```

All the models in the following sections will be presented using the Darwin formalism and some corresponding parts of the Promela code will also be shown.

3.2 Physical units architecture

In the First Steam Boiler model (Fig. 2), all the physical unit are modelled (Valve, Water_Level, Steam_Level and Pumps). No breakdowns are taken into account so we do not specify pump controllers. Some processes have been used to manage the evolution of the boiler (water input and steam output). The system can run in two different modes (INITIALIZATION and NORMAL). With this model, we have checked the following properties :

- WL : When the system is either in NORMAL or DEGRADED mode then the water level must always be maintained between N1 and N2.
- IM : When the System is in the INITIALIZATION mode, the water level will eventually reach the valid range (N1, N2).

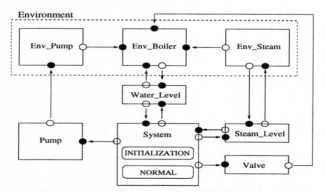

FIGURE 2. Architecture of the model without failures

The *Env_boiler* process is used to calculate the quantity of water in the boiler by using the quantity of steam which exits (*Env_Steam process*), the quantity of water which enters in (*Env_Pump*) and the quantity of water which may exits using the Valve (*Valve*). All communications are performed through synchronous channels.

In this model, we have three kinds of processes. Firstlyl, there is the system which controls and takes decisions. Secondly, there are all processes which simulate all the physical units like pumps, controllers, valve, steam sensor, water sensor. Finally, we have processes which simulate the evolution of the system. For example, the quantity of steam evolves (increase or decrease) according to the power which is used to heat the boiler.

3.3 System with Pumps and Controllers Failures

At this level, failures of pumps and pump-controllers are taken into account. Therefore, new processes are needed to model pump controllers (Fig. 3) and we need at least two pumps to allows failures for one of them. Pump Controller processes receive the state (OPEN, CLOSE) from their respective pump and send a control code to the System process which can be FLOW, NO_FLOW. The system will then compare states from pumps and states from pump controllers to detect eventual failures using the failure detection table (Table1). If an error is detected, the system goes into the DEGRADED mode until the defective unit is repaired.

We have also modelled the Stop Request in this model. We have added the process *Stop* which counts the number of requests which come from the user interface and when three stop requests have been sent it puts the system into Emergency mode.

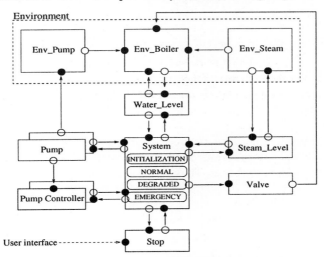

FIGURE 3. Architecture of the model with pumps and controllers failures

Many others properties can be checked on this model. We have the following requirements :

- WL :   When the system is either in NORMAL or DEGRADED mode then the water level must always be maintained between N1 and N2.
- IM :   When the System is in the INITIALIZATION mode, the water level will eventually reach the valid range (N1, N2)
- PF :   We never try to use a pump which is in a failure state unless it is repaired

- NM : When the system is in the Normal Mode no units are in a failure state
- PR : If a pump is working correctly and it receives an order from the system then it can either realize correctly the request or leads in a failure state
- PL: If mode is not EMERGENCY then the water level must always evolve, except in the INTIALIZATION mode.

3.4 System with Steam and Water Failures

This model is more detailed, it can manage pump failures, pump-controller failures, steam sensor and water sensor failures. Because water sensor failures are taken into account in this model we need to model the RESCUE mode. The system can now evolve according to the figure 4. It takes information which is coming from the physical units and by taking this information into account it can evaluate the new situation and send orders to the physical units.

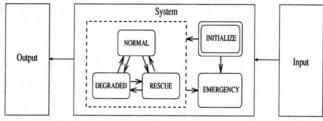

FIGURE 4. Running modes of the system

The architecture of this model is the same as the previous one. The only difference is that the steam and the water sensor units can now send bad values. Therefore the system has to control the validity of the steam and water values in order to detect failures.

3.4.1 Degraded mode

When a steam failure is detected, the steam which exits the boiler is evaluated by using the difference between the previous and the actual water level and the number of pumps which provide water to the boiler. Thus, we have the following estimation :

$$Steam = \left(\sum_{i=1}^{Nb\_Open\_Pump} p[i] \right) + \frac{pre(WaterLevel) - WaterLevel}{Dt}$$

with :
+/- : (+) : when Prev_Water > Water, (-) : else
Nb_Open_Pump : Number of pumps which provide water to the boiler
P : Nominal capacity of one pump (litre / sec)
Cycle : Quantity of time for one cycle (per sec.).

We need to know the quantity of steam which exits in order to decide how many pumps have to be open. This number of pumps is calculated as follows :

nb_pump = (Steam_Outcome / P)

3.4.2 Rescue mode

When a water failure is detected, the level of water in the boiler is evaluated at each cycle of the system by using the previously calculated values of water (qc), the amount of water provided by the pumps and the quantity of steam which comes out of the boiler. For this we have the following estimation :

$$qc = pre(qc) - Steam \times Dt + \sum_{i=1}^{N\_pump} p[i] \times Dt - V \times Dt$$

with :
Nb_Open_Pump : Number of pumps which provide water to the boiler
P : Nominal capacity of one pump (litre / sec)
Cycle : Quantity of time for one cycle (per sec.).
Steam : Adjusted value of steam which comes from the physical unit.
V : Valve Status (1 if open, 0 if closed)

The system cannot manage the boiler if both steam and water sensors are in a failure state. We have thus simplified the calculus proposed in (see Chapter A5, this book) for the system dependencies. When the water sensor is in a failure state the system estimates the water level by using the read steam value and when the steam sensor is broken, the quantity of steam which exits of the boiler is evaluated by using the read water level value. When the water sensor is working correctly the calculated water value is equal to the read value.

On this model, we still have checked the previous properties. Only a new property concerning the steam sensor unit is added. This gives the following requirements :

- SF : The system is always in Degraded mode when the steam unit or a pump or a pump-controller has failed and the water sensor is working correctly.
- WF : The system is always in Rescue mode when water unit is in a failure state.

3.5 System speed requirement

To avoid problems, the time between two cycles of the system (input, analysis and output) has to be evaluated. The shortest time is between two cycles the more the system is able to compensate for failures. For example, assume that time between two cycles is five seconds and we need to open a pump. We first try to open one of the four pumps which could possibly fail. We must wait for the next cycle to know if the order has been executed. If an error has occur, the system has to open an other pump. When the three first pumps fail, we need four cycles to compensate (Fig. 5). So the time needed to pass from N1 to M1 or N2 to M2 has to be at least equivalent to four cycles to insure the best control of the boiler.

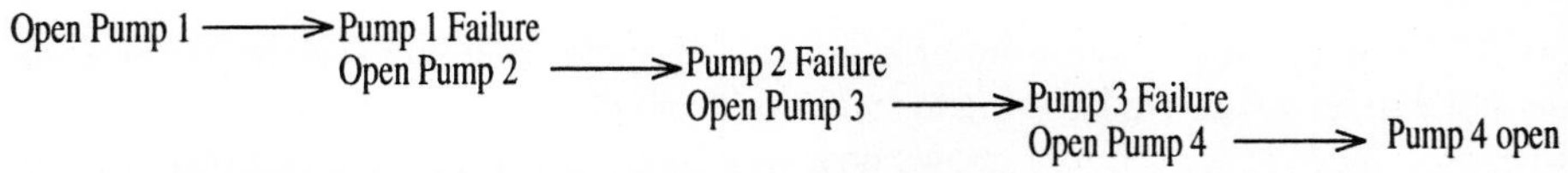

Open Pump 1 ⟶ Pump 1 Failure
　　　　　　　　Open Pump 2 ⟶ Pump 2 Failure
　　　　　　　　　　　　　　　　Open Pump 3 ⟶ Pump 3 Failure
　　　　　　　　　　　　　　　　　　　　　　　　Open Pump 4 ⟶ Pump 4 open

FIGURE 5. Critical scenario for pump failures

4. Modeling and Verification

4.1 Properties

All the properties which have been checked on models are expressed using Linear Temporal Logic. The SPIN tool allows us to use the Linear Temporal Logic to model these properties. SPIN will then automatically transform these LTL properties into Buchi automata and will include them into the model. The properties to be checked are as follows :

WL: $\Box((NORMAL \lor DEGRADED) \Rightarrow ((WaterLevel > N1) \land (WaterLevel < N2)))$

IM: $\Box((INITIALIZE) \Rightarrow \Diamond ((WaterLevel > N1) \land (WaterLevel < N2)))$

PF: $Fail[n] \Rightarrow ((Unused[n]) \; W(Repaired[n])) \quad$ n : pump indice

NM: $\Box(NORMAL \Rightarrow NoFail)$

PR: $\Box((NoFail[n] \land RcvOrder[n]) \Rightarrow \Diamond (ExecOrder[n] \lor Fail[n]))$

PL: $\Box((\neg EMERGENCY \land \neg INIT) \Rightarrow (WaterLevel \neq PreviousWaterLevel))$

SF: $\Box((SteamFail \land \neg WaterFailure) \Rightarrow (DEGRADED))$

WF: $\Box(WaterFail \Rightarrow (RESCUE))$

All these requirements have been successfully checked on the model to insure the correct behaviour of the controller. When a property is checked on the model, the SPIN tool returns a set of date which says whether the property is correct or not. Fig. 6 shows the SPIN report for the verification of the WL property. The information of most interest is shown in bold.

```
Search completed
(Spin Version 2.8.5)
Bit statespace search for:
        never-claim             +
        assertion violations    + (if within scope of claim)
        non-progress cycles     - (not selected)
        acceptance   cycles     + (fairness disabled)
        invalid endstates       - (disabled by never-claim)

State-vector 444 byte, depth reached 1872495, errors: 0
6.85181e+06 states, stored
7.71564e+06 states, matched
1.45675e+07 transitions (= stored+matched)
4.43317e+07 atomic steps
hash factor: 9.79433 (best coverage if >100)
(max size 2^26 states)

3.0902e+09    equivalent memory usage in bytes (stored*vector + stack)
2.68434e+08   actual memory usage
```
FIGURE 6. SPIN verification report

4.2 Description of the model

One of the objectives was to minimize the loss of cycle when the system passes from one mode to an other. We have made an interleaving diagram to optimize the changing of mode. The following figure shows how to pass from one mode to another without losing a cycle. With such a structure, when a failure is detected in the NORMAL mode, the appropriate action is immediately taken in the DEGRADED or RESCUE mode.

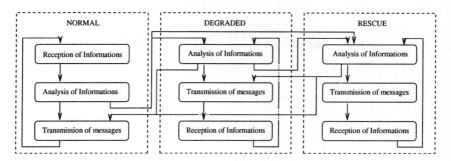

FIGURE 7. Sequence of running modes

The Promela language does not provide the possibility to structure the model using procedures. For this reason we have used C macros (#define) to structure our models. All of the important actions like (*OPEN_PUMP(i), CONTROL_PUMPS(error)*) have been described. Fig. 8 shows the C macro to check that all pumps and all controllers are working correctly. When a failure is detected, it sends the failure detection message and switches the system into the degraded mode.

```
#define CONTROL_PUMPS(error)\
i=0;\
do\
:: (i < N_pump) -> \
   if\
   :: (PUMP_FAILURE(i)) ->\
      PUMP_SEND_MESG(i, PUMP_FAILURE_DETECTION);\
      error=DEGRADED;\
      Sys_Pump_State[i]=FAILURE_DETECTED;\
   :: else;\
   fi;\
   if\
   :: (PUMP_CONTROLER_FAILURE(i)) ->\
      PUMP_CONTROL_SEND_MESG(i,PUMP_CONTROL_FAILURE_DETECTION);\
      error=DEGRADED;\
      Pump_Control_State[i]=FAILURE_DETECTED;\
   :: else;\
   fi;\
   i=i+1;\
:: (i >= N_pump) -> break;\
od
```
FIGURE 8. Pump and Controller failure detection

We present some of the modules of the controller to show how simple the SPIN solution is. The entire program is given elsewhere (see CD-ROM, Annex DC.1). Our actual boiler system has been designed as a single block process (Fig. 9). So each cycle is equivalent to a loop which contains a switch to choose the right mode which has to be executed. However in the future extension we will change this structure and divide it into several subprocesses as we will see in the next section. We can see at the beginning the stop request management which allows the system to stop when a STOP message comes from the user interface. At the end of the process, the Calculated_Q process is called to estimate the dynamics of the system.

```
proctype Boiler_system(byte pid; chan Pump_Order,
Valve_Order,Valve_Reply, Water_Level_Info, Water_Level_Reply,
Steam_Level_Info, Steam_Level_Reply, Stop_Line, Stop_Reply){
   byte P_state, F_state;
   byte  error, nb_open=0;
   bit stop_mesg = FALSE;
   do
   :: /***Environment evolves***/
   Env_Water_Sync!SYNC;
   Env_Water_Sync?k;
      /******Stop request******/
   Stop_Line!STOP;
   Stop_Reply?stop_mesg;
      /******System Cycle******/
   if
   :: (stop_mesg==TRUE) -> Mode=EMERGENCY;
   :: else;
   fi;
   if
   :: (Mode == INITIALIZATION) ->
      Initialization();
      End_Init:
      skip;
   ::(Mode == NORMAL) ->
      Normal();
      End_Normal:
      skip;
   ::(Mode == DEGRADED) ->
      Degraded();
      End_Degraded:
      Mode=error;
   ::(Mode == RESCUE) ->
      Rescue();
      End_Rescue:
      Mode=error;
   ::(Mode == EMERGENCY) ->
      skip;
   fi;
   Calculated_Q();
   od;
}
```
FIGURE 9. Boiler System Process

The mode DEGRADED is one of the most complicated since it has to take decisions by taking into account all failures. It also has to wait for REPAIRED messages from units which were in failure state and it has to check that no new failures occurred. Fig. 10 shows our actual structure of the degraded mode.

```
#define Degraded()\
     /*  Emission of Orders */\
```

```
/* steam failure Ack or Repaired */\
...
if\
:: (Steam_State==FAILURE_DETECTED)->\
   Steam_Level_Reply?STEAM_OUTCOME_FAILURE_ACK;\
   ...
fi;\
if\
:: (Steam_State==FAILURE_ACKNOWLEDGED)->\
   Steam_Level_Reply?STEAM_REPAIRED;\
   Steam_Level_Info!STEAM_REPAIRED_ACK;\
   ...
fi;\
...
/* iteration for N pumps and Controllers */\
do\
:: (i < N_pump) -> \
   /* pumps failure Ack or Repaired */\
   if\
   :: (Sys_Pump_State[i]==FAILURE_DETECTED) ->\
      PUMP_RECV_MESG(i,PUMP_FAILURE_ACK);\
      ...
   fi;\
   if\
   :: (Sys_Pump_State[i]==FAILURE_ACK) ->\
      ...
      PUMP_RECV_MESG(i,PUMP_REPAIRED);\
      PUMP_SEND_MESG(i,PUMP_REPAIRED_ACK);\
   fi;\
   /* Pump-Controlers Acknow or Repaired */\
   if\
   :: (Pump_Control_State[i]==FAILURE_DETECTED) ->\
      PUMP_CONTROL_RECV_MESG(i,PUMP_CONTROL_FAILURE_ACK;\
      ...
   fi;\
   if\
   :: (Pump_Control_State[i]==FAILURE_ACKNOWLEDGED) ->\
      PUMP_CONTROL_RECV_MESG(i,PUMP_CONTROL_REPAIRED);\
      PUMP_CONTROL_SEND_MESG(i,PUMP_CONTROL_REPAIRED_ACK);\
      ...
   fi;\
   i=i+1;\
:: (i >= N_pump) -> break;\
od;\
...
/* pumps, pump-ctrl. steam and water check */\
CONTROL_N_PUMP(error);\
CONTROL_STEAM(error);\
CONTROL_WATER(error)
```
FIGURE 10. Degraded Mode

At the end of the Degraded Mode, the system checks that no new failures have occurred. If a new failure is detected, it will stay in Degraded Mode even if acknowledgements for previous failures have been received.

5. Implementation

Our Promela model has been implemented in Synchronous C++ [1] and linked with the graphical simulation written by FZI in Tcl/Tk. The complete Synchronous C++ program is given elsewhere (see CD-ROM, Annex DC.2). On UNIX, commands are sent via *stdout*, and sensor values are read from *stdin*. In this implementation a particular process called *Sampler* regularly samples the environment and forwards the values of the sensors to the controllers. A graphical user interface with two buttons *Stop* and *Exit* allows the system to be put into an emergency stop state or the simulation to quit. This graphical interface also contains a text window where execution messages such as the current operation mode are displayed.

Synchronous C++, is an extension of C++ that is also very similar to Ada to given extents. The following source code (Fig. 11) shows a part of the implementation of the Boiler System. A process template is declared as an active class and instances are created with *new* as instances of usual C++ classes. Synchronous rendezvous are declared as methods of an active class. These correspond to Ada task entries. For awaiting events, the *accept* statement is used. This may be achieved on a local rendezvous but also with a call to a rendezvous of an other active object, in that sense Synchronous C++ is more symmetrical than Ada.

```
int Boiler::NORMAL(){
  accept GetBoilerState;
  if ((boilerstate.LEVEL > M2) || (boilerstate.LEVEL < M1))
    Mode = EMERGENCY
  else
    {
    if (CONTROL_N_PUMP()||CONTROL_STEAM())
      Mode=DEGRADED;
    else
      {
      nb_pump_needed=(int)(boilerstate.STEAM)/(int)(P);
      if ((boilerstate.LEVEL > N2)&&(boilerstate.LEVEL <= M2)
          &&(NB_OPENED() > 0))
        {
        i=NB_OPENED()-(int)(floor(nb_pump_needed));
        for(j=N_pump;j>0 && i>0;j--){
          if((boilerstate.PUMP_STATE[j]==OPEN)\
            &&(Sys_Pump_State[j]==OPEN))
          {
          PUMP_CLOSE(j);
          i=i-1;
          }
        };
        }
      else if ((boilerstate.LEVEL<N1)\
          &&(boilerstate.LEVEL>=M1)&&(NB_OPENED()<N_pump))
        {
        i=(int)(floor(nb_pump_needed)+1-NB_OPENED());
        for(j=1;j<=N_pump && i>0;j++){
          if ((boilerstate.PUMP_STATE[j]==CLOSED)&&
            (Sys_Pump_State[j]==CLOSED))
          {
          PUMP_OPEN(j);
          i=i-1;
          }
        };
```

```
            };
        };
    };
};
```
FIGURE 11. Normal Mode

Fig. 11 shows the NORMAL mode of the system. At the beginning it receives information from the physical units and then checks that the water level is correct. It calculates the number of needed pumps and sends apropriate orders to the pumps.

Some experiments have been made using our controller and it appears to drive the system correctly. We exercised the simulation for a number of hours without finding any problems.

6. Future extensions

The Steam boiler problem is still studied to be decomposed into several subprocesses as advocated in an other contribution (see Chapter LM, this book). As shown on Fig. 12, the system reads all the values which come from the physicals units and dispatches them to the required processes.

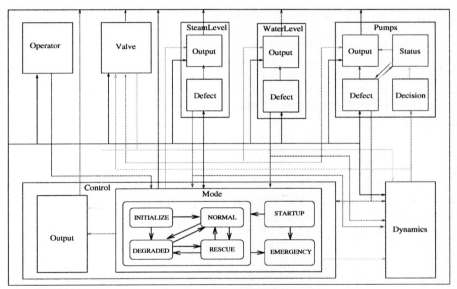

FIGURE 12. Decomposed architecture of the system

Each action is now managed by one process which computes intermediary results as follows :
- **Operator**: It waits for *NB_stop* stop events coming from the operator desk and sends a stop request to the *ControlMode* process if it occurs in a row.
- **Valve**: It opens the valve in Initialization mode if the water level is too high.
- **Steam**:This is composed of two processes. **SteamDefect** for the failure detections and **SteamOuput** for the emission of messages to the physical unit.

- **Level**: This is composed of two processes. **LevelDefect** for the failure detections and **LevelOutput** for the emission of messages to the physical unit.
- **Pumps**: This is composed of four processes. **PumpsDefect** detects the failures on pumps or Controllers. **PumpsDecision** determines how many pump should be providing water to the boiler. **PumpsStatus** calculates the new status of each pump and **PumpsOutput** sends messages to the pumps and controllers.
- **Dynamics**: This estimates the water and steam levels taking into account the possible water and steam failures and the values coming from the physical units.
- **Control**: This is composed of two processes. **ControlMode** manages the operation mode of the system by taking into account errors detected and the levels calculated. **ControlOutput** manages the protocol with the plant.

To simplify all models, transmission failures have not been taken into account although this could be added with the addition of an extra process. This process would just receive all values coming from the physical units and check them to detect any transmission failures. If no failures are detected, it could forward values to the system otherwise it could put the system in Emergency mode.

7. Evaluation and Conclusions

We attempt to answer the evaluation questions included in (see Chapter AS, this book) and then conclude.

1. The sections 1 and 3 comprise the requirement specifications. The solution includes specifications for the system components and the safety and the liveness of the system, but does not deal with the performance of the system.

1'. The requirement specification of the solution is presented in Linear Temporal Logic, from which SPIN will automatically produce Buchi automaton to realize the verification

2. The functional design has been made for the control program, the water level, the steam sensor, the pump actuators, the pumps and the drain specification. However it does not deal with the transmission failures.

2'. The functional design of the solution has been formally verified against the requirements specification using the model-checker SPIN.

3. The architecture has been designed in terms of dataflow using the Darwin tool. This has been done for the control program specification, for the message passing specification, for the steam sensor specification, for the water level sensor specification, for the pump actuator specification, for the pumps sepcification and for the drain specification. It does not deal with the transmission failures.

3'. The architectural design served for rigorously deriving the Promela skeletons. This is done manually but it could be automatized.

4. The solution comprises an implementation of a control program which has been linked to the FZI simulator. We also spent some time on the experimentation of the controller.

5. Only a few contributions use an automatic verifier or include an implementation of the controller (see Chapter DC, CD, WS, this book).

6. Regarding the other contributions, they all complement the SPIN solution to some extent. Some contributions deal with qualitative properties (performance, response time, ...) (see Chapter AL, HW, LW, RS, this book). Others rely on refinement (see

Chapter AL, CW1, VH, this book) which would be an interesting new possibility for the SPIN tool. Using refinement leads automatically to performing compositional verification, although compositional verification may be used without refinement.

7. In terms of effort, one person spent approximately one month to produce the solution. Maybe two or three weeks training should be sufficient to become able to produce such a solution.

8. A detailed knowledge of the used formalism is not necessary for understanding the proposed solution. An overview of the SPIN tool through a tutorial is enough for an average programmer to understand the solution in 4 to 6 days.

As a conclusion, we could say that SPIN is very well adapted for specifying process control applications and for verifying some properties. Different models were produced and many properties were verified at each step. The final model gives an adapted answer to the steam boiler problem. Since all communications are synchronous, the implementation can easily be realised in a concurrent programming language.

Concerning the time spent on producing the solution, we can estimate that it would take two people less than one month. But the most important thing is that the whole approach could be accessible to any engineer in two to three months.

Since all elements of the system may fail, fairness consideration have to be taken into account. However SPIN can not easily manage fairness concerns. No formal verification of refinement relationship between our different models has been checked as SPIN does not provide that possibility at the moment. We hope that SPIN will keep on growing, in particular toward verification of process equivalencies and fairness possibilities.

More details on this case study (reports, models, demos and code) may be found on http://diwww.epfl.ch/w3lti/

8. References

1. G. Caal, A. Divin, C. Petitpierre, Active Objects: a Paradygm for Communications and Event Driven Systems, Globecom'94, San Francisco.
2. G. Duval, J. Jullian. Modeling and Verification of the RUBIS micro-Kernel with SPIN. Proc. of SPIN Workshop 95, INRS-Telecom, Montreal, October 1995.
3. Holzmann G.J., What's new in SPIN version 2, AT&T Bell Laboratories, May 1995.
4. Holzmann G.J., Design and Validation of Computer Protocols, 512 pgs, ISBN 0-13-539925-4, Publ. Prentice Hall, (c) 1991 AT&T Bell Laboratories.
5. Holzmann G.J., Design and validation of protocols : a tutorial, Computer Networks, 25(9), April 93, pp. 981-1017.
6. Manna Z., Pnueli A., The Temporal Logic of Reactive and Concurrent Systems - Specification. Springer-Verlag, 1992.
7. Manna Z., Anuchitanukul A, ... STeP : the Stanford Temporal Prover. Department of Computer Science. Stanford University, California 94395.

TRIO Specification of a Steam Boiler Controller

Angelo Gargantini and Angelo Morzenti
Politecnico di Milano, Dipartimento di elettronica e Informazione, Milano, ITALY
{garganti, morzenti}@elet.polimi.it

Abstract
We specify a controller for a steam boiler starting from an informal descriptions of its requirements. The specification is formalized in the temporal logic TRIO and its object-oriented extension TRIO+. To obtain a maximum of abstraction and reuse we make the specification parametric with respect to all equipment and hardware features, and we avoid to impose any particular strategy in the management of the available resources and in the control of the critical physical quantities.

1 Introduction

Computers are finding increasing applications in the fields of the control of real-time and safety-critical systems (avionic systems, medical systems, plant control systems, etc.). The development of such systems requires appropriate well-structured methods to master their high complexity. A particular importance is ascribed to the specification phase, since very often the errors encountered during their development can be traced back to inaccuracies or ambiguities in the description of the requirements. It is therefore particularly important that requirements specifications be precise (to avoid ambiguities), formal and mathematically well founded (to allow mechanized support in their analysis) and transparent (to serve as a common reference and a means of communication among humans).

The present report presents the specification of a steam boiler controller proposed in [AS] as a benchmark to assess the adequacy of specification methods to cope with practical non-trivial time- and safety-critical systems. The specification is written in TRIO, a temporal logic with metric on time that is particularly well suited to the specification of real time systems. TRIO is a logical language and therefore it favors a descriptive style where properties, rather than procedures or mechanisms, are specified, and the requirements are stated abstractly, avoiding any unnecessary bias with respect to particular design choices or implementation strategies. TRIO is the result of a long term cooperation among industry and academia, and in recent years a specification, validation, and verification environment has been built around the language, to support the development of industrially-sized time critical applications. The environment includes the definition of TRIO+, an object oriented extension of the language that effectively supports the modularization and the reuse of specifications of highly complex systems.

The report is organized as follows. Section 2 contains the formal specification of the steam boiler: it is organized in subsections according to the modular structure of the TRIO+ classes describing the overall system. Section 3 briefly illustrates how the specification can be usefully employed in the subsequent validation and verification activities with the support of automated tools developed around the language, and discusses the notion of safety assessment of the specified system. Appendix 1 "(see

CD-ROM Annex GM.1)" provides a brief overview of TRIO: to make the succeeding presentation reasonably self contained we illustrate the syntax of the language and the definition of the used derived operators; for the sake of brevity, the main features of the TRIO+ language are just recalled, referring the interest reader to the literature.

2. The Specification

To facilitate understanding by the reader, the presentation of the steam boiler formal specification in TRIO+ will follow a top-down approach. First, in Section 2.1, we illustrate the main assumptions and choices that we took in developing our specification. This should provide a rational to help the reader in obtaining a clear overall picture and a general understanding of our specification. Then in Section 2.2 we illustrate the modular structure of the specification describing informally how the various aspects of the requirements are separated and located in the specification components. At this point the reader should have precise and exact expectations on what will be found inside the modules at the lowest level of the part-of hierarchy determined by the modular structure, those containing the TRIO axioms that formalize the requirements. The detailed presentation of TRIO axioms is in Appendices from 2 to 11. "(see CD-ROM Annex GM.2)"

2.1 Assumptions and Choices

For the sake of abstraction the description of the steam boiler in the informal specification document [AS] deliberately leaves undetermined, and thus open to interpretation, several aspects of the control strategy and of the criteria to be used in the interpretation of messages coming from the equipment. Furthermore, each adopted specification formalism provides a particular notation to characterize the desired properties of the specified system, and different ways to obtain a model by abstracting away from irrelevant details. Most of the remarks listed below will be discussed in more depth in the subsequent paragraphs where the specification is presented in complete detail.

Representation of time. The informal specification document describes the operation of the program in terms of a possibly infinite iteration of a cycle that take place each five seconds. It is also assumed that: data transmission among the controller and the equipment is instantaneous and all messages are emitted (and received) simultaneously; that during every cycle the program can receive messages, analyze them, and send (response) messages. We model all these assumptions by choosing for our specification a temporal domain consisting of a discrete set, e.g., the set of integers, where each instant is intended to represent one distinct cycle time for the control program. As a consequence of this choice, the control program appears to have instantaneous reactions times, which is clearly a simplification of reality but is consistent with the abstraction level of the informal specification document [AS]. The main advantage of this choice is that the temporal properties and requirements can be described by means of very simple and transparent TRIO formulas. The description of the steam boiler at a more detailed level is obviously possible by choosing a finer time granularity to represent time instants between consecutive program activations and inside each activation, but this would require to consider information regarding the Hw/Sw architecture of the implemented system and thus would involve the design phase, which we consider to be out of the scope of the present exercise.

Management of the pumps. The informal specification document [AS] does not describe any particular policy in the management of pumps (i.e., how to alternate the usage of the functioning pumps) and provides only a very simple criterion for the diagnosis of faults (i.e., how to establish that a pump and/or its controller is operating correctly). As a consequence we leave unspecified this choice when more pumps than necessary are available, by specifying only that, at any time, the controller must choose nondeterministically, among to functioning pumps, exactly those ones needed to cover the current requested throughput. Further refinement of the specification (or appropriate design choices) could specify a particular pump management policy (e.g., minimizing pumps wear by avoiding pump state changes, or balancing the load by alternating them as much as possible); in this case it would be necessary (and possible using the TRIO deductive system) to prove that such a policy is correct w.r.t. the high-level, nondeterministic specification of the present document.

Regarding the diagnosis of faults for pumps and pump controllers, we remark that even very complex and sophisticated diagnostic criteria cannot lead to absolute certainty on the effective state of the various equipment components if such criteria are based on comparisons among measures perceived through sensors and no assumption or estimation is made (as it happens in the document [AS]) on the availability and reliability of such measures coming from the sensors. In other terms, if *all* information coming from the field is equally subject to some uncertainty then all conclusions drawn from them based on comparisons or deductions (even though arbitrarily complex) cannot, in general, be absolutely secure. Keeping this remark in mind, for the diagnosis of pumps and pump controller faults we provide three sample criteria of increasing complexity, from the simplest one outlined in the [AS] document to two other, more sophisticated ones, where one considers the consistence between the state of each pump and that of its controller, or the probability of simultaneous faults. These more elaborate criteria can permit to improve the *average* effectiveness of the plant management but do not provide an absolutely error-free knowledge (and therefore control) of the plant state. These ideas will be illustrated and discussed in more depth in Appendix 7 "(see CD-ROM Annex GM.7)".

Another consequence of the above remarks on the reliability of the information coming from the field is that a primary property such as *safety* (which in our case can be stated as: the controller will always go into the *emergency stop* mode as soon as the water level reaches the minimal or maximal limit quantity M_1 and M_2) will be asserted (and could be proved) only under the condition that the water level sensor only breaks down in a "recognizable" way, i.e., its indications are correct if they are reasonable, i.e., inside the physical limits for the capacity, 0 and C. In section 3 we will also discuss the methodological implications of this approach in a correct and effective design discipline.

Operation during the rescue mode. For reasons related to the above remarks on the reliability of the information provided by the sensors, our specification prescribes, during the *rescue* mode, an operation of the plant that is more restrictive than that indicated by the informal specification document. When operating in the *rescue* mode, the controller abandons any information currently provided by sensors: it takes as reference the last useful value provided by the level sensor and evaluates the minimum time necessary for the plant to reach a dangerous condition considering it as out of control, i.e., it evaluates the minimum between the time needed to reach level M_1 when all pumps are closed with a maximal steam flow and the time needed to reach level M_2 when all pumps are open and no steam exits the boiler. If this time elapses without any event intervening that takes the controller out of the *rescue* mode, then it spontaneously goes into to *emergency stop* mode. The rationale for this stricter requirement on the controller behavior during the critical *rescue* mode is that reaching this mode is a symptom that something unexpected or unnoticed and potentially dangerous has happened, so the most prudent choice is not to rely on the sensors and actuators and work under the most pessimistic assumptions. Not surprisingly, this specification choice allows one to ensure the safety property under conditions that are particularly simple and easy to implement, as it will be shown in section 3. An alternative, less prudent plant operation during the *rescue* mode (such as the one described in the informal specification) would make the property of safety more difficult to obtain in practice, and also to prove formally as a property of the modeled system.

Management of the water level. Like the management policy for pumps, also the policy for keeping the water level within the prescribed limits is left undetermined in [AS]. An addendum to the informal specification simply suggests to open the pumps (without indicating the measure of such opening) if the water level is estimated to be below N_1, and to close them (again without indicating how much) when it is above the limit N_2. To obtain maximum generality we provide a framework where a variety of strategies for managing the water level can be described: at any time a quantity called "requiredThroughput" is defined as a non-negative real quantity whose actual value is determined by the adopted policy in water management. Then, in the present specification, we adopt a compromise between simplicity and effectiveness of the control policy, assuming that the control aims at keeping the water level as close as possible to the median level $(N_1+N_2)/2$, and consequently the pumps are opened (resp., closed) by a quantity which is essentially proportional (considering also the current estimated steam output) to the difference between the current and the desired water level. The definition of an optimum control algorithm for the water level falls in the area of control theory and is therefore considered out of the scope of the present exercise. As with many other features of the modeled system, the specification can be made parametric w.r.t. the policy for controlling the water level by means of the previously mentioned constructs of genericity and inheritance.

Errors in measuring. Every measurement is subject to some error, and the values obtained by the sensors for the water level and the steam flow can be no exception. The informal specification document [AS], however, does not mention possible inaccuracies in these measures. Consequently, we assume that the minimal and maximal limit quantities M_1 and M_2 for the water level are chosen in a conservative way as to account for any possible inaccuracy in the measurement of the controlled quantities, and thereafter we reason under the assumption that such measures are exact.

Modeling the environment. In the initial phase of requirements elicitation and formalization it is often very useful to model in the adopted formalism not only the device or system to be designed but also the environment where it will be put into operation when implemented. Therefore in our specification we model not only the controller but also the operator, the transmission system, the equipment, and the interactions among them. Then for the sake of brevity we mostly concentrate on the controller, since this component will be the actual object of the design activity. We point out, however, that the specification language can be usefully employed to describe relevant properties of the environment or of its interaction with the control program. In section 3 we provide an example thereof by describing a hypothesis on the functioning of the water level measuring device.

Parameters of the steam boiler plant. Several significant physical quantities of the equipment (such as boiler capacity, maximal and minimal limit and normal water quantity, maximal quantity and maximal gradient of increase or decrease in steam flow, nominal capacity of pumps) are mentioned in the informal specification document [AS], and the obvious fact that they may change from one plant to another is dealt with by indicating their value symbolically through suitable symbolic constants. Other parameters such as the number of pumps in the equipment, or the period of the operation cycle of the control program are probably assumed to be less likely to change and therefore are indicated as fixed values (there are 4 pumps and the cycle period is 5 seconds). To obtain maximum generality, flexibility and reusability of specifications (we could mention a "specifying for change" attitude) our specifications are generic and therefore parametric with respect to all the above mentioned quantities, which are subject to change due to the physical dimensioning of the various plants or are determined by design choices influenced by technological factors (e.g., the cycle period).

Description of the initialization mode. We found it convenient to distinguish three phases in the initialization mode (i.e., *waiting*, *adjustingWaterLevel*, *programReady*) to describe in a simpler and more explicit way the sequence of actions carried out by the control program when operating in this mode. To fully comply with the informal specification document [AS], however, this separation of phases is confined in an inner module of the controller component and does not emerge in the communication between the controller and the equipment.

2.2 Modular Structure of the Specification

Object oriented methodology comprises classical modularization criteria, therefore our specification is divided into modules according to the principles of encapsulation and information hiding, separation of concerns, maximization of intra-module cohesion and minimization of inter-module coupling (and hence of module interfaces). Moreover, the typical object-centered view (emphasizing the physical and logical components of the system) is blended with a more functional view (modules called *controller, management, diagnosis* are introduced) so that system functions can be described in a general and abstract way. As it often happens, the modular structure

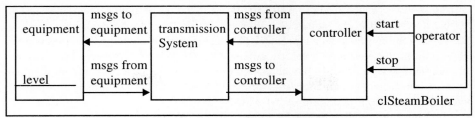

Fig. 1 Graphical representation of class clSteamBoiler

supports abstraction and reuse, but it also facilitates presentation and understanding, therefore the specification is highly structured, especially in the part regarding the pumps.

The highest level in the module hierarchy is shown in Fig. 1, representing class *steamBoiler* that includes modules for the system physical components (*equipment*, *transmissionSystem*, *controller*, and *operator*) and the connections among them consisting of the information exchanged and the delivered commands. In Appendix 2

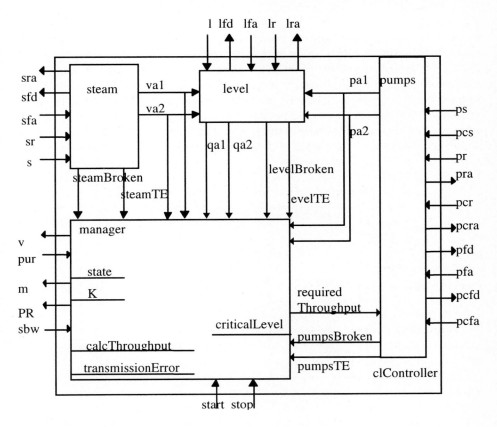

Fig. 2 The class clController

"(see CD-ROM Annex GM.2)" we report the detailed graphic representation and the textual declaration of the same class. It can be noticed that in the *equipment* module the local item *level* models the actual physical water level, which is in principle distinct from the measured level as perceived by the controller through the sensors. Modeling as separate entities the actual and the measured water level will allow us, in Section 4, to formalize some remarks on the reliability of performed measures and on the safety of the control algorithms based on them.

For brevity we do not model other features of the environment, and in the remaining parts of the present specification we focus on the controller, whose structure is shown in the graphical representation of class *clController* of Fig. 2, and whose textual declaration is in Appendix 3 "(see CD-ROM Annex GM.3)". This figure shows the components *steam* and *level*, which concern the measurement and control of the quantity of water in the boiler and of the exiting steam. Module *pumps* specifies control and management of the pumps. Module *manager* specifies the operation of the control program, as described in Section 3 of the informal specification document [AS], with all actions to keep the water level within the required bounds and to face sensor and actuator faults by operating in the *degraded* or *rescue* mode.

Fig. 3 depicts the clPumps: its textual version is in Appendix 5 "(see CD-ROM Annex GM.5)": it includes an array of modules, called *pumpSet*, containing instances of class *clPumps* in a number equal to the number of pumps actually present in the plant (notice that the specification is generic with respect to this number). The *pumpManager* component includes the specification of how to govern pumps, i.e., the indication of pumps opening or closing depending on the current required throughput and on the estimated current state of the pumps. Let us anticipate that the commands to the pumps, the diagnosis of their state, and the messages sent to the equipment are

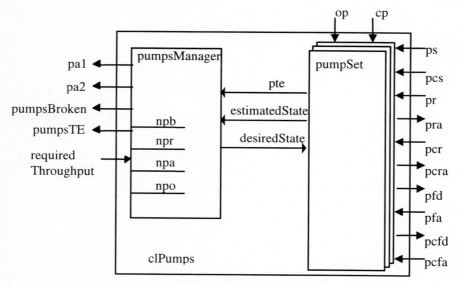

Fig. 3 The class clPumps

2.3 The *Pump* Module

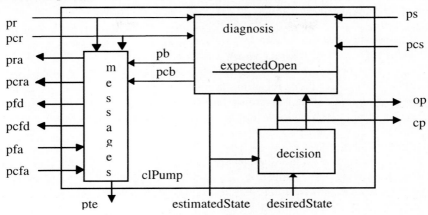

Fig. 4 The class clPump

Fig. 4 reports the graphical representation of the class *clPump*. The textual declaration can be found in Appendix 6 "(see CD-ROM Annex GM.6)". This class models a single pump of the plant, and there are many instances of it in the array of modules included in class clPumps (4 in the case considered by [AS]). The class is organized into three modules: module *diagnosis* specifies how faults of the pump or of its controller are determined; module *decision* characterizes the opening or closing commands to the pump according to its estimated state and to its desired state as determined by the pumpManager module; module *messages* defines the messages to be exchanged with the equipment regarding faults and repairs, in interaction with the diagnosis module.

Performing diagnosis on the state of the pumps and its controller is a crucial operation because the correct plant operation and control depends on the accuracy with which the actual state of the various devices can be estimated. For this reason we specify three possible ways of performing this operation. Here we report only the first one, that is simply a formalization of the criterion reported in [AS].

pumpDiagnosis:

$$pb \leftrightarrow \left(\begin{array}{c} UpToNow(pb) \wedge \neg pr \vee \\ expectedOpen \wedge ps(closed) \vee \\ \neg expectedOpen \wedge ps(open) \vee \\ \left(\begin{array}{c} UpToNow(\neg pcb) \vee \\ pcr \end{array} \right) \wedge \left(\begin{array}{c} ps(closed) \wedge pcs(open) \vee \\ ps(open) \wedge pcs(open) \end{array} \right) \end{array} \right)$$

pumpControlDiagnosis:

$$pcb \leftrightarrow \left(\begin{array}{c} UpToNow(pcb) \wedge \neg pcr \vee \\ expectedOpen \wedge pcs(closed) \vee \\ \neg expectedOpen \wedge pcs(open) \vee \\ \left(\begin{array}{c} UpToNow(\neg pb) \vee \\ pr \end{array} \right) \wedge \left(\begin{array}{c} ps(closed) \wedge pcs(open) \vee \\ ps(open) \wedge pcs(open) \end{array} \right) \end{array} \right)$$

We propose other two ways for pump fault diagnosis in Appendix 7 "(see CD-ROM Annex GM.7)". The complete specifications of *decision* and *messages* modules are respectively in Appendix 8 "(see CD-ROM Annex GM.8)" and in Appendix 9 "(see CD-ROM Annex GM.9)".

2.4 The *Level* and *Steam* Modules

The classes *clSteam* and *clLevel*, which we report in Appendix 4 "(see CD-ROM Annex GM.4)", specify operations similar to those described for the pumps by the class *clPumps*, i.e., operations regarding monitoring of the device state, exchange of messages with the equipment regarding faults, detection of transmission faults, and computation of estimated values for the water level and the exiting steam.

2.5 The *manager* Module

The class *clManager,* reported Appendix 11 "(see CD-ROM Annex GM.11)", specifies the module *manager* of Fig. 2 and describes general operations regarding the various modes of operation, the detection of transmission errors, and the government of the water level.

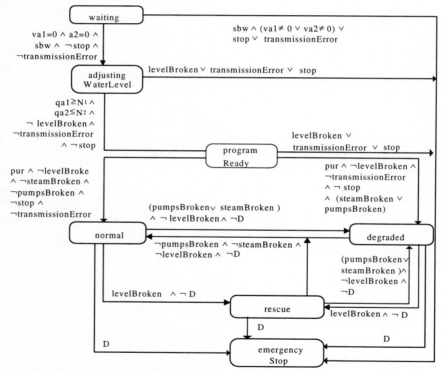

Fig. 5 Finite state automaton for the transitions among modes of the controller

We adhere to the description of the *modes* of operation employed in the informal specification document, therefore we model the principal structure of the control program as a finite state machine, whose states and transitions are represented in Fig. 5. The transitions of this automaton can be formally described in TRIO in a rather obvious and uniform way, which we just exemplify with this axiom:

normalToRescue

$$\begin{pmatrix} UpToNow(mode(normal)) \wedge \\ levelBroken \wedge \\ \neg D \end{pmatrix} \rightarrow \begin{pmatrix} mode(rescue) \wedge \\ Until(mode(rescue), \neg levelBroken \vee D) \end{pmatrix}$$

3. Use of the Specification

The obvious purpose of any specification is to express requirements and to serve as a reference for the successive phases of design, implementation, verification, and maintenance. Before these are undertaken, a very useful activity is often performed (especially when the specified systems are particularly complex or critical), namely the validation activity, which consists of establishing whether the actual requirements were indeed captured and correctly expresses by the specification. Formal methods, being based on a solid mathematical foundation, have a clear and unambiguous semantics, so that the validation and verification activities can be effectively supported by (semi)automatic software tools that can greatly enhance the effectiveness and the practical impact of such activities. This is the case with TRIO, where an environment of tools for editing specification, validating them and verifying design and implementation has been developed in recent years at Politecnico di Milano.

Broadly speaking, the validation activity can take in TRIO the form of *history checking*, *history generation* (i.e., *simulation*), and *property proving*.

When performing history checking [F&M94] the designer invents (with the aid of a suitable tool) histories of the modeled system (i.e., sequences of events, system configurations, and values for the significant quantities that represent a hypothetical trace of a system execution) that in his/her view correspond to a possible behavior of the specified system where the requirements and properties are apparent. For instance, possible histories of the steam boiler could include sequences of faults in the pumps, and the actions of the controller to deal with them, possibly in presence of particularly high (or reduced) steam production. Such histories are then checked, i.e., confronted for consistency with the specification, by considering each history as the frame of an interpretation structure for the TRIO formulas. The results of history checking are useful both to the final user, who verifies that his/her expectations on the system behavior are sensible, and to the specifier, who controls that his/his understanding of the requirements are correct and have been effectively formalized by means of the formal notation.

A more sophisticated method of validation consists of simulating the modeled system by generating (with the support of suitable specialized interpreters, see [MMM95]) histories of the specified systems under particular constraints that may represent an initial system configuration or particular combination of input events coming from the environment and are assumed to stress particular system functionalities that the designer wants to explicitly visualize.

The most complex, general, and effective validation activity is obtained by proving properties that are supposedly ensured by the requirements as expressed in the formal specification. From a logical viewpoint, as it happens in the case of TRIO, such properties are theorems that are derived in a theory consisting of the TRIO general axioms augmented with the axioms that are included the specification document. The derivation of such theorems can be made manually using the axiomatic system presented in [FMM94] or with the support of a theorem proover, such as PVS, where the TRIO semantics and axiomatic system have been suitably encoded, as it was done in [Jef95]. Typical properties that one would like to derive for a time critical system would be liveness, absence of deadlock, or the ability of the system to control the environment by maintaining invariant in time a given configuration or relation among components or physical quantities. As an example thereof, we would like to express in TRIO a property of (physical) safety that was never explicitly formalized in the specification presented in the preceding sections, but is clearly implied as the main purpose of the designed controller. As anticipated in the remarks of section 2.1, we state such safety requirement under suitable assumptions regarding the correct functioning of the measuring and transmission devices that the controller uses during operation. A first assumption is that the transmission system component is correct (although not necessarily permanently available), that is, any received data are equal to those transmitted. This assumption is easily formalized through the following simple TRIO formula

$$l(v) \rightarrow lE(v)$$

asserting that if a value is received by the controller then it is the same value that has been measured (and then sent) by the physical equipment.

The second assumed property regards the water level measuring device and asserts that when it is broken (i.e., its measurement is significantly different from the actual water level) then it gives a value that is out of the possible range of measures. This is formalized is TRIO as follows

$$lE(v) \wedge v\_equipment.level \rightarrow (v<0 \vee v>C)$$

or equivalently, and perhaps in a more intuitive manner, as

$$lE(v) \wedge 0 \leq v \leq C \rightarrow v = equipment.level$$

This assumption ensures that the measurements from the water level sensor are reliable, in that if they are incorrect then they are out of range and thus can be immediately recognized as such. Under the two above hypotheses the fundamental property of physical safety can be stated as follows.

$$(\ddagger)$$

$$\forall v \begin{pmatrix} (l(v) \rightarrow lE(v)) \\ \wedge \\ (lE(v) \wedge 0 \leq v \leq C \rightarrow v = equipment.level) \end{pmatrix} \rightarrow \begin{pmatrix} \neg(M_1 < equipment.level < M_2) \\ \rightarrow m(emergencyStop) \end{pmatrix}$$

A Formal Specification of the Steam-Boiler Control Problem by Algebraic Specifications with Implicit State

Marie-Claude Gaudel, Pierre Dauchy, Carole Khoury
LRI, Université de Paris-Sud & CNRS, Orsay, France

> This paper concentrates on the specification of the requirements for the Steam-Boiler Control System. It contains a detailed analysis and specification of the sent and received messages, the different states the system can be in, etc. Special emphasis is put on the analysis of checking the physical devices and detecting failures in the system environment.
>
> The paper uses an original approach that combines algebraic specification techniques with a notion of implicit state. State changes are achieved by so-called modifiers.

1 Introduction

This paper concentrates on the specification of the requirements for the Steam-Boiler Control System. It contains a detailed analysis and specification of the sent and received messages, the different states the system can be in, etc. Special emphasis is put on the analysis of checking the physical devices and detecting failures in the system environment.

The paper uses an original approach that combines algebraic specification techniques with a notion of implicit state. This approach follows the state-as-algebra paradigm introduced in [Gau 80], [Gau 83]. The changes of states are achieved by some so-called modifiers, which perform transformations on algebras.

This formalism aims at describing the states of a system and its dynamic behaviour in an algebraic framework [DG 94]. It was first introduced in a case study on a formal specification of the embedded safety part of an automatic subway pilot [DM 91], [Dau 92], [DGM 93], in order to cope with the large number of variables whose values made up the state of the system at a given time. A good analysis of the motivations of such formal approaches can be found in [EO 94]: the idea is to enrich algebraic specifications in order to make them more convenient for the description of dynamic behaviours. A comparison with similar works can be found in [DG 94] and [EO 94].

We have used the PLUSS specification language (the variant with positive conditional axioms) [BGM 89], [Bid 89], as a basis on which the formalism is built, but our approach is actually independent from the used specification language, as long as the semantics of a specification is a class of many-sorted algebras.

The paper begins with a brief survey of the implicit state approach (part 2). Then, we present the methodology followed for the development of the specification and provide a "road map" of the various specification parts which appear in the paper (part 3).

In part 4, we analyse the fundamental cycle of the system and elaborate successive sketches of its specification. In part 5, we complete the formal specification of the state and of the modifiers. In part 6, we present the specification of the failure detections and repairs of the physical units. In part 7, we give the specification of the detection of transmission failures. Then the specification of the changes of mode can be stated in part 8.

During the development of the formal specification, several questions arose on the informal specification and we had to make some choices and assumptions on the system. These questions, choices and assumptions are explained all along the paper. They are summarized in the conclusion: among the main interests of the development of such a formal specification is the discovery of such problems in the informal requirements.

2 A Brief Survey of the Implicit State Approach

In this part, we briefly present the formalism we use.

In this approach, the specification of a system is composed of four parts $<<\Sigma, Ax>$, $<\Sigma_{ac}, Ax_{ac}>$, $<\Sigma_{mod}, Def_{mod}>$, $Ax_{Init}>$ which we will present in sequence.

First, a classical algebraic specification $<\Sigma, Ax>$ describes the data types used by the specified system.

Second, there is a specification $<\Sigma_{ac}, Ax_{ac}>$ of some *access functions*; this specification is a persistent enrichment of $<\Sigma, Ax>$ with no new sort. The access functions are the observers of an *implicit state*; that is, they are syntactically analogous to normal operations, but their values can be modified. The specified state is thus a finitely generated ($\Sigma \cup \Sigma_{ac}$)-algebra. Among the access functions, some characterize the state of the system and are called *elementary*, while the others are defined in terms of them via the access axioms of Ax_{ac}.

The admissible initial states of the system are characterized by the set of axioms Ax_{Init}.

The specification of the elementary access functions automatically makes available to the specifier a set of corresponding *elementary modifiers* in the following way: given an elementary access function ac, with profile $s_1 * \ldots * s_n \to s$, the corresponding elementary modifier is $\mu\text{-}ac$ with domain $s_1 * \ldots * s_n * s$. For patterns (terms with variables) $\pi_1, \ldots, \pi_n$ of sorts $s_1, \ldots, s_n$ and a term t of sort s, the meaning of the statement $\mu\text{-}ac(\pi_1, \ldots, \pi_n, t)$ is a modification of ac. More precisely, it transforms a state (a ($\Sigma \cup \Sigma_{ac}$)-algebra) A into a state B such that:

- $ac^B(v_1, \ldots, v_n) = (\sigma t)^A$ if there exists a ground substitution σ (mapping the variables of the patterns into the ground terms) such that $v_1 = (\sigma \pi_1)^A, \ldots, v_n = (\sigma \pi_n)^A$;
- $ac^B(v_1, \ldots, v_n) = ac^A(v_1, \ldots, v_n)$ otherwise;
- access functions that depend on ac are updated according to Ax_{ac};
- all the other carriers and operations are unchanged.

For instance, if a specification of a state contains the elementary access function

$$\text{pump-failure}: \{1, 2, 3, 4\} \to \text{Bool}$$

then the elementary modifier $\mu\text{-}pump\text{-}failure$ is available and its domain is $\{1, 2, 3, 4\}*\text{Bool}$. The statement $\mu\text{-}pump\text{-}failure(1, true)$ leads to a state where *pump-failure(1)* has value true and *pump-failure(i)* is unchanged for i=2..4; the statement $\mu\text{-}pump\text{-}failure(PN,$

true) (where *PN* is a variable of sort {1, 2, 3, 4}) leads to a state where *pump-failure(i)* has value true for all values of *i* in {1, 2, 3, 4}.

All the access functions are visible outside the specification module, unless specified otherwise. Elementary modifiers are hidden: they are not visible outside of the specification module where the elementary access functions are defined.

In the third part <Σ_{mod}, Def_{mod}> of the specification, some *composite modifiers* which can possibly be exported are defined. The profiles of these modifiers have no range. The axioms in Def_{mod} are positive conditional and their preconditions are built on $\Sigma \cup \Sigma_{ac}$. They define the modifiers using statements built from the elementary modifiers and the following primitives:

- the statement **nil** corresponds to the identity on states;
- the semicolon stands for sequential composition;
- the construction **and** stands for indifferent composition; this means that the modifications connected with **and** can be done in any order (it is the responsibility of the specifier to make sure that all possible execution orders lead to the same state);
- the big dot • indicates modifications done on the same state; it means that all preconditions and arguments of the connected modifiers must be evaluated in the initial state prior to doing all the corresponding modifications.

We also allow a special form of conditional definition of a modifier:

$$\text{mod} = \textbf{cases } c_1 \Rightarrow \text{mod}_1 \mid \ldots \mid c_n \Rightarrow \text{mod}_n \textbf{ end cases}$$

This construction means that for all values matching the patterns involved in this expression and satisfying one of the conditions $c_1 \ldots c_n$, the corresponding modification is done. The specifier must make sure that if more than one condition is valid, all corresponding modifications lead to the same state.

For instance, given the following elementary access functions (which could correspond to messages to some pump in the specification of the boiler control system):

$$\text{to-pump: } \{1, 2, 3, 4\} \to \{\text{open, close, nothing}\}$$

and an access function q which gives the water level, it is possible to define the following modifier:

act-on-pumps = **cases**
$q \leq N1 = \text{true} \Rightarrow \mu\text{-to-pump(PN, open)} \mid$
$q \geq N2 = \text{true} \Rightarrow \mu\text{-to-pump(PN, close)} \mid$
$q \geq N2 = \text{false} \wedge q \leq N1 = \text{false} \Rightarrow \mu\text{-to-pump(PN, nothing)}$
end cases

In this definition, PN is a (rather simple) pattern: the elementary access *to-pump* is updated for all the values corresponding to a term matching the pattern, here all the pump numbers. Let us assume that this modifier is exported and is the only one, it means that it would be the only way to act on the specified system: for instance, it would be impossible to open or close one specific pump from outside the system.

Both for the **cases** construction and for plain conditional axioms, it is clear that if no substitution can make the condition(s) valid, no action is performed: the state remains unchanged. This constitutes a *frame assumption* and shortens the specification by avoiding to have to cover all cases with explicit axioms. Of course, it by no means alleviates the specifier's duty of considering all possible cases...

We must stress that this is still ongoing research and that a few rough edges remain to be polished.

3 Methodology and Description of the Specification Development

We use some aspects of the methodology for specification development which has underlied the PLUSS specification language [Gau 92]: it is basically an incremental approach, starting with some *sketches* of the future specification where some sorts, operations and properties are missing. It is not necessary to know all the details of this approach to follow this paper. It is sufficient to know that the first sketch is enriched until the specifier is convinced of having all the necessary sorts and operations. This stage is expressed by transforming the sketch into a specification module which has a different semantics from a sketch, namely a class of finitely generated algebras, and which cannot be enriched as liberally as a sketch. During the evolution of a sketch, the profile of some operations must be often revised: generally, some operands are added.

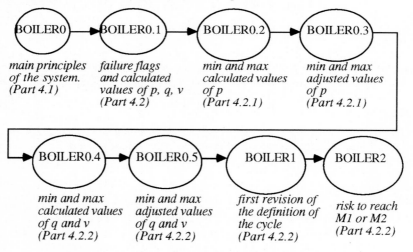

Fig. 1. Successive sketches of the specification of the cycle

This approach is convenient for the kind of problem addressed in this paper, where the data types are not especially complex, but the processing is. In this case, there is no strong need of different abstraction levels: we start with some incomplete specification and make it more and more precise, the abstraction level remaining the same. This methodology is supported by PLUSS, but not yet stated for algebraic specifications with implicit state. It turned out that it was rather easy to use it informally in this framework. Of course, some work remains to be done to actually transpose it: more elegant constructs should be defined to support the evolution of the specification.

Figure 1 above illustrates the evolution of the specification as presented in the paper ; ellipses denote sketches and horizontal arrows stand for enrichments or revisions.

In part 5.2, the BOILER2 sketch is transformed into a specification module named STEAM_BOILER_CONTROL: all the sorts, operations and accesses characterising a state of the control program are specified. This module uses the specification of the received messages which is stated in part 5.1.1, the specification of the sent messages which is given in 5.1.2, and some other elementary specifications of measuring units, modes, etc which are given in appendix (see CD-ROM, Annex GDK.1). The overall structure of the specification is summarized in Figure 2. Rectangular boxes stand for specification modules, and ascending arrows represent the use relation.

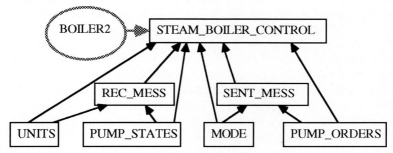

Fig. 2. The general structure of the specification

The STEAM_BOILER_CONTROL module is then completed with the definitions of the modifiers: in part 5.3, the definition of the cycle; in part 6, the checks of the devices; in part 7, the checks of the transmissions; in part 8, the changes of modes. The definitions of the modifiers corresponding to the initialization mode are left in appendix (see CD-ROM, Annex GDK.2).

4 Sketches of the Specification of the Steam-boiler Control System

4.1 Main Principles of the System

The essential points of the specification of the system can be sketched in the following way. From section 2.6 of the informal specification[1], it appears that there is a state with the following access functions (elementary or not):

q: $\to$ Litres   *{quantity of water in the steam-boiler}*
pump-state: $\{1, 2, 3, 4\} \to \{$on, off$\}$ *{states of the four pumps}*
v: $\to$ Litres/sec   *{quantity of exiting steam}*
p: $\to$ Litres/sec   *{throughput of the pumps}*

At every cycle, first the system receives some new values for q, every pump state, and v. Then, as stated in section 4.2 of the informal specification, as soon as q is below a quantity $N1$, the program sends a message for "switching the pumps on"; as soon as q is above $N2$,

---

[1]We refer to "sections" of the informal specification (chapter A in this book), and to "parts" of our paper.

the program sends a message for "switching the pumps off". This message can be considered as a component of the state, thus it is modelled as an access function[2]:

   to-pumps: → {open, close, nothing}

It is updated at every cycle by a modifier (let us call it *update-to-pumps* at this stage), which is defined by:

   update-to-pumps = **cases**
      $q \leq N1$ = true ⇒ μ-to-pumps(open) |
      $q \geq N2$ = true ⇒ μ-to-pumps(close) |
      $q \geq N2$ = false ∧ $q \leq N1$ = false ⇒ μ-to-pumps(nothing)
   end cases

When all the pumps are OK, the following property holds (where the *through* operation returns *P* for *on* and *0* for *off*):

   p = through(pump-state(1)) + through(pump-state(2)) + through(pump-state(3)) + through(pump-state(4))

An imperative requirement of the informal specification (still from section 4) is that *q* must remain above a quantity *M1* and below *M2*. In the informal specification, it is said that "if the water level is risking to reach one of the limits... the program enters the emergency stop mode. This risk is evaluated on the basis of the maximal behaviour of the physical units". Thus it is necessary to specify this notion of risk.

Let us call *Δt* the duration of a cycle in seconds. During the next cycle, the value of *p* will be *0* when *to-pumps* is equal to *close*, *4*P* when it is equal to *open*, and unchanged otherwise. Let us call this value *next-p* . The quantity of water entering the boiler during one cycle is either *next-p*Δt* or *next-p*(Δt – 5)* if the pumps have just been switched on (cf. section 2.3 of the informal specification; we make the assumption that *Δt* is greater than or equal to 5)[3]. The quantity of water exiting the boiler during one cycle is *K*v*Δt* when the steam measurement device is OK (*K* is a physical constant which relates a quantity of produced steam to a quantity of consumed water under the conditions holding in the boiler; it is not an actual constant since its value depends on pressure and temperature; however, since these quantities are not available from the informal specification, we consider *K* a constant). Thus, provided that the water measuring device is OK, one can estimate the water level at the end of the next cycle as:

   q + next-p * Δt – K * v * Δt or *q + next-p * (Δt – 5) – K * v * Δt*.

If this estimation is above or below the limits, there is a risk, and the mode must be changed to *emergency-stop*. This is formally specified in the *update-mode* modifier which is partially defined below.

The specification below summarizes the various points addressed in this first sketch of the formal specification. In a sketch, the axioms and the modifier definitions can be incomplete,

---

[2] In our model, the same message is always sent to all the pumps; then we simplify into this new version the example given in part 2 to introduce the use of patterns.

[3] According to section 3 of the informal specification Δt is equal to 5. We keep Δt in our specification for consistency with the addendum (see part 4.2 of this paper).

but all the used symbols must be declared. Besides, the distinction between elementary accesses and accesses is not definitive.

This specification uses some auxiliary specifications of the units and of the states of the pumps which are given in appendix (see CD-ROM Annex GDK.1). Basic specifications such as BOOL for booleans, NAT for natural numbers and RAT for rational numbers are assumed to exist and to be standard. Besides, the module below contains an implicit definitions of a discrete sort for the numbers of the pumps. Implicit sort definitions are a facility of PLUSS [BGM 89, p. 8] which makes it possible to omit obvious specification modules.

---

sketch-system BOILER0
  use UNITS *{defines the constants and their sorts, see APPENDIX 1}*,
PUMP_STATES *{defines on, off and through}*,MODE *{defines Mode}*,
PUMP_ORDERS *{defines open, close, nothing}*
    elementary accesses
        q: $\to$ Litres   *{quantity of water in the steam-boiler}*
        pump-state: {1, 2, 3, 4} $\to$ PumpState
        v: $\to$ Litres/sec   *{quantity of exiting steam}*
        to-pumps: $\to$ PumpOrder
        mode: $\to$ Mode
    accesses
        p: $\to$ Litres/sec   *{throughput of the pumps}*
        next-p: $\to$ Litres/sec   *{throughput of the pumps at the next cycle}*
        estimated-q: $\to$ Litres   *{estimated level at the end of the cycle}*
    access axioms
        mode = normal $\Rightarrow$ p = through(pump-state(1)) + through(pump-state(2)) +
                          through(pump-state(3)) + through(pump-state(4))
        mode = normal $\wedge$ to-pumps $\ne$ open $\Rightarrow$ estimated-q = q + next-p * Δt − K*v*Δt
        mode = normal $\wedge$ to-pumps = open $\Rightarrow$
                              estimated-q = q+ next-p * (Δt − 5) − K*v*Δt
        mode = normal $\wedge$ to-pumps = close $\Rightarrow$ next-p = 0
        mode = normal $\wedge$ to-pumps = open $\Rightarrow$ next-p = 4 * P
        mode = normal $\wedge$ to-pumps = nothing $\Rightarrow$ next-p = p
  modifiers
        cycle: PumpState x PumpState x PumpState x PumpState x Litres x Litres/sec
        receive: PumpState x PumpState x PumpState x PumpState x Litres x Litres/sec
        update-to-pumps:
        update-mode:
        emission: PumpOrder x Mode
  modifier definitions
        cycle(PS1, PS2, PS3, PS4, Q, V) = receive(PS1, PS2, PS3, PS4, Q, V);
                update-to-pumps; update-mode; emission(to-pumps, mode)
        receive(PS1, PS2, PS3, PS4, Q, V) = µ-pump-state(1, PS1) **and**
                µ-pump-state(2, PS2) **and** µ-pump-state(3, PS3) **and**
                µ-pump-state(4, PS4) **and** µ-q(Q) **and** µ-v(V)

> update-to-pumps = **cases**
>
> > $q \leq N1$ = true $\Rightarrow$ μ-to-pumps(open) |
> >
> > $q \geq N2$ = true $\Rightarrow$ μ-to-pumps(close) |
> >
> > $q \geq N2$ = false $\wedge$ $q \leq N1$ = false $\Rightarrow$ μ-to-pumps(nothing)
>
> **end cases**
> update-mode = **cases**
>
> > mode = normal $\wedge$
> > > or^4($q \leq M1$, $q \geq M2$, estimated-$q \leq M1$, estimated-$q \geq M2$) = true
> > > $\Rightarrow$ μ-mode(emergency-stop) |
> >
> > { ... }
>
> **end cases**
> emission(PO, M) = **nil**
> *{no change of state: the order to the pumps and the mode are sent}*
>
> **with** PS1, PS2, PS3, PS4: PumpState; Q: Litres; V: Litres/sec; PO: PumpOrder;
> M: Mode
>
> **end sketch-system** BOILER0

We chose to keep "multiple or " in prefix notation since we find it more readable. This incomplete specification deals with the case where there is no failure and the mode is normal. The following part deals with the estimations of *p*, *q* and *v* in the presence of failures; it may be skipped at first reading: part 5, 6 and 7 are understandable without the details below.

4.2 The Addendum to the Informal Specification

It is explained in the addendum of the informal specification that when there are some failures of the pumps or of the controllers, the minimal and maximal values of *p* must be calculated and used to make a decision. It is the same for the values of *v* and *q* when the steam measurement device or the water level measurement device are broken. There is a need for some flags in the state, i.e. some elementary access functions, to indicate whether a device is failing. Thus we enrich the sketch above by several access functions:

> **sketch-system** BOILER0.1
> > **enrich** BOILER0   *{introduction of failure flags and calculated values for p, q, v}*
> > **elementary accesses**
> > > pump-failure: {1, 2, 3, 4} $\rightarrow$ Bool
> > >
> > > pump-controller-failure: {1, 2, 3, 4} $\rightarrow$ Bool
> > >
> > > water-level-measuring-unit-failure: $\rightarrow$ Bool
> > >
> > > steam-level-measuring-unit-failure: $\rightarrow$ Bool
> > >
> > > transmission-failure: $\rightarrow$ Bool
> >
> > **accesses**
> > > min-calculated-p: $\rightarrow$ Litres/sec*{minimum possible throughput of the pumps}*
> > >
> > > max-calculated-p: $\rightarrow$ Litres/sec*{maximum possible throughput of the pumps}*
> > >
> > > min-calculated-q: $\rightarrow$ Litres   *{minimum possible calculated water level}*
> > >
> > > max-calculated-q: $\rightarrow$ Litres   *{maximum possible calculated water level}*
> > >
> > > min-calculated-v: $\rightarrow$ Litres/sec *{minimum calculated quantity of exiting steam}*
> > >
> > > max-calculated-v: $\rightarrow$ Litres/sec *{maximum calculated quantity of exiting steam}*
>
> **end** BOILER0.1

```
        or(pump-failure(PN),pump-controller-failure(PN)) = true ⇒
                        max-est-through(PN) = P
 modifier
     update-mode:
 modifier definition
     update-mode = cases
         or⁴(min-adj-q≤M1, max-adj-q≥M2,
             next-min-est-q≤M1, next-max-est-q≥M2) = true
                                        ⇒ μ-mode(emergency-stop) |
         { ... }
     end cases
with PN: {1, 2, 3, 4}
end BOILER2
```

This sketch must now be enriched with the detections, tolerances and recoveries of failures, and the different modes mentioned in section 4 of the informal specification.

5 The Specification of the Steam-boiler Control System

5.1 Specification of the Messages

Before specifying the program itself, we describe the messages received and sent by the program, following respectively the sections 6 and 5 of the informal specification. Since the program simultaneously receives messages as the first action of its cycle, and simultaneously transmits messages as the last action of its cycle (cf. section 3 of the informal specification), we have chosen to have one sort for the collection of received messages, and one for the collection of sent messages. These sorts will be used for the parameters of the *cycle* modifier in the specification of the system.

These sorts are not completely specified: it turns out that we are just interested in the way received messages can be observed, for instance: is the STOP message present in the received messages? At this stage, we are not concerned by the way such messages are generated. Thus the following specification module is a sketch. However, the situation is different for the sent messages, since it is an essential role of the program to build these messages; thus the specification includes operations for the generation of sent messages.

5.1.1 Received Messages

```
sketch REC_MESS {this module closely follows section 6 of the informal specification}
  use BOOL, PUMP_STATES, UNITS
  sort RecMess
  operations
      stop: RecMess → Bool
      steam-boiler-waiting: RecMess → Bool
      physical-units-ready: RecMess → Bool
      pump-state: RecMess x {1, 2, 3, 4} → PumpState      {i. e. {on, off}}
      pump-state-pres: RecMess x {1, 2, 3, 4} → Bool
      pump-control-state: RecMess x {1, 2, 3, 4} → {flow, noflow}
      pump-control-state-pres: RecMess x {1, 2, 3, 4} → Bool
      level: RecMess → Litres
```

```
        level-pres: RecMess → Bool
        steam: RecMess → Litres/sec
        steam-pres: RecMess → Bool
        pump-repaired: RecMess x {1, 2, 3, 4} → Bool
        pump-control-repaired: RecMess x {1, 2, 3, 4} → Bool
        level-repaired: RecMess → Bool
        steam-repaired: RecMess → Bool
        pump-failure-ack: RecMess x {1, 2, 3, 4} → Bool
        pump-control-failure-ack: RecMess x {1, 2, 3, 4} → Bool
        level-failure-ack: RecMess → Bool
        steam-outcome-failure-ack: RecMess → Bool
        all-present: RecMess → Bool
  axioms
        all-present (RM) = and¹⁰(pump-state-pres(RM, 1), pump-state-pres(RM, 2),
pump-state-pres(RM, 3), pump-state-pres(RM, 4), pump-control-state-pres(RM, 1),
pump-control-state-pres(RM, 2), pump-control-state-pres(RM, 3), pump-control-state-
pres(RM, 4), level-pres(RM), steam-pres(RM))
    with RM: RecMess
end REC_MESS
```

All the operations with names such as "xxx-pres" provide a way to check that the xxx message is present in the received messages. The *all-present* operation returns true when all the messages which must be present are present. When it is false, a transmission failure will be detected by the program (see part 7).

5.1.2 Sent Messages

```
draft SENT_MESS
  use BOOL, MODE, PUMP_ORDERS, UNITS
  sort SentMess
  generator
        set-mode: Mode x SentMess → SentMess
        set-program-ready: Bool x SentMess → SentMess
        set-valve: Bool x SentMess → SentMess
        set-pump-order: {1, 2, 3, 4} x PumpOrder x SentMess → SentMess
        {OPEN_PUMP(n) or CLOSE_PUMP(n)}
        set-pump-failure-detection: {1, 2, 3, 4} x Bool x SentMess → SentMess
        set-pump-control-failure-detection: {1, 2, 3, 4} x Bool x SentMess → SentMess
        set-level-failure-detection: Bool x SentMess → SentMess
        set-steam-failure-detection: Bool x SentMess → SentMess
        set-pump-repaired-ack: {1, 2, 3, 4} x Bool x SentMess → SentMess
        set-pump-control-repaired-ack: {1, 2, 3, 4} x Bool x SentMess → SentMess
        set-level-repaired-ack: Bool x SentMess → SentMess
        set-steam-repaired-ack: Bool x SentMess → SentMess
        message: → SentMess
```

```
operations
    sent-open-pump: SentMess x {1, 2, 3, 4} → Bool
    sent-close-pump: SentMess x {1, 2, 3, 4} → Bool
{The two observers above are used to check whether the message OPEN_PUMP(n) or
CLOSE_PUMP(n) was among the sent messages during the previous cycle}
    sent-pump-failure-detection: SentMess x {1, 2, 3, 4} → Bool
    sent-pump-control-failure-detection: SentMess x {1, 2, 3, 4} → Bool
{The observers above are used in the checks that the acknowledgement arrives}
{Operations to access to all parts of a sent message are defined similarly. All their names
begin with 'sent-'}
axioms
    sent-open-pump(set-pump-order(PN,open,SM),PN) = true
    sent-open-pump(set-pump-order(PN,close,SM),PN) = false
    sent-open-pump(set-pump-order(PN,nothing,SM),PN) = false
    PN ≠ PN' ⇒
        sent-open-pump(set-pump-order(PN',O,SM),PN) = sent-open-pump(PN,SM)
    {etc: for all the other operations the value of sent-open-pump is unchanged}
with  PN,PN': {1, 2, 3, 4}; O: PumpOrder; SM: SentMess
end SENT_MESS
```

5.2 Specification of the States of the System

In this part, we assume that the REC_MESS and SENT_MESS modules have been completed into some specifications of the same names and we use them.

Some information must be kept by the system from one cycle to the following ones: for instance, the number of STOP messages received in a row, since after three times the program must go into the *emergency-stop* mode (cf. section 6, item 1 of the informal specification); the fact that an OPEN_PUMP(n) or CLOSE_PUMP(n) message has been sent, since the calculations of q can be different. Moreover, the program has to check something during the second following cycle (cf. section 7, item 2 of the informal specification), etc. In order to keep the formulation of the formal specification close to the informal one, we chose to keep the two last collections of sent and received messages as components of the state: they are named *mess-in-minus-2, mess-in-minus-1, mess-out-minus-2* and *mess-out-minus-1*; accordingly, the sets of current messages are named *mess-in-0* and *mess-out-0*: the first one is the input of the current cycle, the second one its result.

These choices imply some changes with respect to the sketches given in part 4: it means that v and q are no more elementary access functions, but access functions since they are observed from *mess-in-0* as it appears in the specification below.

Similarly, the profile of *pump-state* is changed (see the REC_MESS module in part 5.1.1). Everywhere *pump-state(PN)* must be replaced by *pump-state(mess-in-0, PN)*. In the current version of our specification language, the only possibility is to forget the *pump-state* access from the sketches, and to rewrite all the axioms and modifier definitions of the sketches where it occurs, mainly the definitions of *p* and *min-through* and *max-through*, which is a bit tedious. This corresponds to a specification development operator which already exists in the GLIDER language [Lev 90, SL 93] and that we plan to introduce in our language.

At every cycle, the program computes the new messages to be sent: thus these messages are elementary access functions of the state, and *mess-out-0* is a non elementary access function constructed from them (see the last access axiom below). Conversely, the input of a cycle is *mess-in-0*, which is an elementary access function; a specific received message is a component of it, thus a non elementary access function.

The specification is no more a sketch, since it contains all the definitions. It is built from the BOILER2 sketch, forgetting all the items which need to be redefined. The only exported modifier is *cycle*, since it is an atomic action (from the last sentence of section 3).

system STEAM_BOILER_CONTROL
   from BOILER2 **forget** q, v, pump-state, p, min-through, max-through, receive,
                failure-detections, update-mode, emission
  exports cycle
  use UNITS, REC_MESS, SENT_MESS, MODE, PUMP_ORDERS,
     PUMP_STATES
  elem-accesses
{messages received, previously received, previously sent:}
       mess-in-0: $\rightarrow$ RecMess
       mess-in-minus-1: $\rightarrow$ RecMess
       mess-in-minus-2: $\rightarrow$ RecMess
       mess-out-minus-1: $\rightarrow$ SentMess
       mess-out-minus-2: $\rightarrow$ SentMess
{messages to be sent (the existing elementary accesses are recalled as comments):}
       {mode: $\rightarrow$ Mode}
       program-ready: $\rightarrow$ Bool
       valve: $\rightarrow$ Bool
       {to-pumps $\rightarrow$ PumpOrder}
       pump-failure-detection: $\{1, 2, 3, 4\} \rightarrow$ Bool
       pump-control-failure-detection: $\{1, 2, 3, 4\} \rightarrow$ Bool
       level-failure-detection: $\rightarrow$ Bool
       steam-failure-detection: $\rightarrow$ Bool
       pump-repaired-ack: $\{1, 2, 3, 4\} \rightarrow$ Bool
       pump-control-repaired-ack: $\{1, 2, 3, 4\} \rightarrow$ Bool
       level-repaired-ack: $\rightarrow$ Bool
       steam-repaired-ack: $\rightarrow$ Bool
  accesses
       mess-out-0: $\rightarrow$ SentMess
       q: $\rightarrow$ Litres
       v: $\rightarrow$ Litres/sec
       p: $\rightarrow$ Litres/sec
       min-through: $\{1, 2, 3, 4\} \rightarrow$ Litres/sec
       max-through: $\{1, 2, 3, 4\} \rightarrow$ Litres/sec
       {some auxiliary accesses related to the failure flags will be introduced in part 8 to make the expression of the changes of mode easier}

The six possibly embarrassing cases are derived from the preconditions of the various occurrences of μ-pump-failure:

sent-open-pump(mess-out-minus-1, PN) = true ∧ pump-state(mess-in-0, PN) = off ∧
  pump-just-repaired (mess-in-0, PN) = true ∧ pump-failure-detection(PN) = false ⇒ ??

sent-open-pump(mess-out-minus-1, PN) = true ∧ pump-state(mess-in-0, PN) = off ∧
  pump-just-repaired (mess-in-0, PN) = true ∧ pump-failure-detection(PN) = true ∧
  pump-failure-ack(mess-in-0,PN) = true ⇒ ??

sent-close-pump(mess-out-minus-1, PN) = true ∧ pump-state(mess-in-0, PN) = on ∧
  pump-just-repaired (mess-in-0, PN) = true ∧ pump-failure-detection(PN) = false ⇒ ??

sent-close-pump(mess-out-minus-1, PN) = true ∧ pump-state(mess-in-0, PN) = on ∧
  pump-just-repaired (mess-in-0, PN) = true ∧ pump-failure-detection(PN) = true ∧
  pump-failure-ack(mess-in-0,PN) = true ⇒ ??

sent-open-pump(mess-out-minus-1, PN) = false ∧
  sent-close-pump(mess-out-minus-1, PN) = false ∧
  pump-state(mess-in-0, PN) ≠ pump-state(mess-in-minus-1, PN) ∧
  pump-just-repaired (mess-in-0, PN) = true ∧ pump-failure-detection(PN) = false ⇒ ??

sent-open-pump(mess-out-minus-1, PN) = false ∧
  sent-close-pump(mess-out-minus-1, PN) = false ∧
  pump-state(mess-in-0, PN) ≠ pump-state(mess-in-minus-1, PN) ∧
  pump-just-repaired (mess-in-0, PN) = true ∧ pump-failure-detection(PN) = true ∧
  pump-failure-ack(mess-in-0,PN) = true ⇒ ??

The first four cases may occur since in our specification, the program emits commands to all the pumps even when some of them are known as failing. A way to explicitly avoid these cases is to introduce the condition that the pump was not already detected as failing in the first two conditional modifications:

---
pump-failure(PN) = false ∧ sent-open-pump(mess-out-minus-1, PN) = true ∧
  pump-state(mess-in-0, PN) = off ⇒ μ-pump-failure(PN, true)
pump-failure(PN) = false ∧ sent-close-pump(mess-out-minus-1, PN) = true ∧
  pump-state(mess-in-0, PN) = on ⇒ μ-pump-failure(PN, true)

The two last cases (spontaneous change of state of a pump) are quite possible: a spontaneous change of state may be compatible with the fact that the pump has been repaired and that pump-failure must be switched to false. However, it does not correspond to a new failure. Thus we decided to ignore spontaneous changes of failing pumps; the third axiom finally becomes:

pump-failure(PN) = false ∧ sent-open-pump(mess-out-minus-1, PN) = false ∧
   sent-close-pump(mess-out-minus-1, PN) = false ∧
   pump-state(mess-in-minus-1, PN) ≠ pump-state(mess-in-0, PN) ⇒
   μ-pump-failure(PN, true)

In conclusion, the update of the four *pump-failure* flags is specified by :

update-pump-failures = **cases**
   pump-failure(PN) = false ∧ sent-open-pump(mess-out-minus-1, PN) = true ∧
         pump-state(mess-in-0, PN) = off ⇒ μ-pump-failure(PN, true) |
   pump-failure(PN) = false ∧ sent-close-pump(mess-out-minus-1, PN) = true ∧
         pump-state(mess-in-0, PN) = on ⇒ μ-pump-failure(PN, true) |
   pump-failure(PN) = false ∧ sent-open-pump(mess-out-minus-1, PN) = false ∧
      sent-close-pump(mess-out-minus-1, PN) = false ∧
      pump-state(mess-in-minus-1, PN) ≠ pump-state(mess-in-0, PN) ⇒
         μ-pump-failure(PN, true) |
   pump-just-repaired (mess-in-0, PN) = true ∧ pump-failure-detection(PN) = false ⇒
         μ-pump-failure(PN, false) |
   pump-just-repaired (mess-in-0, PN) = true ∧ pump-failure-detection(PN) = true ∧
         pump-failure-ack(mess-in-0, PN) = true ⇒ μ-pump-failure(PN, false)
end cases

6.2 Messages and Acknowledgements

The issue of pump failures is not completely specified without a description of the related issued messages and acknowledgements, and the way they are dealt with by the program. These points are discussed in sections 5 and 6 of the informal specification.

The PUMP_FAILURE_DETECTION(n) message must be sent when a failure of a pump is detected. In the formal specification, this sending is expressed by *pump-failure-detection(PN)* being true. Thus the three cases where *pump-failure(PN)* is switched to true also appear in the definition of the *update-pump-failure-detections* modifier as conditions where the message is sent, *(i.e. pump-failure-detection(PN)* is switched to true). Moreover, in section 5, it is said that this message is sent until reception of the corresponding acknowledgement; thus *pump-failure-detection(PN)* is switched to false when such a message is present in *mess-in-0* and kept true when there is no acknowledgement. We get a first version of *update-pump-failure-detection* :

update-pump-failure-detections = **cases**
   pump-failure(PN) = false ∧ sent-open-pump(mess-out-minus-1, PN) = true ∧
         pump-state(mess-in-0, PN) = off ⇒ μ-pump-failure-detection(PN, true) |
   pump-failure(PN) = false ∧ sent-close-pump(mess-out-minus-1, PN) = true ∧
         pump-state(mess-in-0, PN) = on ⇒ μ-pump-failure-detection(PN, true) |
   pump-failure(PN) = false ∧ sent-open-pump(mess-out-minus-1, PN) = false ∧
      sent-close-pump(mess-out-minus-1, PN) = false ∧
      pump-state(mess-in-minus-1, PN) ≠ pump-state(mess-in-0, PN) ⇒
         μ-pump-failure-detection(PN, true) |

> sent-pump-failure-detection(mess-out-minus-1, PN) = true ∧
>   pump-failure-ack(mess-in-0, PN) = true ⇒
>   μ-pump-failure-detection(PN, false) |
> sent-pump-failure-detection(mess-out-minus-1, PN) = true ∧
>   pump-failure-ack(mess-in-0, PN) = false ⇒
>   μ-pump-failure-detection(PN, true)
> **end cases**

In the last case above, one could specify the detection of a transmission failure since the program does not receive an expected message. However, this could contradict the informal specification, where it is said that the detection message must be sent again: the detection of a transmission failure causes an emergency stop and would prevent that... Thus we decided not to do it.

Moreover, one can remark that this last case, as it is, is useless because of our frame assumption (see part 2). It is included here for similarity with the informal specification.

We prudently decided that the system detects a transmission failure when it receives an acknowledgement without having sent the *pump-failure-detection* message:

> sent-pump-failure-detection(mess-out-minus-1, PN) = false ∧
>   pump-failure-ack(mess-in-0, PN) = true ⇒ μ-transmission-failure(true)

This case is included in the forthcoming definition of the *check-transmission* modifier, in part 7.

When checking consistency we noticed a risk in *update-pump-failure-detection* if the fourth case is not disjoint of the others. Such cases may occur when *sent-pump-failure-detection(mess-out-minus-1,PN) = true* and one of the three first cases is true. Thus we add *sent-pump-failure-detection(mess-out-minus-1,PN) = false* to the first three preconditions. This does not restrict the considered cases, since the case where *sent-pump-failure-detection(mess-out-minus-1,PN) = true* is covered exhaustively.

Thus the final version of *update-pump-failure-detection* is:

> update-pump-failure-detections = **cases**
>   pump-failure(PN) = false ∧ sent-open-pump(mess-out-minus-1, PN) = true ∧
>   pump-state(mess-in-0, PN) = off ∧
>   sent-pump-failure-detection(mess-out-minus-1, PN) = false ⇒
>   μ-pump-failure-detection(PN, true) |
>   pump-failure(PN) = false ∧ sent-close-pump(mess-out-minus-1, PN) = true ∧
>   pump-state(mess-in-0, PN) = on ∧
>   sent-pump-failure-detection(mess-out-minus-1, PN) = false ⇒
>   μ-pump-failure-detection(PN, true) |
>   pump-failure(PN) = false ∧ sent-open-pump(mess-out-minus-1, PN) = false ∧
>   sent-close-pump(mess-out-minus-1, PN) = false ∧
>   pump-state(mess-in-minus-1, PN) ≠ pump-state(mess-in-0, PN) ∧
>   sent-pump-failure-detection(mess-out-minus-1, PN) = false ⇒
>   μ-pump-failure-detection(PN, true) |

> sent-pump-failure-detection(mess-out-minus-1, PN) = true ∧
>   pump-failure-ack(mess-in-0, PN) = true ⇒ μ-pump-failure-detection(PN, false) |
> sent-pump-failure-detection(mess-out-minus-1, PN) = true ∧
>   pump-failure-ack(mess-in-0, PN) = false ⇒ μ-pump-failure-detection(PN, true)
> **end cases**

6.3 Pump Repairs

In section 5, item PUMP_REPAIRED_ACKNOWLEDGEMENT(n), it is said that this acknowledgement is sent by the program after a PUMP_REPAIRED(n) message was received. It is formally specified by:

> pump-failure(PN) = true ∧ pump-repaired(mess-in-0) = true ⇒
>   μ-pump-repaired-ack(PN, true)
> pump-repaired (mess-in-0, PN) = false ⇒ μ-pump-repaired-ack(PN, false)

Such a reception is normal only if the corresponding pump is failing. As stated above in part 6.1, a transmission failure is detected when it is not true.

Checking completeness reveals that the cases above are not sufficient, since it is said in section 6, item PUMP_REPAIRED(n), that this message is sent by the physical unit until it receives an acknowledgement message. The choice made in part 6.1, to detect a transmission failure when an unjustified *pump-repaired* message arrives, is consistent with this point of the informal specification. But it may be useful to add some emission of an acknowledgement message to this detection to try to stop the erroneous emission of the *pump-repaired* message. It leads to:

> pump-failure(PN) = false ∧ pump-repaired (mess-in-0, PN) =true ⇒
>   μ-pump-repaired-ack(PN, true)

Thus the definition of the update of *pump-repaired-ack(PN)* is simply:

> update-pump-repaired-acks = **cases**
>   pump-repaired (mess-in-0, PN) = true ⇒ μ-pump-repaired-ack(PN, true) |
>   pump-repaired (mess-in-0, PN) = false ⇒ μ-pump-repaired-ack(PN, false)
> **end cases**

This terminates the specification of the cases where the *check-pumps* modifier changes the state. In all other cases, following the frame assumption mentioned in part 2, the state is unchanged. The final definition of *check-pumps* is:

> check-pumps = [update-pump-failures • update-pump-failure-detections
>   • update-pump-repaired-acks]

It would be possible to present *check-pumps* in another way, namely by enumerating all the cases and their global effects on the failure flags, the detection messages and the acknowledgement messages; we have found it convenient to organize the presentation elementary access by elementary access: it ensures a good separation of concerns, and makes it easier to detect inconsistencies and incompleteness.

The checks of the other devices (the water level measurement unit, the pump controllers, the steam measurement unit) follow the same principles as the check of the pumps. Their formal specification is not given here.

7 Checking Transmissions

The informal specification mentions numerous cases where a transmission failure must be detected. Moreover, other cases have been identified in the previous parts of this paper.

The case of the absence of some necessary messages in the set of received messages was mentioned in part 5.1.1:

> all-present(mess-in-0) = false $\Rightarrow$ µ-transmission-failure(true)

We mentioned in the part on the pump failure detections (6.1) the following cases:

> pump-just-repaired(mess-in-0, PN) = true $\wedge$ pump-failure-detection(PN) = true $\wedge$
>   pump-failure-ack(mess-in-0, PN) = false $\Rightarrow$ µ-transmission-failure(true)
> pump-failure(PN) = false $\wedge$ pump-repaired (mess-in-0, PN) = true $\Rightarrow$
>   µ-transmission-failure(true)

Moreover, in the part on pump failure messages and acknowledgements (6.2), we decided to have:

> sent-pump-failure-detection(mess-out-minus-1, PN) = false $\wedge$
>   pump-failure-ack(mess-in-0, PN)= true $\Rightarrow$ µ-transmission-failure(true)

There are three similar cases for the pump controllers:

> pump-controller-failure(PN) = true $\wedge$ pump-control-failure-detection(PN) = true $\wedge$
>   pump-control-failure-ack(mess-in-0, PN) = false $\wedge$
>   pump-control-repaired(mess-in-0, PN) = true $\Rightarrow$ µ-transmission-failure(true)
> pump-controller-failure(PN) = false $\wedge$ pump-control-repaired (mess-in-0, PN) = true $\Rightarrow$
>   µ-transmission-failure(true)
> sent-pump-control-failure-detection(mess-out-minus-1, PN) = false $\wedge$
>   pump-control-failure-ack(mess-in-0, PN) = true $\Rightarrow$ µ-transmission-failure(true)

For the water level measuring unit they become:

> water-level-measuring-unit-failure = true $\wedge$ level-failure-detection = true $\wedge$
>   level-failure-ack(mess-in-0) = false $\wedge$ level-repaired(mess-in-0) = true $\Rightarrow$
>   µ-transmission-failure(true)
> water-level-measuring-unit-failure = false $\wedge$ level-repaired (mess-in-0) = true $\Rightarrow$
>   µ-transmission-failure(true)
> sent-level-failure-detection(mess-out-minus-1) = false $\wedge$
>   level-failure-ack(mess-in-0) = true $\Rightarrow$ µ-transmission-failure(true)

For the steam level measuring unit, there is similarly:

> steam-level-measuring-unit-failure = true $\wedge$ steam-failure-detection = true $\wedge$
>   steam-outcome-failure-ack(mess-in-0) = false $\wedge$ steam-repaired(mess-in-0) = true $\Rightarrow$
>   µ-transmission-failure(true)
> steam-level-measuring-unit-failure = false $\wedge$ steam-repaired (mess-in-0) = true $\Rightarrow$
>   µ-transmission-failure(true)
> sent-steam-failure-detection(mess-out-minus-1) = false $\wedge$
>   steam-outcome-failure-ack(mess-in-0) = true $\Rightarrow$ µ-transmission-failure(true)

There is no mention in the informal specification of a way of recovering from a transmission failure. Thus the *transmission-failure* elementary access is set to false in the initial state, and the *update-transmission-failure* modifier either leaves it unchanged or switches it to true, inducing a change into the *emergency-stop* mode and then the stop of the program.

Other cases of transmission failures have been considered by other authors in this book: in chapters BCPR and OKW, the reception of messages specific to the initialization mode in another mode raises a transmission failure. Moreover, in chapter BCPR, the authors consider the possibility of "junk" messages. The informal specification is not prescriptive on these cases and we have not considered them.

8 The Functioning Modes

We come back here to the various functioning modes which depend on the failure cases whose detection has been specified above. In order to simplify the specification, some access functions have been added. They correspond to some compositions of failure cases which occur in the specification of the changes of mode.

---

accesses
  failure-phys-unit: $\to$ Bool
  rescue-possible: $\to$ Bool
  three-stops: $\to$ Bool
  emergency-in-any-case: $\to$ Bool
access-axioms
  failure-phys-unit = or^9(pump-failure(1), pump-failure(2),pump-failure(3),
        pump-failure(4), pump-controller-failure(1), pump-controller-failure(2),
        pump-controller-failure(3), pump-controller-failure(4),
        steam-level-measuring-unit-failure)
  rescue-possible = and^5(not(steam-level-measuring-unit-failure),
          not(pump-controller-failure(1)),
          not(pump-controller-failure(2)),
          not(pump-controller-failure(3)),
          not(pump-controller-failure(4)))
{Receiving three stops in a row must stop the program}
  three-stops = and^3 (stop(mess-in-0), stop(mess-in-minus-1), stop(mess-in-minus-2))
{In any mode, the program goes into emergency-stop mode when it detects a transmission failure, when the water level risks to reach M1 or M2, or when it receives three stops in a row}
  emergency-in-any-case = or^3 (transmission-failure,
          or^4(min-adj-q$\leq$M1, max-adj-q$\geq$M2,
          next-min-est-q$\leq$M1, next-max-est-q$\geq$M2),
          three-stops)

The following transition table summarizes the changes of mode described in section 4 of the informal specification:

	normal	degraded	rescue	emergency
normal	C1	C2	C3	C4
degraded	C1	C2	C3	C4
rescue	C1	C2	C3	C4
initialization	C5	C6	false	C7

where C1, C2, C3, C4 correspond to the following cases:

(C1) failure-phys-unit = false ∧ water-level-measuring-unit-failure = false ∧
emergency-in-any-case = false
(C2) failure-phys-unit = true ∧ water-level-measuring-unit-failure = false ∧
emergency-in-any-case = false
(C3) water-level-measuring-unit-failure = true ∧ rescue-possible = true ∧
emergency-in-any-case = false
(C4) emergency-in-any-case = true

The transition rules are the same for three modes: *normal, degraded* and *rescue*; as seen in part 5.3, we have called these modes "intermediate" (see the MODE specification in CD-ROM Annex GDK.1). The definition of the *update-mode* conditional elementary modifier is then:

---

update-mode = **cases**
{Cases where the mode is normal, degraded or rescue:}
  intermediate(mode) = true ∧ emergency-in-any-case= true ⇒ μ-mode(emergency-stop)|
  intermediate(mode) = true ∧ failure-phys-unit = false ∧
    water-level-measuring-unit-failure = false ∧ emergency-in-any-case = false ⇒
                                    μ-mode(normal) |
  intermediate(mode) = true ∧ failure-phys-unit = true ∧
    water-level-measuring-unit-failure = false ∧ emergency-in-any-case = false ⇒
                                    μ-mode(degraded) |
  intermediate(mode) = true ∧ water-level-measuring-unit-failure = true ∧
    rescue-possible = true ∧ emergency-in-any-case = false ⇒ μ-mode(rescue) |
  intermediate(mode) = true ∧ water-level-measuring-unit-failure = true ∧
    rescue-possible = false ∧ emergency-in-any-case = false ⇒
                                    μ-mode(emergency-stop) | ...

The complete definition of *update-mode*, including the initialization, is given in appendix (see CD-ROM Annex GDK.3).

9 Conclusion on this Case Study

This paper illustrates two points:

- the use of the Dauchy-Gaudel formalism of algebraic specifications with implicit state for a second realistic example;
- the stepwise development of a formal requirement specification following the PLUSS methodology, i.e. using sketches of specifications.

We were able to use our "implicit state" approach on this case study without any major modification. However, an important methodological point is worth being mentioned: we chose to structure the modifiers according to the access functions modified rather than the different cases; it turned out to be very useful during the development of the specification, since it helps to detect inconsistencies and tricky cases. It turned out to be a very good basis for the validation of the specification, as shown in part 6.

The use of sketches has proved to be convenient for the development of the specification (it was no surprise for us). Moreover, it is interesting to present the different development steps of such a large specification as an element of justification.

The only new point here was that elementary access functions could evolve into non elementary ones during the development of the specification. The solution presented here is not definitive: the use of "forget x" followed immediately by a new definition of x is not very elegant... All that is needed is a new construction making possible the change of the profile and the properties of an item in the specification.

An interesting point is that the development of the specification was not sequential, as presented in the paper. Actually, once the basic decisions were made on the state, we were able to work concurrently on:

- the fault tolerance aspects of the devices;
- the estimations of levels (addendum to the informal specification).

The integration was then straightforward.

This work emphasizes requirement capture rather than system design. We think that continuing the development of the system, following the PLUSS methodology (i.e. transforming sketches into system modules) will raise no problem. Actually, the specification we get states a partial order on the treatments during a cycle. Thus, it provides a precise guideline for the programming task.

One important feedback of the writing of a formal specification is the discovery of ambiguities, incompletenesses or inconsistencies in the original informal specification (see for instance [CGR 93]). It has been the case here, and we list below some of the questions which arose during the development of the specification and the decisions we made when it was necessary. This should make easier the comparison with other formal specifications of the same problem. We follow the order of the paper, not the order of discovery.

It is not clear from the informal specification that all the pumps have the same characteristics, i.e. the same P. However, this is explicitly stated in the document [Bau 92] which is at its origin, where a unique value is given for the four pumps. It is also explicitly assumed in the definitions of the calculated values of p in the addendum. Similarly, it can be understood that "the pumps" are switched on or off all together, as we did, or that they can be switched independently, e.g. as in chapters BCPR and OKW in this book.

The informal specification is mute on what happens when a PUMP_REPAIRED(n) message arrives and the pump n is supposed to work correctly... We decided to detect a transmission failure in this case (in part 6.1) since it is a message whose presence is aberrant. Besides, nothing is said on what happens when a PUMP_REPAIRED(n) message and a failure occur at the same time. When a pump is broken, it is not clear that the failure cases are still meaningful as we explain in part 6.1. We decided to ignore the *pump-state* messages from a pump which is failing at the previous cycle and to put the failure flag to true as soon as a *pump-repaired* message arrives. It was a significant decision, and in a real project it should be discussed with the specialists of the domain.

In the addendum to the informal specification, it is said: "Note that an equipment (that is not already considered broken) becomes broken when the corresponding raw quantity is not a member of the interval of quantities calculated at the previous cycle." In the case of p, this sentence is unclear since p is always calculated and never actually measured. Moreover, as noted in part 4, p is not estimated from the previous cycle. We have considered that this sentence concerned q and v and not p. It is not completely satisfactory: when q is outside the acceptance interval, it may come from a pump failure as well as from a water level measuring device failure.

In the description of the passage from the normal mode into the rescue mode, we check that the rescue mode is possible, which is not explicitly specified in section 4.2: from this section, in any case, if there is a failure of the water level measuring unit, the system goes into rescue mode; then, if something else is broken, it goes into the emergency stop mode. We decided to go directly to the emergency stop mode; in chapter BCPR of this book, the same remark and the same choice were made.

We think that this list of remarks on the informal specification is an important byproduct of the writing of this formal specification.

10 Evaluation and Comparison

1. The paper provides a formal requirements specification of the control program.

2. The answer is "no" for a), b), c).

3. The comparable solutions are chapter BCPR, chapter BBDGR, and chapter OKW. But they don't address exactly the same aspects of the problem, thus the comparison is difficult. The closest specification is in chapter BCPR. It is purely algebraic, and can be considered more abstract than ours: it defines the set of sent messages for a given state and a given set of received messages, without decomposing the cycle as we do. However, our decomposition provides a guideline for the sequel of the development.

4.a) The time spent has been globally 1.5 person-month.

4.b) Our estimate is 8 weeks of personal experience, including a 2-week course on logic, algebraic specifications, and implicit states (Carole Khoury joined our group after a DEA, in september 1995, with only a few notions on algebraic specifications; she was able to actively participate after $\pm$ 2 weeks).

5. Logic is the main premise for a good understanding of our solution. We believe that a detailed knowledge of the formalism is not needed. We think that a good programmer can understand it. We estimate the time necessary for the average programmer without knowledge of the used specification method to learn what is needed to be able to understand the solution to be 1 or 2 weeks.

References

[BGM 89] Bidoit, M., Gaudel, M-C. and Mauboussin, A., How to make algebraic specifications more understandable? An experiment with the PLUSS specification language. Science of Computer Programming, vol. 12, n° 1, pp. 1-38, June 1989.

[Bau 92] Bauer J.C., Generic Problem Competition, a component of the International Symposium Design and Review of Software Controlled Safety-Related Systems, Institute for Risk Research, University of Waterloo, 1992.

[Bid 89] Bidoit M., PLUSS, un langage pour le développement de spécifications algébriques modulaires, Thèse d'Etat, Université de Paris-Sud, Orsay, May 1989.

[CGR 93] Craigen D., Gerhart S., Ralston T., On the use of formal methods in industry — an authoritative assessment of the efficacy, utility, and applicability of formal methods to systems design and engineering by the analysis of real industrial cases, Report to the US National Institute of Standards and Technology, March 1993.

[Dau 92] Expériences de spécification formelle d'un pilote automatique de métro. Thèse, Université de Paris-Sud, 1992.

[DG 94] Dauchy P., Gaudel M.-C., Algebraic Specifications with Implicit State, Rapport LRI n°887, Feb. 1994.

[DGM 93] Dauchy P., Gaudel M-C, Marre B., Using Algebraic Specifications in Software Testing: a case study on the software of an automatic subway, Journal of Systems and Software, vol. 21, n° 3, June 1993, pp. 229-244.

[DM 91] Dauchy P., Marre B., Test data selection from algebraic specifications: application to an automatic subway module, 3rd European Software Engineering Conference, (ESEC'91), Milan, 1991, LNCS n° 550.

[EO 95] Ehrig H., Orejas F., Dynamic Abstract Data Types: an informal proposal, EATCS bulletin, Sept. 1994.

[Gau 80] Gaudel M.-C., Génération et Preuve de Compilateurs basées sur une Sémantique Formelle des Langages de Programmation, Thèse d'état, INPL (Nancy), 1980.

[Gau 83] Gaudel M.-C., Correctness Proofs of Programming Language Translations, IFIP-TC2 Working Conference on Formal Description of Programming Concepts, Garmisch-Partenkirchen, June 83, North-Holland 1983, pp.25-43.

[Gau 92] Gaudel M-C., Structuring and Modularizing Algebraic Specifications: the PLUSS specification language, evolutions and perspectives, STACS'92, Cachan, Feb. 1992, LNCS n°577, pp. 3-18, Springer-Verlag.

[Lev 90] Lévy N., Definition of "add-a-component", a Specification Construction Process Operator, Technical Report ForSem-005-R, ICARUS ESPRIT Project, 1990.

[SL 93] Souquières J., Lévy N., Description of Specification Developments, 1st IEEE Symposium on Requirements Engineering, San Diego, January 1993.

Annex GDK.1

APPENDIX 1: Auxiliary Specification Modules

The following specification modules define the units (litres, seconds, etc) and the constants used by the specification (cf. the table of part 2.6 of the informal specification).

```
spec LITRES
  use RAT renaming Rat as Litres
  operations
       C, M1, M2, N1, N2: → Litres
  axioms
{definition of the constants C, M1, M2, N1, N2, cf. part 2.1 of the informal
specification}
end LITRES

spec SECONDS
  use RAT renaming Rat as Seconds
  operations
       Δt: → Seconds
  axioms
{definition of Δt: cf. part 2.1 of the informal specification}
end SECONDS
spec LITRES/SECOND
  use RAT renaming Rat as Litres/sec
  operations
       W, P: → Litres/sec
  axioms
{definition of W, cf. part 2.1 of the informal specification}
end LITRES/SECOND

spec LITRES/SECOND/SECOND
  use RAT renaming Rat as Litres/sec/sec
  operations
       U1, U2: → Litres/sec/sec
  axioms
{definition of U1, U2}
end LITRES/SECOND/SECOND
spec UNITS
  use LITRES, SECONDS, LITRES/SECOND, LITRES/SECOND/SECOND
  operations
       _ * _: Litres/sec/sec x Seconds → Litres/sec
       _ * _: Seconds x Litres/sec/sec → Litres/sec
       _ * _: Litres/sec x Seconds → Litres
       _ * _: Seconds x Litres/sec → Litres
       _ / _: Litres/sec x Seconds → Litres/sec/sec
       _ / _: Litres x Seconds → Litres/sec
end UNITS
```

The following module defines the five different modes of operation of the program (cf. section 4 of the informal specification).

```
basic spec MODE
  use BOOL
  sort Mode
  generators
      initialization, normal, degraded, rescue, emergency-stop: → Mode
  operations
      intermediate: Mode → Bool
  axioms
      intermediate(initialization) = false
      intermediate(normal) = true
      intermediate(degraded) = true
      intermediate(rescue) = true
      intermediate(emergency-stop) = false
end MODE
```

The following module specifies the two functioning modes of the pumps. The through operation is useful for specifying easily the throughput of the pumps in the main specification module.

```
spec PUMP_STATES
  use   UNITS
  sort PumpState
  generators
      on, off: → PumpState
  operations
      through: PumpState → Litres/sec
  axioms
      through (on) = P
      through (off) = 0
end PUMP_STATES
```

The following module specifies the orders which can be sent to a pump.

```
basic spec PUMP_ORDERS
  sort PumpOrder
  generators
      open, close, nothing: → PumpOrder
end PUMP_ORDERS
```

Using HYTECH to Synthesize Control Parameters for a Steam Boiler*

Thomas A. Henzinger[1]   Howard Wong-Toi[2]

[1] Department of Electrical Engineering and Computer Sciences
University of California, Berkeley, CA
[2] Cadence Berkeley Labs, Berkeley, CA

Abstract. We model a steam-boiler control system using hybrid automata. We provide two abstracted linear models of the nonlinear behavior of the boiler. For each model, we define and verify a controller that maintains safe operation of the boiler. The less abstract model permits the design of a more efficient controller. We also demonstrate how the tool HYTECH can be used to automatically synthesize control parameter constraints that guarantee safety of the boiler.

1 Introduction

A description of an industrial steam boiler has been proposed as a benchmark problem for the formal specification and verification of embedded reactive systems [1, 2]. Our approach to the problem is unique in that we use *algorithmic* techniques to analyze *directly*, without discretization, the mixed discrete-continuous components of the system. In this way we are able to fully automatically synthesize safe values for the parameters that control the continuous behavior of the boiler.

We describe the steam boiler and its controller using *hybrid automata* [4, 3]. These automata model nondeterministic continuous activities within a nondeterministic discrete transition structure. Since the nonlinear behavior of the steam-boiler system is not directly amenable to automatic analysis, we provide an approximating model using *linear hybrid automata*. Linear hybrid automata are a subclass of hybrid automata with linearity restrictions on continuous activities (inequalities between linear combinations of first derivatives) and discrete transitions (inequalities between linear combinations of transition sources and targets). Model-checking based analysis techniques [5] for this subclass have been implemented in HYTECH [16, 17] and used to verify numerous distributed real-time systems [14, 19]. Borrowing ideas from the field of abstract interpretation [9, 15], we choose our approximations such that if a desired property holds for an approximating linear automaton, then it holds also for the original nonlinear automaton.

---

* This research was supported in part by the ONR YIP award N00014-95-1-0520, by the NSF CAREER award CCR-9501708, by the NSF grant CCR-9504469, by the AFOSR contract F49620-93-1-0056, and by the ARPA grant NAG2-892.

We also take algorithmic analysis a step beyond the checking of system properties. Given a parametric description of a controller, we use HyTech to automatically synthesize constraints on the safe values for the control parameters [6, 10]. These constraints are necessary and sufficient for the correctness of the approximating linear automaton, and because of our choice of approximations, they are also sufficient (though not necessary) for the correctness of the original nonlinear automaton. More accurate approximations thus provide less restrictive constraints on the controller, which can be used to control the system more efficiently.

Steam-boiler description. The steam boiler consists of a water tank, four pumps, and sensors that measure the pumping rates, the steam evacuation rate, the water level, and the operational status of each component (see Figure 1). The entire physical system operates under the guidance of a controller. The controller must keep the water level between the extreme values M_1 and M_2 at all times, and it should try to keep the water level between the normal operating levels of N_1 and N_2 as much as possible. All communication between the controller and the physical plant occurs in discrete rounds, once every Δ seconds. In each round, all units send information to the controller, and the controller responds by sending messages to the units. All communication is assumed to take place instantaneously.

The controller operates in five modes: initialization (waiting for the steam boiler to signal its readiness for operation), normal, rescue (the water-level sensor has failed), degraded (other components have failed, but the water-level sensor is working correctly), and emergency stop. In the initialization mode, the controller receives a signal that the boiler is ready, and then tests the amount of steam escaping from the boiler. If this is nonzero, it enters the emergency stop mode. Otherwise, it either drains the water level to N_2 or activates a pump to raise the water level to N_1. Once the range of normal water levels has been reached, the controller sends a signal to the physical units, waits for acknowledgements, and then proceeds with normal operation. In normal mode, the controller makes its decisions to turn pumps on or off based on the current water level, the states of the physical units, and the rate at which steam is being emitted. No action is taken if the water level lies in the range $[N_1, N_2]$. In degraded mode, some unit, other than the water-level sensor, has failed. Messages are sent to repair the faulty components, and the controller attempts to maintain correct water levels with the operational components. The reader is referred to [2] for a description of the rescue mode.

Our goal is not to provide a detailed model of the message passing between system components and the controller; to do so would result in state explosion. Rather, we focus on the high-level interactions between discrete control decisions and the continuous aspects of the underlying physical plant. To this end, we restrict most of our discussion to two fault-free pumps in normal operating mode. This simplifies the discrete control space of the system and allows us to concentrate on the continuous evolution of variables modeling the water level, steam and pumping volumes. Only in a later section do we briefly consider four

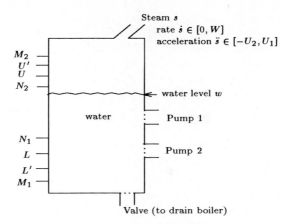

Fig. 1. Overview of the steam boiler

pumps and a simple fault-tolerant system with a degraded mode.

Steam-boiler analysis. We provide two models of the system, at two different levels of abstraction. For each model, we design controllers that ensure that whenever all physical components operate correctly, then the water level is maintained within the desired bounds, and the emergency stop mode is never entered. Our controllers rely on sensor values to determine how many pumps should be operating. The simpler the model of a system, the more complex the questions we are able to answer using HYTECH. Our first model ignores all information about the second derivative of the steam output. For this simple model, we provide a controller whose decisions are based solely on the water level. We verify the controller, determine the minimal and maximal water levels that can occur, determine a safe upper bound on the time period Δ that separates consecutive rounds of communication, and determine constraints on the water-level thresholds that trigger decisions to turn a pump on or off. All of this is done completely automatically using HYTECH.

The simpler the model, the more extraneous behaviors it admits. Therefore, a simple model may lead to an unnecessarily restrictive choice of control-parameter values such as water-level thresholds. Our second model achieves a closer approximation of the original system by taking into account the steam rate at the beginning and end of each round, and inferring bounds on the steam volume emitted during a round. A controller that utilizes this information is strictly superior to the previous one in that it requires the pumps to be on less often, while still maintaining the required water levels.

2 Hybrid Automata

We define *hybrid automata*, which are used to model mixed discrete-continuous systems [4]. Informally, a hybrid automaton consists of a finite set X of real-valued variables and a labeled multigraph (V, E). The edges E represent discrete

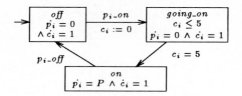

Fig. 2. Automaton for pump i

transitions and are labeled with guarded assignments to X. The vertices V represent continuous activities and are labeled with constraints on the derivatives of the variables in X. The state of the automaton changes either instantaneously through a discrete transition or, while time elapses, through a continuous activity. We use the pump automaton in Figure 2 as an example accompanying the formal definition.

2.1 Definition

Given a set X of variables, a *predicate* over X is a boolean combination of inequalities between algebraic terms with free variables from X. A *hybrid automaton* A consists of the following components.

Variables. A finite ordered set $X = \{x_1, \ldots, x_n\}$ of real-valued *variables*, and a subset $Y = \{y_1, \ldots, y_k\} \subseteq X$ of *controlled variables*. For example, the pump automaton has two controlled variables, p_i and c_i, representing the pumping volume and a clock. A *valuation* is a point $\mathbf{a} = (a_1, \ldots, a_n)$ in the n-dimensional real space $\mathbb{R}^n$, or equivalently, a function mapping each variable x_i to its value a_i.

Control modes. A finite set V of vertices called *control modes*. The pump automaton has three control modes, *on*, *off*, and *going_on*, used for modeling the pump when it is on, off, and in the process of going on. A *state* $(v, \mathbf{a})$ of the hybrid automaton A consists of a control mode $v \in V$ and a valuation $\mathbf{a} \in \mathbb{R}^n$.

Invariant conditions. A labeling function *inv* that assigns to each control mode $v \in V$ an *invariant condition* $inv(v)$, which is a predicate over X. A state $(v, \mathbf{a})$ is *admissible* if $inv(v)[X := \mathbf{a}]$ is true. The automaton control may reside in control mode v only while the invariant condition $inv(v)$ is true; so the invariant conditions can be used to enforce progress. For example, in the pump automaton, the invariant condition $c_i \leq 5$ of the control mode *going_on* ensures that the automaton control must leave at the latest when the monotonically increasing clock c_i reaches the value 5.

Flow conditions. A labeling function *flow* that assigns to each control mode $v \in V$ a *flow condition* $flow(v)$, which is a predicate over the set $X \cup \dot{X}$ of variables, where $\dot{X} = \{\dot{x}_1, \ldots, \dot{x}_n\}$. Each dotted variable $\dot{x}_i$ represents the first derivative of the variable x_i with respect to time. While the automaton control resides in control mode v, the variables change along differentiable trajectories whose first derivatives satisfy the flow condition. Formally, for each real $\delta \geq 0$, we define the binary *flow relation* $\xrightarrow{\delta}$ on the admissible states such that

$(v, \mathbf{a}) \xrightarrow{\delta} (v', \mathbf{a}')$ if $v' = v$, and there is a differentiable function $\rho\colon [0, \delta] \to \mathbb{R}^n$ such that

- the endpoints of the flow match those of ρ, *i.e.* $\rho(0) = \mathbf{a}$ and $\rho(\delta) = \mathbf{a}'$;
- the invariant condition is satisfied throughout the flow, *i.e.* for all reals $t \in [0, \delta]$, $inv(v)[X := \rho(t)]$ is true; and
- the flow condition is satisfied throughout the flow, *i.e.* for all reals $t \in (0, \delta)$, $flow(v)[X, \dot{X} := \rho(t), \dot{\rho}(t)]$ is true, where $\dot{\rho}(t) = (\rho_1(t)/dt, \ldots, \rho_n(t)/dt)$.

For example, in the pump automaton, the flow condition $\dot{p}_i = 0 \wedge \dot{c}_i = 1$ of the control mode *going_on* ensures that the pumping volume stays unchanged (at 0) and that the clock c_i measures the amount of elapsed time.

Initial conditions. A labeling function *init* that assigns to each control mode $v \in V$ an *initial condition* $init(v)$, which is a predicate over Y. The automaton control may start in control mode v when $init(v) \wedge inv(v)$ is true, where $\mathbf{a}[Y]$ denotes the restriction of $\mathbf{a}$ to the variables in Y. A state $(v, \mathbf{a})$ is *initial* if it is admissible and $init(v)[Y := \mathbf{a}[Y]]$ is true. In the graphical representation of automata, we omit initial conditions of the form *false*. For example, in the pump automaton, the initial condition of the control mode *off* is *true*, and the initial conditions of the other control modes are *false*.

Control switches. A finite multiset E of edges called *control switches*. Each control switch (v, v') has a source mode $v \in V$ and a target mode $v' \in V$. For example, the pump automaton has three control switches.

Jump conditions. A labeling function *jump* that assigns to each control switch $e \in E$ a *jump condition* $jump(e)$, which is a predicate over the set $X \cup Y'$ of variables, where $Y' = \{y_1', \ldots, y_k'\}$. The unprimed symbol y_i refers to the value of the controlled variable before the control switch, and the primed symbol y_i' refers to its value after the control switch. Only controlled variables are updated by a control switch. Formally, for the control switch $e = (v, v')$, we define the binary *jump relation* $\xrightarrow{e}$ on the admissible states such that $(v, \mathbf{a}) \xrightarrow{e} (v', \mathbf{a}')$ iff

- $jump(e)[X, Y' := \mathbf{a}, \mathbf{a}'[Y]]$ is true; and
- for all uncontrolled (environment) variables $x_i \in X \setminus Y$, $\mathbf{a}_i' = \mathbf{a}_i$.

The control switch e is *enabled* in the valuation $\mathbf{a}$ if there exists a state $(v', \mathbf{a}')$ such that $(v, \mathbf{a}) \xrightarrow{e} (v', \mathbf{a}')$. We use nondeterministic guarded interval-valued assignments to write jump conditions. For example, we write $\phi \to y_i := [l, u]$ for the jump condition $\phi \wedge l \leq y_i' \leq u \wedge \bigwedge_{j \neq i} y_j' = y_j$, where l and u are predicates over X. Intuitively, a control switch is enabled in the valuation $\mathbf{a}$ if the guard is satisfied, *i.e.* $\phi[X := \mathbf{a}]$ is true. Then, the controlled variable y_i is updated nondeterministically to any value in the interval $[l[X := \mathbf{a}], u[X := \mathbf{a}]]$. In the graphical representation of automata, guards of the form *true* and identity assignments are omitted. For example, in the pump automaton, the control switch from *going_on* to *on* has the jump condition $c_i = 5 \wedge p_i' = p_i \wedge c_i' = c_i$.

Events. A finite set Σ of *visible events*, and a labeling function *event* that assigns to each control switch $e \in E$ either a visible event from Σ or the internal event τ. The *internal event* τ is contained neither in Σ, nor in the set of visible events of any other automaton. The event labels are used to define the parallel composition of automata. Internal events are omitted in the graphical representation of automata. For example, in the pump automaton, $event(off, going\_on) = p_i\_on$ and $event(going\_on, on) = \tau$.

2.2 Parallel composition

Nontrivial systems consist of several interacting components. We model each component as a hybrid automaton, and the components coordinate with each other through both shared variables and events. For example, the controller communicates with the i-th pump by synchronizing control switches with the events p_i_on and p_i_off. A hybrid automaton that models the entire system is obtained from the component automata using a product construction.

Let A_1 be the hybrid automaton $(X_1, Y_1, V_1, inv_1, flow_1, init_1, E_1, jump_1, \Sigma_1, event_1)$, and define A_2 similarly. In the product automaton $A_1 \times A_2$, two control switches e_1 and e_2 from the two component automata A_1 and A_2 occur simultaneously if $event_1(e_1) = event_2(e_2)$. They are interleaved if $event_1(e_1) \neq event_2(e_2)$ and neither $event_1(e_1)$ is a visible event of A_2, nor $event_2(e_2)$ is a visible event of A_1. Internal events may occur simultaneously or interleaved. Formally, provided Y_1 and Y_2 are disjoint, the *product* $A_1 \times A_2$ of A_1 and A_2 is the following hybrid automaton $A = (X_1 \cup X_2, Y_1 \cup Y_2, V_1 \times V_2, inv, flow, init, E, jump, \Sigma_1 \cup \Sigma_2, event)$.

Control modes. Each control mode (v_1, v_2) in $V_1 \times V_2$ has the invariant condition $inv(v_1, v_2) = inv_1(v_1) \wedge inv_2(v_2)$, the flow condition $flow(v_1, v_2) = flow_1(v_1) \wedge flow_2(v_2)$, and the initial condition $init(v_1, v_2) = init_1(v_1) \wedge init(v_2)$.

Control switches. E contains the control switch $e = ((v_1, v_2), (v_1', v_2'))$ if

(1) $e_1 = (v_1, v_1') \in E_1$, $v_2' = v_2$, and $event_1(e_1) \notin \Sigma_2$; or
(2) $e_2 = (v_2, v_2') \in E_2$, $v_1' = v_1$, and $event_2(e_2) \notin \Sigma_1$; or
(3) $e_1 = (v_1, v_1') \in E_1$, $e_2 = (v_2, v_2') \in E_2$, and $event_1(e_1) = event_2(e_2)$.

In case (1), $event(e) = event_1(e_1)$ and $jump(e) = jump_1(e_1) \wedge \bigwedge_{y \in Y_2} y' = y$.
In case (2), $event(e) = event_2(e_2)$ and $jump(e) = jump_2(e_2) \wedge \bigwedge_{y \in Y_1} y' = y$.
In case (3), $event(e) = event_1(e_1) = event_2(e_2)$ and $jump(e) = jump_1(e_1) \wedge jump_2(e_2)$.

2.3 Verification

Let A be a hybrid automaton with n variables and the control graph (V, E). A subset of the state space $S = V \times \mathbb{R}^n$ is called a *region*. We define the binary transition relation $\to$ on S as $\bigcup_{e \in E} \overset{e}{\to} \cup \bigcup_{\delta \in \mathbb{R}_{\geq 0}} \overset{\delta}{\to}$. For a region R, the *successor region* $post(R)$ is the set of states that are reachable from some state in R via a single transition, i.e. $post(R) = \{s' \mid \exists s \in R. s \to s'\}$. The *reachable region* $reach(A)$ is the set of states that are reachable from some initial state via any finite number of transitions, i.e. $reach(A) = \bigcup_{i \geq 0} post^i(R)$ for the set R of initial states of A.

Safety and timing analysis. Safety and timing verification problems can be posed in a natural way as reachability problems. For this purpose, the system is composed with a special monitor automaton that "watches" the execution of the system and enters a violation state whenever the system violates a given safety or timing requirement. The automaton A is correct with respect to the region $T \subseteq S$ of violation states if $reach(A) \cap T$ is empty.

Parameter synthesis. A system description often contains symbolic (unknown) constants, which we call *system parameters*. We are interested in the problem of finding the values of the parameters for which the system is correct. When the correctness criterion is a safety or timing requirement, expressed via a monitor with violation states, then necessary and sufficient conditions on the parameter values can be discovered automatically using reachability analysis. We model a parameter as a controlled variable whose derivative is 0 in every control mode, and whose value is left changed by every control switch. Then, the system is correct for precisely those parameter values for which the region $reach(A) \cap T$ is empty.

2.4 Linear hybrid automata and the tool HyTech

Linear hybrid automata are a subclass of hybrid automata that can be analyzed algorithmically. A *linear term* is a linear combination of variables with rational coefficients. A *convex linear predicate* is a conjunction of inequalities between linear terms. A hybrid automaton A is *linear* if (1) all invariant, flow, initial, and jump conditions of A are convex linear predicates, and (2) no flow condition of A contains undotted variables, *i.e.* all flow conditions are convex linear predicates over the set $\dot{X}$ of first derivatives. Hence, in flow conditions, linear dependencies between the rates of variables can be expressed, but the current flow must be independent of the current state. The convexity restrictions can be achieved by splitting control modes and control switches, if necessary. For example, the pump automaton is a linear hybrid automaton.

A region is *linear* if it can be defined by a disjunction of convex linear predicates. The computation of the successor region $post(R)$ is effective for a linear region R, and yields again a linear region [5]. Therefore, the reachability problem, which can be solved by iterating successor computations, is semidecidable for linear hybrid automata. Our analysis of linear hybrid automata is performed using the symbolic model checker HyTech [17].[3]

2.5 Approximation

While it may be difficult to reason automatically about a complex automaton, in order to establish a particular property, it may suffice to prove a related property for a simpler approximating automaton. If the correctness criterion for an automaton A is a safety property, expressed via a set of violation states, then it suffices to consider an approximating automaton B such that $reach(A) \subseteq reach(B)$. Then, the safety property holds for A if it can be verified for B. The same idea applies to parametric analysis. Suppose that we can find necessary and sufficient parameter constraints for the correctness of the approximating automaton B. Since more states are reachable in B than in A, all parameter values that cause

---

[3] HyTech can be obtained on the web at http://www.eecs.berkeley.edu/~tah.

Fig. 3. Valve and boiler automata

violations in A also cause B to be incorrect. Hence, the parameter constraints for the correctness of B are sufficient for the correctness of the original automaton A. However, it may not be the case that the conditions are necessary, and a closer approximation may yield more permissive conditions.

The use of HYTECH demands approximation if the hybrid automaton A to be analyzed is nonlinear. In this case, we approximate A by a linear hybrid automaton B [13, 18]. For a system with several components, it may suffice to approximate only the nonlinear components. The steam-boiler system contains only one nonlinear component —the automaton that models the steam— and we replace it first by a very rough and then by a more accurate linear approximation.

3 Steam-boiler Description

We model the steam-boiler system at various levels of detail. The primary modeling issues are (1) modeling the nonlinear behavior of the steam exiting the boiler, (2) how many of the failure modes are considered, and (3) the design of the controller. To facilitate algorithmic analysis, we first restrict the state space by considering only two pumps and omitting the failure modes. We use the following variables: p_i for the volume of water pumped by pump i, c_i for a clock of pump i, v for the volume of water drained through the valve, w for the water level in the boiler, s for the volume of steam emitted from the boiler, r for the rate at which steam is emitted from the boiler, t_s for an auxiliary clock for modeling steam emission, and t for the clock of the control program.

Pumps. The automaton for pump i appears in Figure 2. Control is initially in the mode *off*, where the flow condition $\dot{p}_i = 0$ indicates that the pump is idle. The pump synchronizes with the controller automaton on the events p_i_on, through which the controller commands the pump to be turned on, and p_i_off, through which the pump is turned off. The pump takes 5 seconds to respond to the command p_i_on to begin pumping, modeled as follows. The variable c_i is a clock, with first derivative equal to 1 at all times. When the switch from mode *off* to mode *going_on* occurs, the clock c_i is set to 0 and measures the delay before water is actually pumped. The invariant condition $c_i \leq 5$ together with the guard $c_i = 5$ on the switch to mode *on* ensures that the delay is exactly 5 seconds. The flow condition $\dot{p}_i = P$ in mode *on* reflects our assumption that when the pump is on, it operates at its maximal capacity P.

Valve. The valve automaton appears on the left of Figure 3. The controller automaton communicates with the valve through the events *open_valve* and *close_valve*. We assume that the valve opens and closes instantaneously, and from the normal range above $N_1 = 100$ liters down to below 60 liters. Suppose that the controller has to make a decision when the water level is exactly L liters with both pumps active. It may leave both pumps running through the next round. Minimal steam emission over this period causes the next water-level reading to lie in the normal range [100, 150], because the net input of the two pumps is 8 liters per second. Within the range of normal water levels, no control action is taken, so both pumps are still active. It is therefore possible for a future water-level reading to occur at exactly $N_2 = 150$, and again both pumps remain operating. Minimal steam emission causes the water level to rise 40 more liters to 190 liters. Now if U' is not above 190, then the controller must enter the emergency-stop mode.

$\phi_0 : L > L' + 20$: The explanation here is similar to that for the constraint $U' > U + 20$ above. The controller may decide to leave exactly one pump running when it reads the water level as L. The earliest time that the second pump could be active is 10 seconds later, because the next decision is made after 5 seconds, and there is a delay of 5 seconds before an activated pump begins pumping water. Within 10 seconds, the water level may drop by 20 liters.

$\phi_1 : L < 40 \lor L' < 30 \lor L > L' + 40$: Intuitively, this condition addresses the spacing between L and L' when the parameters are set close to the minimal normal water level N_1. Assume that $L' \geq 30$. We show that $L > L' + 40$. Consider the following scenario. The controller reads the water level when it is exactly $N_1 = 100$ with no pumps active, reads it again at 70 liters and activates only one pump. By the end of the next round, the water level drops to 40 liters, and then the first pump is active in the round after that, leading to a water level of 30 liters. If $L' \geq 30$, then the controller will enter the emergency-stop mode. To avoid this kind of situation, it is necessary for both pumps to be active before the water level drops to L'. In particular, the above scenario shows that L must be above 70. It is therefore possible for the controller to read the water level as L with no pumps active, and for the water level to drop to $L - 40$ over the next two rounds. Thus we conclude that $L' < L - 40$.

$\phi_2 : L' < 40 \lor L < 70$: First, observe that the conjunction $\phi_0 \land \phi_1 \land \phi_2$ is equivalent to $\phi_0 \land \phi_1 \land L' < 40$, because $\phi_0 \land \phi_1 \land L < 70$ implies $L' < 40$. The condition ϕ_2 can therefore be simplified to $L' < 40$, but currently HyTech does not perform minimality checks on expressions in conjunctive normal form. It is easy to see that $L' < 40$ is a necessary condition. From a water level of $N_1 = 100$, it may take two rounds before a pump is activated, and during that time a total of 60 liters of water may be lost.

In hindsight, the necessity of each of the above conditions is not difficult to explain. However, it would have been nontrivial to manually generate each one, and justify that their conjunction is sufficient for the safe operation of the boiler. As explained in Subsection 2.5, the conditions are necessary and sufficient for the approximating linear system, but only sufficient for the correctness of the approximated nonlinear system.

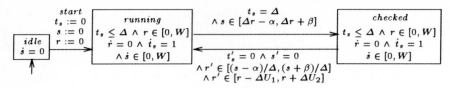

Fig. 8. Second steam approximation A_B^S

4.2 Model B: linear approximation of the steam acceleration

We now give a closer approximation of the steam automaton, which allows us to prove the correctness of a controller that lets the water level drop lower than previously before operating the pumps. We use the variable r to represent the steam rate at the end of each round. The consistency between the steam volume emitted during a round and the steam rate is maintained by ensuring that (1) the emitted steam volume is consistent with the steam rate at the beginning of the round and the possible steam acceleration, and (2) the steam rate at the end of the round is consistent with the emitted steam volume and the possible steam acceleration. The linear hybrid automaton for this steam model appears in Figure 8. The derivative of the variable r is always 0. If the steam rate at the beginning of a round is $\dot{s} = r_0$, and the steam emission accelerates constantly at k liters per second per second, then the volume of steam emitted over the next Δ seconds is $\int_0^\Delta (r_0 + kt)dt = \Delta r_0 + \frac{1}{2}k\Delta^2$. Thus the steam emitted lies in the range $[\Delta r_0 - \alpha, \Delta r_0 + \beta]$, where $\alpha = \frac{1}{2}\Delta^2 U_1$ and $\beta = \frac{1}{2}\Delta^2 U_2$. The jump condition on the control switch from mode *running* to mode *checked* enforces the consistency condition (1), and the jump condition from *checked* to *running* enforces the consistency condition (2).

Improved controller. If the controller has access to the measured steam rate r, in addition to the water level w, then it can bound more tightly the possible future behaviors of the boiler. It may therefore be able to maintain the water level correctly in situations where this is not possible without a steam-rate measurement. In this subsection, we assume that the controller keeps the pumps either both active or both idle at any given time. The key control decision is when to turn on the pumps. The controller (see the appendix) makes this decision at the end of each round based on the water level w and the steam rate r. As before, when the water level falls below the control parameter L', or rises above U', then the controller enters the emergency-stop mode. For water levels between L' and N_1, the controller activates both pumps if $w \leq \varphi(r)$, and determines that the pumps should be idle if $w > \varphi(r)$, where the values of the function φ are sufficiently high to avoid the water level dropping to L' over the next two rounds. The controller turns off both pumps whenever the water level is between N_2 and U', and aborts when the level is above U'.

We define the function φ. A decision not to turn on the pumps means that at least two rounds will pass before the pumps begin actively filling the boiler tank. The controller must ensure that no disaster can occur during this pe-

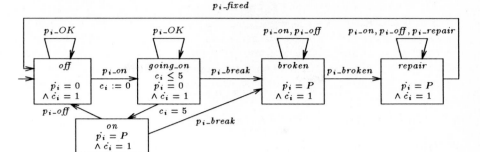

Fig. 9. Automaton for pump i, assuming possible failure

riod. Leaving the pumps inactive will not cause the water level to drop to L' if $w - max2(r) > L'$, where $max2(r)$ is the maximal amount of steam emitted during the next 2Δ seconds, given a current steam rate of r. However, $max2(r)$ is quadratic, and therefore cannot be expressed in a linear hybrid automaton. We instead use the upper bound $Max2(r)$ on $max2(r)$, which is defined as $Max2(r) = \min(2W\Delta, 2r\Delta + 2\Delta^2 U_2)$. Thus the function φ is defined by $\varphi(r) = L' + Max2(r)$.

Verification of the improved controller. Given the bounds $U_1 = U_2 = 2/5$ on the steam acceleration, HyTech verifies that the controller C_B, with the control parameters set to $L' = 25$ and $U' = 200$, guarantees the safety requirements. This controller is more flexible than a comparable controller that relies on the simple steam model A_A^S. Consider a simple controller C'_A (see the appendix) that activates both pumps between L' and L, and neither pump between L and N_1 and between N_2 and U'. Parametric analysis of C'_A for $L' = 25$ and $U' = 200$ determines that the safe values of L are characterized by $L \geq 85$. The improved controller C_B is more relaxed than the simple controller C'_A, in that it does not activate the pumps as often. For a given steam rate r, the threshold value for turning on the pumps is $\varphi(r) = L' + Max2(r)$. When the steam rate is high (*i.e.* above 4), then $\varphi(r) = 85$, and both pumps are activated just as in C'_A. But when the steam rate r is lower than 4, then the threshold value $\varphi(r)$ is only $L' + 10r + 50U_2 = 45 + 10r$. In particular, if $r = 0$, then the improved controller allows the water level to drop as low as 45 before turning on the pumps.

4.3 A fault-tolerant system

Finally, we describe a simple fault-tolerant model of a steam-boiler system where pumps may fail. The new model of pump i appears in Figure 9. It includes a *broken* mode and a *repair* mode. The pump may fail at any time other than when it is in the *off* mode. It remains in the *broken* mode until the controller is informed of the status of the pump via the events p_i_OK and p_i_broken, at which point the pump may be repaired and restored to the *off* mode. While broken or in the process of being repaired, the pump delivers water at any rate between 0 and P.

The controller (see the appendix) attempts to maintain the water level within the required bounds by using operational pumps wherever possible. It aborts to the emergency-stop mode if it cannot guarantee a safe water level for the next two rounds, assuming that any active pump may break at any time. The control decisions are based on five control parameters: if the water level is below *lo_abort*, then the controller aborts because of a danger of reaching the minimal allowable level; it activates all functional pumps if the water level is between *lo_abort* and L, and at most one functional pump if the water level is between L and N_1; if the water level is above *hi_abort_both_off* with both pumps off, or it is above *hi_abort_one_off* with exactly one pump on, or it is above *hi_abort_none_off* with both pumps on, then the controller also aborts.

It makes no sense to check the same safety requirements as before, because in the presence of pump faults it is impossible to avoid extreme water levels. We instead show two properties of the fault-tolerant system. First, if no pumps ever break (*i.e.* we remove the p_i_break edges), then the system never enters the emergency-stop mode, and the water level remains within the required bounds. Second, if the system can be completely shutdown (*i.e.* no steam being emitted and no water being pumped) within 5 seconds of entering the emergency-stop mode, then the water level is always safe. HyTech verifies both these properties for the system with two pumps and the simple steam model A_A^S.

4.4 Computational data

The analysis reported in this paper was performed on a Sun Sparcstation 5 with 32 MBytes of main memory and additional swap space. For the simple steam model A_A^S, HyTech requires 5 seconds and 3 MBytes of memory for verification, 7s (5MB) to generate bounds on the water level, 11s (7MB) for synthesizing the sampling time, and 62s (8MB) for analyzing the control parameters. For the more sophisticated steam model B, HyTech requires 119s (9MB) for verification.

The complexity of our models has been restricted by HyTech's computational capacity. For a boiler with four pumps and the simple steam model A, HyTech completes the verification in 33s (15MB). The fault-tolerant system with a degraded mode strains resources, taking 123 seconds and 39 MBytes. However, for neither system does a parametric analysis of the allowable sampling rates complete.

5 Evaluation and Comparison

Hybrid automata enable a natural, yet mathematically precise, modeling of the steam-boiler system. Both discrete and continuous phenomena are modeled directly in one integrated formalism. The theory of abstractions for this formalism ensures that the automatic analysis results apply to the original mixed discrete-continuous system. Because our modeling language includes continuous information about the physical behavior of the boiler, we were able to discover nontrivial control constraints, *e.g.* a lower bound on the safe delay between rounds of communication, and parametric correctness criteria for the water thresholds used by the controller. The advantage of an expressive model comes at the cost of computationally expensive automated analysis. We presented fault-free components as well as a simple fault-tolerant system because of restrictions on the size of the model that HyTech can currently handle.

Designing the nonlinear hybrid automata was reasonably straightforward, even for the full fault-tolerant system (not shown in this paper). We expect an average programmer with some background in a state-machine formalism to have little difficulty understanding our model, and to be able to produce a similar solution after a couple of weeks' training[6]. The majority of time developing the solution was spent in the iterative process of abstracting the nonlinear model to linear models, designing a suitable controller, testing whether HyTech is able to produce useful results, and then redesigning the abstraction and the controller. The simplest abstraction adn its controller was easy to derive, but produces coarse results that may be too conservative to establish system correctness for safe parameter combinations. On the other hand, more accurate approximations sometimes cause HyTech to fail, due to either memory overflow, arithmetic overflow in solving linear constraints, or a nonterminating sequence of successor computations. Finding a useful balance between detail and abstraction, and modeling the system to optimize the use of HyTech's analysis algorithms, is still somewhat of an art.

Our goal is the *direct* and *automatic* analysis of mixed discrete-continuous systems such as the steam boiler. Among the papers in this volume, many abstract the continuous behavior into a discrete system (*e.g.* [8, 11, 22]) or into a simple timed system (*e.g.* [21, 24]), which may then be subjected to automated analysis. Others analyze the continuous behavior deductively with the possible assistance of a mechanical proof checker (*e.g.* [7, 12, 20, 23]). This paper is unique in that we retain continuous information about the boiler's physical quantities during the automated analysis. While computational complexities force us to keep our models simpler and smaller compared to discrete-state based methods, our approach allows a fully automatic synthesis of constraints on parameters that control continuous behavior.

References

1. J.-R. Abrial, E. Börger, and H. Langmaack. The steam-boiler case study project. An introduction. This volume.
2. J.-R. Abrial. Steam-boiler control specification problem. This volume.
3. R. Alur, C. Courcoubetis, N. Halbwachs, T.A. Henzinger, P.-H. Ho, X. Nicollin, A. Olivero, J. Sifakis, and S. Yovine. The algorithmic analysis of hybrid systems. *Theoretical Computer Science*, 138:3–34, 1995.
4. R. Alur, C. Courcoubetis, T.A. Henzinger, and P.-H. Ho. Hybrid automata: an algorithmic approach to the specification and verification of hybrid systems. *Hybrid Systems I*, Lecture Notes in Computer Science 736, pp. 209–229. Springer-Verlag, 1993.

---

[6] A U.C. Berkeley graduate student took three weeks to learn the formalism from scratch, acquaint herself with HyTech, and generate a solution to a nonlinear control problem of similar complexity.

5. R. Alur, T.A. Henzinger, and P.-H. Ho. Automatic symbolic verification of embedded systems. *IEEE Trans. Software Engineering*, 22:181–201, 1996.
6. R. Alur, T.A. Henzinger, and M.Y. Vardi. Parametric real-time reasoning. *Proc. Symp. Theory of Computing*, pp. 592–601. ACM Press, 1993.
7. R. Buessow and M. Weber. A steam-boiler control specification with STATECHARTS and Z. This volume.
8. T. Cattel and G. Duval. The steam-boiler problem in LUSTRE. This volume.
9. P. Cousot and R. Cousot. Abstract interpretation: a unified lattice model for the static analysis of programs by construction or approximation of fixpoints. *Proc. Symp. Principles of Programming Languages*. ACM Press, 1977.
10. P. Cousot and N. Halbwachs. Automatic discovery of linear restraints among variables of a program. *Proc. Symp. Principles of Programming Languages*. ACM Press, 1978.
11. G. Duval and T. Cattel. Specifying and verifying the steam-boiler problem with SPIN. This volume.
12. G. Leeb and N. Lynch. Proving safety properties of the steam-boiler controller. This volume.
13. T.A. Henzinger and P.-H. Ho. Algorithmic analysis of nonlinear hybrid systems. *Computer-aided Verification*, Lecture Notes in Computer Science 939, pp. 225–238. Springer-Verlag, 1995.
14. T.A. Henzinger and P.-H. Ho. HYTECH: The Cornell Hybrid Technology Tool. *Hybrid Systems II*, Lecture Notes in Computer Science 999, pp. 265–293. Springer-Verlag, 1995.
15. T.A. Henzinger and P.-H. Ho. A note on abstract-interpretation strategies for hybrid automata. *Hybrid Systems II*, Lecture Notes in Computer Science 999, pp. 252–264. Springer-Verlag, 1995.
16. T.A. Henzinger, P.-H. Ho, and H. Wong-Toi. HYTECH: The next generation. *Proc. Real-time Systems Symp.*, pp. 56–65. IEEE Computer Society Press, 1995.
17. T.A. Henzinger, P.-H. Ho, and H. Wong-Toi . A user guide to HYTECH. *Tools and Algorithms for the Construction and Analysis of Systems*, Lecture Notes in Computer Science 1019, pp. 41–71. Springer-Verlag, 1995.
18. T.A. Henzinger and H. Wong-Toi. Linear phase-portrait approximations for nonlinear hybrid systems. *Hybrid Systems III*, Lecture Notes in Computer Science 1066, pp. 377–388. Springer-Verlag, 1995.
19. P.-H. Ho and H. Wong-Toi. Automated analysis of an audio control protocol. *Computer-aided Verification*, Lecture Notes in Computer Science 939, pp. 381–394. Springer-Verlag, 1995.
20. X.-S. Li and J. Wang. Specifying optimal design of a steam-boiler system. This volume.
21. P.C. Olveczky, P. Kosiuczenko, and M. Wirsing. An object-oriented algebraic steam-boiler control specification. This volume.
22. C. Schinagl. VDM specification of the steam-boiler control using RSL notation. This volume.
23. J. Vitt and J. Hooman. Assertional specification and verification using PVS of the steam-boiler control system. This volume.
24. A. Willig and I. Schieferdecker. Specifying and verifying the steam-boiler control system with time extended LOTOS. This volume.

A VDM Specification of the Steam-Boiler Problem

Yves LEDRU and Marie-Laure POTET*

Laboratoire Logiciels Systèmes et Réseaux - Institut IMAG (UJF - INPG - CNRS)
BP 53 38041 Grenoble Cedex 9 (FRANCE) - Tel + 33 76827214 - Fax + 33 76827287
e-mail: Yves.Ledru@Imag.fr, Marie-Laure.Potet@imag.fr

Abstract. A model-based formal specification of the steam boiler problem is presented, using VDM-SL. The development of the specification follows an environment-based approach. First, the physical boiler is specified, then its interface with the control system and finally, the control system itself. The integrations of the interface and of the controller in the specification of the boiler take the form of successive refinements. The approach has several advantages: it provides a methodological guidance to the specification activity and tends to avoid over-specification and premature design decisions, by focusing later on the control aspects of the problem.

1 Introduction

The requirements specification activity plays a crucial role in the development of software systems. Formal methods are perceived as one way to achieve high quality specifications, mainly because their semantical foundations allow an increased precision in the description of the system under development. In [CGR93], several applications of formal methods are presented. Some of these correspond to complete formal developments where an implementation is proved correct with respect to a specification. The other applications only use formal methods at the specification stage. In these cases, the increased precision of the formal language is perceived as a significant but sufficient improvement to the current practice. The work presented here belongs to this second category, although it does not prevent the subsequent development of a proved implementation. It starts thus from the assumption that, in order to develop code for a system, it is necessary to precisely understand and state the problem.

The steam boiler is not a small problem. The first specification we came up with was about 50 pages long and therefore it is necessary to structure the specification and its development process. In this work, we use two constructs to structure the formal description:

- *modules* cut the specification into components of manageable size;
- *refinements* allow the incremental introduction of details in the description.

But technical constructs are not sufficient, it is also necessary to have guidelines to help structuring the specification activity. In this work, the environment

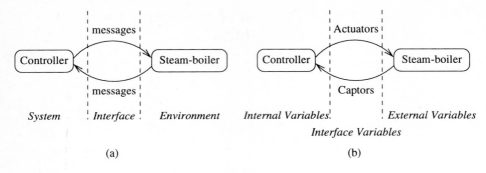

Fig. 1. The system and its environment

based approach has been chosen. This approach first models the environment of the system under development (see Fig. 1.a). Here, this environment is the physical boiler. It then gradually introduces the interfaces between this environment and the system that controls it. Finally, it specifies the behaviour of the control system. This specification process starts thus with the description of an existing system which provides a context where the specification of the controller may be stated. The environment-driven approach is one of the principles of the JSD method [Jac83]. This principle may be adapted to the development of reactive systems [Led91, Led93, LC94] with the following expected benefits:

- it is easier to model an existing reality (the environment) than to specify a non-existing controller;
- the concepts needed to specify the system to develop are concepts of its environment;
- specifying the system in terms of its environment produces a more abstract specification and prevents premature design decisions and overspecification.

The environment-based approach is not linked to a particular specification language. Previous experiments have been based on a combination of VDM and state machines or on the Temporal Logic of Actions [Lam91]. The present specification uses VDM. VDM [Jon90] is a model-based formal method, like Z [Spi92] or B [Abr96] aimed at the specification and development of transformational systems. Its specification language (VDM-SL) is undergoing an ISO standardisation process where it has now reached the "Draft International Standard" status [ABH+95]. This standardisation process has favoured the development of industrial tools to support the formalism. The current version of the VDM-SL standard does not include the structuring facilities mentioned above. But modular extensions are proposed in the IFAD tools [ELL94], and refinements in both data and control structures are defined in VDM [Jon90].

Nevertheless, the specification presented here must face several problems. Since VDM is aimed at describing transformational systems, it does not provide support for reactive aspects of systems (like the specification of input/output activities or real-time aspects), nor does it support parallel activities. The steamboiler problem features three such problems:

- the program to develop must communicate with the steamboiler ;
- the program and the steamboiler evolve in parallel;
- the physical laws that rule the boiler are related to time; it is thus necessary to take the real time into account to express several properties of the system.

In this case study, several technical solutions have been adopted to express these aspects in the standard framework of VDM. VDM specifications are based on the abstract machine paradigm. They link a series of state variables to several operations on this state. Therefore, the environment and the interface will be modeled by several state variables. Three kinds of variables are distinguished here (see figure 1.b):

- The *internal variables* are the variables of the controller. The controller has an unrestricted access to these variables, i.e. it may read or write from these.
- The *external variables* model the state of the environment, i.e. the steamboiler. For example, the level of water, the outcome of steam, are such variables. These variables can not be accessed directly by the controller. For example, it is impossible to modify the level of water by a simple assignment within the controller.
- The *interface variables* provide a connection between the controller and the boiler. Several properties link the evolution of these variables and the evolution of the environment. The controller has limited access to these variables: *captors* provide information on the environment and are read-only for the controller, *actuators* have an action on the environment and may be written by the controller.

The environment-based approach starts thus with a specification of the external variables, then introduces captors and actuators, and finally specifies the controller. This document is organised as follows. Sect. 2 gives an informal description of the steamboiler and states the major requirements of the problem. The subsequent sections present a VDM specification of the boiler structured by two refinements. The first specification (Part I) states the properties of the physical uncontrolled boiler, i.e. the possible behaviours of the boiler when arbitrary actions are performed by the actuators. The first refinement introduces the actuators to control the behaviour of the boiler (Part II). Finally, the captors are introduced and the relationship between the controller and its interface is more precisely defined (Part III). Sect. 15 proposes several evolutions of the specification in order to take into account the possible failures of components. Finally Sect. 16 draws the conclusions of this study and Sect. 17 evaluates and compares it to the other contributions.

2 Informal model of the boiler

In the environment-driven approach, the first focus is on the physical boiler. The steamboiler (Fig. 2) may be modeled by the following variables:

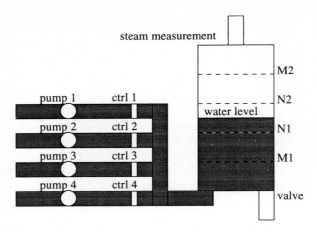

Fig. 2. The steamboiler (similar to the graphical appearance of the simulator)

- q_e the quantity of water in the steam-boiler;
- s_e the quantity of steam exiting the steam-boiler;
- $p_e(i), 1 \leq i \leq 4$, four variables which denote the status of the corresponding pumps: *stopped*, *pumping* and the transitory state *starting* (stopping the pump is instantaneous);
- v_e the *opened/closed* status of the valve;
- h_e the quantity of heat provided to the boiler.

These variables vary with time. The e subscript stands for "environment".

The physical behaviour of this system is ruled by several laws. One of the most important laws in the perspective of this study is the evolution of q_e with respect to time. It is ruled by the following differential equation:

$$dq_e/dt = (\sum_{i=1}^{4} p_e(i) * P) - v_e * V - s_e \qquad (1)$$

where P is the nominal capacity of each pump, and V the nominal capacity of the valve. In this equation, p_e is equivalent to 0 if it is *stopped* or *starting* and to 1 if it is *pumping*; similarly, v_e is 0 when *closed* and 1 when *opened*. This conversion could easily be stated more formally.

Variables p_e, v_e, and s_e vary with time, following a function that depends, among others, on the controller's actions. Therefore, it is impossible to solve this differential equation. For completeness sake, it should also be stated that:

$$s_e = f(h_e) \qquad (2)$$

This means that the amount of steam produced by the boiler is a function of the heat provided to it. This function is not linear since there exists a maximum quantity W of steam at the exit of the boiler.[2]

$$s_e \leq W \qquad (3)$$

There also exist maximum gradients on the increase (U_1) and decrease (U_2) of steam outcome[3].

$$-U_2 \leq ds_e/dt \leq U_1 \qquad (4)$$

The statement of the problem also introduces several constants of the boiler:

- C the maximal quantity of water in the steam-boiler,
- M_1 and M_2 the critical levels in the boiler,
- N_1 and N_2 the minimum and maximum amounts of water during normal operation.

To complete this physical model of the boiler, the notion of "mode" must be introduced. The statement of the problem details the following modes for the controller: initialization, normal, degraded, rescue, and emergency stop. These correspond to the controller viewpoint of the boiler. From the point of view of the physical boiler, this notion of mode also makes sense:

- in the emergency stop mode (*Stopped*), the physical environment is responsible to take appropriate actions;
- another physical, but undesired, mode is *Explosion*;
- normal operation takes place in a mode that is distinct from *Stopped* and *explosion*, it will be called *Running*;
- finally, an initialization mode (*Init*) is introduced which corresponds to the starting operations; it corresponds to a different physical behaviour than the *Running* mode, e.g. the environment eventually issues the STEAM-BOILER-WAITING message.

The notion of mode is thus not specific to the controller but also makes sense in the description of the steamboiler, since it impacts on its behaviour. In particular, one rule links this mode to the physical variables of the boiler:

$$\Box(M_1 < q_e < M_2) \Rightarrow \Box(mode \neq Explosion) \qquad (5)$$

where $\Box$ is the "always" temporal operator. This guarantees that the boiler never explodes if the water level remains non-critical. The correctness of this law is a basic assumption of the development of the controller. Actually the principal requirement on the controller is that:

Requirement 1 *mode $\neq$ Explosion*

Combined with this law, this requirement may be reached by:

Requirement 2 $M_1 < q_e < M_2$

---

[2] Obviously, if the heat increases until s reaches W, the exceeding heat will no longer be evacuated and there might be a danger for the boiler.

[3] As stated in appendix A, $U1$ and $U2$ are positive values.

Unfortunately, it is impossible to guarantee that the controller will always be able to keep q_e within these limits. In this case, the controller should give up and switch to *Stopped* mode. The requirement may thus be stated as:

Requirement 3 $M_1 < q_e < M_2 \vee (mode = Stopped)$

In normal operating mode, it is recommended that the water level remains between N_1 and N_2. This correspond to a "weak" requirement:

Requirement 4 *(weak)* $N_1 < q_e < N_2$

This weak requirement should be met most of the time but is not critical. Therefore, one may demonstrate it is met by means of informal arguments combined with a convincing series of tests.

PartI
Abstract Model of the Boiler

Starting from this informal model, the initial VDM specification describes the behaviour of the boiler without any controller. Fig. 3 shows the modular structure of this specification. Module *BOILER* (Sect. 3) defines several types and constants. Module *NCT-PUMPS-VALVE* (Sect. 4) defines the possible evolutions of the pumps and the valve within a given duration. Module *PHYSICAL-LAWS* (Sect. 5) defines the possible evolutions of all variables of the boiler, with the exception of the mode, for a given duration. It also computes if a given instant of time is safe for given initial values of the boiler. This module takes the evolutions of the pumps and the valve as parameters. *NCT-LAWS* corresponds to the instantiation of *PHYSICAL-LAWS* by *NCT-PUMPS-VALVE*. Module *NON-CONTROLLED* (Sect. 6) models the evolution of the boiler mode and specifies the evolution of all boiler variables. Finally, *PHYSICAL-BOILER* specifies an abstract machine which associates the boiler variables to one single operation which models the effect of the real time on these variables.

3 Types and constants of the boiler

A first module, named *BOILER*, introduces constants and types.

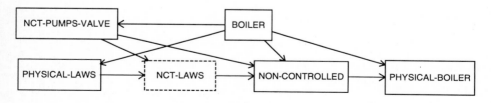

Fig. 3. The modular structure of the initial specification

Fig. 4. Graphical conventions for module diagrams

module *BOILER*
    exports all
definitions
    The constants of this module are the ones defined in Sect. 2. The values given here are the ones of the steamboiler simulator [L]. They are given because VDM mandates to link constants to values.
values

1.0 $C : kilograms = 1000$;
2.0 $M1 : water\text{-}level = 200$;
3.0 $M2 : water\text{-}level = 800$;
4.0 $N1 : water\text{-}level = 400$;
5.0 $N2 : water\text{-}level = 600$;
6.0 $W : kilograms\text{-}psec = 25$;
7.0 $U1 : kilograms\text{-}psec2 = 1$;
8.0 $U2 : kilograms\text{-}psec2 = 1$;
9.0 $P : kilograms\text{-}psec = 10$;
10.0 $V : kilograms\text{-}psec = 50$;

Additional constants in the statement of the problem are the number of pumps, the set of valid pump numbers, and the time needed to start a pump.

11.0 $Pump\text{-}nb\text{-}max : \mathbf{N}_1 = 4$;
12.0 $Pump\text{-}nb\text{-}set : pump\text{-}nb\text{-}\mathbf{set} = \{1, \ldots, Pump\text{-}nb\text{-}max\}$;
13.0 $Pump\text{-}starting\text{-}time : seconds = 5$

A first series of type definitions introduces the units of the problem: *kilograms* measures quantities of water or steam, *seconds* is the time unit, *kilograms-psec* measures flows of water or steam, *kilograms-psec2* measures gradients of flows of water or steam and *watts* is a power unit. These units are real numbers; the gradients and *watts*, aimed at measuring the heat provided to the boiler, are considered positive reals. Invariants on these types enforce this property.

types

14.0 $kilograms = \mathbf{R}$;
15.0 $seconds = \mathbf{R}$;
16.0 $kilograms\text{-}psec = \mathbf{R}$;
17.0 $kilograms\text{-}psec2 = \mathbf{R}$
   .1   **inv** $k \triangleq 0 \leq k$;
18.0 $watts = \mathbf{R}$
   .1   **inv** $w \triangleq 0 \leq w$;

As far as the semantics of VDM are concerned, these definitions do not express a difference between *kilograms* and *seconds*. Both are equivalent to reals!

Nevertheless, the introduction of these types is a meaningful way to document the specification. Three units are constructed on these: *water-level* corresponds to a level of the boiler, *steam-outcome* corresponds to the flow of steam and *duration* corresponds to a positive time interval.

19.0   *water-level* = *kilograms*

.1   **inv** $wl \triangleq 0 \leq wl \wedge wl \leq C$;

20.0   *steam-outcome* = *kilograms-psec*

.1   **inv** $so \triangleq 0 \leq so \wedge so \leq W$;

21.0   *duration* = *seconds*

.1   **inv** $d \triangleq d \geq 0$;

Further types are associated to pumps:

- *pump-nb* is the type of valid pump numbers;
- *pump-state* is an enumerated type that lists the possible states of the pump: state *Starting* corresponds to the intermediate state where a pump is switched on but has not balanced the pressure of the boiler.
- *pumps* is a map from pump numbers to pump states; an invariant enforces that it is total, i.e. its domain is the set of all pump numbers. In other words, each pump has a state.

22.0   *pump-nb* = $\mathbf{N}_1$

.1   **inv** $p \triangleq p \leq Pump\text{-}nb\text{-}max$;

23.0   *one-pump* = STOPPED | STARTING | PUMPING;

24.0   *pumps* = *pump-nb* $\xrightarrow{m}$ *one-pump*

.1   **inv** $pm \triangleq \mathbf{dom}\ pm = Pump\text{-}nb\text{-}set$;

The specification of the pumps could have been more precise: the last "switch-on/off" time could be recorded for each pump and the invariant could enforce the relation between the state of the pump and the corresponding time.

Type *valve* is an enumerated type that lists the possible states of the valve.

25.0   *valve* = OPENED | CLOSED;

Type *bmode* lists the possible modes of the boiler defined in Sect. 2.

26.0   *bmode* = STOPPED | RUNNING | INIT | EXPLOSION;

A composite type *bvalues* groups the physical variables of the boiler. Finally, *bstate* groups these variables with the boiler mode.

27.0   *bvalues* :: *qe* : *water-level*

.1   *se* : *steam-outcome*

.2   *pe* : *pumps*

.3   *ve* : *valve*

.4   *he* : *watts*;

28.0   *bstate* :: *val* : *bvalues*

.1   *mode* : *bmode*

end *BOILER*

Several physical constraints are enforced by the invariants associated to the types of these variables, e.g. an upper bound is set on *se*.

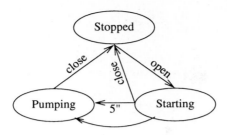

Fig. 5. The state diagram for the pumps

4 Pumps and Valve evolutions

The next sections will characterize the possible evolutions of the boiler variables. In this section, the possible behaviours of the pumps and valve are characterized. When the steam boiler is considered in isolation, every possible evolution of these state variables is allowed, in particular the pumps may be switched on or off and the valve state may be modified at any time. These evolutions may both result from actions from the controller or defects of the units.
module *NCT-PUMPS-VALVE*

      imports

29.0   **from** *BOILER* **all**

      exports all

definitions

As shown in figure 5, a pump may only be in one of three states: *Stopped*, *Starting*, and *Pumping*. The allowed transitions are *Stopped* to *Starting*, *Starting* to *Pumping*, *Starting* to *Stopped* and *Pumping* to *Stopped*. The transition from *Starting* to *Pumping* may take up to 5 seconds (*Pump-starting-time*). Actually, nothing prevents the pump from instantaneously reaching state *Pumping*. For example, it is reasonable to think that a "stop and pump" sequence, where the pump is only stopped for a very small amount of time, will not allow the pressure to diminish in the pump and that it will restart instantaneously. Our interpretation of the original statement is thus that the pump is guaranteed to reach state *Pumping* within 5 seconds, but may reach it before this time is elapsed. This is expressed in the diagram by two transitions between these states. The unlabelled transition may be taken at any time, while the transition labelled "5 seconds" will be taken after this time has elapsed. Function *reachable-pumps* computes the set of states that can be reached in dt from state p:

functions

   30.0   *reachable-pumps* : $BOILER`one\text{-}pump \times BOILER`duration$

    .1 $\rightarrow BOILER`one\text{-}pump\text{-}\mathbf{set}$

    .2   *reachable-pumps* $(p, dt) \triangleq$

    .3   {STOPPED, STARTING, PUMPING};

Actually, all states may always be reached instantaneously. Although the transition from *Starting* to *Pumping* may take some time, if the pump is *Starting* and no information is available about the time it went into this state, it may reach

Pumping at any time! In the refinements of the abstract model, this function will be more elaborated in order to take into account the action of the controller on the pumps.

A similar function is now defined for the valve. Since it may be opened or closed at arbitrary times, its behaviour is not constrained.

31.0   *reachable-valve* : $BOILER`valve \times BOILER`duration$
  .1 $\rightarrow BOILER`valve$-**set**
  .2   *reachable-valve* $(ve, dt) \triangleq$
  .3   {OPENED, CLOSED}

end *NCT-PUMPS-VALVE*

5 Approximations of physical laws

The physical laws that rule the behaviour of the boiler have been expressed in section 2 as differential equations. Unfortunately, VDM does not allow to constrain the behaviour of its variables by such elaborate mathematical formulae. In this section, several approximations of these physical laws are presented. They define upper and lower bounds on the physical variables for an evolution ranging over a given duration dt and starting from a given state. Within this dt interval, several evolutions of the pumps state and valve state are allowed. These evolutions may both result from actions by the controller or defects of the units. Thus the *PHYSICAL-LAWS* module is parametrized by two functions *reachable-pumps* and *reachable-valve* which characterize these evolutions. *Reachable-pumps* takes an initial pump state and a duration and returns the set of states which can be reached after this duration. The module exports two functions: *evolution* which states whether a final configuration of the boiler can be reached within a given time from an initial configuration, and *safe* which states whether it is safe to let the boiler evolve for a given duration from an initial configuration.

module *PHYSICAL-LAWS*
       parameters
32.0   **functions** *reachable-pumps* : $BOILER`one\text{-}pump \times BOILER`duration$
  .1 $\rightarrow BOILER`one\text{-}pump$-**set**,
  .2   *reachable-valve* : $BOILER`valve \times BOILER`duration$
  .3 $\rightarrow BOILER`valve$-**set**
       imports
33.0   **from** *BOILER* **all**
       exports
34.0   **functions** *evolution* : $BOILER`bvalues \times BOILER`duration$
  .1 $\times BOILER`bvalues \rightarrow \mathbf{B}$,
  .2   *safe* : $BOILER`bvalues \times BOILER`duration \rightarrow \mathbf{B}$

definitions

5.1 Boiler evolution

Function *evolution* takes as parameters two states of the boiler and a duration. It returns true if the final state $s1$ can be reached from the initial state $s0$ after the duration dt. It is decomposed into more elementary functions which rule the evolutions of qe, se, ve, and pe. Since no information is available about the evolution of he, no constraint is put on this element of the state. Function *evolution* expresses the upper and lower bounds on the final level of water, the steam outcome, and the sets of possible pump and valve states. As far as pumps and valves are concerned, it refers to the functions given as parameters to the module. As far as the water level and steam outcome are concerned, they must remain within minimum and maximum levels computed by the *min-water*, *max-water*, *min-steam*, and *max-steam* respectively. These min and max functions are defined in the rest of this module.

functions

35.0   *evolution* : $BOILER`bvalues \times BOILER`duration \times BOILER`bvalues$
.1 $\rightarrow \mathbf{B}$
.2   *evolution* $(s0, dt, s1) \triangleq$
.3   *min-water* $(s0, dt) \leq s1.qe \wedge s1.qe \leq$ *max-water* $(s0, dt) \wedge$
.4   *min-steam* $(s0.se, dt) \leq s1.se \wedge s1.se \leq$ *max-steam* $(s0.se, dt) \wedge$
.5 $s1.ve \in$ *reachable-valve* $(s0.ve, dt) \wedge$
.6 $\forall i \in BOILER`Pump\text{-}nb\text{-}set \cdot s1.pe(i) \in$ *reachable-pumps* $(s0.pe(i), dt)$;

5.2 Safe time

Based on the model of the physical evolution, a boolean function, *safe-instant*, is defined which estimates if the configurations that can be reached after dt from initial values bv are all safe. This function corresponds to a worst case analysis. It returns *true* if all possible evolutions of the water level remain inside the allowed limits $M1$ and $M2$ after dt, following requirement 2 of Sect. 2.

36.0   *safe-instant* : $BOILER`bvalues \times BOILER`duration \rightarrow \mathbf{B}$
.1   *safe-instant* $(bv, dt) \triangleq$
.2 $\forall bv2 : BOILER`bvalues \cdot$
.3   *evolution* $(bv, dt, bv2) \Rightarrow$
.4 $(bv2.qe < BOILER`M2 \wedge bv2.qe > BOILER`M1)$;

The previous function characterises one particular moment in time. The *safe* function is true if every instant between 0 and its parameter dt is safe.

37.0   *safe* : $BOILER`bvalues \times BOILER`duration \rightarrow \mathbf{B}$
.1   *safe* $(bv, dt) \triangleq$
.2 $\forall t : BOILER`duration \cdot t \leq dt \Rightarrow$ *safe-instant* (bv, t);

5.3 Steam

The next function definitions provide estimates on the evolution of the boiler variables. The first physical law is an estimate on the evolution of the steam. The following function computes the maximum flow of steam that can be reached after dt starting from a given flow $s0$. Fig. 6 shows that the evolution of the steam outcome (s) is a function of dt. *max-steam* and *min-steam* define an envelope

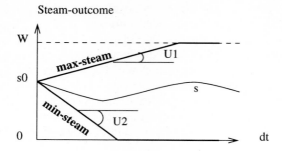

Fig. 6. The evolution of the steam outcome

on the set of physical functions of s (which does not mean that any function that fits into this envelope is admissible). They give thus a first estimate on s. Obviously, this estimate is more accurate if dt is small. Function $max\text{-}steam$ increases linearly until it reaches the maximum steam outcome W. The minimal value may be computed similarly (see appendix C).

38.0 $max\text{-}steam : BOILER`steam\text{-}outcome \times BOILER`duration$
.1 $\to BOILER`steam\text{-}outcome$
.2 $max\text{-}steam\ (s0, dt) \triangleq$
.3   **if** $s0 + dt \times BOILER`U1 \leq BOILER`W$
.4   **then** $s0 + dt \times BOILER`U1$ **else** $BOILER`W$

end $PHYSICAL\text{-}LAWS$

5.4   Quantity of water

The remaining functions of this module similarly compute upper and lower bounds on the quantity of water that goes inside the boiler. The specification of the module is given in appendix C.

6   Specifying the evolution of the boiler state

Module $NON\text{-}CONTROLLED$ specifies the evolution of every boiler variable without a controller. Most evolutions of variables have been defined in module $PHYSICAL\text{-}LAWS$. Therefore, $NON\text{-}CONTROLLED$ imports this module and instantiates it with the behaviour of the pumps and valve. This provides the *evolution* and *safe* functions.
module $NON\text{-}CONTROLLED$

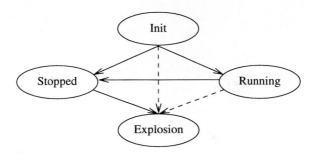

Fig. 7. The evolution of the mode

       imports
39.0   **from** *BOILER* **all** ,
40.0   **from** *NCT-PUMPS-VALVE* **all**
       instantiations
41.0   *NCT-LAWS* **as** *PHYSICAL-LAWS*
  .1   (*reachable-pumps* → *NCT-PUMPS-VALVE'reachable-pumps*,
  .2   *reachable-valve* → *NCT-PUMPS-VALVE'reachable-valve*) **all**
       exports all
definitions

6.1 Mode evolution

The evolution of the physical mode of the boiler is not described in module *PHYSICAL-LAWS*. Based on the notion of safe time, it is possible to express the logical laws that rule the evolution of the mode. The evolution of the mode is not arbitrary but must correspond to figure 7. For example, *Stopped* may be reached from *Init* or *Running* but may only be left to lead to *Explosion*.

Several assumptions are made here:

1. The possibility to restart the boiler is not taken into account. Once in *Stopped* mode, it is not possible to go back to *Init* mode. This is justified by the fact that this process is not specified in the problem statement. Another view of this choice is to consider that the specification corresponds to one complete cycle in the life of the boiler; restarting the boiler is thus to restart the whole system.
2. *Explosion* can be reached in *Stopped* mode: the engineers in charge of saving the boiler while in *Stopped* mode may fail.

Equation 5 of section 2 guarantees that the boiler may not explode if the water level remains within the critical limits *M1* and *M2*. This is expressed here by the dashed arrows. In *Stopped* mode, this assumption is not taken into account, mainly because the statement of the problem is primarily concerned with *Init* and *Running* and does not really detail *Stopped*. This evolution is

captured in the following function which expresses the set of modes that can be reached from an initial configuration of the state variables:
functions

    42.0   *reachable-modes* : $BOILER`bstate \times BOILER`duration$
      .1 $\rightarrow BOILER`bmode$-**set**
      .2   *reachable-modes* (bl, dt) $\triangleq$
      .3   **let** $explosion =$ **if** $NCT\text{-}LAWS`safe\,(bl.val, dt)$
      .4   **then** $\{\}$ **else** $\{\text{EXPLOSION}\}$ **in**
      .5   **cases** $bl.mode$:
      .6   INIT $\rightarrow$ {INIT, RUNNING, STOPPED} $\cup$ *explosion*,
      .7   RUNNING $\rightarrow$ {RUNNING, STOPPED} $\cup$ *explosion*,
      .8   STOPPED $\rightarrow$ {STOPPED, EXPLOSION},
      .9   EXPLOSION $\rightarrow$ {EXPLOSION}
     .10   **end**;

It must be noted that this function is not transitive: *Stopped* is a quite dangerous mode because explosion may take place at any time. As a consequence of Fig. 7, if the boiler is safe and in mode *Running*, and if it switches to emergency stop, it "forgets" about its safe status and may explode at any time. The *reachable-mode* function takes the assumption that if the boiler is safe for dt it remains safe even if it switches to emergency stop.

6.2 Boiler evolution

This *reachable-modes* function is then combined with $NCT\text{-}LAWS`evolution$. It states that the new physical mode is one of the reachable modes and that the physical variables are bounded according to the physical and logical laws.

    43.0   *boiler-evolution* : $BOILER`bstate \times BOILER`duration \times BOILER`bstate$
      .1 $\rightarrow \mathbb{B}$
      .2   *boiler-evolution* $(bl1, dt, bl2)$ $\triangleq$
      .3 $bl2.mode \in$ *reachable-modes* $(bl1, dt)$ $\wedge$
      .4 $(bl2.mode \in \{\text{INIT, RUNNING}\} \Rightarrow$
      .5 $NCT\text{-}LAWS`evolution\,(bl1.val, dt, bl2.val))$
end *NON-CONTROLLED*

6.3 The Physical Boiler

The module *PHYSICAL-BOILER* is the abstract machine of this first model of the boiler. An abstract machine is a collection of variables associated to operations on these variables. Here, the variables are bl of type $bstate$, the collection of variables of the physical boiler and te, the real time. A given configuration of the state variables corresponds thus to a snapshot of these values at time te.

module *PHYSICAL-BOILER*
       imports
    44.0   **from** *BOILER* **all** ,
    45.0   **from** *NON-CONTROLLED* **all**
       exports all
definitions

46.0 **state** *BOILER* **of**
.1   *bl* : *BOILER'bstate*
.2   *te* : *BOILER'seconds*
.3   **end**

Operations A single operation is provided which specifies the evolution of the system over a duration *dt*. This specification corresponds to an abstract machine with a single button (operation). Each time the button is pressed, the time evolves by an arbitrary duration *dt* and the state variables are updated, according to *boiler-evolution*.
operations

47.0 *Advance-time* ()
.1   **ext wr** *bl* : *BOILER'bstate*
.2   **wr** *te* : *BOILER'seconds*
.3   **pre true**
.4   **post let** *dt* : *BOILER'duration* **in**
.5 $te = \overleftarrow{te} + dt \land NON\text{-}CONTROLLED\text{'}boiler\text{-}evolution\,(\overleftarrow{bl}, dt, bl)$

end *PHYSICAL-BOILER*

The post-condition of the *Advance_time* operation states that the real time has increased by an arbitrary *dt* and that the boiler variables have evolved. VDM does not allow to specify what happens during the action, only the initial and final states are described by the pre- and post-conditions. But since *dt* is arbitrary, this provides a good approximation of the behaviour of the steam-boiler. This specification of the physical boiler completes the specification of the environment which is the first stage of the environment-driven approach.

PartII
Introducing the Controller

This first refinement introduces the notion of controller in the specification. Two major notions appear at this level: the notion of discrete time, i.e. the control takes place periodically in time, and the notion of actuator, i.e. the means available for the controller to influence the evolution of the boiler. Fig. 8 gives the structure of modules and refinements for this controlled version of the boiler. Module *D-TIME* (Sect. 7) defines several types, constants and functions related to the discrete view of time. Module *CT-PUMPS-VALVE* (Sect. 8) is a refinement of *NCT-PUMPS-VALVE* which takes into account the fact that the evolution of the pumps and valve are controlled. A new instantiation of *PHYSICAL-LAWS* is generated with the functions of this module. Module *CONTROLLED* (Sect. 9) is the refined version of *NON-CONTROLLED*, which takes into account this new instantiation. Module *ACTUATORS* (Sect. 10) introduces the actuators and their action on the environment. *SAFE-ACTIONS* (Sect. 11) defines, for a given configuration of the boiler, the actions of the controller that do not lead to a critical state. Finally *CONTROLLED-BOILER*

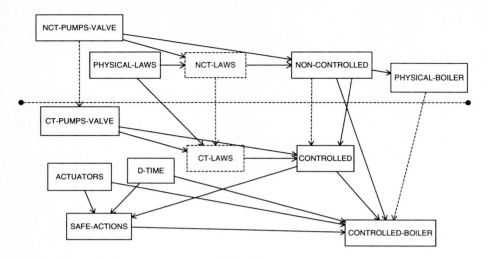

Fig. 8. The first refinement (*BOILER* has been omitted)

(Sect. 12) is the abstract machine which refines *PHYSICAL-BOILER* and takes into account the influence of the controller.

7 The sampling clock

An important characteristic of the controller is that it gets information from its captors every five seconds and computes actions for the actuators to be executed before the next cycle takes place. It thus relies on the assumption that the whole reaction is ended before the next cycle. Module *D-TIME* introduces this discrete notion of time.

module *D-TIME*
     imports
48.0   **from** *BOILER* **all**
     exports all
definitions

    Two constants are defined in this module:

- *Sampling-period* defines the periodicity of the controller, i.e. five seconds;
- *React-time* defines an upper bound on the time needed by the controller to get the values, compute a reaction, and transmit this reaction; here, we do not have any estimation on this time and it has been *arbitrarily* defined as 0.1 seconds, which corresponds to an order of magnitude under the sampling period; it is obvious that in a realistic application, this value should be the result of a preliminary study.

An important hypothesis is that *React-time* is smaller than the *Sampling-period*. Type *time-const* is a composite which will encapsulate these time constants. It enforces this property as an invariant.

types

49.0 $time\text{-}const :: sp : BOILER\text{`}duration$
.1 $rt : BOILER\text{`}duration$
.2   **inv mk-**$time\text{-}const\ (sp, rt) \triangleq rt < sp$

values

50.0   **mk-**$time\text{-}const\ (Sampling\text{-}period, React\text{-}time)$
.1 $: TIME\text{-}const = $ **mk-**$time\text{-}const\ (5, 0.100000)$

Based on these constants, two functions are defined on the real time. A boolean function, *sampling*, is defined to test if an instant of time is a sampling instant. Actually, the origin of time has been conveniently defined to be one of the sampling instants, so that sampling instants are multiples of the sampling period. Finally, a function of time provides the next sampling instant.

functions

51.0 $sampling : BOILER\text{`}seconds \to \mathbf{B}$
.1 $sampling\ (t) \triangleq$
.2   **floor** $(t/Sampling\text{-}period) \times Sampling\text{-}period = t;$

52.0 $Next\text{-}sampling : BOILER\text{`}seconds \to BOILER\text{`}seconds$
.1 $Next\text{-}sampling\ (t) \triangleq$
.2   **floor** $(t/Sampling\text{-}period) \times Sampling\text{-}period + Sampling\text{-}period$

end *D-TIME*

8   Pumps and Valve change

One of the characteristics of the controller is that it spends most of its time doing nothing. Actually, it only acts on the boiler once every five seconds. It is therefore interesting to instantiate the *PHYSICAL-LAWS* module to this kind of behaviour. The *CT-PUMPS-VALVE* module restricts the evolutions of the pump and valve states in the context where no action is performed. Figure 9 shows the possible transitions in this context. Actually, the only possible transition is *Starting* to *Pumping*.

module *CT-PUMPS-VALVE*

       imports

53.0   **from** *BOILER* **all**

       exports all

definitions
functions

54.0 $reachable\text{-}pumps : BOILER\text{`}one\text{-}pump \times BOILER\text{`}duration$
.1 $\to BOILER\text{`}one\text{-}pump\text{-}\mathbf{set}$
.2 $reachable\text{-}pumps\ (s, dt) \triangleq$
.3   **if** $s = $ STARTING
.4   **then if** $dt > BOILER\text{`}Pump\text{-}starting\text{-}time$
.5   **then** {PUMPING} **else** {STARTING, PUMPING}
.6   **else** $\{s\}$;

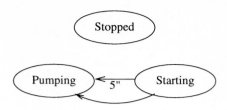

Fig. 9. The state diagram for the pumps, when no action takes place

This function is a refinement of the one presented in *NCT-PUMPS-VALVE* because the sets of resulting states are always included in the ones of the abstract function. The next function is a similar refinement for the behaviour of the valve. Obviously, when no action is performed, the state of the valve does not change.

55.0   *reachable-valve* : $BOILER`valve \times BOILER`duration \to BOILER`valve$-**set**

    .1   *reachable-valve* $(ve, dt) \triangleq$
    .2 $\{ve\}$

end *CT-PUMPS-VALVE*

9   Refined evolution of the boiler state

Module *CONTROLLED* is nearly an updated version of *NON-CONTROLLED* which takes into account *CT-PUMPS-VALVE* to instantiate *PHYSICAL-LAWS*. It exports a *boiler-evolution* function, and the *evolution* and *safe* functions of *CT-LAWS*. Its definition is given in appendix D.

10   Introducing Actuators

Actuators correspond to the available means for the controller to act on the boiler. In this case study, these are the pumps, the valve, and the ability to switch to emergency stop mode.

module *ACTUATORS*

       imports

56.0   **from** *BOILER* **all**

       exports all

definitions

The controller modifies these actuators by performing *actions*: on the pumps and the valve (to switch it on or off, and to do nothing) and on the mode (to activate the emergency stop or to do nothing). For each of the actuators, it was decided to introduce a "do nothing" action. This was motivated by the fact that the choice to do nothing is a decision of the controller. A *pumps-action* type is also introduced to assign an action to each pump.

types

57.0   *valve-action* = OPEN | CLOSE | NOTHING;

58.0   *pumps-action* = $BOILER`pump\text{-}nb \xrightarrow{m} valve\text{-}action$

.1   **inv** $pa \triangleq$ **dom** $pa = BOILER`Pump\text{-}nb\text{-}set$;

59.0 $mode\text{-}action = \text{STOP} \mid \text{NOTHING}$;

A composite type *actions* groups the simultaneous actions of the controller: pa is a map that assigns an action to each pump, ea is the emergency action performed, va is the action performed on the valve.

60.0 $actions :: pa : pumps\text{-}action$
.1 $ea : mode\text{-}action$
.2 $va : valve\text{-}action$

Actions modify the state of actuators. Functions *mode*, *valve*, and *pumps*, given in appendix E, express state machines which model the response of the actuators to the actions. These functions are used by *boiler-actuators*. It is a boolean function that establishes the consistency between some initial state of the boiler, bi, the actions performed on this state and the immediately resulting boiler state, bs. It is *true* if the actuators react to the actions as defined in the above listed functions.

functions

61.0 $boiler\text{-}actuators : BOILER`bstate \times actions \times BOILER`bstate$
.1 $\to \mathbf{B}$
.2 $boiler\text{-}actuators\,(bi, ac, bs) \triangleq$
.3 $bs.val.pe = pumps\,(ac.pa, bi.val.pe) \wedge$
.4 $bs.val.ve = valve\,(ac.va, bi.val.ve) \wedge$
.5 $bs.val.qe = bi.val.qe \wedge bs.val.se = bi.val.se \wedge$
.6 $bs.val.he = bi.val.he \wedge bs.mode = mode\,(ac.ea, bi.mode)$

end $ACTUATORS$

11   Safe actions

Having introduced the notion of action and their effects on the boiler state, the notion of "safe action" may be specified. A safe action is such that the boiler will remain safe for the next sampling period.

module $SAFE\text{-}ACTIONS$
    imports
62.0   **from** $ACTUATORS$ **all** ,
63.0   **from** $BOILER$ **all** ,
64.0   **from** $CONTROLLED$ **all** ,
65.0   **from** $D\text{-}TIME$ **all**
    exports all
definitions

In $CONTROLLED$, the notion of safe time corresponds to the absence of actions. Here, an additional notion of safe time is defined. It corresponds to the notion of safe time when an action takes place before the reaction time elapsed. Function $ct\text{-}safe\text{-}instant$ is true if the boiler is safe dt seconds after the actions pa and va have been performed on bv within the reaction time. This definition involves three instants in time:

- the initial time which corresponds to the values bv of the boiler variables;

- some intermediate time *date* where the actions take place; this intermediate time is less than the reaction time and less than dt.
- the final time, dt after the initial time.

The underlying assumption is that there is some uncertainty on the exact value of *date*, i.e. the time where the reaction takes place. It may be any time between 0 and *React-time*. Two states correspond to time *date*. The state before the actions take place, $bv2$, and the state after, $bv3$. They differ by the new state of the pumps and the valve and are linked by the *pumps* and *valve* functions to the values of the actuators (pa, va). The state before the actions took place must be consistent with the initial state and the *date* according to the *evolution* function. The final instant dt is safe if the time (*date*) when the actions took place was already safe, and if the time interval (dt - *date*), elapsed since the action took place, is safe with respect to $bv3$. It is interesting to notice that no action takes place within both periods of time (0 to *date* and *date* to dt), which justifies the use of CONTROLLED.

functions

66.0   *ct-safe-instant* : $BOILER`bvalues \times ACTUATORS`pumps\text{-}action$
.1 $\times ACTUATORS`valve\text{-}action \times BOILER`duration \to \mathbf{B}$
.2   *ct-safe-instant* $(bv, pa, va, dt) \triangleq$
.3 $\forall\, date : BOILER`duration \cdot$
.4 $date \leq D\text{-}TIME`React\text{-}time \wedge date \leq dt \Rightarrow$
.5 $(CONTROLLED`safe\,(bv, date) \wedge$
.6 $\forall\, bv2 : BOILER`bvalues, bv3 : BOILER`bvalues \cdot$
.7 $(CONTROLLED`evolution\,(bv, date, bv2) \wedge$
.8 $bv3.qe = bv2.qe \wedge$
.9 $bv3.se = bv2.se \wedge bv3.he = bv2.he \wedge$
.10 $bv3.pe = ACTUATORS`pumps\,(pa, bv2.pe) \wedge$
.11 $bv3.ve = ACTUATORS`valve\,(va, bv2.ve)) \Rightarrow$
.12 $CONTROLLED`safe\,(bv3, dt - date));$

The next function is a copy of *safe* of Sect. 5 where the actions are taken into account into the computation of the safe time.

67.0   *ct-safe* : $BOILER`bvalues \times ACTUATORS`pumps\text{-}action$
.1 $\times ACTUATORS`valve\text{-}action \times BOILER`duration \to \mathbf{B}$
.2   *ct-safe* $(bv, pa, va, dt) \triangleq$
.3 $\forall\, t : BOILER`duration \cdot t \leq dt \Rightarrow ct\text{-}safe\text{-}instant\,(bv, pa, va, t);$

This revised notion of safe time which takes into account the actions of the controller leads to define the notion of safe action: 1) emergency stop is always a safe action, 2) actions on the pumps and valve that ensures a safe time larger than the sampling period added to the reaction time are safe. A safe action allows thus to reach the next cycle and to react.

68.0   *safe-actions* : $BOILER`bstate \times ACTUATORS`actions \to \mathbf{B}$
.1   *safe-actions* $(bl, ac) \triangleq$
.2 $ACTUATORS`mode\,(ac.ea, bl.mode) = \text{STOPPED} \vee$
.3   *ct-safe* $(bl.val, ac.pa, ac.va,$
.4 $D\text{-}TIME`Sampling\text{-}period + D\text{-}TIME`React\text{-}time)$

.5 **pre** $bl.mode \in \{\text{RUNNING}, \text{INIT}\} \land$
.6 $CONTROLLED\text{'}safe\,(bl.val,\,D\text{-}TIME\text{'}React\text{-}time)$

end *SAFE-ACTIONS*

A pre-condition for the action to be safe is that the current safe time is larger than the reaction time. In other words, the reaction time is safe with respect to the initial state of the boiler. Otherwise, the action could take place too late. Also, the laws stated here correspond to the normal operating modes *Running* and *Init*.

12   Specification of a boiler under control

To end this first refinement, an abstract machine of the controlled boiler is specified. It refines the *PHYSICAL-BOILER* module.
module *CONTROLLED-BOILER*
   imports
69.0 **from** *ACTUATORS* **all** ,
70.0 **from** *BOILER* **all** ,
71.0 **from** *CONTROLLED* **all** ,
72.0 **from** *D-TIME* **all** ,
73.0 **from** *NON-CONTROLLED* **all** ,
74.0 **from** *SAFE-ACTIONS* **all**
   exports all
definitions

12.1   State variables

The state variables are the ones of *PHYSICAL-BOILER*. The initial state condition is defined. It merely expresses that, if the initial mode is *Init* or *Running*, the initial instant is a sampling instant and the state of the boiler allows some reaction to take place.

75.0 **state** *BOILER* **of**
.1 $bl : BOILER\text{'}bstate$
.2 $te : BOILER\text{'}seconds$
.3 **init** $\text{mk-}BOILER\,(bs,\,te)\ \triangleq$
.4 $bl.mode \in \{\text{INIT}, \text{RUNNING}\}\ \Rightarrow\ (D\text{-}TIME\text{'}sampling\,(te) \land$
.5 $CONTROLLED\text{'}safe\,(bl.val,\,D\text{-}TIME\text{'}React\text{-}time))$
.6 **end**

12.2   Operations

The single *Advance-time* operation of *PHYSICAL-BOILER* is now structured into two sub-operations :

– *Safe-advance-time* corresponds to the *Init* and *Running* modes. It is decomposed into two operations: *React* that takes place at every sampling instant and *Wait* which lets the boiler evolve until the next sampling instant, if *React* did not switch to emergency stop mode.

– *Unsafe-advance-time* corresponds to the *Stopped* and *Explosion* modes.

Advance-time groups these operations into a single operation that chooses the adequate evolution of the system depending on *mode*. While the mode remains *Init* or *Running*, this operation may only take place at sampling instants and it must be guaranteed that the reaction time is safe. This is expressed by the precondition.

operations

    76.0   *Advance-time* : () $\xrightarrow{o}$ ()
       .1   *Advance-time* () $\triangleq$
       .2   **cases** *bl.mode*:
       .3   RUNNING, INIT $\rightarrow$ *Safe-advance-time*() ,
       .4   EXPLOSION, STOPPED $\rightarrow$ *Unsafe-advance-time*()
       .5   **end**
       .6   **pre** $bl.mode \in \{\text{INIT}, \text{RUNNING}\} \Rightarrow D\text{-}TIME\text{'}sampling\,(te) \wedge$
       .7 $CONTROLLED\text{'}safe\,(bl.val,\,D\text{-}TIME\text{'}React\text{-}time)$;

Safe operations. *React* takes place in modes *Init* and *Running* at the sampling instants, as expressed in its pre-condition. The pre-condition also requires the boiler to remain safe for at least the reaction time.

    77.0   *React* ()
       .1   **ext wr** $bl : BOILER\text{'}bstate$
       .2   **wr** $te : BOILER\text{'}seconds$
       .3   **pre** $bl.mode \in \{\text{INIT}, \text{RUNNING}\} \wedge D\text{-}TIME\text{'}sampling\,(te) \wedge$
       .4 $CONTROLLED\text{'}safe\,(bl.val,\,D\text{-}TIME\text{'}React\text{-}time)$
       .5   **post let** $dt : BOILER\text{'}duration$ **be st**
       .6 $dt \leq D\text{-}TIME\text{'}React\text{-}time \wedge dt > 0$ **in**
       .7   **let** $ac : ACTUATORS\text{'}actions$ **be st**
       .8 $SAFE\text{-}ACTIONS\text{'}safe\text{-}actions\,(\overleftarrow{bl},\,ac)$ **in**
       .9   **let** $bi : BOILER\text{'}bstate$ **be st**
      .10 $CONTROLLED\text{'}boiler\text{-}evolution\,(\overleftarrow{bl},\,dt,\,bi)$ **in**
      .11 $ACTUATORS\text{'}boiler\text{-}actuators\,(bi,\,ac,\,bl) \wedge te = \overleftarrow{te} + dt$;

To reach the final state of the operation, three local variables are used:

– dt is the duration of the reaction; it is chosen to be less than the reaction time. It is also greater than 0, so that the second conjunct of the pre-condition of *Wait* is valid at the end of *React*.
– ac is the collection of actions taken; these actions must be safe with respect to the initial state. Moreover, if the controller does not switch to emergency stop mode, this guarantees that the boiler is safe for the next sampling period and the next reaction.
– bi is the state reached before the actions are performed. The underlying idea is that actions take place precisely at the final instant of *React*. bi corresponds thus to an evolution without action during the time of the operation.

React only involves two instants in time: the initial and final instants. This final instant corresponds to two states of the boiler: bi and bl. The last part of the

.4 $captors := Get\text{-}captors\,()$;
.5   **if** $bl.mode \in \{\text{INIT}, \text{RUNNING}\}$
.6   **then** $\|\ (actions := Compute\text{-}reaction\,(captors),\ Evolve()\,)$ **else skip**;
.7   **if** $bl.mode \in \{\text{INIT}, \text{RUNNING}\}$
.8   **then** $Put\text{-}actuators(actions)$ **else skip**)
.9   **pre** $bl.mode \in \{\text{INIT}, \text{RUNNING}\} \wedge D\text{-}TIME`sampling\,(te) \wedge$
.10 $CONTROLLED`safe\,(bl.val, D\text{-}TIME`React\text{-}time)$;

Get-captors returns the information stored in the captors variables *cp*. Since it takes some *Acquire-time* to perform the operation, it also affects the boiler state. This operation models its evolution while the operation takes place.

99.0 $Get\text{-}captors\,()\ r : CAPTORS`cstate$
.1   **ext wr** $bl : BOILER`bstate$
.2   **wr** $te : BOILER`seconds$
.3   **rd** $cp : CAPTORS`cstate$
.4   **pre** $bl.mode \in \{\text{INIT}, \text{RUNNING}\} \wedge$
.5 $CONTROLLED`safe\,(bl.val, Acquire\text{-}time)$
.6   **post let** $dt : BOILER`duration$ **be st** $dt \leq Acquire\text{-}time \wedge dt > 0$ **in**
.7 $r = cp \wedge te = \overleftarrow{te} + dt \wedge$
.8 $CONTROLLED`boiler\text{-}evolution\,(\overleftarrow{bl}, dt, bl)$;

The post-condition states that the result is equal to the values of the captors variables, that the time has increased by less than *Acquire-time* and that the boiler has evolved, without action, during the operation.

Put-actuators is the dual operation to *Get-captors*. It modifies the actuators variables according to the collection of actions given as input. The evolution of the boiler takes place during the operation. The action of the actuators on the state is supposed to take place at the final instant of the operation.

100.0 $Put\text{-}actuators\,(actions : ACTUATORS`actions)$
.1   **ext wr** $bl : BOILER`bstate$
.2   **wr** $te : BOILER`seconds$
.3   **wr** $ac : ACTUATORS`actions$
.4   **pre** $bl.mode \in \{\text{INIT}, \text{RUNNING}\} \wedge CONTROLLED`safe\,(bl.val, Act\text{-}time)$
.5   **post let** $dt : BOILER`duration$ **be st** $dt \leq Act\text{-}time \wedge dt > 0$ **in**
.6   **let** $bi : BOILER`bstate$ **be st**
.7 $CONTROLLED`boiler\text{-}evolution\,(\overleftarrow{bl}, dt, bi)$ **in**
.8 $te = \overleftarrow{te} + dt \wedge actions = ac \wedge$
.9 $ACTUATORS`boiler\text{-}actuators\,(bi, ac, bl)$;

The post-condition states the maximal bound on the duration of the operation, the new values of the interface variable *ac*, and the evolution of the boiler during the operation. This evolution is slightly more complex than for *Get-captors*. It includes an intermediate state *bi* which corresponds to the state reached after *dt* but before the actions are applied. Then, *boiler-actuators* expresses the instantaneous effect of the actuators on the boiler at the last instant of the operation.

Controlling the boiler. The control part of *React* is an operation that has no access to either the boiler state or the captors/actuators. The interactions with the interface only take place through its input/output parameters. It may only modify its internal variables *local-ve* and *local-mode*.

101.0   *Compute-reaction* $(cp : CAPTORS`cstate)\ ac : ACTUATORS`actions$
   .1   **ext wr**   *local-ve* : $BOILER`valve$
   .2   **wr**   *local-mode* : $BOILER`bmode$
   .3   **pre**   *local-mode* $\in$ {INIT, RUNNING}
   .4   **post** *captors-safe-actions* $(cp, local\text{-}ve, local\text{-}mode, ac) \land$
   .5 $local\text{-}ve = ACTUATORS`valve\,(ac.va, \overleftarrow{local\text{-}ve}) \land$
   .6 $local\text{-}mode = ACTUATORS`mode\,(ac.ea, \overleftarrow{local\text{-}mode})$;

The pre-condition of this operation is not absolutely necessary, it reminds the fact that the notion of safe action only makes sense in modes *Init* and *Running*. Therefore, it is reasonable to assume that the operation will be started with these values for *local-mode*. The post-condition expresses that the computed actions are safe with respect to the captors values and that the local variables should be updated with respect to these actions. As such, The *Compute-reaction* operation takes no time, because it may not access the real-time variable *te*. To model its duration, *Evolve* is defined. It expresses the evolution of the boiler during the computation of the reaction.

102.0   *Evolve* ()
   .1   **ext wr** $bl : BOILER`bstate$
   .2   **wr** $te : BOILER`seconds$
   .3   **pre** $bl.mode \in$ {INIT, RUNNING} $\land$
   .4 $CONTROLLED`safe\,(bl.val, Compute\text{-}time)$
   .5   **post let** $dt : BOILER`duration$ **be st** $dt \leq Compute\text{-}time \land dt > 0$ **in**
   .6 $te = \overleftarrow{te} + dt \land CONTROLLED`boiler\text{-}evolution\,(\overleftarrow{bl}, dt, bl)$

end *ARCHITECTURE*

The pre- and post-conditions are similar to the ones of *Get-captors* as far as the boiler variables are concerned.

14.5   Correspondence with $CONTROLLED\text{-}BOILER`React$

In order to prove that this operation is actually a refinement of the one defined in the previous module, several points should be examined:

— The definition of the time constants guarantees that the reaction takes place before *React-time*.
— It should be proven that the actions taken are safe. In fact, the reaction takes place at a sampling instant and the state invariant guarantees the correspondence of the captors values with the boiler state at sampling instants. Therefore, the state reconstructed by *captors-safe-actions* corresponds to the initial state of the boiler in the abstract version of *React*.
— It should be proved that boiler evolutions are compatible. Actually, the boiler evolution of the abstract operation is refined as a sequence of three evolutions. For this sequence to be a valid refinement, it may be proved that if

$dt = dt_1 + dt_2$, every possible evolution over dt_1 followed by an evolution over dt_2 remains in the envelope of the evolution over dt.
- Each sub-operation requires a safe time in its pre-condition. Since the pre-condition of *React* guarantees a safe time greater than *React-time*, since the sum of the sub-operations times is smaller than *React-time*, and since no change is performed on the actuators before the end of the sequence of sub-operations, it is reasonable to assume that all three sub-operations are undertaken in safe conditions.

15   Extensions

The specification presented up to now does not cover every aspect of the original problem. This section briefly presents how several characteristics of the problem can be modeled. These include the failures of the captors, a more precise description of the behaviour of the pumps, the specification of the initialization mode, the communication protocols between the boiler and its controller, and the weak requirement on the water level.

Failures of the captors   One of the major challenges of the problem is to take into account the possible failures of the physical units. This aspect could easily be taken into account in this specification. The defective state of a physical unit can be stored in a boolean variable which is part of the environment and hence can not be read by the program. A new definition of *boiler-captors* and *boiler-actuators* would take these boolean variables into account to express the consistency between the interface (captors and actuators) and the environment (the boiler). For example, in *boiler-captors* instead of $qe = qc$, one would state $qc\text{-}captor\text{-}ok \Rightarrow qe = qc$. This means that a defective captor may return any value.

A further look should be taken at the definition of the notion of safe time. If the pumps can break at any time, the corresponding safe time is no longer *ct-safe* but is closer to the more general *safe* instantiated in *NON-CONTROLLED*.

Behaviour of the pumps   The behaviour of the pumps, and in particular the transition from *Starting* to *Pumping*, could be modeled more precisely. In particular, the time when the pump enters the *Starting* mode could be added to its state variables. In the present specification, this information is not given. So that once a pump is *Starting*, the model does not provide a very precise information on the maximal time remaining before it starts pumping.

Since the *Sampling-period* is equal to the *Pump-starting-time*, this information is not really necessary. But if one wanted to decrease the sampling period (e.g. to 1 second), it would be required to model this particular pump behaviour. Adding this information to the model is straightforward: for example, a map could be defined between the pumps and their starting times.

Initialization mode and communication protocols The model given sofar is valid in both initialization and running modes. Initialization is a mode where the heat, and thus the steam outcome, is null. Nevertheless, all the requirements on the initialization mode are not captured in this model: the controller must enter a protocol with the "physical units". The description of this protocol corresponds to the definition of a state machine which corresponds to this global state. To describe a state machine in VDM is not very complex: for example it may be modeled by a state variable and operations corresponding to the transitions. It must be noted that the emission of the *PROGRAM-READY* signal is a supplementary action on the physical mode of the boiler.

Similarly, the treatment of the communication protocols, e.g. reporting that a captor has failed, can be described by state variables and operations, or by a state transition function that returns the new state and the emitted messages in function of the input messages and the initial state. The current description of the boiler abstracts from these communication details. For example, the state *Init* abstracts from the initialization protocol, and the operation *Get-captors* abstracts from the details of the communication with the captors (e.g. the actual format of the messages). Incorporating these details corresponds to a refinement of the specification.

Weak requirement on the water level In section 2, the fourth requirement expresses that the level should remain between $N1$ and $N2$. It is weak because if should not be fulfilled at all times. This requirement has not been taken into account in this specification. It should rather be considered as a criterion to decide between possible refinements of the controller. Its fulfillment can be checked by informal arguments or by a series of tests which evaluate the percentage of time where the requirement is met.

16 Conclusions

This study was merely concerned with the development of a specification of the steam boiler. The use of a formal specification language fostered a precise description of the system which led to further clarify some aspects of the boiler (see appendix A). This specification process has improved our understanding of the problem which is a prerequisite for a safe development of a controller. Moreover, the result of this study has been reported in this document to provide a precise communication support for discussions with either engineers responsible for the boiler or developers in charge of designing a controller.

The environment-based approach supported this specification activity. It started with a very liberal specification of the boiler (*PHYSICAL-BOILER*) which included every possible behaviour, taking into account the assumption that an explosion may not happen if the water level remains within $M1$ and $M2$. The first refinement gives a first specification of the controller where the notion of discrete time and the notion of actuator are introduced. The safe character of this controller has been demonstrated informally. Finally, it is refined into *ARCHITECTURE* which structures the reaction of the controller. Operations enforce the read-only access to the captors, the write-only access to the actuators, and the absence of direct access to the variables of the boiler. Also, a fourth kind of state variable is added: the internal variables of the controller.

The environment-based approach definitely directed the specification process. It provided a guideline to help us answer the "What to do next?" question. This constrasted with our previous attempts to directly focus on the specification of the controller where several questions were popping up constantly: "Did I choose the right level of abstraction?", "What properties of the boiler do I rely on?", "Is this specification feature a premature design decision?" A major benefit of this specification approach is that it does not restrict the freedom of the designer by selecting a set of solutions. The task assigned to the controller is clear: to keep the water level within limits or to give up and go into emergency stop. It corresponds to the safety requirements of the boiler. It is the designer's responsibility to meet these objectives while ensuring the maximum availability of the boiler.

The standard VDM-SL [ABH+95] has been used throughout this specification process. Its use was supported by state of the art tool technology for pretty-printing, syntax and type checking, i.e. the IFAD toolbox[ELL94]. Also, the (non-standard) modular constructs supported by the toolbox structure the specification and allow to reuse parts of it by instantiations of the *PHYSICAL-LAWS* module. Obviously the size of the specification mandates the use of such contructs. One of the limitations of these modular constructs is that neither imported functions nor the functions from instantiated modules may be exported from the module. For example, module *BOILER* had to be explicitly imported into every module.

The main limitation with the standard VDM-SL encountered in this case study was the lack of constructs to express parallel activities. This is particularly obvious in *ARCHITECTURE*, where the parallel evolutions of the controller and the boiler are not straightforward to express. One possible evolution of the semantics of the language could be to consider modules as independent state machines which interact through their operations. Separate state machines could then be defined for the boiler and the controller. In fact, the semantics associated to modules is still a matter of discussion for VDM-SL.

Validation of the specification. In order to get high quality specifications, it is not sufficient to write these in a formal language. In this case study, several techniques were used to improve the quality of the specification. The syntax and type checkers have ensured the conformance with the standard notation. The specification is documented by this paper and the attached appendixes. This documentation process has forced to understand the specification precisely and to informally prove some properties. The document has been read by its authors but also by independent readers. Finally, rework of the specification simplified and restructured it. Each of these activities forces to take a different viewpoint on the specification. Further validation activities could have been undertaken:

the IFAD toolbox allows to test the executable parts of the specification, proof obligation generators and theorem provers help demonstrating the consistency of the specification.

In summary The environment-based specification approach has provided us with an intellectual tool to address the steam boiler problem, in a structured and rigorous way. Its result is an abstract specification that improved the understanding of the problem by the specifiers. We wish it brought similar benefits to the readers of this document.

17 Evaluation and Comparison

1. The specification covers the behaviour of the boiler, how it can be perceived through the available captors, or influenced by the available actuators. These provide a context for the implicit specification of the controller. These specifications are formal. The specification of the possible failures of the boiler has not been expressed formally, but sketched verbally. Several consistency properties of the specification have been considered verbally.
2. (a) No, it only addresses the specification activity. Still some parts of the specification are executable, but they correspond to a modelisation of the boiler, and not of the controller.
   (b) No, but the simulator helped clarifying some imprecisions in the original statement of the problem (see appendix A).
   (c) No.
3. The main characteristics of our work are: a model-based language, the modelisation of the environment, and the fact that it is limited to the specification phase. The solutions that use B, Z, or RSL also use a model-based specification language. Abrial's solution [A] mainly focuses on design and verification activities and less addresses the problem of stating the specification. To some extent, it addresses the activity that follows the specification activity covered by our work and can be considered as complementary. Also, the fact that it uses a model-based formalism may ease the transition between methods. Buessow and Weber [BW] are closer to our contribution. The use of the Statecharts formalism in combination with Z is a good way to address the reactive aspects of the problem. Schinagl's approach, using RSL [S], appears more object-oriented than ours. Also, it tends to be more closer to an implemented solution. One objective of our approach was to avoid over-specification and premature design decisions which may appear in the very explicit style adopted in Schinagl's work.
4. (a) Two persons worked part-time on the problem during 3 months. The total amount of time spent is estimated to 1.5 man*month, but it definitely benefted from the "idle" time to mature the solution.
   (b) An average programmer needs 2 weeks of training for the VDM method and tools, but the production of a similar specification needs a deep understanding of the method which can only be reached after several months of practice under the supervision of an expert.

5. (a) No. The paper assumes a basic knowledge of the principles and concepts of VDM, corresponding to the level of a one-day tutorial.
   (b) The goal of the document is mainly to specify the problem. Not to propose a solution. An average programmer can understand the specification provided he has a basic knowledge of VDM.
   (c) One week to understand the specification.

Acknowledgements We wish to thank Sten Agerholm and the anonymous reviewers of this book for their comments on the first version of this document.

References

[ABH+95] D.J. Andrews, H. Bruun, B.S. Hansen, P.G. Larsen, N. Plat, et al. *Information Technology — Programming Languages, their environments and system software interfaces — Vienna Development Method-Specification Language Part 1: Base language*. ISO, 1995.

[Abr96] J.R. Abrial. *The B-Book*. Cambridge University Press, to appear in 1996.

[CGR93] D. Craigen, S. Gerhart, and T. Ralston. An international survey of industrial applications of formal methods. Technical Report NISTGCR 93/626, U.S. National Institute of Standards and technology, 1993.

[ELL94] R. Elmstrom, P. G. Larsen, and P. B. Lassen. The IFAD VDM-SL toolbox: a practical approach to formal specifications. *ACM SIGPLAN Notices*, 29(9):77–80, 1994.

[Jac83] M.A. Jackson. *System development*. Prentice-Hall, 1983.

[Jon90] C. B. Jones. *Systematic Software Development Using VDM (Second Edition)*. Prentice-Hall, London, 1990.

[Lam91] L. Lamport. The temporal logic of actions. Technical Report SRC-79, DEC Systems Research Center, Palo Alto, december 1991.

[LC94] Y. Ledru and P. Collette. Environment-based development of reactive systems. In D. Till, editor, *Sixth BCS-FACS Refinement Workshop, London*, Workshops in Computing. Springer Verlag, 1994.

[Led91] Y. Ledru. *Towards the formal development of terminating reactive systems*. PhD thesis, Université Catholique de Louvain, Unité d'Informatique, 1991.

[Led93] Y. Ledru. Developing reactive systems in a VDM framework. *Science of Computer Programming*, 20(1-2):51–71, 1993.

[Spi92] J.M. Spivey. *The Z notation - A Reference Manual (Second Edition)*. Prentice Hall, 1992.

Proving Safety Properties of the Steam Boiler Controller

Formal Methods for Industrial Applications: A Case Study

Gunter Leeb
leeb@auto.tuwien.ac.at
Vienna University of Technology
Department for Automation
Treitlstr. 3, A-1040 Vienna, Austria

Nancy Lynch
lynch@lcs.mit.edu
Massachusetts Institute for Technology
Laboratory for Computer Science
Technology Square 545, Cambridge, MA

Abstract

In this paper we model a hybrid system consisting of a continuous steam boiler and a discrete controller. Our model uses the Lynch-Vaandrager Timed Automata model to show formally that certain safety requirements can be guaranteed under the described assumptions and failure model. We prove incrementally that a simple controller model and a controller model tolerating sensor faults preserve the required safety conditions. The specification of the steam boiler and the failure model follow the specification problem for participants of the Dagstuhl Meeting "Methods for Semantics and Specification."

1 Introduction

The number of different formal methods for specifying, designing, and analyzing real-time systems has grown difficult to survey. For the purpose of comparison, some problems have been defined or borrowed from real-life applications. One such benchmark problem is the Steam Boiler Controller problem discussed in this paper. Another representative of this kind of problem is the Generalized Railroad Crossing (GRC) [Hei93]. Various approaches have been applied to the latter, e.g., [Cle93,Jah86,Sha93,Hoa93]. Many steps of the approach described here are similar to the steps described in [Hei94].

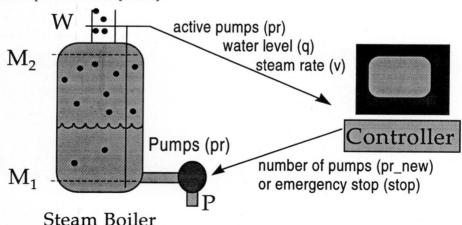

Figure 1: The steam boiler system. This picture shows the information flow between the controller and the steam boiler. It also gives some notion about the capacities of a pump (*P*), the limits for the steam rate (*W*) and the boundaries for the water level (*M_1* and *M_2*).

However, the Steam Boiler Controller represents a different kind of problem. Basically, it consists of a discrete control loop where several components may fail. The full version of this paper on the CD-ROM contains a condensed and informal description of the Steam Boiler Controller specification. The original specification can be found in [AS96]. Since even the original specification is informal and ambiguous, the condensed version on the CD-ROM summarizes our interpretation of the described problem.

The rest of this paper is organized as follows: After presenting an outline of our formal methods (Section 2), we state the assumptions we make for our model and show how the model is related to the physical model (Section 3). The following two sections describe the model of the boiler and a simple controller. In Section 6, we show some key model invariants. In Section 7, we present a similar controller which allows for sensor faults and we show its correctness incrementally based on the simpler controller model.

2 The Formal Framework

Applying formal methods to a system involves three steps: the system requirements specification, the design of an implementation, and the verification that the implementation satisfies the specification. The system requirements specification describes all acceptable system implementations [Hei94]. It has three parts:

1. A formal model describing the environment (e.g., the steam boiler) and its interface
2. A formal model describing the controller system and its interface at an abstraction level
3. Formal statements of the properties that the system must satisfy

The formal method we used to specify the steam boiler problem and to develop and verify a solution represents both the controller and the system environment as Timed Automata, according to the definition of Lynch and Vaandrager [Lyn91]. A Timed Automaton is a very general automaton, i.e., a labeled transition system. It is not finite-state: for example, the state can contain real-valued information, such as the current time or the current steam rate. This characteristic makes Timed Automata suitable for modeling not only discrete computer systems but also real-world entities such as the steam boiler. We base our work directly on an automaton model rather than on any particular specification language, programming language, or proof system, so that we may obtain the greatest flexibility in selecting specification and proof methods. The formal definition of a Timed Automaton appears in the CD-ROM version in Appendix A. Appendix B describes the Simulation Mapping method used for incremental reasoning about other increasingly specific instances of the model.

The Timed Automaton model supports the description of systems as collections of Timed Automata, interacting by means of common actions. In our example, we define separate Timed Automata for the steam boiler and the controller system; the common actions are sensors reporting the current state of some parameters of the boiler and actuators controlling the pumps of the boiler.

Actions change the state and, in particular, some variables of the state of an automaton. As a distinction between variables of the pre-state and the post-state, we write variables of the post-state (or the representation of the whole post-state) with a prime. In changing the state, actions perform a step or transition. Such a step or transition defines the change from one state s to another state s' by an action a, which is formally written as (s, a, s') or $s_A \xrightarrow{a}_A s'_A$, where the subscript A stands for the name of the particular automaton.

For the communication with other automata, we define input, output and internal actions. Such input actions will be enabled by output actions of another automaton. For example, the actuator output action in the controller model is synchronized with the actuator input action of the steam boiler model. The inherent flexibility of the method allows, for example, the introduction of a new automaton representing channel and message transfer characteristics to be employed in-between the boiler automaton and the controller automaton, interfacing with an input action from the controller and an output action to the steam boiler model. This allows us to model more complex systems without major changes to the previous automata. Furthermore, with this composition, we can reuse information we gained about the separate automata.

We describe the Timed Automata using precondition-effect notation. The precondition identifies particular states in which the system performs some actions. For any state fulfilling the precondition, the effect part describes how the state is changed by the particular action. This has several advantages. First of all, it is easy to understand. Even more important is that implementations can follow the abstract model description and even allow for simple validity checks in the code. In addition, all the invariants proved represent useful checks to be validated while running the final application. This approach will help to identify rare kinds of faults that are not even considered in the model. In this view, formal verification with Timed Automata is a constructive approach to systems development.

3 Further Considerations for Our Model

For our model, we need to know some more information about the physical behavior. Some of the following assumptions follow the informal specification of [AS96] or are intended to resolve some ambiguity. As suggested by [AS96], to simplify reasoning about the model, we ignore second order effects like the volume expansion of water when heated. This reasoning implies that a unit of water measured as steam can be replaced by pumping in exactly one unit of water.

Most important is some knowledge about how fast the steam rate may change over time. We assume a reasonable worst case situation where the steam rate increases at most with U_1 liters per second per second. In other words, the maximum gradient of increase of the steam rate is U_1 l/s^2. Symmetric to this, we know that the fastest decrease of the steam rate is denoted with U_2 l/s^2.

Furthermore, no pump supplies water unless activated and then it supplies a constant, exactly known amount of water per second denoted with P liters per second. The delay between reading the sensors and consequently changing the active pumps, denoted with S, is caused mainly by the slow reaction of the physical pumps. As a minor difference to the specification in [AS96], we assume the same delay for the activation and the deactivation of pumps. Since the pumps cause most of the delay S, we assume any boiler shut down is activated instantaneously and the whole process of shutting down the steam boiler is left to a later phase which we do not consider in this model. In the same way, we omit the initialization phase, which should force the boiler state into a particular acceptable set of start states before the boiler becomes fully operational. We assume all parameters of the start state for this model are already in their correct operational ranges. Moreover, we assume that the controller may decide to shut down the boiler any time it sets the new pumps. This assumption includes the possibility that the operator initiates an emergency stop, and it provides the flexibility to incorporate other reasons to shut down the boiler.

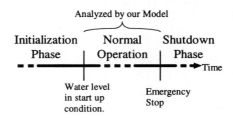

Figure 2: Our model only considers the time of normal operation. At the beginning, the initialization phase provides all parameters in the correct range and the shutdown phase is activated through setting parameter *stop* to *true*.

Other helpful assumptions are correct and accurate sensor values or the detection of a sensor fault. Perfect fault detection and identification are necessary for our model but will not be available in reality. In this aspect our model might need improvement if it is necessary to study such general cases. For example, the techniques developed for probabilistic Timed Automata [Seg94] seem to be appropriate for a problem requiring the analysis of such probabilistic properties. Probabilistic Timed Automata would allow one to assign probabilities to certain actions, e.g., for a successful error detection, and to prove the probability of a certain system behavior.

As a further simplification, we choose a very simple fault model which, in fact, includes or is close to most common fault conditions. The fault model assumes that every pump may fail and stop pumping water into the boiler. As a minor simplification, we assume for our model that any pump fault only occurs at times when pumps may be activated or stopped. This happens periodically whenever the parameter *set* equals the current time (*now*). Thus, pumps, when successfully activated, supply water at least to the next instant where pumps might change their behavior. Moreover, we assume that the activation delay, i.e., the time from reading the sensor values until consequently the pumps change their behavior, is smaller than the time between two successive sensor readings ($S < I$).

The goal of modeling the steam boiler and the controller with Timed Automata is to show certain important properties. In this case, we want to verify that our controller model does not violate safety. Therefore, we have to show that neither the steam rate nor the water level crosses its critical limits.

Next, we summarize the information we have about the physical model.

3.1 The Physical Model

We assume the steam rate expressed as a function over time $(sr(t) \geq 0)$ is differentiable. Furthermore, we know that

$$-U_2 \leq \dot{sr}(t) \leq U_1$$

and

$$wl(t) = wl(0) + \int_0^t pr(x)dx - \int_0^t sr(x)dx$$

where $\dot{sr}(t)$ represents the derivative of the steam rate function and $wl(t)$ the amount of water in the boiler at the time t and $pr(t)$ (≥ 0) the (discrete) pump rate function over time. We apply the following transformation to this information to make our model easier to follow.

We know $-U_2 \leq \dot{sr}(t)$, which implies $0 \leq \dot{sr}(t) + U_2$ and in general

$$\int \dot{sr}(t) + U_2 \, dt = sr(t) + t*U_2 + C.$$

Thus, we know that for all Δt,

$$sr(t + \Delta t) + U_2 * \Delta t \geq sr(t)$$

and symmetrically

$$sr(t + \Delta t) - U_1 * \Delta t \leq sr(t).$$

In the following, we use s for $sr(t)$ and s_{new} for $sr(t + \Delta t)$. With a similar straightforward calculation as before, we get

$$wl(t + \Delta t) \geq wl(t) + \int_t^{t+\Delta t} pr(x)dx - \delta_{HIGH}(s, s_{new}, \Delta t)$$

and symmetrically

$$wl(t + \Delta t) \leq wl(t) + \int_t^{t+\Delta t} pr(x)dx - \delta_{LOW}(s, s_{new}, \Delta t)$$

with

$$\delta_{HIGH}(s, s_{new}, \Delta t) = \left(\frac{2\Delta t U_2 s + 2\Delta t U_1 s_{new} + \Delta t^2 U_1 U_2 - (s - s_{new})^2}{2U_1 + 2U_2} \right)$$

and

$$\delta_{LOW}(s, s_{new}, \Delta t) = \begin{cases} \left(\frac{2\Delta t U_1 s + 2\Delta t U_2 s_{new} - \Delta t^2 U_1 U_2 + (s - s_{new})^2}{2U_1 + 2U_2} \right) & \text{if } \left(\frac{s}{U_2} + \frac{s_{new}}{U_1} \right) > \Delta t \\ \left(\frac{s^2}{2U_2} + \frac{s_{new}^2}{2U_1} \right) & \text{otherwise} \end{cases}$$

δ_{HIGH} describes the maximum amount of water that could evaporate and δ_{LOW} the minimum amount of water. Obviously, δ_{LOW} depends on whether the steam rate might drop to 0 in the interval Δt. Figure 3 represents δ_{HIGH} and δ_{LOW} graphically for an arbitrary interval t. Figure 3 ignores the pump rate, and the shaded areas represent the water evaporated into steam until a certain point in time. In other words, δ_{HIGH} and δ_{LOW} represent the worst case amount of water that could evaporate into steam in interval Δt. Both depend on the knowledge of the steam rate at the beginning and the end of the interval. The basic dependencies shown in the following Lemma 1 are sufficient for all further proofs.

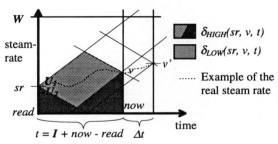

Figure 3: Example of what δ_{HIGH} and δ_{LOW} represent. For different intervals the maximum and minimum amount of water evaporated into steam depends on the steam rate at the beginning of the interval and at the end.

The following Lemma lists all necessary relations about the steam development functions δ_{HIGH} and δ_{LOW}. The intuition for this lemma can be gained from Figure 3. Obviously, two consecutive intervals can be joined and the minimum and maximum amount of water is smaller and bigger respectively or equal to the minimum/maximum water evaporated in both subintervals.

Lemma 1: For all $a, b, c \geq 0$, all constants > 0 and $t, u > 0$:

1) $\delta_{LOW}(a, b, u) \leq \delta_{HIGH}(a, b, u)$
2) $\delta_{LOW}(a, b, u) \geq \max(a * u - U_2 * u^2/2, a^2/(2*U_2)) \geq 0$
3) $\delta_{LOW}(a, b, u) \geq \max(b * u - U_1 * u^2/2, 0)$
4) $\delta_{LOW}(a, b, u) + \delta_{LOW}(b, c, t) \geq \delta_{LOW}(a, c, t + u)$
5) $(a + b)*u/2 \geq \delta_{LOW}(a, b, u)$
6) $\delta_{HIGH}(a, b, u) \leq (b * u + U_2 * u^2/2)$
7) $\delta_{HIGH}(W-U_1, W, I) = W*I - U_1*I^2/2$
8) $\delta_{HIGH}(a, b, u) + \delta_{HIGH}(b, c, t) \leq \delta_{HIGH}(a, c, u + t)$
9) $\delta_{HIGH}(a, b, u) \geq (a + b)*u/2$
10) $\delta_{HIGH}(a, b, u) \leq (a * u + U_1 * u^2/2)$

Proof: 1. - 10.: By calculus. ∎

Based on this information, we can now model the steam boiler as a Timed Automaton.

4 The Boiler Model

Constants

Name	Type	Unit	Description
I	real, > 0	s	time in-between periodical sensor readings
S	real, > 0	s	delay to activate pumps after the last sensor reading
U_1	real, > 0	l/s^2	maximum gradient of the increase of the steam rate
U_2	real, > 0	l/s^2	maximum gradient of the decrease of the steam rate
M_1	real, > 0	l	minimum amount of water before boiler becomes critical
M_2	real, > 0	l	maximum amount of water before boiler becomes critical
W	real, > 0	l/s	maximum steam rate before boiler becomes critical
P	real, > 0	l/s	exact rate at which one active pump supplies water to the boiler
#pumps	int, > 0		number of pumps that can supply water to the boiler in parallel
C	real, > 0	l	capacity of the boiler

Table 1: Constants for the boiler and controller models

Variables

Name	Initial Value	Type	Values Range	Unit	Description
now	0	real	$[0 \ldots \infty)$	s	current time
pr	0	integer	$\{0, \ldots \text{\#pumps}\}$		number of pumps actively supplying water to the boiler
q	$\gg M_1$, $\ll M_2$	real	$[0 \ldots C]$	l	actual water level in the boiler
v	0	real	$[0 \ldots \infty)$	l/s	steam rate of the steam currently leaving the boiler
pr_new	0	integer	$\{0, \ldots \text{\#pumps}\}$		number of pumps that are supposed to supply water after the activation delay
error	0	integer	$\{0, \ldots pr\_new\}$		number of pumps that fail to supply water to the boiler after activation
do_sensor	**true**	boolean	{*true*, *false*}		enable a single sensor reading
set	S	real	$[0 \ldots \infty)$	s	next time the pumps change to the new settings
read	0	real	$[0 \ldots \infty)$	s	next time the sensors will be read
stop	**false**	boolean	{*true*, *false*}		flag that determines whether emergency shut-down is activated

Table 2: Variables of the steam boiler model. Together they represent the (initial) state of the steam boiler.

For providing a formal description of the steam boiler, we first define all constants and the state (Table 1 and Table 2). For all variables of the state, we provide the type, value range and description. Moreover, we describe the initial state which immediately forces the automaton to read the current sensor values and forwards them to the controller. The controller will provide an appropriate pump setting. The checks in the controller, which is described in the following section, require that there is a certain minimal amount of water between the critical limits or otherwise the controller would stop the steam boiler at once. Thus, a valid start condition of the water level and steam rate must be far enough from the critical boundaries not to force the controller to execute an emergency stop.

4.1 The Boiler Automaton

Expressing our interpretation of the informal specification more precisely leads to the following Timed Automaton:

Input Action

actuator (e_stop, pset)
Effect:
$pr\_new' = pset$
$stop' = e\_stop$
$do\_sensor' = \textbf{true}$
$read' = now + I$

Output Action

sensor (s, w, p)
Precondition:
$now = read$
$do\_sensor = \textbf{true}$
$stop = \textbf{false}$
$w = q$
$s = v$
$p = pr$
Effect:
$do\_sensor' = \textbf{false}$

Internal Actions

activate
Precondition:
$now = set$
$stop = \textbf{false}$
Effect:
$set' = read + S$
$0 \leq error' \leq pr\_new$
$pr' = pr\_new - error'$

v(Δt)
Precondition:
$stop = \textbf{false}$
$now + \Delta t \leq read$
$now + \Delta t \leq set$
Effect:
$v - U_2 * \Delta t \leq v' \leq v + U_1 * \Delta t$
$q + pr * P * \Delta t - \delta_{HIGH}(v, v', \Delta t) \leq q'$
$q' \leq q + pr * P * \Delta t - \delta_{LOW}(v, v', \Delta t)$
$now' = now + \Delta t$

This formal description of the steam boiler is easily readable: The steam boiler reads periodically the current water level and the current steam rate and forwards these values to the controller. In the addition, the controller learns about the number of pumps that currently actually supply water to the boiler. The controller evaluates the data and through the actuator supplies a new pump setting or enables the shut-down phase. After the activation delay, all non-faulty pumps of the new setting supply water to the boiler. In the meantime, water evaporates into steam unpredictably but limited by its worst case rules.

With the ***actuator*** action the boiler receives the new pump setting requested by the controller and learns whether the controller shuts down the boiler. Furthermore, it schedules and enables the next reading of the sensor values. After an emergency stop is executed by setting the variable *stop* to *true*, our model ignores any further development.

As an internal action, the boiler changes the steam rate and the water level unpredictably over time. The purpose of the ***time-passage action*** denoted with $v(\Delta t)$ is to provide a method for describing formally a time-dependent process. Δt represents an arbitrary, non-empty interval of time. A possible value for the parameter Δt depends on the precondition. Obviously, Δt may be arbitrary as long as the next activation of the pumps and the next sensor reading occur. Formally, the time-passage action must follow some rules as described in CD-ROM version in the Appendix A, which we are going to verify in the next section.

The ***activate action*** occurs after the pump activation delay. It sets the new pump rate with respect to an arbitrary number of pumps that fail, expressed as *error*. We chose this rather strong fault model where all pumps might fail at the activation time regardless whether such a pump was already supplying water before. This can be as much as all pumps that should supply water for the next cycle. Finally, it schedules the next activation time. Periodically, the ***sensor action*** forwards the current amount of water, the current steam rate and the number of active pumps to the controller. To prevent the sensor action from happening multiple times, it disables itself by setting *do_sensor = false*.

4.2 Checking the Model

As described formally in the CD-ROM version in Appendix A (the complete definition can be found in [Lyn91]), each Timed Automaton has to follow five axioms. We need to show that the Boiler Model satisfies these axioms. Overall, these axioms are used to define the concept of time in Timed Automata. Axiom [A1] says that the current time is always 0 in a start state. Axiom [A2] says that non-time-passage steps do not change the time; that is, they occur "instantaneously", at a single point in time. Axiom [A3] says that time-passage steps must cause the time to increase. The fourth axiom [A4] enforces transitivity in the representation of time, i.e.; transitivity of the time passage action. Whenever it is possible to describe a development over time with several succeeding time-passage steps it must be possible to describe this change in a single time-passage step. The fifth axiom [A5] describes trajectory consistency: Whenever the change from one state to another with the time-passage action can be expressed as a trajectory (or function), the change between any two states in this interval follows the same trajectory.

The CD-ROM version contains the details to these proofs. Basically, with these axioms fulfilled the Timed Automaton model allows us to combine automata through their input and output actions. We will combine the boiler model with a controller model, which we present in the next section.

4.3 Properties of the Boiler

Based on the automaton description, we can derive useful information about the boiler system. These intermediate results can be favorably employed for fault detection and consistency checks in any actual boiler implementation based on this model. This information is expressed in the form of logic expressions invariant in all possible executions of this boiler model. Therefore, these expressions are called *invariants*. In other words, no order of steps will produce a state in which any of these logical expressions is not true. All the statements and their proofs can be found in the CD-ROM version of this report. All proofs are by induction on the steps of the automaton.

5 The Controller Model

In order to solve the steam boiler problem, we have to find a controller that guarantees the required safety properties. For this purpose, we take advantage of a characteristic of the Timed Automaton model. First, we will show that a simple controller that cannot tolerate sensor faults guarantees the safety properties under described assumptions. Then, the Simulation Mapping technique is used to show incrementally that a different controller which allows for sensor failures preserves the safety properties.

Obviously, it is most important that the controller identifies water levels and steam rates that might cross their critical limits before the next sensor values arrive. In case such sensor values are identified the controller will enable the shut-down phase. In a non-critical case, the controller chooses an appropriate new setting for the pumps to adjust the water level and compensate for the amount of steam leaving the boiler.

5.1 The Controller Model

Definitions

Name	Type	Unit	Value	Description
$max\_pumps\_after\_set$	integer		#pumps	maximum number of pumps that can supply water to the boiler after the delay considering the pump failure model
$min\_pumps\_after\_set$	integer		0	minimum number of pumps that can supply water to the boiler after the delay considering the chosen pump failure model. For a different pump failure model, e.g., in which pumps might fail when activated or stopped, this constant may actually be a function of the change in the number of pumps.
$min\_steam\_water(sr)$	real	1	$max(0, (sr - U_2 * I/2)*I$	minimum amount of water that can evaporate into steam until the next sensor reading
$max\_steam\_water(sr)$	real	1	$(sr + U_1*I/2)*I$	maximum amount of water that can evaporate into steam until the next sensor reading

Table 3: Definitions and abbreviations for the controller model

Variables

Name	Initial Value	Type	Value Range	Unit	Description
do_output	**false**	boolean	{**true, false**}		flag that enables the output. This represents a kind of program counter.
stopmode	**false**	boolean	{**true, false**}		flag to activate the shut down
wl	q	real	[0 ... C]	l	current water level reading
sr	0	real	[0 ... W]	l/s	current steam rate reading
now	0	real	[0 ... ∞)	s	current time
pumps	0	integer	{0 ...#pumps}		number of currently active pumps supplying water to the boiler
px	0	integer	{0 ...#pumps}		number of pumps that shell supply water next

Table 4: The state of the controller including all variables and their initial values

5.2 The Simple Controller Automaton

The input and output actions are complementary to the input and output actions of the steam boiler model.

Input Action

sensor (s, w, p)
Effect:
$sr' = s$
$wl' = w$
$pumps' = p$
$do_output' = $ **true**

\# safety checks:
if $sr' \leq W - U_1 * I$ or $wl' \geq M_2 - P * (pumps' * S + (max\_pumps\_after\_set) * (I - S)) + min\_steam\_water(sr)$ or $wl' \leq M_1 - P * (pumps' * S + (min\_pumps\_after\_set) * (I - S)) + max\_steam\_water(sr)$
then
    $stopmode' = $ **true**
else
    $stopmode' = \{true, false\}$ arbitrary

Internal Actions

controller
Precondition:
    true
Effect:
    $0 \leq px' \leq \#pumps$

ν(Δt)
Precondition:
    true
Effect:
    $now' = now + \Delta t$

Output Action

actuator (e_stop, pset)
Precondition:
    $do_output = $ **true**
    $pset = px$
    $e\_stop = stopmode$
Effect:
    $do_output' = $ **false**

With the **sensor action**, the controller receives periodically the current steam rate, water level and number of activated pumps. Its primary purpose is to test if the current sensor values are "close" to either critical limit. In such a case the sensor action sets a flag for the actuator to initiate the shut-down. Likewise, external critical conditions are modeled by non-deterministically setting *stopmode* to true. Furthermore, the sensor action enables the actuator action. The test for what is "close" depends on the particular fault model used and controller capabilities. The controller can try to start all pumps every period and our fault model allows up to all pumps to fail. The point in time for the decision how many pumps actually supply water to the boiler is every *set* time. Therefore, we must choose all pumps for *max_pumps_after_set*. On the other hand, all pumps could fail and therefore *min_pumps_after_set* equals 0. Similarly, *min_steam_water* and *max_steam_water* express the minimum and maximum amounts of water that can evaporate into steam in the following period starting with given current steam rate, respectively. The test simply calculates the worst case situations for the water level and steam rate and compares the results with the critical limits M_1, M_2 and W.

The **controller action** chooses an appropriate new pump setting. Actually, it can choose any pump setting. For our approach, we are not particularly interested in the performance of the controller. On the other hand, we are interested in generality. Therefore, we chose a controller model that can incorporate any possible control algorithm for setting the pumps. As a consequence, our results concerning the safety are valid for an arbitrary control algorithm. Although the choice of a new setting for the pumps is irrelevant to the safety of the steam boiler system, for a performance analysis the pump setting would be of major importance. The **time-passage action** ($v(\Delta t)$) allows time to pass. For the following proofs, we ignore these two actions, since they do not provide additional information and are irrelevant to the proofs.

Finally, the **actuator action** forwards the new pump setting and whether the boiler must be stopped to the boiler environment. Furthermore, it disables itself, by setting *do_output* back to false.

As suggested in the original specification, this controller model acts instantaneously. Therefore, the time-passage action is trivial and all five axioms for Timed Automata are satisfied. Moreover, there is no useful information gained from the controller model alone. So far the proofs have involved only either the steam boiler model or the controller model. Next, we use the composition property of Timed Automata for combining the two automata, and we prove the required safety properties.

6 Properties of the Combined Steam Boiler System

Following, we show in several steps that the combined model (formally a composition), consisting of the steam boiler model and the simple controller model together, guarantee the safety conditions. The first safety property requires that the steam rate must always stay below W. Before the steam rate can cross this limit, the boiler must be shut down. Expressing this in terms of the state of the steam boiler system, we have to show

S1) $v < W$ or *stop* = **true**

The second safety property requires that the water level must always stay between its critical limits M_1 and M_2. Before the water level can cross either limit, the boiler must be stopped. Thus, we have to show

S2) $M_1 < q < M_2$ or *stop* = **true**

6.2 Steam Boiler System Properties

The following lemmas lead us step-by-step toward proving the safety conditions. While we provide only some important statements in this version of this paper, all the statements (lemmas and theorems) and their proofs can be found on the CD-ROM. The numbering is consistent.

Coming up with the right invariants that lead to showing the safety properties is the most complicated task in working with Timed Automata. On the other hand, the proofs themselves are usually straightforward and follow well-established, stylized methods and the usual pattern for proving by induction. The main work for proving the safety properties is done by means of these invariants. All the proofs for our model are by induction on the model and can easily be verified using current mechanical proof technology.

Lemma 5 concludes that the next time the pumps will be activated can only be either the constant delay after or before the next sensor reading. This lemma depends on *if do_output then now = read* (Lemma 4) and *now ≤ set* (Lemma 2).

Lemma 5: *set = read + S or set = read - I + S*

The following lemma helps us later to show that whenever the sensors are read (or, at the same instant, the new pumps settings sent to the boiler) the pumps are activated exactly after the delay *S*, as specified. It depends on Lemma 4 (*if do_output then now = read*) and Lemma 5.

Lemma 6: *now ≤ read - I + S or set = read + S*

Lemma 10 expresses that at the time the sensors are read the representation of the active pumps in the controller are equal to the pumps actually supplying water to the boiler. During the entire operation of the boiler system the number of pumps supplying water is either the number requested by the controller minus some faulty pumps or equal to the status sensed at the last reading point after the pumps were activated. It depends on Lemma 2, Lemma 5, Lemma 6, *if now < read then do_output = false* (Lemma 7), *if do_output then pumps = pr* (Lemma 8) and *do_output xor do_sensor* (Lemma 9).

Lemma 10: *if set = read + S and do_output = false then pr = pr_new - error else pr = pumps*

Using the test conditions in Lemma 5, we can now prove that the actual steam rate will stay under a certain limit depending on how long it takes until the next sensor reading. This lemma depends on $sr + U_1 * I < W$ or *stopmode = true* (Lemma 3.3) and *if do_output then now = read and sr = v* (Lemma 4).

Lemma 11: $v + U_1*(read - now) < W$ or *stop = true*

Similar to the previous lemma, we can find an upper limit on the water level depending on the pump activity since the last sensor reading. This lemma depends on Lemma 1.2, *now ≤ set* (Lemma 2), $wl < M_2 - P * (pumps * S + \#pumps * (I - S)) + max(0, sr - U_2 * I^2/2)$ or *stopmode = true* (Lemma 3.1), *if do_output then now = read and wl = q* (Lemma 4), Lemma 5, Lemma 6, *do_output = false* from *if now < read* then $do\_output = \textbf{\textit{false}}$ (Lemma 7), and if $set = read + S$ and $do\_output = \textbf{\textit{false}}$ then $pr = pr\_new - error$ else $pr = pumps$ (Lemma 9).

Lemma 12: $q < M_2 - psa$ or $stop = \textbf{\textit{true}}$

With
$$psa = \begin{cases} P*pr*(read - now) & \text{if} \quad set = read + S \\ P*(pumps*(set - now) + pr\_new*(I - S)) & \text{otherwise} \end{cases}$$

The following lemma describes the lower limit of the water level depending on the current steam rate, the maximum amount it can increase and amount of time left until the next sensor reading. Lemma 13 depends on Lemma 1.10, $wl > M_1 - P * pumps * S + (sr * I + U_1 * I^2/2)$ or $stopmode = \textbf{\textit{true}}$ (Lemma 3.2) and if do_output then $now = read$ and $wl = q$ and $sr = v$ (Lemma 4).

Lemma 13: $q > M_1 + v * (read - now) + U_1 * (read - now)^2/2$ or $stop = \textbf{\textit{true}}$

6.3 Summarizing Theorems

The following theorems summarize the previous lemmas and translate them into the form in which the required safety properties were expressed.

Theorem 1: In all reachable states of boiler system, $v < W$ or $stop = \textbf{\textit{true}}$.

Theorem 2: In all reachable states of boiler system, $M_1 < q < M_2$ or $stop = \textbf{\textit{true}}$.

With above proofs, we have shown that the steam boiler model together with the controller model meets all the safety requirements. As a further step, we must modify the controller model to allow sensor faults. This is presented in the following section.

7 Sensor Fault-tolerant Controller

In this section, we extend the model of the controller to be tolerant to sensor faults. Rather than proving the safety properties all over again, we use a technique called *Simulation Mapping*. This technique is used to show consistency between abstraction levels. In particular, it provides a means to show that properties proved for an abstract model are preserved in a particular implementation. In this case, the previously described boiler system represents the specification and a new controller that tolerates sensor faults represents a possible implementation.

The two lemmas which supply additional information about the boiler system with the previous controller can be found in the CD-ROM version of this text. Next, we present the Timed Automaton model of the sensor fault-tolerant controller.

7.1 The Controller Model Allowing Sensor Faults

Variables

Name	Initial Value	Type	Value Range	Unit	Description
do_output	**false**	boolean	{*true, false*}		flag that activates the output; This parameter represents a kind of program counter.
stopmode	**false**	boolean	{*true, false*}		flag to activate the emergency stop
wll	*q*	real	[0 ... C]	l	lower bound of the estimation of the current water level
srl	0	real	[0 ... W]	l/s	lower bound of the estimation of the current steam rate
wlh	*q*	real	[0 ... C]	l	upper bound of the estimation of the current water level
srh	0	real	[0 ... W]	l/s	upper bound of the estimation of the current steam rate
sr_ok	**true**	boolean	{*true, false*}		flag that tells whether the steam rate sensor has failed
wl_ok	**true**	boolean	{*true, false*}		flag that tells whether the water level sensor has failed
now	0	real	[0 ... ∞)	s	current time
pumps	0	integer	{0 ... #*pumps*}		number of currently active pumps supplying water to the boiler
px	0	integer	{0 ... #*pumps*}		number of pumps that shall supply water next

Table 5: The initial state of the fault-tolerant controller including all variable declarations

7.2 The Fault-tolerant Controller Automaton

Input Action

sensor (s, w, p)
Effect:
  $pumps' = p$
  $do\_output' = \textbf{true}$

\# estimate steam rate
if sr_ok' then $srh' = srl' = s$
else $srh' = srh + U_1 * I$
  $srl' = srl - U_2 * I$

\# estimate water level
if wl_ok' then $wlh' = wll' = w$
else $wlh' = wlh - (max(0, srl' - U_1*I/2)) *I +$
  $P * (pumps * S + pumps' * (I - S))$
  $wll' = wll - (srh' + U_2*I/2)*I + P *$
  $(pumps * S + pumps' * (I - S))$

\# safety checks
if $srh' \geq W - U_1 * I$ or
  $wlh' \geq M_2 - P *(pumps' * S +$
  $(max\_pumps\_after\_set) * (I- S))+$
  $min\_steam\_water(srl)$ or
  $wll' \leq M_1 + P *(pumps' * S +$
  $(min\_pumps\_after\_set) * (I - S)) -$
  $max\_steam\_water(srh)$
then $stopmode' = \textbf{true}$
else $stopmode' = \{true, false\}$ arbitrary

Internal Actions

bad
Precondition:
  true
Effect:
  $sr\_ok' = \{\textbf{true}, \textbf{false}\}$ arbitrary
  $wl\_ok' = \{\textbf{true}, \textbf{false}\}$ arbitrary

controller
Precondition:
  true
Effect:
  $0 \leq px' \leq \textbf{\#pumps}$

v(Δt)
Precondition:
  true
Effect:
  $now' = now + \Delta t$

Output Action

actuator (e_stop, pset)
Precondition:
  $do\_output = \textbf{true}$
  $pset = px$
  $e\_stop = stopmode$
Effect:
  $do\_output' = \textbf{false}$

The controller model that allows sensor faults has the same structure as the simple controller. An additional action **bad** tell the controller whether a sensor has failed. The fault model allows arbitrary combinations of sensor break downs and fast or slow repairs. The **sensor** action expresses the strategy of the controller to cope with sensor faults. Basically, the strategy is to calculate an upper and lower limit for the missing value of the failed sensor, using its last recent value and the remaining sensor values. Even in the case that both sensors break, the controller still may allow the operation of the boiler and guarantee safety. In this respect, our controller definition is better than the one suggested in [AS96], since he suggests to shut down the boiler system whenever both steam rate and water level sensors fail.

The various operational modes (normal, degraded and rescue) as specified in [AS96] can be inferred from the variables sr_ok, wl_ok and the difference between $pumps$ and px. In our model, these modes are not relevant to the safety of the boiler system and have therefore been ignored.

7.3 Proving the Safety Properties by Simulation Mapping

After composing the steam boiler automaton with the new fault-tolerant controller, we have to prove that the safety properties are satisfied in the new model.

We use a **Simulation Mapping** for proving that one Timed Automaton "implements" another. This technique shows that all possible traces[*] of the new automata are included in the traces of the already proven model. Therefore, all safety properties involving the states of the steam boiler with the simple controller are valid for the system with the fault-tolerant controller, too. A Simulation Mapping is most useful to show that an implementation actually preserves properties of the specification. This method can be applied repeatedly to get from a very abstract model, which is proven to fulfill the required properties, to a detailed implementation (maybe even the final implementation). Like invariants, the Simulation Mappings involve time deadline information, in particular, they include inequalities between time deadlines. Therefore, they are suitable for showing timing properties, too.

We apply a Simulation Mapping from states of the steam boiler system with the fault-tolerant controller (in short "fault-tolerant controller system") to the system with the simple controller ("simple controller system"). Appendix B of the CD-ROM version contains a formal definitions of the Simulation Mapping technique and the correctness properties it guarantees.

7.3.1 Simulation Relation

Theorem 3: The relation f as defined below is a Simulation Mapping from the states of the fault-tolerant controller system to the states of the simple controller system.

Let s denote a state of the simple controller system and i denote a state of the fault-tolerant controller system. We define s and i to be related by the relation f provided that:

1) $i.Boiler = s.Boiler$[†]
2) $i.do\_output = s.do\_output$, $s.px = i.px$, $s.pumps = i.pumps$, $s.now = i.now$
3) $i.srl \leq s.sr \leq i.srh$
4) $i.wll \leq s.wl \leq i.wlh$
5) $s.stopmode = i.stopmode$

Proof sketch: Let i lead to i' via action a in the fault-tolerant controller. We must find an s' such that $s' f i'$ and there exists an execution fragment from s to s' with the same trace as a. Usually, we break by cases on the type of a. In the initial state f is fulfilled. For this proof it remains to show the case for the sensor action because all other actions are identical in the specification and implementation. It remains to show that there is an equivalent sensor step enabled in s, and s' relates to i' following the definition of f. In particular, we must show the three conditions in the definition of a Simulation Mapping in Appendix B on the CD-ROM. The first condition, preservation of the *now* value, is immediate from the definition of f. The second

---

[*] The exact meaning of "traces" is defined in Appendix A on the CD-ROM.
[†] This relation expresses that the entire boiler state is preserved.

condition, correspondence of the start states, is also immediate, because f is fulfilled between the start states. The interesting condition is the induction step. As before, the detailed proof can be found in the full version of this text. The proof depends on Lemma 1.3&6, $s.now \leq s.read - I + S$ or $s.set = s.read + S$ (Lemma 6), $-U_2*(I + s.now - s.read) \leq s.v - s.sr \leq U_1*(I + s.now - s.read)$ (Lemma 14) and *if s.do_output = false* then $ps - \delta_{HIGH}(s.sr, s.v, t) \leq s.q - s.wl \leq ps - \delta_{LOW}(s.sr, s.v, t)$ with $ps = $ *if* $s.set = s.read + S - I$ *then* $P * s.pumps * t$ *else* $P * (s.pumps * S + s.pr * (t - S))$ and $t = (I + s.now - s.read)$ (Lemma 15).

This Simulation Mapping maps every reachable state of the boiler system with the fault-tolerant controller to a corresponding reachable state in the system with the simple controller by the relation f. Therefore, the safety properties involving the states of the specification (simple controller) are valid for the implementation (fault-tolerant controller), too. Thus, we have shown that the steam boiler system with the fault-tolerant controller satisfies the required safety properties.

8 Conclusion

We have applied a formal method based on Timed Automata, invariant assertions and Simulation Mappings to the steam boiler model, and verified that our controller fulfills the required safety properties. In doing so we have made it possible to compare our techniques to other approaches.

Summarizing, the *Timed Automata, composition* and *Simulation Mapping* techniques present an excellent combination for system analysis. The main advantage of Timed Automata is their flexibility in modeling a hybrid system. Timed Automata allow us to combine a continuous environment that is fairly unpredictable over time with a discrete control system such as a computer. The composition and Simulation Mapping techniques supplement this specification tool for formal verification, for more flexibility in how to search for a solution and for the reuse of already gained knowledge. The composition technique lets you combine different automata and scale incrementally solutions from smaller problems to more complex ones. The Simulation Mapping technique provides a consistent transition between different abstraction layers.

This method seems to scale better than other formal verification techniques because of the possibility of applying this method to different abstraction layers, and applying various decomposition techniques [Wei96]. A Simulation Mapping can be used to prove that two abstraction layers preserve certain properties. Decomposition techniques provide modular and incremental verification. For instance, suppose that you have proved that a certain implementation of a shared register provides mutual exclusion. The automaton model together with already proved properties may then be composed into a bigger application without having to prove the mutual exclusion property again.

Constructing the proofs, though not difficult, requires significant work. The hardest parts were getting the details of the models right and finding the right invariants. Unfortunately, this seems to be an art rather than an automatic procedure. Nevertheless, our experience in this paper and others (e.g., [Hei94]) shows that this art is easily learnable even for application engineers. The techniques are very systematic and understandable. The description allows for much flexibility and is very powerful in describing the possible progression of a system.

The actual proofs of the invariants were tedious but routine work. Much work can be avoided by proving the required properties on a general model and using *Simulation Mappings* for more specialized models. Moreover, the characteristics of these techniques make them amenable for mechanical generation and verification of proofs. Related to this, we are currently considering the use of automatic provers such as Larch [Soe93] or PVS [Sha93] with the described techniques.

The only major disadvantage we encountered while working with Timed Automata and the Simulation Mapping technique is that we could not gain any information or any measurement towards the optimality of parameters of a solution. Although our controllers preserve provable safety, there are obviously better implementations. For example, on a steam rate sensor failure, the steam rate estimation could take into account the amount of water which has evaporated since the last sensor reading. Moreover, we like to note that more of the reality could be modeled formally with a more relaxed pump failure model and diverse pump controller algorithms. The latter might lead to interesting performance comparisons and tighter parameters such as the distance between M_1 and M_2.

Future work includes applying this method to larger and more complex examples, and developing the appropriate computer assistance for carrying out and checking the proofs. On-going research in our group shows that the timed-automata method provides high potential for automating the generation of the proofs [Sha93], [Arc96].

Acknowledgments

We thank Anya Pogosyants and Roberto Segala for several useful comments as well as Angelika Leeb and Dave Evans for comments and proofreading.

References

[AS96]   Abrial, J.-R.: A B-solution for the steam-boiler problem. Contains: Steam-boiler control specification problem for the meeting Methods for Semantics and Specification, Dagstuhl; See chapter AS in this LNCS volume.

[Arc96]   Archer, M.; Heitmeyer, C.: Mechanical Verification of Timed Automata: A Case Study, To appear in the proceedings of RTAS, 1996

[Cle93]   Cleaveland, R.; Parrow, J.; Steffen, B.: The concurrency workbench: A semantics-based tool for verification of concurrent systems. ACM Trans. on Prog. Lang. and Sys., 15(1):36-72, Jan. 1993

[Hei93]   Heitmeyer, C.; Jeffords, R.; Labaw, B.: A benchmark for comparing different approaches for specifying and verifying real-time systems. In Proc., 10th Intern Workshop on Real-Time Operating Systems and Software, May, 1993

[Hei94]   Heitmeyer, C.; Lynch, N.: The Generalized Railroad Crossing: A Case Study in Formal Verification of Real-Time Systems. In Proceedings of the 15th IEEE Real-Time Systems Symposium, San Juan, Puerto Rico, IEEE Computer Society Press, pages 120 -131, December 1994 real-time behavior in the specifications of the individual controller modules and model all modules of the controller as performing synchronous transitions when the variable *round* changes value. This assumption is reflected in the form of the TLA formula that specifies a module, which will be of the form

$$Init \wedge \Box[\mathcal{N} \wedge (round' \neq round)]_{\langle v, round \rangle}$$

The formulas *Init* and $\mathcal{N}$ describe the initial conditions and the next-state relation of the module, v is a tuple of the "output" variables controlled by the module. The formula asserts that the next-state relation has to hold whenever *round* changes value and, conversely, that the variables in v may only change value if *round* does, too. Taken together, these conditions ensure that all state changes happen simultaneously.

The structure of this paper is as follows: We specify a simplified controller in section 2. Sections 3 and 4 introduce a formal model of the physical steam boiler and relate the physical state to the approximations maintained by the controller. We give a more refined controller specification in section 5, while sections 6 and 7 are concerned with detecting sensor failures and modelling message transmissions. We discuss the implementability of our specification in section 8 and derive a data flow graph. Finally, section 9 concludes with answers to the questions posed by the organizers of this case study. The complete specification, together with further explanations and the proofs, which had to be omitted in the main text due to space constraints, can be found in the appendix.

2 Abstract controller specification

Our first specification concerns an abstract version of the controller for the steam boiler, it appears as module *Abstract* in figure 2. The main purpose of the controller is to maintain a satisfactory water level in the steam boiler. We assume that the abstract controller receives indications concerning the current water level in the steam boiler (normal, low, high or dangerous) at each cycle, as well as information about catastrophic system failures. It may also receive the signals *units_ready*, and *stop_req*, which are abstractions of the messages PHYSICAL_UNITS_READY and STOP described in the problem description [AS]. We model signals by the values T (true) and F (false); formally, the value T represents the presence of a signal, while any other value means that the signal is absent.

Based on its input, the steamboiler determines its mode of operation and may react by opening and closing the valve and the pumps, as modelled by the following variables *ctl_mode*, *prog_ready*, *valve*, and *pumps*. The current mode of operation of the abstract controller may be "initialize", "operating" or "emergency". Mode "initialize" corresponds to the second phase of the initialization described in the problem statement, where the controller tries to ensure a normal water level before steam production starts. Mode "operating" subsumes the "normal", "degraded", and "rescue" modes described in the problem statement, which are not distinguished in the abstract specification. The variable

┌──────────────────────── module *Abstract* ────────────────────────┐
parameters
  round : VARIABLE
  units_ready, *stop_req*, *system_failure* : VARIABLE
  normal_level, *low_level*, *high_level*, *dangerous_level* : VARIABLE
  ctl_mode, *prog_ready*, *valve*, *pumps* : VARIABLE
├───┤
$critical\_failure \triangleq \ \vee\ (system\_failure' = T)$
$\qquad\qquad\qquad\quad\ \vee\ (ctl\_mode = \text{``operating''}) \wedge (dangerous\_level' = T)$
$initialization\_complete \triangleq\ (ctl\_mode = \text{``initialize''}) \wedge (normal\_level' = T)$
$open\_valve \triangleq\ (ctl\_mode' = \text{``initialize''}) \wedge (high\_level' = T)$
$open\_pumps \triangleq\ (ctl\_mode' \neq \text{``emergency''}) \wedge (low\_level' = T)$
$close\_pumps \triangleq\ (ctl\_mode' = \text{``emergency''}) \vee (high\_level' = T)$
$Init \triangleq\ \wedge\ (ctl\_mode = \text{``initialize''}) \wedge (prog\_ready = F)$
$\qquad\quad\ \wedge\ (valve = \text{``closed''}) \wedge (pumps = \text{``off''})$
$\mathcal{N} \triangleq\ \wedge\ ctl\_mode' = \textbf{if}\ critical\_failure \vee (stop\_req' = T) \vee (ctl\_mode = \text{``emergency''})$
$\qquad\qquad\qquad\quad\ \textbf{then}\ \text{``emergency''}$
$\qquad\qquad\qquad\quad\ \textbf{elsif}\ (ctl\_mode = \text{``initialize''}) \wedge (units\_ready' = T)$
$\qquad\qquad\qquad\quad\ \textbf{then}\ \text{``operating''}$
$\qquad\qquad\qquad\quad\ \textbf{else}\ ctl\_mode$
$\qquad\ \wedge\ (prog\_ready' = T) \equiv initialization\_complete$
$\qquad\ \wedge\ \vee\ open\_valve \wedge (valve' = \text{``open''})$
$\qquad\qquad\ \vee\ \neg open\_valve \wedge (valve' = \text{``closed''})$
$\qquad\ \wedge\ (pumps' \in \{\text{``on''}, \text{``off''}\}) \wedge (pumps' = \text{``on''}) \Rightarrow open\_pumps$
$\qquad\qquad\qquad\qquad\qquad\qquad\quad\ \wedge\ (pumps' = \text{``off''}) \Rightarrow close\_pumps$
$\Phi \triangleq\ Init \wedge \Box[\mathcal{N} \wedge round' \neq round]_{\langle ctl\_mode, prog\_ready, valve, pumps, round \rangle}$
└───┘

Fig. 2. Specification of an abstract controller.

prog_ready represents a signal sent from the controller to the steam boiler that indicates that the initialization is complete. This signal is an abstraction of the message PROGRAM_READY of the problem statement.

The variables *valve* and *pumps* indicate the state of the actuators during the subsequent control cycle. The abstract controller does not distinguish between different pumps.

The behavior of the abstract controller is specified by the formula Φ, which appears at the end of module *Abstract*. It has the expected form

$$Init \wedge \Box[\mathcal{N} \wedge (round' \neq round)]_{\langle v, round \rangle}$$

of a module specification. We now explain the action formula $\mathcal{N}$, which asserts the next-state relation of the four variables *ctl_mode*, *prog_ready*, *valve*, and *pumps* that represent the controller's output state. Figure 3 illustrates the tran-

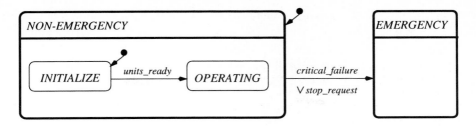

Fig. 3. Abstract controller: control modes.

sitions between the control modes of the abstract controller, using a Statechart-like notation. The specification uses the auxiliary action formula *critical_failure* to determine when the controller should enter "**emergency**" mode. This happens when the environment signals that the water level might be dangerous or that a system failure has occurred.

The second conjunct of formula $\mathcal{N}$ states that the signal *prog_ready* is emitted iff the initialization is complete. The third and fourth conjuncts of formula $\mathcal{N}$ concern opening and closing the valve and the pumps. The valve should be "**open**" iff condition *open_valve* holds, which requires the controller to be in mode "**initialize**" and the signal for high water level to be present. The behavior of the pumps is specified rather loosely: We only require that the environment has signalled high water level whenever the controller wants to switch "**on**" some pump, and similarly for the "**off**" state. Specifically, we do not rule out situations where both signals *high_level* and *low_level* are present; the controller may then behave in either way (the problem statement does not prescribe how the system should react in such a situation).

The specifications of module *Abstract* are a little unusual in that they contain primed versions of variables that represent input to the controller, such as the variables *low_level* and *high_level*. This peculiarity is due to our model of the abstract controller operating in synchrony with its environment: we think of the input variables as changing at the very moment that the controller performs a step; hence the specification refers to the new (primed) values of the input. Note that changes of the input variables are completely unconstrained by the formula Φ. However, some of the signals will actually be produced by (different modules of) the controller itself when we introduce a refinement of the controller later, and the refinement proof will become a little simpler if we adopt the model of synchronous computation right from the beginning. We will discuss the notion of primed input variables from the point of view of implementability in section 8.

3 Steam boiler physics

The abstract controller specification of module *Abstract* relies on signals that indicate normal, low, high or dangerous water levels. The problem description

$$q_{min}[q, v, p, pst, e, p\_cmd, e\_cmd] =$$
$$\max(0, q - v \cdot \Delta - \tfrac{1}{2} \cdot U_1 \cdot \Delta^2 - e_{max}[q, v, p, pst, e, p\_cmd, e\_cmd]$$
$$+ \sum_{i=1}^{4} p_{min}[q, v, p, pst, e, p\_cmd, e\_cmd][i])$$
$$q_{max}[q, v, p, e, p\_cmd, e\_cmd] =$$
$$\min(C, q - v \cdot \Delta + \tfrac{1}{2} \cdot U_2 \cdot \Delta^2 - e_{min}[q, v, p, pst, e, p\_cmd, e\_cmd]$$
$$+ \sum_{i=1}^{4} p_{max}[q, v, p, pst, e, p\_cmd, e\_cmd][i])$$

Fig. 4. Steam boiler physics: possible definitions (incomplete).

states that the controller receives information about the water level and other important data from unreliable sensors. In order to cope with sensor failures, the controller maintains a model of the state of the steam boiler, based on physical laws that underly the behavior of the steam boiler.

We use the following entities to describe the state of the steam boiler at any given moment:

- the amount of water q in the steam boiler,

- the amount of steam v exiting the steam boiler,

- the amount of water pumped into the steam boiler by each pump during the preceding control cycle, represented as a function $p : \{1, 2, 3, 4\} \to [0, P]$,

- the state of the pumps, which we describe by a function $pst : \{1, 2, 3, 4\} \to \{\text{"off"}, \text{"switching"}, \text{"on"}\}$,

- and the amount of water $e \in [0, V]$ that has exited through the evacuation valve during the previous control cycle.

The pump state "switching" indicates that the pump has been switched on during the previous cycle. The problem statement indicates that a pump needs a full controller cycle to start pouring water into the boiler. In particular, the pump state cannot be inferred from the amount of water delivered by the pump.

Given the current boiler state and the commands sent to the acutators by the controller as represented by a function $p\_cmd : \{1, 2, 3, 4\} \to \{\text{"on"}, \text{"off"}\}$ and $e\_cmd \in \{\text{"open"}, \text{"close"}\}$, we assume given functions to compute lower and upper bounds for the value of each entity after Δ seconds (that is, at the following controller cycle). The precise definitions of these functions are unimportant, but they should yield results within the static bounds for the entity such that the result of the lower-bound function is below that of the upper-bound function. Moreover, we assume certain monotonicity conditions. For example, the functions that compute bounds for the water level should be monotonic in the arguments q, p, and p_cmd and anti-monotonic in v, e, and e_cmd, where we let pump and valve commands are ordered by "off" < "on" and "close" < "open",

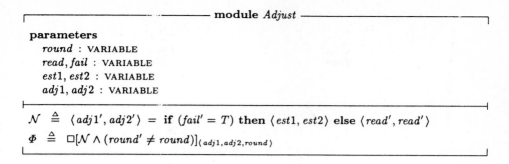

Fig. 5. Adjusting sensor readings.

respectively, and let arrays be ordered component-wise. A precise statement of these assumptions is given in module *Dynamics*, shown in figure 8.

Figure 4 contains possible definitions of functions q_{min} and q_{max} that are inspired by the information in the problem description. We do not state these definitions in the form of a TLA+ module, because they are not part of our specification. Similar definitions of the remaining functions are given in figure 18 in the appendix (see CD-ROM annex LM.A.4).

4 Relating component and environment

Based on the physical laws discussed in the previous section, we now introduce specifications that concern both the physical evolution of the steam boiler state and the approximations maintained by the controller. This section contains three specifications: module *Adjust* describes how bounds for the relevant data are computed from the sensor readings, the system's projections, and information about sensor failures. Module *Estimate* specifies how the system arrives at its projections, and module *Environment* states assumptions on the evolution of the physical steam boiler. Finally, we assert a theorem that states that the system's estimates are bounds for the physical state of the steam boiler as long as failure information is accurate.

Figure 5 shows the specification of a generic module that computes "adjusted" values for an entity based on the reading of the corresponding sensor, information about the failure of the sensor, and estimations for the expected range of sensor values. As in the case of the abstract controller specification, we assume that environment and system operate synchronously. Therefore, the new information about sensor failures and sensor readings (represented by primed variables) are used to update the adjusted values, whereas the old estimates (presumably computed in the previous cycle) are used in case of failure.

```
┌─────────────────────── module Estimate ───────────────────────┐
│ parameters                                                     │
│   lwb, upb : CONSTANT                                          │
│   round : VARIABLE                                             │
│   qa₁, qa₂, va₁, va₂, pa₁, pa₂, psta₁, psta₂, ea₁, ea₂, p_cmd, e_cmd : VARIABLE │
│   est1, est2 : VARIABLE                                        │
├────────────────────────────────────────────────────────────────┤
│ N ≜ ∧ est1' = lwb[qa'₁, qa'₂, va'₁, va'₂, pa'₁, pa'₂, psta'₁, psta'₂, ea'₁, ea'₂, p_cmd', e_cmd']│
│     ∧ est2' = upb[qa'₁, qa'₂, va'₁, va'₂, pa'₁, pa'₂, psta'₁, psta'₂, ea'₁, ea'₂, p_cmd', e_cmd']│
│ Φ ≜ □[N ∧ (round' ≠ round)]⟨est1,est2,round⟩                   │
└────────────────────────────────────────────────────────────────┘
```

Fig. 6. Computing estimates for the next cycle.

We now discuss how one can compute estimates for the state of the steam boiler at the next controller cycle and how these estimates relate to the actual (physical) values. To state the relationship between the actual state of the steam boiler and the model of the steam boiler state maintained by the controller, we use the variables q, v, p, pst, and e to denote the actual water level, amount of steam, pump throughput, pump state, and valve throughput, the variables qr, vr, pr, $pstr$, and er to denote the "readings" of the values transmitted from the respective sensors,[1] and the variables qa_1, qa_2, va_1, va_2, pa_1, pa_2, $psta_1$, $psta_2$, ea_1, and ea_2 to denote lower and upper bounds for these entities maintained by the controller. (This nomenclature follows the suggestions given in the "additional information" part of the problem statement, where they are called "adjusted" values.) Using the assumptions on the monotonicity of the functions q_{min}, q_{max}, etc., they can be generalized to compute bounds for the state of the steam boiler at the next cycle given *bounds* for the present state of the steam boiler instead of actual values. For example, the generalized version qa_{min} of function q_{min} can be defined as

$$qa_{min}[qa_1, qa_2, va_1, va_2, pa_1, pa_2, psta_1, psta_2, ea_1, ea_2, p\_cmd, e\_cmd]$$
$$= q_{min}[qa_1, va_2, pa_1, psta_1, ea_2, p\_cmd, e\_cmd]$$

and similarly for the other functions (all definitions are contained in the complete specification A.5 in the appendix).

Module *Estimate*, whose specification appears in figure 6, defines a generic module to compute estimates for a particular entity of the steam boiler state. Its

---
[1] We are deviating somewhat from the problem statement and assume that the steam boiler transmits the actual throughput of the pumps and the valve instead just binary status information for the pumps and no information for the valve. This assumption makes our specification more uniform.

```
┌──────────────────── module Environment ────────────────────┐
  parameters
     lwb, upb : CONSTANT
     round : VARIABLE
     q, v, p, pst, e, p_cmd, e_cmd : VARIABLE
     actual, read, fail, adj1, adj2, est1, est2 : VARIABLE
├────────────────────────────────────────────────────────────┤
  $\mathcal{N} \triangleq lwb[q, v, p, pst, e, p\_cmd, e\_cmd] \leq actual' \leq upb[q, v, p, pst, e, p\_cmd, e\_cmd]$
  $\Phi \triangleq \Box[\mathcal{N} \wedge (round' \neq round)]_{\langle actual, round \rangle}$
├────────────────────────────────────────────────────────────┤
  $Fail\_reliable \triangleq (actual \neq read) \Rightarrow (fail = T)$
  $Good\_approx \triangleq adj1 \leq actual \leq adj2$
  $Good\_estimate \triangleq \wedge est1 \leq lwb[q, v, p, pst, e, p\_cmd, e\_cmd]$
  $\phantom{Good\_estimate \triangleq} \wedge upb[q, v, p, pst, e, p\_cmd, e\_cmd] \leq est2$
└────────────────────────────────────────────────────────────┘
```

Fig. 7. Behavior of the environment.

parameters include two estimation functions *lwb* and *upb*[2] to compute lower and upper bounds for the particular value. These functions are static in the sense that they do not change over time; this is indicated by the keyword CONSTANT in the parameter declaration. A typical lower-bound function would be the function qa_{min} defined above. Module *Estimate* applies these functions at every cycle to the "adjusted" values and the commands that will be sent to the pumps and the valve during the subsequent control cycle. Again, primed variables appear as inputs because the estimates should be based on the most recent data available.

Module *Environment*, shown in figure 7, will be used to relate the physical evolution of the steam boiler system with the model maintained by the controller. Similar to module *Estimate*, it begins by importing constant functions *lwb* and *upb* to compute lower and upper bounds on the evolution of a physical state variable given the current state of the steam boiler (not its approximation). Typical instantiations will be q_{min} and q_{max}. The specification Φ asserts that the new value of the variable *actual*, which will be used to represent a particular entity of the steam boiler state, must fall within the lower and upper bounds computed by the functions *lwb* and *upb* (we let $x \leq y \leq z$ be an abbreviation for $x \leq y \wedge y \leq z$). To understand this specification, note again that we model the environment and the system as evolving synchronously, which may seem somewhat counterintuitive—after all, the water level in the steam boiler is a continuous function whose evolution cannot be described accurately with a

---

[2] Since TLA is an untyped logic, the parameter declaration does not express that *lwb* and *upb* are expected to be functions or specify their functionality.

model of discrete time as that underlying TLA. However, think of the variable *actual* as a "probe" that is taken precisely at each controller cycle. The problem statement asserts that the steam boiler would be in danger if the water level exceeded the limit values for more than Δ seconds. We interpret this statement as implying that the controller does not have to care about peak values outside the limit values that do not persist for at least Δ seconds and may therefore pass unnoticed.

Module *Environment* also defines three state predicates for later use. Predicate *Fail_reliable* holds if the sensor value agrees with the actual value unless a sensor failure is signalled. Predicate *Good_approx* holds if the actual value falls within the interval $[adj1, adj2]$ while predicate *Good_estimate* asserts a similar relationship between the estimates and the bounds computed from the current state.

Module *Dynamics*, part of which is shown in figure 8 (the complete module appears in figures 22, 23 and 24 in the appendix, see CD-ROM annex LM.A.5), assembles instantiations of the specifications defined in the modules discussed in this section. It declares the following parameters:

- functions $q_{min}, \ldots, pst_{max}$ that compute bounds for the respective subcomponents of the state of the steam boiler as discussed in section 3,

- the variable *round* used for synchronization,

- variables $q, \ldots, e$ that represent the actual state of the steam boiler and variables *p_cmd* and *e_cmd* that represent the commands sent to the actuators as described in section 3,

- variables $qr, \ldots, er$ that represent the readings of the sensors associated with the different steam boiler state components,

- variables $qa_1, \ldots, ea_2$ that represent the "adjusted" values used by the controller, and

- variables $qc_1, \ldots, ec_2$ that represent the estimations of the steam boiler state made by the controller.

The module goes on to state assumption *MonotonicityAssumption*, which formally asserts the assumptions on the functions $q_{min}, \ldots, e_{max}$ discussed in section 3. A TLA+ module that contains an assumption may only be instantiated with parameters that satisfy the assumption. Any theorems asserted in the module (such as theorem *Good_model* of module *Dynamics*) need only hold if the assumption is satisfied.

Next, module *Dynamics* gives definitions for the functions $qa_{min}, \ldots, ea_{max}$ that compute bounds for the steam boiler state at the next cycle given bounds for its current state. These functions are obtained by supplying lower or upper bounds to the functions $q_{min}, \ldots, e_{max}$, according to the monotonicity of the function for the respective argument. Module *Dynamics* then includes instantiated versions of the modules *Adjust*, *Estimate*, and *Environment* for each

---------- module *Dynamics* ----------

parameters
  $q_{min}, q_{max}, v_{min}, v_{max}, p_{min}, p_{max}, pst_{min}, pst_{max}, e_{min}, e_{max}$: CONSTANT
  $round$: VARIABLE
  $q, v, p, pst, e, p\_cmd, e\_cmd$: VARIABLE
  $qr, vr, pr, pstr, er$: VARIABLE
  $qf, vf, pf, pstf, ef$: VARIABLE
  $qa_1, qa_2, va_1, va_2, pa_1, pa_2, psta_1, psta_2, ea_1, ea_2$: VARIABLE
  $qc_1, qc_2, vc_1, vc_2, pc_1, pc_2, pstc_1, pstc_2, ec_1, ec_2$: VARIABLE

---

assumption
  $MonotonicityAssumption \triangleq$
    $(q1 \leq q2) \land (v1 \leq v2) \land (p1 \leq p2) \land (pst1 \leq pst2) \land (e1 \leq e2)$
    $\Rightarrow \land\ 0 \leq q_{min}[q1, v1, p1, pst1, e1, pc, ec] \leq q_{max}[q1, v1, p1, pst1, e1, pc, ec] \leq C$
          $\land\ q_{min}[q1, v2, p1, pst1, e2, pc, ec] \leq q_{min}[q2, v1, p2, pst2, e1, pc, ec]$
          $\land\ q_{max}[q1, v2, p1, pst1, e2, pc, ec] \leq q_{max}[q2, v1, p2, pst2, e1, pc, ec]$
          $\land\ \ldots$ (* similar assumptions for $v_{min}, \ldots, e_{max}$ omitted *)

$qa_{min}[q1, q2, v1, v2, p1, p2, pst1, pst2, e1, e2, pc, ec] = q_{min}[q1, v2, p1, pst1, e2, pc, ec]$
$qa_{max}[q1, q2, v1, v2, p1, p2, pst1, pst2, e1, e2, pc, ec] = q_{max}[q2, v1, p2, pst2, e1, pc, ec]$
$\ldots$ (* similar definitions for $va_{min}, \ldots, ea_{max}$ omitted *)

---

include *Adjust* **as** QA **with**
  $read \leftarrow qr, fail \leftarrow qf, adj1 \leftarrow qa_1, adj2 \leftarrow qa_2, est1 \leftarrow qc_1, est2 \leftarrow qc_2$
include *Estimate* **as** QC **with**
  $lwb \leftarrow qa_{min}, upb \leftarrow qa_{max}, est1 \leftarrow qc_1, est2 \leftarrow qc_2$
include *Environment* **as** Q **with**
  $lwb \leftarrow q_{min}, upb \leftarrow q_{max}, actual \leftarrow q, read \leftarrow qr, fail \leftarrow qf, adj1 \leftarrow qa_1, adj2 \leftarrow qa_2$
$\ldots$ (* similar **include** clauses for other entities omitted *)

$All\_fail\_reliable \triangleq\ \land\ Q.Fail\_reliable \land V.Fail\_reliable \land P.Fail\_reliable$
                         $\land\ PST.Fail\_reliable \land E.Fail\_reliable$
$All\_good\_approx \triangleq\ \land\ Q.Good\_approx \land V.Good\_approx \land P.Good\_approx$
                         $\land\ PST.Good\_approx \land E.Good\_approx$
$EnvDynamics \triangleq\ Q.\Phi \land V.\Phi \land P.\Phi \land PST.\Phi \land E.\Phi$
$ModDynamics \triangleq\ \land\ QA.\Phi \land VA.\Phi \land PA.\Phi \land PSTA.\Phi \land EA.\Phi$
                 $\land\ QC.\Phi \land VC.\Phi \land PC.\Phi \land PSTC.\Phi \land EC.\Phi$

---

theorem
  $Good\_model \triangleq EnvDynamics \land ModDynamics \Rightarrow$
                             $(All\_good\_approx \land \Box All\_fail\_reliable \twoheadrightarrow \Box All\_good\_approx)$

Fig. 8. Module *Dynamics* (incomplete).

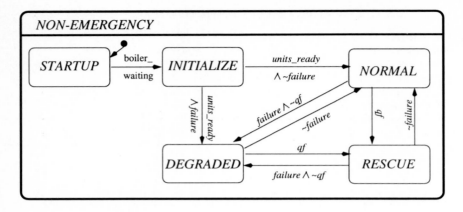

Fig. 9. Control modes for the refined controller.

component of the steam boiler state. The *actual*, *read*, *fail*, *adj*1, *adj*2, *est*1, and *est*2 parameters of these modules are instantiated with the variables representing the actual, read, adjusted or estimated values or the failure information for the respective state component. Any parameter that is not explicitly instantiated in an **include** clause is instantiated with the parameter of the same name of the including module.

Finally, module *Dynamics* states theorem *Good_model*, which asserts that the actual value of every state component falls within the bounds given by the corresponding "adjusted" values in any run of the environment (the physical steam boiler) and the controller, as long as the information about failures is reliable, and assuming that the initial values fall within the bounds. The theorem is expressed as a formula of the form $P \Rightarrow (Q \twoheadrightarrow R)$. The formal definition of the operator $\twoheadrightarrow$ has been given in [AL95]. Intuitively, formula $Q \twoheadrightarrow R$ asserts that R holds for at least as long as Q holds. We sketch a proof of theorem *Good_model* in the appendix, see CD-ROM annex A.5.

5 A refined controller

5.1 Refining control modes

We now refine the specification of the abstract controller of section 2, providing for different modes of operation as described in the problem statement. The initialization mode is split into a "startup" mode where the controller waits for the signal *boiler_waiting* and an "initialize" mode where it tries to ensure a normal water level by operating the pumps and the valve. The mode "operating" of the abstract controller is subdivided into several submodes, and the variable *system_failure* that was used as an "oracle" in the abstract specification is now defined in terms of sensor and transmission failures. Figure 9 illustrates the different control modes (except for the emergency mode, which stays unchanged) of

―――――――――――――――― module *Control* ――――――――――――――――
import *Naturals*
parameters
  round : VARIABLE
  boiler_waiting, *units_ready*, *stop_req* : VARIABLE
  transmission_failure, *qf*, *vf*, *pf*, *pstf*, *ef*, *dangerous_level*, *normal_level* : VARIABLE
  system_failure, *ctl_mode*, *prog_ready* : VARIABLE
―――

$failure \triangleq (qf' = T) \lor (vf' = T) \lor \exists i \in \{1,2,3,4\} : (pf'[i] = T) \lor (pstf'[i] = T)$
$critical\_failure \triangleq$
  $\lor (system\_failure' = T)$
  $\lor (ctl\_mode \in \{\text{"normal"}, \text{"degraded"}, \text{"rescue"}\}) \land (dangerous\_level' = T)$
$initialization\_complete \triangleq (ctl\_mode' = \text{"initialize"}) \land (normal\_level' = T)$
$Init \triangleq (ctl\_mode = \text{"startup"}) \land (prog\_ready = F)$
$\mathcal{N} \triangleq \land (system\_failure' = T) \equiv$
       $\lor (transmission\_failure' = T)$
       $\lor (ctl\_mode = \text{"startup"}) \land (vf' = T)$
       $\lor (ctl\_mode = \text{"initialize"}) \land (qf' = T \lor vf' = T)$
       $\lor (ctl\_mode = \text{"rescue"}) \land \lor (dangerous\_level' = T) \lor (qf' = T)$
                          $\lor \forall i \in \{1,2,3,4\} : (pf'[i] = T) \lor (pstf'[i] = T)$
  $\land ctl\_mode' =$ **if** $critical\_failure \lor (stop\_req' = T) \lor (ctl\_mode = \text{"emergency"})$
                  then "emergency"
                  elsif $(ctl\_mode = \text{"startup"})$
                  then if $(boiler\_waiting' = T)$ **then** "initialize" **else** "startup"
                  elsif $(ctl\_mode = \text{"initialize"}) \land \neg (units\_ready' = T)$
                  then "initialize"
                  elsif $(qf' = T)$ **then** "rescue"
                  elsif *failure* **then** "degraded"
                  else "normal"
  $\land (prog\_ready' = T) \equiv initialization\_complete$
$\Phi \triangleq Init \land \Box[\mathcal{N} \land round' \neq round]_{\langle system\_failure, ctl\_mode, prog\_ready, round\rangle}$
―――

Fig. 10. Module *Control*.

the refined controller and the state transitions between these modes of control. The TLA specification of the refined controller is given in module *Control* of figure 10. The component specification defines next-state relations for the variables *system_failure*, *ctl_mode*, and *prog_ready*. The definitions of *failure* and *critical_failure*, are virtually literal transcriptions from the problem description. The definition of *system_failure* reflects the failure conditions that are considered to be critical in the informal problem statement, depending on the current mode of operation. The transition relation for *ctl_mode* is easily read off the statechart-like illustration in figure 9. Unlike module *Abstract* of figure 2, module *Control* only describes the state transitions of the controller: the operation of the pumps and the valve is specified in module *Actuators*, discussed in the following section.

5.2 Operating the pumps and valve

Module *Actuators*, shown in figure 11, refines the abstract controller's decisions about the operation of the pumps and the valve. In particular, the specification decides on the number of pumps the controller wants to operate during the next cycle, but defers the decision about which specific pumps should be switched on or off to a later module. The reason for this separation is that we do not want to be concerned with pump failures at this stage to simplify the situation. In particular, we do not want to worry here about the number of pumps that are currently operational or about pump latency.

The specification is parameterized by two functions *lwb* and *upb* that compute lower and upper bounds for the water level in the steam boiler, given the current water level, the current amount of steam, the amount of water exiting through the valve, and assuming that k pumps were operating throughout the following cycle. For concreteness, we give possible definitions of these functions as suggested by the problem description:

$$lwb[q1, q2, v1, v2, e, k] = q1 - v2 \cdot \Delta - \tfrac{1}{2} \cdot U_1 \cdot \Delta^2 - e + k \cdot P$$
$$upb[q1, q2, v1, v2, e, k] = q2 - v1 \cdot \Delta - \tfrac{1}{2} \cdot U_2 \cdot \Delta^2 - e + k \cdot P$$

The module states obvious monotonicity assumptions about these functions.

Action $\mathcal{N}$ describes the opening and closing of the valve and the pumps. The behavior of the valve is specified exactly as in the abstract specification of figure 2, except that the condition *high_level* is made explicit. The decision concerning the number of pumps to operate is based on the water level estimated by the functions *lwb* and *upb*. Ideally, the controller should choose a number such that the estimated water level falls within the interval $[N_1, N_2]$ of normal operation. Otherwise, the specification asserts that at least one pump should be open if the water level is guaranteed to be below N_1, while all pumps should be closed if the water level is guaranteed to be above N_2. This is of course still a very loose specification that would have to be refined during the design stage. Besides, module *Actuators* gives specifications for the signals *normal_level* and *dangerous_level* that are used in module *Control*. We consider the water level to be dangerous not only if it may exceed the limit values M_1 or M_2, but also if the information given by the adjusted values is so imprecise that we cannot tell whether the level is above or below the limits for normal operation. (The reaction of the controller in such a state has been left open in the problem description.)

In the appendix (see CD-ROM annex LM.A.7) we prove that the specifications of modules *Control* and *Actuators* refine the abstract controller of module *Abstract* for suitable substitutions. The appendix also contains a module *PumpAssignment* (see CD-ROM annex LM.A.8) that specifies which pumps should be switched on or off, given the number of pumps the controller wants to operate.

of failures. It should then compute the "adjusted" values, decide on the new mode of operation, make decisions about the actuators, and finally estimate the new bounds for the state of the steam boiler as well as send the outputs to the physical units.

9 Evaluation and Comparison

1. We have given formal specifications of all parts of the system in the specification language TLA+, starting from an abstract specification of the controller and adding detail in successive steps of refinement. We have also considered the evolution of the physical system and its relation to the model maintained by the controller. Our specification is given further structure by grouping related requirements into modules. The input/output dependencies between these modules can be made explicit, yielding a dataflow analysis of our solution to the problem.

2. Our solution does not include an implementation, nor has it been linked to the simulator. There does not presently exist a prototyping tool for TLA specifications. However, the contributions [CD] and [DC] give implementations for the control program based on (a previous version of) our specifications, which have been linked to the simulator.

3. Many formalisms are based on concepts similar to those found in TLA. These include TLT [CW1, CW2], action systems [BSS], Timed Automata [LL], system B [A], and evolving algebras [BBDGR]. The basic setup of the TLT and evolving algebra solutions are quite similar to ours, although they differ in scope. For example, the TLT solution gives a much more detailed account of failure detection, while the contributions [A, BBDGR] are less detailed than our solution and do not model the environment.

   Our solution is best complemented by the solutions [CD] and [DC] that give implementations based on our specification of the control program in the synchronous languages Lustre and SPIN.

4. We spent about four weeks to write the initial solution and another two weeks to produce the version presented in this paper. TLA is based on very few, but elementary and powerful concepts. Specifications have a distinctly operational flavor, which should help programmers write TLA specifications. The framework is flexible enough to accomodate various specification styles. Although it is sometimes non-trivial to find the right fairness conditions, this is not an issue in a real-time specification such as the present one. In our experience, programmers can begin to write specifications after a few days of exposure to the method. Of course, the difficult part in writing specifications is to find an adequate abstraction and decomposition of the problem.

5. We believe that our solution should be understandable to programmers after a few days of training. TLA formulas use flexible variables, which are

straightforward abstractions of ordinary program variables. The notation uses primed and unprimed variables to refer to the value of a variable before and after an action is executed. Our solution models several components executing in synchrony—a familiar abstraction used in process control languages.

Acknowledgements

We would like to thank Thierry Cattel, Pierre Collette, and Jorge Cuéllar for insightful comments on a previous version of this specification.

References

[AL94]   Martín Abadi and Leslie Lamport. An old-fashioned recipe for real time. *ACM Transactions on Programming Languages and Systems*, 16(5):1543–1571, September 1994.

[AL95]   Martín Abadi and Leslie Lamport. Conjoining specifications. *ACM Transactions on Programming Languages and Systems*, 17(3):507–534, May 1995.

[A]   Jean-Raymond Abrial. A B-solution for the steam-boiler problem. This volume (see CD-ROM Annex.A).

[AS]   Jean-Raymond Abrial. Steam-boiler control specification problem. This volume (see CD-ROM Annex.AS).

[BBDGR]   Christoph Beierle, Egon Börger, Igor Durdanović, Uwe Glässer, Elvinia Riccobene. An evolving-algebra solution to the steam-boiler control specification problem. This volume (see CD-ROM Annex.BBDGR).

[BSS]   Michael Butler, Emil Sekerinski, Kaisa Sere. An Action System approach to the steam boiler problem. This volume (see CD-ROM Annex.BSS).

[CD]   Thierry Cattel, Gregory Duval. The steam-boiler problem in Lustre. This volume (see CD-ROM Annex.CD).

[CW1]   Jorge Cuéllar, Isolde Wildgruber. The steam boiler problem—a TLT solution. This volume (see CD-ROM Annex.CW1).

[CW2]   Jorge Cuéllar, Isolde Wildgruber. The real-time embedding of the steam boiler. This volume (see CD-ROM Annex.CW2).

[DC]   Gregory Duval, Thierry Cattel. Specifying and verifying the steam-boiler problem with SPIN. This volume (see CD-ROM Annex.DC).

[L96]   Leslie Lamport. TLA—temporal logic of actions. At URL http://www.research.digital.com/SRC/tla/ on the World Wide Web.

[L94]   Leslie Lamport. The temporal logic of actions. *ACM Transactions on Programming Languages and Systems*, 16(3):872–923, May 1994.

[LL]   G. Leeb, Nancy Lynch. Proving safety properties of the steam boiler controller. This volume (see CD-ROM Annex.LL).

Specifying Optimal Design for a Steam-Boiler System

Li XiaoShan[1] * and Wang JuAn[2]

[1] International Institute for Software Technology
United Nations University, P.O.Box 3058, Macau
E-mail: Xiaoshan.Li@newcastle.ac.uk
[2] Software Engineering Programme, Faculty of Science and Technology
University of Macau, P.O.Box 3001, Macau
E-mail: fstjaw@sftw.umac.mo

Abstract. Mean Value Calculus is a real-time interval logic which can be used to specify and reason about timing and logical constraints of real-time systems. In this paper, we apply it to specify and verify the steam-boiler control problem for Dagstuhl seminar. In addition to specifying safety requirement, emphasis is put on capturing non-functional requirements which reflect the qualities of designs, such as performance and optimization. Especially the fault-tolerant optimal design is presented to make the system run as long as possible while the water sensor is broken down.
Keywords: Specification, Verification, Mean Value Calculus, Steam-boiler System, Hybrid Systems, Fault-tolerant Design.

1 Introduction

Hybrid systems contain both continuous and discrete variables of time. Generally, only physical plants of control systems contain continuous variables, and the variables in decision makers i.e., the computer control programs are discrete. Plant and decision maker are connected by the sensors and actuators. Computer control programs in decision makers first obtains the sampling data from the sensors before making control decisions. The plant is controlled through actuators while it provides feedback to control programs via sensors.

Mean Value Calculus [4] is a real-time interval logic. In this paper, we apply it to specify and verify the steam-boiler control problem described in Chapter AS, this book. In addition to specifying safety requirement, emphasis is put on capturing non-functional requirements which reflect the qualities of designs, such as performance and optimization. From the case study, we present a general methodology of specifying and verifying this kind of sampling control systems. That is, firstly giving the specification for the mathematics model of the physical plant; then presenting the specification of the mathematical model of the

---

* Present address is Department of Electrical and Electronic Engineering, University of Newcastle upon Tyne, Newcastle upon Tyne NE1 7RU, UK.

sampling system; thirdly making design decisions by analyzing the mathematical models and requirements; finally verifying the design satisfying the requirements of the system.

Specially we discussed the fault-tolerant optimal design problem which is interesting for the theoretical scientists and practical engineerers. Taking the steam-boiler control system as a running example, this paper presents an approach to specifying optimal design of hybrid control system based on Mean Value Calculus(MVC).

A steam-boiler control problem was used as a testing case for the existing formal methods. It represents a typical hybrid system which is a challenge for formalization. In order to provide a concise solution, the original problem is simplified as follows:

The system comprises the main units:

- the steam-boiler;
- a sensor to measure the quantity of water in the steam-boiler;
- one pump to provide the steam-boiler with water;
- a sensor to measure the quantity of steam coming out of the steam-boiler.

The total capacity of the steam-boiler is C liters. There are two dangerous bounds for the water quantity in the steam-boiler: the minimal limit M_1 (in liters) and maximal limit M_2. When the water quantity keeps below M_1 or above M_2 five seconds, the system will be in danger. This is the safety requirement of the system. The water quantity is denoted by a continuous variable q.

There are also two normal water quantity bounds N_1 and N_2. The preferable water quantity ($N_1 < q < N_2$) should be maintained during regular operation. This is a performance requirement for the system. Obviously we have the following inequation:

$$0 < M_1 < N_1 < N_2 < M_2 < C$$

For simplification, it is assumed that the pump can supply water immediately when it is opened. The throughput of the pump, denoted by p (in liters/sec), can be adjusted from 0 (i.e. , the pump is closed) to its maximal capacity P (in liters/sec).

The control program communicates with the physical units (plant, sensors and actuators) through message passing. A sampling event occurs every five seconds. When it happens, the program first receive messages from the physical units, i.e., getting the values of q, v(steam quantity), and two Boolean states R(representing the system is running) and S(representing the water sensor works well). Then the control program analyzes the information and calculate the suitable pump control value p by the control decisions. Finally, the control value p is transmitted to the physical pump by the controller.

The time for message transmission is assumed to be zero. For more details of steam-boiler system, please see Chapter AS in this book.

By using MVC, the present study provides a technique to specify and verify the steam-boiler control system. It can also be applied to similar type of control systems. In the following we first provide a brief introduction to MVC in section 2. Then the specification of the steam-boiler system is in MVC. In section 4, We analyze the optimal rescue time of system while water sensor is broken down. An optimal control design is analyzed and the verifications of the safety requirement, performance and optimization is provided in section 5. In section 6 we summarize the formalization of the steam-boiler control system. Finally, evaluation and comparison are given in section 7.

2 Mean Value Calculus

Mean Value Calculus [4] is an extension of Duration Calculus [3] which is a real-time interval logic, by adding a different structure to interval temporal variables, which are mean values of Boolean functions instead of integrals. Here we just give a brief introduction.

Boolean functions are piecewise continuous, denoted F, G, etc., and 0 and 1 are constant Boolean functions. With the usual Boolean operator $\neg, \wedge, \vee, \cdots$ Boolean functions can be combined to form composite Boolean functions. When a Boolean function is a step function, it represents a *state*, denoted P, Q, S, etc., and it represents an *event*, denoted e_1, e_2, etc., when it is a δ-function (only to be 1 at isolated time points).

For any Boolean function F, the mean value of F, denoted $\bar{F}$, is a *term* in MVC, and is defined, for an arbitrary interval $[a, b]$,

$$\bar{F} \triangleq \begin{cases} \frac{1}{b-a} \int_a^b F & \text{if } b > a \\ F(b) & \text{if } b = a \end{cases}$$

Another *term* in MVC is the special symbol l, denoting the length of an interval. For an arbitrary interval $[a, b]$, the length of the interval is defined by

$$l \triangleq b - a \qquad (a \leq b)$$

The integral of Boolean function F then can be defined as:

$$\int F \triangleq \bar{F} * l$$

Term in mean value calculus has the form:

$$t ::= \bar{F} \mid l \mid r \mid t_1 \oplus t_2$$

where r is a real number, and $\oplus \in \{+, -, *\}$.

Formulae are constructed from such terms with operators in real arithmetics and logical connectives, and has the form:

$$A ::= t_1 = t_2 \mid t_1 \leq t_2 \mid A_1 \wedge A_2 \mid \neg A \mid A_1; A_2 \mid A^+$$

where ';' is a temporal modality, called *chop* and '+' is called chop plus. For an arbitrary interval $[a, b]$, $[a, b]$ satisfies $(A_1; A_2)$ iff there exists c such that

$a \leq c \leq b$, $[a,c]$ satisfies A_1, and $[c,b]$ satisfies A_2, and $[a,b]$ satisfies (A^+) iff there exists a non negative natural number n and there $[a,b]$ satisfies (A^n), where $A^1 = A$ and $A^{i+1} = A^i; A$.

The conventional modalities can be defined by *chop* operator as follows:

$$\Diamond A \triangleq true; A; true \quad \text{(for some subinterval } A \text{ holds)}$$
$$\Box A \triangleq \neg \Diamond \neg A \quad \text{(for all subintervals } A \text{ holds)}$$

We define some abbreviations :

$\lceil \rceil \triangleq l = 0$ (*true* for any point interval)
$\lceil F \rceil^0 \triangleq \lceil \rceil \wedge (\overline{F} = 1)$ (F has value 1 at a point)
$\lceil F \rceil \triangleq (l > 0) \wedge \neg((l > 0); (\lceil \rceil \wedge \overline{F} = 0); (l > 0))$
  (F has value 1 everywhere inside a non-point interval)
$\lceil F \rceil^* \triangleq \lceil F \rceil \vee \lceil F \rceil^0$

The axioms of Mean Value Calculus are:

(A1) $\overline{0} = 0$
(A2) $\overline{F} \geq 0$
(A3) $\overline{F \vee G} = \overline{F} + \overline{G} - \overline{F \wedge G}$
(A4) $(\overline{F} * l = r + s) \Leftrightarrow (\overline{F} * l = r); (\overline{F} * l = s)$ $r, s \geq 0$
(A5) $\lceil \rceil \Rightarrow (\lceil F \rceil^0 \vee \lceil \neg F \rceil^0)$
(A6) $\lceil F \rceil^* \Rightarrow (\overline{F} = 1)$
(A7) $\Box(\lceil \rceil \vee \lceil F \rceil; true \vee \lceil \neg F \rceil; true)$
(A8) $\Box(\lceil \rceil \vee true; \lceil F \rceil \vee true; \lceil \neg F \rceil)$

For every given state P there are two corresponding primitive events: the rising edge and the falling edge of P

- $\uparrow P$ holds at the point where there is a changing from $\neg P$ to P
- $\downarrow P$ holds at the point where there is a changing from P to $\neg P$

$\uparrow P$ and $\downarrow P$ are called *germ-ships*. A transition of a system from state P to state Q can be considered as a composite germ-ship:

$$\downarrow P \wedge \uparrow Q$$

which means, leaving state P and at the same time entering state Q.

There are a group of axioms of germships and corresponding germship calculus in [4]. Here we omit them.

States are assumed *finitely variable*. That is, a state can have only finite alternations of its presence and absence in a finite interval.

For dealing with the functions of real value we extend MVC by adding the part of Extended Duration Calculus [5], where point properties of real-valued functions are taken to be states, and lifted to interval properties by the ceiling operator $\lceil \cdot \rceil$, such as $\lceil Continuous(f) \rceil$, $\lceil f1 \geq f2 \rceil$, $\lceil \dot{f} = w \rceil$, etc. $\lceil Continuous(f) \rceil$ means that real-valued function f is continuous everywhere *within* a non-point interval, $\lceil f1 \geq f2 \rceil$ means that function $f1$ is greater than or equal to function $f2$ everywhere inside a non-point interval, etc.

3 Specification of the Steam-Boiler Problem

The steam-boiler problem is a typical hybrid system which has both continuous and discrete variables. Its behavior and requirements are captured by MVC in this section.

3.1 Mathematical Model of Physical Plant

The system physical plant behavior can be characterized by the following functions q, v, p. q and v are continuous functions; p is a piecewise continuous function. Let $Time$ represent the observation time domain, which is a non-negative real number set $[0, \infty)$.

$q : Time \rightarrow [0, C]$ (water quantity (liters))
$v : Time \rightarrow [0, W]$ (quantity of steam(liters/sec))
$p : Time \rightarrow [0, P]$ (throughput of pump(liters/sec))

The mathematical model of the steam-boiler problem can be established through the following laws:

$$\Box(\lceil 0 \leq q < C \rceil^* \wedge \lceil 0 \leq p \leq P \rceil^* \wedge \lceil 0 \leq v \leq W \rceil^*) \qquad (Law_1)$$

$$\Box(\lceil -U_2 \leq \dot{v} \leq U_1 \rceil^*) \qquad (Law_2)$$

$$\Box(\lceil \dot{q} = p - v \rceil^*) \qquad (Law_3)$$

The meaning of these laws are obvious. Law_1 says that the water quantity q can never exceed the total capacity of the steam-boiler; the maximal throughput of the pump is P; and the maximal quantity of steam coming out of steam-boiler is W. Law_2 says nothing but U_1, U_2 are maximal gradient of increase and decrease, respectively, of steam coming out of the steam-boiler. Law_3 specifies that the change speed of water quantity is decided by p and v.

We assume that the variables q and v are continuous, and p is a piece-wise continuous

$$\Box(\lceil\ \rceil \vee (\lceil Continuous(q) \rceil \wedge \lceil Continuous(v) \rceil \wedge \lceil Continuous(p) \rceil^+)) \; (Assum_1)$$

3.2 Mathematical Model of Computer Sampling

Since the program follows a cycle which takes place every five seconds to sample data and make decisions, a mathematical model for sampling is needed to capture the requirements formally. In each sampling cycle, the program receives messages from physical units, makes decisions based on current status of physical units, and sends out control command to physical units. We use the following functions to characterize sampling data.

$\bar{q} : Time \rightarrow [0, C]$ (the sampling value of water quantity)
$\bar{v} : Time \rightarrow [0, W]$ (the sampling value of steam quantity)
$\bar{p} : Time \rightarrow [0, P]$ (the sampling value of throughput of pump)

These three sampling functions are piece-wise continuous, which are assumed to be right continuous:

$$\Box((\lceil \ \rceil \vee \lceil Continuous(f) \rceil^+) \wedge (\lceil f = f_0 \rceil \Rightarrow \lceil f = f_0 \rceil^0; l > 0)) \qquad (Assum_2)$$

where $f \in \{\bar{p},\ \bar{v},\ \bar{q}\}$.

We use two Boolean state variables S and R to express the working states of the system.

$S : Time \rightarrow \{0, 1\}$
$R : Time \rightarrow \{0, 1\}$

If the water sensor is working properly at time t, then we have $S(t) = 1$, otherwise $S(t) = 0$ represents the system is in the rescue mode. Similarly $R(t) = 1$ represents the system is running. Otherwise $R(t) = 0$ denotes the system was stopped. Meanwhile the Boolean state variable S is assumed to be right continuous, and R is assumed to be 1 at the beginning and end points of interval while $\lceil R \rceil$ holds.

$$\Box((\lceil R \rceil \Rightarrow \lceil R \rceil^0; l > 0; \lceil R \rceil^0) \wedge (\lceil S \rceil \Rightarrow \lceil S \rceil^0; l > 0)) \qquad (Assum_3)$$

Since the program follows a cycle which takes place every five seconds to sample data and make decisions, a mathematical model for sampling is needed to capture the requirements formally. We use $\bar{q}$, $\bar{v}$ and $\bar{p}$ to characterize sampling data in control program, which are piece-wise continuous functions and are assumed to be right continuous.

The events are introduced for the model, which are δ-functions in MVC. sp represents the sampling event. Event $\uparrow R$ starts the steam-boiler. Event $\downarrow R$ stops the steam-boiler. Event $\uparrow S$ represents the water sensor is repaired, and event $\downarrow S$ means the water sensor is broken at sampling time.

The sampling behavior can be formalized by the following sampling law.

$$\Box((\lceil R \rceil \wedge (l = 5) \Rightarrow \Diamond sp) \wedge (sp; \lceil \neg sp \wedge R \rceil; sp \Rightarrow l = 5)) \qquad (Law_4)$$

It means a sampling event occurs in every five seconds during the system operation and there are exact five seconds between two adjacent samplings.

When the water sensor works perfectly, it provides correct water quantity at sampling points.

$$\Box(\lceil S \rceil^0 \wedge sp \Rightarrow \lceil \bar{q} = q \rceil^0) \qquad (Law_5)$$

If the water sensor is broken down, the water quantity is estimated by the following calculation:

$$sp \wedge \lceil \bar{q} = a_0 \wedge \bar{p} = b_0 \rceil^0; l = 5; sp \wedge \lceil \neg S \rceil^0 \Rightarrow l = 5; \lceil \bar{q} = a_0 + 5b_0 - \widehat{\int v} \rceil^0 \quad (Law_6)$$

where $\widehat{\int v}$ is the estimation of $\int v$, which is the steam quantity emitted from the steam-boiler within one sampling cycle. It is a function of the two sampling steam values in a sample cycle. The value of $\widehat{\int v}$ in a sample cycle depends on the concrete situation of steam-boiler. The analysis and calculation of $\widehat{\int v}$ will be discussed later. Law_6 is derived from Law_3. If the water sensor works well, $\int v$ can be calculated precisely by Law_3.

Because we assume that the steam and pump sensors always work properly and the pump is controlled by the control program, we have

$$\Box(sp \Rightarrow \lceil \bar{v} = v \rceil^0) \wedge \Box(\lceil \bar{p} = p \rceil^*) \quad (Law_7)$$

3.3 Assumptions and Requirements

For simplification, we make the following assumptions about the system in addition to $Assum_1$, $Assum_2$, and $Assum_3$:

$$\Box(\uparrow R \Rightarrow \lceil M_1 < q < M_2 \rceil^0 \wedge \lceil S \rceil^0) \quad (Assum_4)$$

$$\Box(\lceil N_2 - N_1 \geq 10P \rceil^* \wedge \lceil W < P \rceil^*) \quad (Assum_5)$$

$$\Box(\uparrow S \vee \downarrow S \vee \downarrow R \vee \uparrow R \Rightarrow sp) \quad (Assum_6)$$

$Assum_4$ assumes that the system starts with its water quantity q within the safe range $[M_1, M_2]$ and the water sensor works well at the beginning. $Assum_5$ says there is enough space between N_1 and N_2 so that it takes at least 10 seconds to fill up such space, and W is less than P. $Assum_6$ prescribes that events can only occur at sampling points. We use Law and $Assum$ to represents all laws and assumptions about the system hereafter.

$$Law \stackrel{\frown}{=} Law_1 \wedge Law_2 \wedge \cdots \wedge Law_7$$

$$Assum \stackrel{\frown}{=} Assum_1 \wedge Assum_2 \wedge \cdots \wedge Assum_6$$

From the discussion above, the reasonable design of the steam-boiler system should satisfy the following requirements.

Safety The water quantity q in the steam-boiler can never exceed continuously the safe bounds for more than 5 seconds.

$$\Box(((\lceil q \leq M_1 \rceil \wedge \lceil R \rceil) \Rightarrow l \leq 5) \wedge ((\lceil q \geq M_2 \rceil \wedge \lceil R \rceil) \Rightarrow l \leq 5)) \quad (\text{Req})$$

This is the basic requirement for the system since any design must guarantee it. For simplification without loosing generality, the safety requirement (Req) is changed into the sampling safety requirement ($Safe$) as follows

$$\Box(sp \wedge \lceil R \rceil^0 \Rightarrow \lceil M_1 < q < M_2 \rceil^0) \tag{Safe}$$

It means that at each sampling time we must guarantee the quantity of water q in the safe range $[M_1, M_2]$ while the system is running. Obviously, it implies the safety (Req). Meanwhile it also implies that the event $\downarrow R$ must occur when q is found to be out of the safe range $[M_1, M_2]$ at a sampling time, i.e., $sp \wedge (\lceil q \leq M_1 \rceil^0 \vee \lceil q \geq M_2 \rceil^0) \Rightarrow \downarrow R$.

Performance If the water sensor works perfectly, the whole system should not be stopped and the water quantity should be kept within normal bounds continuously and stably.

$$\Box(\lceil N_1 + 5W < q < N_2 - 5P \rceil^0; \lceil S \wedge R \rceil) \Rightarrow \lceil N_1 < q < N_2 \rceil) \tag{Perf}$$

A better design should have this good performance property.

Optimization If the water sensor is broken down, the whole system should not be stopped until it has to. In other words, the fault-tolerant design should make the rescue time as long as possible. By analyzing the system, we find the longest rescue sampling time, and give a fault-tolerant design to satisfy the following optimal property.

$$\Box(\downarrow S; \lceil \neg S \wedge R \rceil; \downarrow R \Rightarrow l \geq 5m) \tag{Opt}$$

in which $m = min\{n \mid (\sum_{i=1}^{n-1} S_\Box^i + S_\triangle^n) \geq (M_2 - M_1)\}$,
$S_\Box^i = [25U_1U_2 + 5(U_1 - U_2)(v_i - v_{i-1}) - (v_i - v_{i-1})^2]/(U_1 + U_2)$ [3] and
$S_\triangle^n = 25(U_1 + U_2)/2$.

4 Analysis of Rescue Phase

The rescue phase is a fault-tolerant period of the system, in which the computer control program tries to maintain a satisfactory water quantity q in the safe range $[M_1, M_2]$ while the water sensor is broken down. The water quantity is estimated by a calculation taking into account the maximum dynamics of the steam quantity from the steam-boiler.

Suppose we have a sequence of sampling values:

$$(\bar{q}_0, \bar{v}_0, \bar{p}_0), \ (\bar{q}_1, \bar{v}_1, \bar{p}_1), \ \cdots, (\bar{q}_i, \bar{v}_i, \bar{p}_i), \ (\bar{q}_{i+1}, \bar{v}_{i+1}, \bar{p}_{i+1}), \cdots$$

---

[3] Here is the general case where v_{i-1} and v_i are two successive sampling values of steam quantity. The detail analysis will be given in the next section.

All physical units work perfectly at the i-th sampling time, but the water sensor is found broken at the $(i+1)$-th sampling time. The water quantity q_{i+1} can be estimated by using q_i, v_i, v_{i+1} and p_i. Then we can make the control decision of p_{i+1} according to the estimated water quantity. However, there may be some errors at each estimating calculation since exactly the intermediate steam values emitted between two adjacent samplings are not known. The key point of rescue is to analyze these possible estimating errors and try to find maximal rescue time for a given sequence of sampling values.

As q_{i+1} is largely dependent on the dynamic behavior of steam $v(t)$, where $t_i \leq t \leq t_{i+1}(t_{i+1} = t_i + 5)$, we depict the possible change of steam as the following diagram:

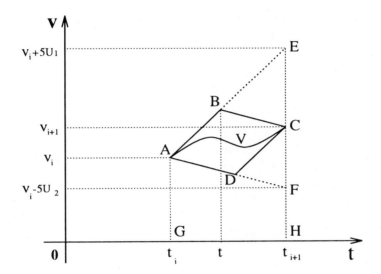

Fig. 1. Possible Changes of Steam

It is obvious from Law_2, Law_3 and Fig. 1, there exists

$$0 \leq (v_i - 5U_2) \leq v_{i+1} \leq (v_i + 5U_1) \leq W$$

So at the sampling time t_i we know that all possible changes of steam quantity from the steam-boiler between two successive sampling points t_i and t_{i+1} will fall into the area of triangle AEF. There are two extreme cases in which steam quantity continuously increases from v_i to $v_i + 5U_1$ or decreases from v_i to $v_i - 5U_2$ on the time interval $[t_i, t_{i+1}]$, i. e., $v_{i+1} = v_i + 5U_1$ or $v_{i+1} = v_i - 5U_2$ as shown by the line AE or AF in Fig. 1.

At the sampling time t_{i+1}, we can know the actual value v_{i+1}. If $v_i - 5U_2 < v_{i+1} < v_i + 5U_1$ holds, we have two extreme cases:

(1) v_i initially increases at maximal velocity U_1 from t_i to t which is a intermediate time within interval $[t_i, t_{i+1}]$, and then decreases at maximal velocity

U_2 from t to t_{i+1}, as shown by ABC in Fig. 1.

(2) v_i initially decreases at maximal velocity U_2 and then increases at maximal velocity U_1, as shown by ADC in Fig. 1.

All possible changes of steam quantity will fall into the area of quadrilateral $ABCD$. A typical curve of steam dynamically change between two successive sampling points is depicted as the curve AVC in Fig. 1.

It is not difficult to understand that the broken line ABC represents the case where the greatest quantity of steam comes from the steam-boiler between two successive samplings. Similarly, the broken line ADC represents the case where the least one.

By Law_3, the following equation holds :

$$q_{i+1} = q_i + 5p_i - \int_{t_i}^{t_{i+1}} v(t)dt.$$

As two instant values v_i and v_{i+1} are known, the total quantity of steam out from t_i to t_{i+1}, i.e. $\int_{t_i}^{t_{i+1}} v(t)dt$, should satisfy:

$$S_{ADCHG} \le \int_{t_i}^{t_{i+1}} v \le S_{ABCHG}$$

Here S_p means the area of polygon p. So S_{ADCHG} and S_{ABCHG} are the least and greatest quantities of steam coming out of the steam-boiler from t_i to t_{i+1}. Hereafter we abbreviate them as S'^{i+1}_{min} and S'^{i+1}_{max} respectively.

Therefore, water quantity in the steam-boiler q_{i+1} at time t_{i+1} satisfies:

$$(q_i + 5p_i - S'^{i+1}_{max}) \le q_{i+1} \le (q_i + 5p_i - S'^{i+1}_{min}).$$

$$(5p_i - S'^{i+1}_{max}) \le q_{i+1} - q_i \le (5p_i - S'^{i+1}_{min})$$

When water sensor is broken down, the $(i+1)$-th sampling value of water quantity $\bar{q}_{i+1}$ is calculated based on $\bar{q}_i, v_i, v_{i+1}$ and an estimated value for $\int_{t_i}^{t_{i+1}} v(t)dt$, denoted as $\widehat{\int v}$. For instance, at time t_i, the behavior of steam between t_i and t_{i+1} can be estimated as the curve AVC Fig. 1. Thus the estimated value of water quantity $\bar{q}_{i+1}$ at next sampling time can be calculated :

$$\bar{q}_{i+1} = \bar{q}_i + 5p_i - \widehat{\int v}, \text{ and then } \bar{q}_{i+1} - \bar{q}_i = 5p_i - \widehat{\int v}$$

So from above we get

$$(\widehat{\int v} - S'^{i+1}_{max}) \le (q_{i+1} - q_i) - (\bar{q}_{i+1} - \bar{q}_i) \le (\widehat{\int v} - S'^{i+1}_{min})$$

Obviously, there might exist an error when we estimate $\int_{t_i}^{t_{i+1}} v(t)dt$ by $\widehat{\int v}$. For example, suppose that the curve AVC in Fig.1 represents the actual dynamic behavior of v and the broken line ABC is the estimated change of v, then

For simplification, let $x_i = M_2 - qa2_i$ and $y_i = qa1_i - M_1$ ($x_i, y_i > 0$). Therefore

$$\begin{aligned}
x_{i+1} &= M_2 - qa2_{i+1} \\
&= M_2 - \bar{q}_{i+1} - maxH_{i+1} \\
&= M_2 - \bar{q}_i - 5p_i + \widehat{\int v_{i+1}} - H_i - maxH_{i+1} \\
&= M_2 - qa2_i - 5p_i + \widehat{\int v_{i+1}} - maxH_{i+1} \\
&= x_i - 5p_i + S'^{i+1}_{min}
\end{aligned}$$

where $S'^{i+1}_{min} = \widehat{\int v_{i+1}} - maxH_{i+1}$
Similarly,

$$\begin{aligned}
y_{i+1} &= qa1_{i+1} - M_1 \\
&= \bar{q}_{i+1} - M_1 - maxL_{i+1} \\
&= \bar{q}_i + 5p_i - \widehat{\int v} - M_1 - L_i - maxL_{i+1} \\
&= qa1_i + 5p_i - \widehat{\int v} - M_1 - maxL_{i+1} \\
&= y_i + 5p_i - S'^{i+1}_{max}
\end{aligned}$$

where $S'^{i+1}_{max} = \widehat{\int v_{i+1}} + maxL_{i+1}$.

If the water quantity q is kept in the safe range $[M_1, M_2]$ at $(i+1)$-th sampling time, then $x_{i+1} > 0$ and $y_{i+1} > 0$ should hold, i.e.,

$$S'^{i+1}_{max} - y_i < 5p_i < S'^{i+1}_{min} + x_i$$

The steam value v_{i+1} is not known at i-th sampling time, but it is within the range of $[max\{0, v_i - 5U_2\}, min\{W, v_i + 5U_1\}]$ by Law_1 and Law_2. When the steam decreases at maximum speed U_2, $S'^{i+1}_{min} = S^{i+1}_{min}(S_{AFHG}$ in Fig. 1); and when the steam increases at maximum speed U_1, $S'^{i+1}_{max} = S^{i+1}_{max}(S_{AEHG}$ in Fig. 1). As well known, from the Fig. 1 we can get

$$S'^{i+1}_{min} \geq S^{i+1}_{min} \quad \text{and} \quad S'^{i+1}_{max} \leq S^{i+1}_{max}$$

Thus the pump control condition (PC) is obtained:

$$max\{0, S^{i+1}_{max} - y_i\} \leq 5p_i \leq min\{5P, S^{i+1}_{min} + x_i\} \tag{PC}$$

Hence if p_i satisfies (PC) at the i-th sampling time, then the water quantity will be able to be guaranteed in the safe range $[M_1, M_2]$ at the $(i+1)$-th sampling time.

However there must be a controllable pre-condition at the i-th sampling time (CC):

$$(x_i > 0) \wedge (y_i > 0) \wedge (x_i + y_i > S^{i+1}_{max} - S^{i+1}_{min}) \tag{CC}$$

This means that only under this situation the water quantity can be possibly guaranteed the water quantity in the safe range $[M_1, M_2]$ at next sampling. From Fig. 1, $S_{max}^{i+1} - S_{min}^{i+1} = S_{AEF}$. If $x_i + y_i \leq S_{max}^{i+1} - S_{min}^{i+1}$, the pump can not be opened properly. No matter how to control the pump, there is a possibility that $qa1 < M_1$ or $qa2 > M_2$ may be violated because v_{i+1} is non-determined. That is to say, at this time the system is out of our control ability. Meanwhile, we can say that we have done our best to control the system when the sensor is broken , and the design is an optimal one. Here $S_{max}^{i+1} - S_{min}^{i+1}$ is abbreviated as $S_\triangle^{i+1}$. Define

$$S_\square \triangleq S'_{max} - S'_{min} \quad \text{and} \quad S_\triangle \triangleq S_{max} - S_{min}$$

The optimal rescue sampling times can be calculated by the following formula:

$$m = min\{n \mid \sum_{i=1}^{n-1} S_\square^i + S_\triangle^n \geq M_2 - M_1\}$$

So actual design rescue sampling times m approachs the theoretical ideal maximal limit $\mathcal{M}$ analysed in section 4.

5.3 Rescue control decisions

First of all, let us define the control condition (C) derived from (CC):

$$C \triangleq \lceil R \land \neg S \rceil^0 \land \lceil x + y > S_\triangle \rceil^0 \land \lceil (x > 0) \land (y > 0) \rceil^0$$

where $x = M_2 - qa2$ and $y = qa1 - M_1$.

If the condition C holds at the sampling time, then the system is controllable by opening the pump properly. Otherwise, the steam-boiler is not controllable, that is to say, the water quantity q may be out of the safe range $[M_1, M_2]$. If we find the system is out of the safe at next sampling time, then the system must be stopped.

According to (CC) and (PC), we give the following rescue decisions :

$$Des_4 \triangleq \square(C \land sp \land \lceil (x > S_\triangle) \land (x \geq y) \rceil^0 \Rightarrow \lceil p = S_{max}/5 \rceil^0)$$
$$Des_5 \triangleq \square(C \land sp \land \lceil (y > S_\triangle) \land (y > x) \rceil^0 \Rightarrow \lceil p = S_{min}/5 \rceil^0)$$
$$Des_6 \triangleq \square(\lceil R \land \neg S \rceil^0 \land sp \land \lceil (0 < x < S_\triangle) \land (0 < y < S_\triangle) \rceil^0$$
$$\Rightarrow \lceil p = (2S_{min} + S_\triangle + x - y)/10 \rceil^0)$$
$$Des_7 \triangleq \square(\lceil R \land \neg S \rceil^0 \land sp \land \lceil (x \leq 0) \lor (y \leq 0) \rceil^0 \Rightarrow \downarrow R)$$

where $S_{max} = 5v + 25U_1/2$, $S_{min} = 5v - 25U_2/2$ and $S_\triangle = 25(U_1 - U_2)/2$ (suppose v is the present sampling value of steam and $0 \leq v - 5U_2 < v + 5U_1 \leq W$).

The auxiliary variables x, y, H and L are introduced for simplification. They satisfy the following assignment formulas whose conjunction called Des_8.

$$[S]^0 \wedge sp \Rightarrow \lceil (H = 0) \wedge (L = 0) \rceil^0$$
$$C \wedge sp \wedge \lceil x \geq y \rceil^0 \wedge \lceil H = H_i \rceil^0; SP \Rightarrow l = 5; \lceil H = H_i + S_\square \rceil^0$$
$$C \wedge sp \wedge \lceil y > x \rceil^0 \wedge \lceil L = L_i \rceil^0; SP \Rightarrow l = 5; \lceil L = L_i + S_\square \rceil^0$$
$$\lceil (\bar{q} = \bar{q}_0) \wedge (p = p_0) \wedge (H = H_i) \rceil^0; SP$$
$$\Rightarrow (l = 5); \lceil \bar{q} = \bar{q}_0 + 5p_0 - S'_{min} - H + H_i \rceil^0$$
$$\lceil (\bar{q} = \bar{q}_0) \wedge (H = H_i) \wedge (L = L_i) \rceil^0 \Rightarrow \lceil x = M_2 - \bar{q}_0 - H_i \rceil^0 \wedge \lceil y = \bar{q}_0 - M_1 - L_i \rceil^0$$

where $SP \triangleq \lceil (v = v_i) \rceil^0; (l = 5); sp \wedge \lceil v = v_{i+1} \rceil^0$. The calculation of $S_\square$, $S_\triangle$ and $S'min$ can be calculated [4] by Fig. 1

$$S_\square = [25U_1U_2 + 5(U_1 - U_2)(v_{i+1} - v_i) - (v_{i+1} - v_i)^2]/(U_1 + U_2)$$
$$S_\triangle = 25(U_1 + U_2)/2$$
$$S'_{min} = 5(v_i + v_{i+1})/2 - [25U_1U_2 + 5(U_1 - U_2)(v_{i+1} - v_i)$$
$$- (v_{i+1} - v_i)^2]/2(U_1 + U_2).$$

So the following theorems can be proved by the proof system of MVC.

Theorem 2

$$Assum \wedge Law \wedge Des_4 \wedge Des_5 \wedge Des_6 \wedge Des_7 \wedge Des_8 \Rightarrow Opt$$

Theorem 3

$$Assum \wedge Law \wedge Des_1 \wedge Des_2 \wedge Des_3 \wedge Des_4 \wedge Des_5 \wedge Des_6 \wedge Des_7 \wedge Des_8 \Rightarrow Safe$$

Des_4 means while the system is in the situation of $(x > S_\triangle) \wedge (x \geq y)$, then the pump is opened at $p = S_{max}/5$. Thus there is S_{max} water which will be pumped into steam-boiler in five seconds. Obviously S_{max} is greater than the quantity of steam coming out, but less than x. Thus the water quantity q can be guaranteed to increase at next sampling time. Meanwhile, the water quantity q increases $S_\triangle$ if no steam comes out during the sampling cycle. Therefore *Lemma* 1 holds.

Lemma 1

$$Dec_4 \wedge C \wedge sp \wedge \lceil (x > S_\triangle) \wedge (x \geq y) \rceil^0; (l = 5) \Rightarrow (l = 5); sp \wedge \lceil (x > 0) \wedge (y > 0) \rceil^0$$

Similarly while the system is in the situation $(y > S_\triangle) \wedge (y > x)$, we can get *Lemma* 2 by using Des_5

Lemma 2

$$Dec_5 \wedge C \wedge sp \wedge \lceil (y > S_\triangle) \wedge (y > x) \rceil^0; (l = 5) \Rightarrow (l = 5); sp \wedge \lceil (x > 0) \wedge (y > 0) \rceil^0$$

---

[4] Here is a general case in which we suppose $0 \leq (v_i - 5U_1) < (v_i + 5U_2) \leq W$. For the particular cases such as the point B touches the line $v = W$ can be similarly calculated. Please refer to [2] for the details

Des_6 means that the pump will be opened at $p = (2S_{min} + S_\triangle + x - y)/10$ while $(x < S_\triangle) \wedge (y < S_\triangle)$. If $(x + y > S_\triangle)$, then we can guarantee that system is still safe at next sampling time(using x' etc. to express the parameters of next sampling time) and since

$$\begin{aligned}
x' &= M_2 - qa2' \\
&= M_2 - (\bar{q} + 5p - \widehat{\int v}) - H' \\
&= x - S_{min} - (S_\triangle + x - y)/2 + \widehat{\int v} - maxH' \\
&= (x + y - S_\triangle)/2 + S'_{min} - S_{min} \\
&> 0 \qquad\qquad\qquad\qquad \text{and} \\
y' &= qa1' - M_1 \\
&= (\bar{q} + 5p - \widehat{\int v}) - L' - M_1 \\
&= (x + y - S_\triangle)/2 + S_{max} - S'_{max} \\
&> 0
\end{aligned}$$

Hence we have *Lemma* 3

Lemma 3

$Dec_6 \wedge C \wedge sp \wedge \lceil (x < S_\triangle) \wedge (y < S_\triangle) \rceil^0; (l = 5) \Rightarrow (l = 5); sp \wedge \lceil (x > 0) \wedge (y > 0) \rceil^0$

From *lemma* 1, 2, 3, the following optimal theorem can be easily proved.

Theorem 4 (Optimal Theorem)

$Dec_4 \wedge Dec_5 \wedge Dec_6 \wedge Des_8 \wedge C \wedge sp; (l = 5) \Rightarrow (l = 5); sp \wedge \lceil (x > 0) \wedge (y > 0) \rceil^0$

If any design of the steam-boiler satisfies this optimal theorem, then the design can achieve the optimal requirement (*Opt*). From *Optimal Theorem*, the *Theorem* 2 can be easily proved.

Up to now, we can find that $Des_1, Des_2, Des_3, Des_4, Des_5, Des_6, Des_7$ and Des_8 consist of an optimal design of steam-boiler.

6 Discussion and Conclusion

In this paper, we used Mean Value Calculus to specify and verify the steam-boiler system. MVC can specify not only the high level abstract requirements of hybrid system, but also specify the low level operational design decisions. The method and techniques of this case in this paper can be used to the similar type of control systems.

From our experience we found MVC is suitable for describing hybrid control systems containing critical events which occur at *instantaneous* time points instead of having *durations* on certain time intervals. Examples of such events include *sample* data, *shut-down* the system etc.

This case study demonstrates the formal method is very important to specify and verify the control system. When we give the specification of the system models, the details of the system can be clarified. The mathematics of analysis is useful to present the design. After the specification and analysis finish, verification is necessary to confirm the correctness of design. The formal verification is the converse procedure of the analysis design. However, it can not be taken the place by the informal mathematical analysis.

The main feature of this paper is the fault-tolerant optimal design. Optimization is a very important and interesting for theoretical research and practical application. For the steam-boiler case, there exist many optimal designs. We use formal method, MVC to specify and verify the optimization. However, how to abstract some general ideas and methodology for optimal control problems still needs future research.

7 Evaluation and Comparison

1. Main parts of the system are specified formally. Especially we presented the optimal design while the water sensor was broken down. Optimization is a new interesting topic for the formal methods.

2. No. However, the decisions of design are operational and it is not difficult to transform them into an implementation of a control program.

3. The paper deeply discusses the fault-tolerant optimal design to the water sensor. The details on the communication, testing failures of system and other problems we omitted can see Chapter RS, CW1, VH and other relative chapters in this book.

4. a. 4 person months. b. 4 weeks.

5. a. Yes. b. No. c. 4 hours.

Acknowledgment Many of these ideas have been improved by discussions with Prof. Zhou ChaoChen. The authors would like to thank him for valuable suggestions and guidance. We would also like to thank Prof. Dines Bjørner for his support and help. Thanks to Dr He Weidong, Phillp Chen, Wang Ji for their comments and discussions. Special thanks are due to two referees for their good comments.

References

1. He Weidong, Zhou Chaochen: A Case Study of Optimization. In *Computer Journal*, Vol.38, No.9, pp734-746, 1995.
2. Li Xiaoshan, Wang Juan: Specifying Optimal Design of a Steam-boiler Control System, *UNU/IIST Technical Report Draft*, February 1995.
3. Zhou Chaochen, C.A.R. Hoare and A.P. Ravn: A Calculus of Durations. In *Information Processing Letters, 40*, 5, pp. 269-276, 1991.

4. Zhou Chaochen and Li Xiaoshan: A Mean Value Calculus of Durations. In *A Classical Mind (Essays in Honour of C.A.R. Hoare)*, Edited by A.W.Roscoe, Prentice-Hall, pp. 431-451,1994.
5. Zhou Chaochen, Anders P. Ravn and Michael R. Hansen: An Extended Duration Calculus for Hybrid Real-Time Systems. In *Hybrid Systems, LNCS 736. Edited by R.L.Grossman, A. Nerode, A.P.Ravn and H. Rischel*, pp. 36-59, Springer Verlag,1993.

An Object-Oriented Algebraic Steam-Boiler Control Specification

Peter Csaba Ölveczky[1]*, Piotr Kosiuczenko[2,3], and Martin Wirsing[2]

[1] Dept. of Informatics, University of Bergen, Norway

[2] Institut für Informatik, Ludwig-Maximilians-Universität München, Germany

[3] Instytut Matematyki, Politechnika Warszawska, Poland

Abstract. In this paper an object-oriented algebraic solution of the steam-boiler specification problem is presented. The solution is written in Timed Maude. Timed Maude is a specification language under development where the static parts of the specified system are described by equational specifications, whereas the behaviour of a process is described by timed term rewriting. Timed Maude is based on Meseguer's Maude language, and its underlying logic is timed rewriting logic, an extension of rewriting logic to deal with hard real-time systems.

The specification focuses on the description of the control program, which is designed as the parallel composition of several objects which communicate using messages. The transmission of such internal messages is assumed to be instantaneous whereas the communication with the environment can be time consuming.

To validate the specification it is shown that only finitely many zero-time transitions are possible in a row implying that Zeno computations cannot happen.

1 Introduction

The steam-boiler control specification problem has been proposed as a challenge for different methods for real-time, safety-critical, and fault-tolerant systems. Such a system consists of a *plant*, a *control program* and some *transmission medium* through which plant and program communicate. The problem is to give a formal specification of the control program and to verify that it behaves in a certain way when composed with its *environment*, the plant and the transmission medium.

As the informal specification of Abrial this paper focuses on the control program for which an operational specification in an object-oriented algebraic style is given. For the environment only a simple specification will be given which can be used for validation and simulation issues. The environment and the control program communicate by exchanging messages. We will loosen the restriction

---

* Work completed while on leave at LMU München.

in Chapter AS, this book, that all messages from the environment in a round are received simultaneously. This will instead be a special case of environment behavior. The fact that messages from the environment can arrive at arbitrary times in a steam-boiler cycle, together with the assumption that the time a message is received reflects the time the message-triggering action took place in the environment, turns a merely reactive steam-boiler system into a really hybrid system. The control program is designed as the parallel composition of several objects which communicate using messages. The transmission of such internal messages is assumed to be instantaneous whereas the communication with the environment can be time consuming.

As part of the validation it is shown that only finitely many zero-time transitions are possible in a row implying that Zeno computations cannot happen. Moreover, it is shown by an example how informal statements about the behavior of the steam-boiler can be formally deduced.

The underlying logic of our specification is *timed rewriting logic* (TRL) [2]. TRL extends algebraic specification techniques and tools to handle dynamic and, in particular, real-time systems like the steam-boiler. In TRL, the static parts of a system are described by equations while the transitions are described by timed rewrite rules. TRL has well-defined initial dynamic algebra semantics.

TRL is based on Meseguer's rewriting logic [4] and allows us to reason about time elapse in real-time systems. In TRL, a time stamp is added to each rewrite step. Timed rewriting logic is a logic about *change* in real-time systems. Terms are considered as processes and proofs as behaviors of a process. Basically, $t \xrightarrow{r} t'$ means that a system could rewrite from state t to state t' in time r.

Our specification is written in the real-time specification language Timed Maude (see also [3]). Timed Maude is based on TRL and Meseguer's language Maude [4, 5], from which Timed Maude borrows its object-oriented features. Rewriting in Timed Maude is nothing but deduction in TRL, which means that a Timed Maude specification defines a logical theory and is at the same time close enough to being an executable specification. We believe that an object-oriented specification consisting of equations and timed rewrite rules yields a specification which is easily understandable for the customer, yet detailed enough to point out loose ends in the informal specification.

This chapter is organized as follows: In Section 2 timed rewriting logic and Timed Maude are defined. Basic notions and notations are introduced. In Section 3 the organization of the specification is explained and motivated. Parts of the specification itself is given in Section 4, while the rest of the specification is given in CD-ROM Annex OKW.A. In Section 5 some properties of the specification are proven informally. The chapter is concluded with an evaluation of using Timed Maude for the specification and verification of real-time systems and a comparison with other solutions to the steam-boiler problem. In CD-ROM Annex OKW.B the full, formal control program specification is given, while CD-ROM Annexes OKW.C and OKW.D provide formal proofs of the claims in Section 5.

2 Timed Rewriting Logic and Timed Maude

In this section timed rewriting logic and Timed Maude are briefly introduced. In Section 2.4 some useful techniques for writing Timed Maude specifications are given. These techniques are used later in the steam-boiler specification. The reader is referred to [2] for a more thorough treatment of the full syntax and semantics of TRL.

2.1 Timed Rewriting Logic

For the sake of simplifying the exposition, we treat the many-sorted case, the order-sorted case can be given a similar treatment.

Notation: The set of free variables in a term t is denoted $\mathcal{V}(t)$. The term $t(u_1/x_1, \ldots, u_n/x_n)$ denotes the term obtained from t by *simultaneously substituting* u_i for x_i, $i = 1, \ldots, n$, provided u_i and x_i have the same sort.

The notion of arithmetical monoid was introduced in [2] to model time abstractly. In this paper, we assume that time is modelled by the natural numbers, but time can as well be modelled by the non-negative real numbers. To abstract away from the particular choice, the sort *time* will denote the time values.

An equational specification (Σ, E) is a *timed specification* if it contains a (finite or infinite) equational axiomatization of the sort *time* with the appropriate functions.

Definition 1 *A timed rewrite specification $\mathcal{R}$ is a triple $\mathcal{R} = (\Sigma, E, R)$ where (Σ, E) is a timed specification and R is a set $R \subseteq T_\Sigma(X)_{time} \times (T_\Sigma(X))_s^2 \times (T_\Sigma(X))_{s_1}^2 \times \cdots \times (T_\Sigma(X))_{s_n}^2, s, s_1, \ldots, s_n \in sorts(\Sigma)$. Elements of R are called timed rewrite rules and we use the notation $u_1 = v_1 \wedge \ldots \wedge u_n = v_n \Longrightarrow t \xrightarrow{r} t'$ for $(r, t, t', u_1, v_1, \ldots, u_n, v_n)$ and $t \xrightarrow{r} t'$ if $n = 0$.*

A timed rewrite specification $\mathcal{R} = (\Sigma, E, R)$ entails a sequent $t \xrightarrow{r} t'$ if and only if $t \xrightarrow{r} t'$ can be obtained from the axioms E and rules R by using the axioms and rules of equational logic and the axioms and deductions rules for timed rewriting given in Figure 1.

This deduction system extends the rules of deduction in (unlabeled) rewriting logic [4] with time stamps as follows:

- Reflexivity is dropped as a general axiom since we also aim at hard real-time systems. Reflexivity would not allow describing hard real-time systems since (parts of) the system could stay idle for an arbitrary long period of time. For specifying soft real-time systems particular reflexivity axioms could be added.
- Transitivity yields the addition of the time stamps. If t_1 evolves to t_2 in time r_1 and t_2 evolves to t_3 in time r_2, then t_1 evolves to t_3 in time $r_1 + r_2$.
- The synchronous replacement rule enforces uniform time elapse in all components of a system: a system rewrites in time r iff all its components do so. Synchronous replacement combined with irreflexivity also induces maximal parallelism, which means that no component of a process can stay idle.

Timed transitivity (TT):

$$\frac{t_1 \xrightarrow{r_1} t_2, \quad t_2 \xrightarrow{r_2} t_3}{t_1 \xrightarrow{r_1+r_2} t_3}$$

Synchronous replacement (SR):

$$\frac{t_0 \xrightarrow{r} t'_0, \quad t_{i_1} \xrightarrow{r'} t'_{i_1}, \ldots, t_{i_k} \xrightarrow{r'} t'_{i_k}}{t_0(t_1/x_1, \ldots, t_n/x_n) \xrightarrow{r'} t'_0(t'_1/x_1, \ldots, t'_n/x_n)}$$

where $\{x_{i_1}, .., x_{i_k}\} = \mathcal{V}(t_0) \cap \mathcal{V}(t'_0)$ and $r' = r(t_1/x_1, \ldots, t_n/x_n)$.

Compatibility with equality (EQ):

$$\frac{t_1 = u_1, \; r_1 = r_2, \; t_2 = u_2, \; t_1 \xrightarrow{r_1} t_2}{u_1 \xrightarrow{r_2} u_2}$$

Renaming of variables (RV):

$$x \xrightarrow{r} x \qquad \text{for all } x \in X, r \in X_{time}$$

Conditional rewrite rules (CR):

$$\frac{u_j(t_1/x_1, \ldots, t_n/x_n) = v_j(t_1/x_1, \ldots, t_n/x_n), \; t_{i_1} \xrightarrow{r'} t'_{i_1}, \ldots, t_{i_k} \xrightarrow{r'} t'_{i_k}}{t_0(t_1/x_1, \ldots, t_n/x_n) \xrightarrow{r'} t'_0(t'_1/x_1, \ldots, t'_n/x_n)}$$

where $\bigwedge_{j=1}^{m} u_j = v_j \Longrightarrow t_0 \xrightarrow{r} t'_0$ is a rule, $\{x_{i_1}, .., x_{i_k}\} = \mathcal{V}(t_0) \cap \mathcal{V}(t'_0)$ and $r' = r(t_1/x_1, \ldots, t_n/x_n)$.

Fig. 1. Deduction rules in timed rewriting logic.

- The renaming rule assures that timed rewriting is independent of names of variables. Observe that the renaming axiom does not imply that $t \xrightarrow{r} t$ holds for all terms t.

2.2 From TRL to Timed Maude

In this section we show how order-sorted timed rewriting logic can be applied for defining an object-oriented specification language for real-time systems. This is done by adding a module facility and syntactic features for defining classes and expressing that terms are dynamic. We base this approach on the language Maude [5] introduced by Meseguer. In our version of Maude, called Timed Maude, the equational part is kept unchanged; only concurrent rewriting is replaced by TRL.

The rewrite rules in Timed Maude are of the form

$$t \xrightarrow{r} t' \text{ if } c_1 \wedge c_2 \wedge \ldots \wedge c_n$$

for $n \geq 0$, where c_i is an equation $t_i = t'_i$ or a term t_i of sort *bool* (and is an abbreviation for $t_i = true$). In Timed Maude, sets of sort, subsort, function, and variable declarations and equations and rewrite rules are preceded by the keywords **sorts, subsorts, ops, vars, eqs** and **rls** respectively. These keywords are omitted in running text. Variables that are not explicitly declared are assumed to have the (greatest) appropriate sort.

We find it natural and convenient to include the axiom $t \xrightarrow{0} t$ for all terms t. This allows for interleaving of actions that take zero time.

Dynamic Sorts. Most values in Timed Maude specifications like the booleans, the natural numbers, etc. do not change in time. To avoid the need for adding for each "static" sort s many rules of the form $t \xrightarrow{r} t$ for all $t \in T_\Sigma(X)_s, r : time$ we introduce the language concept

$$\textbf{dynamic sorts } s_1, \ldots, s_m$$

which defines a set $\{s_1, \ldots, s_m\}$ of dynamic sorts. The set must be closed wrt. supersorts, i. e. if s' is a dynamic sort and $s' \leq s$ then s must also be declared a dynamic sort. Sorts that are not dynamic are called *static*. To avoid observing dynamic behavior into a static domain, the signature must satisfy that whenever $f : s_1, \ldots, s_n \to s$ is a declaration and s a static sort, then $s_1, \ldots, s_{n-1}$ and s_n must be static sorts. The mathematical meaning is that $t \xrightarrow{r} t$ is valid for all term t of static sort, and therefore these values can stay idle.

Features Supporting Object-Oriented Specification. As in Maude, an object can be represented by a term $\langle o : c | a_1 : v_1, \ldots, a_n : v_n \rangle$ where o is the object's name, belonging to a set *oid* of object identifiers, c is its class, a_1 to a_n are the names of the objects attributes and the v_i's are the corresponding values. Given a sort *oid* of object identifiers, a class declaration

$$\textbf{class } c \mid \textbf{atts } a_1 : s_1, \ldots, a_n : s_n.$$

is equivalent to the following translation[4]:

dynamic sorts *object*
op $\quad\quad\quad \langle \_ : c | a_1 : \_, \ldots, a_n : \_ \rangle : oid, s_1, \ldots, s_n \to object.$

Inheritance is treated in the same way as in Maude (see [5]).

A Timed Maude program makes computational progress by rewriting its global state, its *configuration*. A configuration is a multiset of objects and messages. Multiset union is expressed by juxtaposition and corresponds to parallel composition of objects. It is formalized in the following way:

dynamic sorts *msg, object, configuration*
subsort $\quad$ *msg, object* $\leq$ *configuration*
ops $\quad\quad\quad$ $\_\_$: *configuration, configuration* $\to$ *configuration*

---

[4] We adhere to the notational convention that $\_$ denotes the place of an argument in the declaration of a "mixfix" function symbol, as e. g. **if** $\_$ **then** $\_$ **else** $\_$: $bool, s, s \to s$.

vars	$\emptyset : \rightarrow configuration$
	$x, y, z : configuration, \ r : time$
eqs	$x(yz) = (xy)z$
	$xy = yx$
	$x\emptyset = x$
rl	$xy \xrightarrow{r} xy.$

The meaning of the last rule is that if configurations c_1 and c_2 rewrites in time r to configurations c_1' and c_2' respectively, then the configuration $c_1 c_2$ rewrites to $c_1' c_2'$ in time r. The rule does however *not* imply that any configuration can stay idle, but that a configuration can only proceed in time if all its components do so. In particular, since a message is a term of the dynamic sort *msg*, no configuration can proceed in time if it contains a message (that does not proceed in time). In this way strong reactivity is obtained, since a message must be read as soon as it is received.

We follow the Maude convention that attributes not actively involved in a rewrite step need not be mentioned explicitly. For instance

$$\langle o : c | a_1 : v_1, a_4 : v_4 \rangle \xrightarrow{r} \langle o : c | a_1 : f(v_1, v_4) \rangle$$

will be written instead of

$$\langle o : c | a_1 : v_1, \ldots, a_4 : v_4, \ldots, a_n : x_n \rangle \xrightarrow{r} \langle o : c | a_1 : f(v_1, v_4), \ldots, a_4 : v_4, \ldots, a_n : x_n \rangle.$$

If v_i does not occur in the left-hand side, it is by definition a variable x_i. The class name c will be omitted when it can be deduced from the context.

2.3 Some Examples

This subsection presents some constructs that will be used later.

Timer. A timer counts down. When it has reached the value zero it cannot be further rewritten and blocks the configuration within which it appears, forcing a rewrite rule to be applied to "rewind" the timer. It is specified as follows:

dynamic sort *Timer*	
op	$timer : time \rightarrow Timer$
rl	$timer(r + r') \xrightarrow{r} timer(r').$

A timer will be used in objects to force actions to happen at a certain time.

Delay of Messages. Given a message m we construct the message $dly(m, r)$ which means that message m is received r time units after it is sent. The dly operator is specified

op	$dly : msg, time \rightarrow msg$
rls	$dly(m, r + r') \xrightarrow{r} dly(m, r')$
	$dly(m, 0) \xrightarrow{0} m.$

This operator can model the fact that messages need time to travel. An object could also deliberately delay a message if it is ready to send the message but knows that the receiver is not yet ready to receive it. Note that m in the above rules should be a message and not a variable of sort msg.

2.4 Some Timed Maude Specification Techniques

In this section we present some techniques which we use as guidelines writing specifications in Timed Maude. These concern decisions about the choice of the form of the rules.

Synchronous Versus Asynchronous Rules. As in Maude, we support two kinds of communication: synchronous and asynchronous communication. A communication is called synchronous when a rewrite step involves more than one object in the left-hand side; an asynchronous communication is defined by a rewrite rule with only one object (but possibly several messages) in the left-hand side. One important difference between the two styles of communication is the degree of parallelism, which increases in the asynchronous model. Moreover, in the asynchronous style we can model and analyze the behavior of objects independently, whereas in the synchronous style we always have to consider a group of objects. Therefore, we prefer to use the asynchronous style as much as possible. However, this also means that whenever an object needs to know the value of some other object's attribute, this value must be sent (sometimes upon request) by that other object.

In the steam-boiler specification only the two rules of Section 4.1 specifying the elapse of time in the system are synchronous, all other rules are asynchronous.

Using only asynchronous rules in addition to one or two synchronous rules allows us to specify each object alone, thereby enhancing easy specification and confidence in the final specification. The messages can be seen as the abstract interface of objects, such that one object need not know anything about the inner structure and state of the other objects.

0-Time Rules. All state changes that do not involve time aspects are modelled by zero-time transitions. A state change can be forced to happen at a certain time t because a message is received or because a timer reaches the value 0 at this time. Zero-time (asynchronous) rules are therefore of either of two forms. In rules of the form

$$(m_1)\ldots(m_k)\langle o|\ldots s \ldots\rangle \stackrel{0}{\longrightarrow} \langle o|\ldots s' \ldots\rangle(m'_1)\ldots(m'_n)$$

for $k \geq 1, n \geq 0$ the messages $(m_1),\ldots,(m_k)$ are read, the messages $(m'_1),\ldots,(m'_n)$ are sent, and the state of object o changes. In particular, in the steam-boiler specification some rules are of the form

$$\langle o|timer : timer(0),\ldots s \ldots\rangle \stackrel{0}{\longrightarrow} \langle o|timer : timer(\Delta t),\ldots s' \ldots\rangle(m'_1)\ldots(m'_n).$$

Such a rule defines the behavior of o in the situation where the timer "rings": the timer is reset (to Δt in the steam-boiler example), the state of o is changed and the messages $(m'_1), \ldots, (m'_n)$ are sent.

Zero-time transitions have the advantage that they correspond directly to untimed transitions in Maude. However, a problem arises, namely that zero-time transitions may lead to a Zeno behavior with infinitely many transitions in a finite amount of time. To avoid such situations the use of zero-time transitions should be accompanied by a termination proof ensuring that only finitely many such transitions can occur in a row. For an example see Section 5.2, where a proof for the absence of Zeno behavior for the steam-boiler specification is given.

3 Outline of the Specification

In this section, the overall organization of the specification of the steam-boiler system is explained and motivated. First in Section 3.1 time aspects are discussed; in particular it is motivated why we loosen Abrial's assumption of simultaneously sending and receiving messages. In Section 3.2 we present some observations and assumptions about the environment that have influenced our solution. Finally, in Section 3.3 the structure of our solution is explained including the main components of the control program specification.

3.1 Time Aspects

The system consists of two parts, the *control program* and the *environment*, which in turn consists of the *plant* and the *transmission medium* where messages could be created or lost. The control program and the environment communicate by exchange of messages. In the calculations it will be assumed that all messages take the same time to travel.

Denoting the cycle period (5 sec) by Δt, the communication pattern in the nth round between control program and environment is:

- In the *open* time interval $((n-1)\Delta t, n\Delta t)$ messages from the environment are received by the program.
- At time $n\Delta t$ the information is analyzed and the next operation mode of the steam-boiler is decided. All messages from the control program to the environment are sent at this instant.

This communication pattern is more loose than that required in the informal specification in Chapter AS, this book, which states that "in first approximation, all messages coming from (or going to) the physical units are supposed to be received (emitted) *simultaneously* by the program at each cycle".

However, receiving all messages from the environment at the same time causes problems when receiving message combinations reporting a non-sensible water level value and the fact that the water measurement device has been repaired.

The question is whether the device is supposed to work correctly after having sent these messages. Two different physical behaviors could have preceded the sending of the above mentioned messages. In the first case, the measure of the water level was undertaken first (which is why the incorrect value is reported), followed by the repair of the device, which is then supposed to work correctly. In the other case, first the broken device was repaired, then (still within the 5 seconds) broke again and thereafter the level was measured, which means that the device is broken.

The problem is similar for all devices. There are at least three solutions to the problem:

1. Treat the case non-deterministically and allow for further refinement.
2. Make some assumption about the physical device, like "the device is always repaired before the level is measured".
3. Allow messages from the environment to arrive at arbitrary times during the cycle. If the device is repaired before it is measured, then this is reflected by the order in which the messages come: the *level repaired* message is received before the *level(x)* message, and vice versa.

We choose the third solution because we find it the most natural. Messages that arrive at the same time, will be treated independently of each other, which in that case gives a nondeterministic result. Note also that solution 1 is a special case of solution 3, when all messages arrive at the same time. The notion of cycle is still needed for sending the *mode* message to the plant, to check if all "mandatory" messages have arrived in a cycle etc. In order not to deviate too much from the requirements given in the informal specification, all messages *to* the environment are still sent at the same time.

All calculations concerning the mode of steam-boiler operation in round $n+1$ should be based on calculated steam and water levels at time $n\Delta t$, even if the messages *level(x)* and *steam(v)* are received much earlier in the cycle.

We assume that computations take zero time. This is a normal assumption when reasoning about reactive systems. It also means that all rules not explicitly involving change of time (such as counting down a timer or modeling the dynamics of the system) are zero-time rules.

3.2 Observations and Assumptions About the Environment

We realized the need for making the following observations:

1. Messages sent to the plant can disappear, and due to noise in the transmission medium the control program can receive a message that was not sent from the plant, but a message that is assumed sent from the plant and received by the program is never corrupted! This fact is crucial when specifying the pumps and is supported by the problem description, where receiving a water level value way beyond the total capacity C indicates a water device failure and not a transmission failure.
2. All mandatory messages should be present during each transmission, even in the initializing phase before the steam-boiler is actually boiling. During this time the devices behave as usual.

3. We did not see how we could separate the failure of a pump from that of its pump controller. The parts of the control program dealing with the pump controllers will therefore not be specified. Failure of a pump and the corresponding pump controller will be identified.
4. While the steam-boiler is not boiling, water level outside the critical interval M_1 to M_2 poses no danger. E. g., after opening the valve, the level is 0.

The following are our own assumptions:

1. All messages from the environment in the nth cycle are expected to be received in the open time interval $((n-1)\Delta t, n\Delta t)$.
2. At most one message of each kind could arrive from the environment during each cycle[5].
3. In order to avoid too complicated specifications, we assume that a *device repair acknowledged* message (for any device) will be received by the plant. Otherwise, time stamps would be needed to know whether a *device repaired* message originates in the last repair, or in some earlier repair (from which the plant has not received the *device repair acknowledged* message).
4. No steam is exiting the boiler when it is not boiling.
5. The valve is unfailing and empties the boiler instantaneously.
6. When any object discovers a fatal error, the system can stop its computation and just send the *mode(emergency)* message to the environment. In our case, the control program may deadlock after having sent that message.

3.3 The Structure of the System

In this solution, we will focus on the specification of the control program. To make the problem of specifying the control program (which can receive more than 2^{21} different message combinations in each round) more manageable and to increase parallelism, the control program is defined as the parallel composition of the following objects (described informally together with their interface to the environment):

controller object c contains information about the mode of operation the steam-boiler is presumed to be in. It handles *stop*, *steam-boiler waiting*, and *physical units ready* messages. It is responsible for sending *program ready* and *mode(m)* messages to the environment.

water object w contains information about the current assumed state of the water measurement device and the estimated water level of the steam-boiler. The water object handles the messages *level failure acknowledged*, *level repaired*, and the mandatory *level(x)* message coming from the environment, and is responsible for sending *level failure detected*, *level repair acknowledged*, and *valve* messages when appropriate.

---

[5] Our solution can easily be modified to handle the case where more than one message of each kind can be received by the control program in the same round.

pump object p_i (for $i = 1, 2, 3, 4$) handles the messages $pump_i\text{-}mode(b)$, $pump_i$ *failure acknowledged*, and $pump_i$ *repaired*, and is responsible for the possible sending of $open(p_i)$, $close(p_i)$, $pump_i$ *failure detected*, and $pump_i$ *repair acknowledged* to the plant. It contains information about the state of the device (*ok* or *broken*) and its mode of operation (open, just_opened, or closed).

steam object *st* contains information about estimated steam output and the state of the steam measurement device. It handles the messages *steam(v)*, *steam repaired*, and *steam failure acknowledged* from the environment, and is responsible for sending the messages *steam failure detected* and *steam repair acknowledged* to the environment.

pump system object *ps* is a special object not corresponding to any physical part of the steam-boiler. It receives data from the other objects and decides what to do with the pumps in the next round.

Each configuration *conf* of the control program has the form[6]

$$\langle c : controller \rangle \langle w : water \rangle \langle st : steam \rangle \langle p_1 : pump \rangle \ldots \langle p_4 : pump \rangle \langle ps \rangle M$$

where M is a (possibly empty) multiset of messages. A full configuration of the steam-boiler system adds the environment *env* to *conf*:

$$env\ conf.$$

The steam-boiler system proceeds in cycles of length Δt. During a time interval $(n\Delta t, (n+1)\Delta t)$ the control program alternates between the following two behaviors:

- While no new messages are received from the environment, the system evolves according to the dynamics of the system and updates the minimal and maximal expected water and steam levels. This is due to the fact that in our setting messages from the environment can be received at any time, and that a level value message received at time t is supposed to indicate the level at time t, not the level at the end of the round.
- Whenever a message from the environment is sent, the system treats this immediately (since time cannot proceed when a configuration contains a received but unread message). A message is read by the appropriate object, which then
    - updates some of its attributes (e. g. if the received water level is within the current value of the estimated minimal and maximal water level attributes, these attributes are updated accordingly) and/or
    - sends messages to other control program objects.

At the end of each cycle (time $n\Delta t$), timers count down to zero and thereby block the system from proceeding in time. Then

---

[6] For the formal definition of the involved objects see Section 4 and CD-ROM Annex OKW.A.

- the objects check if all messages that must arrive from the environment in each cycle have arrived,
- the control program computes which messages to send to the environment, and
- when an object has finished all computations for this cycle, it resets its timer to Δt. When all timers are reset, the system can again proceed in time and a new cycle can begin.

4 The Control Program Specification

In this section the specification of the control program is given. As mentioned earlier, it is divided into two parts. The first part consists of some synchronous rules which define how the objects proceed in time. The second part consists of the asynchronous rules and specifies the "actions" of the systems.

To avoid a too lengthy exposition, we make the following assumptions about the specification:

- Specifications *NAT*, *INT*, and *BOOL* of natural and integers numbers and boolean values with sorts *nat*, *int* and *bool* and definition of the appropriate function symbols are assumed given.
- Time is represented by the natural numbers.
- For each sort s there is a function $\mathit{if}\_\mathit{then}\_\mathit{else}\_ : \mathit{bool}, s, s \to s$ and a function $eq : s, s \to \mathit{bool}$ which is true if its arguments are equal and false if its arguments are different.
- A constant *error* of sort *nat* is needed and is defined to be greater than any sensible steam-boiler value.
- Unless otherwise stated, we use the same notation and names as the informal description. Messages are not explicitly declared, but will be found in tables.

4.1 Specifying the Dynamics of the System

Since the water level and steam output values can be received at arbitrary times in a cycle, the values qa_1, qa_2, va_1, and va_2 should at every instant of the computation denote the current estimated minimal and maximal water level and steam output respectively. The dynamics of the system is specified by synchronous rules where the water and steam values are updated.

The estimated water and steam values r time units after current moment are given by the following equations:

$$new\_qa_1 = \max(qa_1 - va_2 \cdot r - (U_1(r^2))/2 + pc_1 \cdot r, 0)$$
$$new\_qa_2 = \max(qa_2 - va_1 \cdot r + (U_2(r^2))/2 + pc_2 \cdot r, 0)$$
$$new\_va_1 = \max(va_1 - U_2 \cdot r, 0)$$
$$new\_va_2 = va_2 + U_1 \cdot r.$$

When the steam-boiler is not boiling, all steam values must be zero. We assume that water *could* flow through the ith pump if either the pump is considered broken or if it was opened more than 5 seconds ago (in which case the pump is in pump-mode *open*). Similarly, there is a possibility that the water is *not* flowing through the ith pump either if the pump is considered broken or if it is not in pump-mode *open*. These calculations give the following estimated upper and lower bounds of pump throughput:

$pc_1 = \sum_{i=1}^{4}$ (**if** $not(eq(p_i\_state, ok))$ **or** $not(eq(p_i\_mode, open))$ **then** 0 **else** P)
$pc_2 = \sum_{i=1}^{4}$ (**if** $not(eq(p_i\_state, ok))$ **or** $eq(p_i\_mode, open)$ **then** P **else** 0).

Based on this analysis, we realize the need for the following class attributes: the controller class must have an attribute *state* which is either *boiling* or *not_boiling*, the water class has attributes qa_1 and qa_2 (current calculated minimal and maximal water level), the steam class has attributes va_1 and va_2 (current calculated minimal and maximal steam output), and finally the pump class must have attributes *state* (with values *ok* or *broken*) and *mode* of operation (*open* or not). The Maude convention that attributes not used in a rewrite rule need not be stated explicitly provides for the possibility of developing the classes step by step. We therefore defer the declaration of the classes to the next subsection. Alternatively, the classes could be declared with the above-mentioned attributes, and when the need of more attributes is realized, these could be added by defining subclasses.

When the steam-boiler is not boiling, the steam output is supposed to be zero, so the system proceeds in time r for all $r : time$ by the following rule:

$\langle c | state : not\_boiling \rangle \langle w | qa_1 : wa_1, qa_2 : wa_2 \rangle \langle st | va_1 : 0, va_2 : 0 \rangle$
$\langle p_1 | state : p_1\_ok, mode : p_1\_mode \rangle \ldots \langle p_4 | mode : p_4\_mode, state : p_4\_ok \rangle \langle ps \rangle$
$\xrightarrow{r}$
$\langle c \rangle \langle w | qa_1 : new\_qa_1, qa_2 : new\_qa_2 \rangle \langle st | va_1 : 0, va_2 : 0 \rangle \langle p_1 \rangle \ldots \langle p_4 \rangle \langle ps \rangle$
if $new\_qa_1 = wa_1 + pc_1 \cdot r$
$\bigwedge new\_qa_2 = wa_2 + pc_2 \cdot r$
$\bigwedge pc_1 = \sum_{i=1}^{4}$ (**if** $not(eq(p_i\_ok, ok))$ **or** $not(eq(p_i\_mode, open))$ **then** 0 **else** P)
$\bigwedge pc_2 = \sum_{i=1}^{4}$ (**if** $not(eq(p_i\_ok, ok))$ **or** $eq(p_i\_mode, open)$ **then** P **else** 0).

In case the steam-boiler is boiling, the system proceeds in time r by the following rule:

$\langle c | state : boiling \rangle \langle w | qa_1 : wa_1, qa_2 : wa_2 \rangle \langle st | va_1 : sta_1, va_2 : sta_2 \rangle$
$\langle p_1 | state : p_1\_ok, mode : p_1\_mode \rangle \ldots \langle p_4 | state : p_4\_ok, mode : p_4\_mode \rangle \langle ps \rangle$
$\xrightarrow{r}$
$\langle c \rangle \langle w | qa_1 : new\_qa_1, qa_2 : new\_qa_2 \rangle \langle st | va_1 : new\_va_1, va_2 : new\_va_2 \rangle \langle p_1 \rangle \ldots \langle p_4 \rangle \langle ps \rangle$
if $new\_qa_1 = \max(wa_1 - sta_2 \cdot r - (U_1(r^2))/2 + pc_1 \cdot r, 0)$
$\bigwedge new\_qa_2 = \max(wa_2 - sta_1 \cdot r + (U_2(r^2))/2 + pc_2 \cdot r, 0)$
$\bigwedge new\_va_1 = \max(sta_1 - U_2 \cdot r, 0)$
$\bigwedge new\_va_2 = sta_2 + U_1 \cdot r$
$\bigwedge pc_1 = \sum_{i=1}^{4}$ (**if** $not(eq(p_i\_ok, ok))$ **or** $not(eq(p_i\_mode, open))$ **then** 0 **else** P)
$\bigwedge pc_2 = \sum_{i=1}^{4}$ (**if** $not(eq(p_i\_ok, ok))$ **or** $eq(p_i\_mode, open)$ **then** P **else** 0).

sorts	*cstate, boilingstate, not_boilingstate*	
subsorts	*not_boilingstate, boilingstate* $\leq$ *cstate*	
ops	*startup, ch_w_s, StBR, PrReady, PUReady* : $\rightarrow$ *not_boilingstate*	
	normal, degraded, rescue, emergency : $\rightarrow$ *boilingstate*	
vars	*boiling, newmode* : *boilingstate*, *not_boiling* : *not_boilingstate*	
class	*controller*	
atts	*timer* : *Timer*	– "rings" at time $n\Delta t$.
	state : *cstate*	– states of the controller object.
	stop_v : *nat*	– number of *stop* messages received in row.
	stoprec : *bool*	– *stop* message received in current round?
initially	*timer* := *timer*(Δt), *state* := *startup*, *stop_v* := 0, *stoprec* := *false*.	

Table 1. *The class controller.*

4.2 The Classes and the Asynchronous Rules

Now that the dynamics of the system is specified, the asynchronous zero-time rules for handling the messages to and from the environment are defined.

Messages from the environment are received in the time interval $((n-1)\Delta t, n\Delta t)$, but most of the actions take place at time $n\Delta t$, when all the messages from the nth round have been treated. At this time, the objects decide the next mode of operation of the steam-boiler, which pumps to open or close, and so on. These decisions are all based on estimated values at time $n\Delta t$ (values which are continuously updated by the previously defined synchronous rules).

The Controller Object. The controller object c handles *stop*, *steam-boiler waiting (StBW)*, and *physical units ready* messages from the environment and is responsible for sending the next mode of steam-boiler operation at the end of each cycle. The decision is taken on the basis of received values from the other objects. The declaration of the class *controller* is found in Table 1 and the messages sent and received by the controller are found in Table 2.

The controller object is in *startup* state when the system is started. As soon as it receives a *steam-boiler waiting (StBW)* message, it goes into the state *check water and steam* (*ch_w_s*) and sends *init*-requests to the water and steam objects. Since the next mode of operation must depend on the water and steam values at time $n\Delta t$, a timer is introduced which indicates the amount of time left until $n\Delta t$, and the *init*-messages are delayed to arrive at the appropriate time. Receiving any aberrant message is handled by going to state *emergency*:

$$(StBW)\langle c|timer:timer(r),state:startup\rangle \xrightarrow{0} \langle c|state:ch\_w\_s\rangle$$
$$dly((\text{to } w \text{ init}), r) \; dly((\text{to } st \text{ init}), r)$$

$$(StBW)\langle c|state:s\rangle \xrightarrow{0} \langle c|state:emergency\rangle$$
$$\text{if } not(eq(s, startup)).$$

Receive:	Send:	Abbreviation for	From/To:
(stop)			environment
(StBW)		steam-boiler waiting	environment
(PUR)		physical units ready	environment
(trans_fail)		transmission failure detected	any object
	(to s init)	asks for steam value	steam
(st is x)		x is a steam value, either 0 or error	steam
	(to w init)	ask for water status	water
(w is x)		get water level	water
(w_stat : w_ok, qa_1, qa_2, qc_1, qc_2)		updated water status	water
(st_stat : s_ok)		steam device status	steam
(p_i_stat : p_i_ok)		pump device status	pump$_i$
	(ProgramReady)	program ready	environment
	(mode(x))	mode is x	environment

Table 2. *Messages received or sent by the controller object.*

The controller goes to state *check water and steam (ch_w_s)* and waits for values from the steam and water objects. If any of steam or water devices are broken, then it quits, otherwise it goes to either state *StBR* or *PrReady*:

$$(st \text{ is } error)\langle c|state : ch\_w\_s\rangle \xrightarrow{0} \langle c|state : emergency\rangle$$
$$(w \text{ is } error)\langle c|state : ch\_w\_s\rangle \xrightarrow{0} \langle c|state : emergency\rangle$$
$$(st \text{ is } 0)(w \text{ is } x)\langle c|state : ch\_w\_s\rangle \xrightarrow{0} \langle c|state : PrReady\rangle \text{ if } N_1 \leq x \leq N_2$$
$$(st \text{ is } 0)(w \text{ is } x)\langle c|state : ch\_w\_s\rangle \xrightarrow{0} \langle c|state : StBR\rangle \text{ if } not(N_1 \leq x \leq N_2)$$
$$\text{and } not(eq(x, error)).$$

From the *StBR* state, the *init*-procedure is repeated in each round until the water level is between N_1 and N_2, while from state *PrReady* a *program_ready* message is sent in each round. These actions are performed together with the sending of the mode of operation specified at the end of this subsection. When a *physical units ready (PUR)* message is received in the *PrReady* state, the controller goes to *PUReady* state and then starts the steam-boiler:

$$(PUR)\langle c|state : PrReady\rangle \xrightarrow{0} \langle c|state : PUReady\rangle$$
$$(PUR)\langle c|state : s\rangle \xrightarrow{0} \langle c|state : emergency\rangle \text{ if } not(eq(s, PrReady)).$$

If the controller receives *stop* messages in three consecutive cycles it goes to emergency mode, and if it does not receive a *stop* message in one round it must reset its attribute *stop_v* to zero. Therefore, we must know whether a *stop* message is received in this round. An attribute *stoprec* is set to true if a *stop* message was received in the current cycle (note that, as specified later, some of these attributes are reset at the end of each cycle):

$$(stop)\langle c|stop\_v : v, stoprec : false\rangle \xrightarrow{0} \langle c|stop\_v : v+1, stoprec : true\rangle \text{ if } v \leq 1$$
$$(stop)\langle c|state : s, stop\_v : 2\rangle \xrightarrow{0} \langle c|state : emergency\rangle.$$

A transmission failure in any object puts the controller in *emergency* state:

$$(trans\_fail)\langle c|state : s\rangle \xrightarrow{0} \langle c|state : emergency\rangle.$$

At the end of each computation, the controller receives status information from all objects, computes its new mode and sends the appropriate messages.

When the steam-boiler is boiling, the function

NextMode : *boilingstate, bool, bool, bool, bool, nat, nat, nat, nat* → *boilingstate*

is used to compute the new mode. It takes as arguments the current state, the condition of the water device, the condition of the steam device, the condition of the pump devices and the water level values qa_1, qa_2, qc_1 and qc_2, which is enough to decide the next mode of the steam-boiler. Its definition is a straightforward formalization of the informal description:

NextMode(normal, w_ok, st_ok, pumps_ok, qa_1, qa_2, qc_1, qc_2) =
  if *water_crisis(qa_1, qa_2, qc_1, qc_2)* **then** *emergency*
  else (**if** *not(w_ok)* **then** *rescue*
    else (**if** *not(st_ok)* **and** *pumps_ok*) **then** *degraded* **else** *normal*))

NextMode(degraded, w_ok, st_ok, pumps_ok, qa_1, qa_2, qc_1, qc_2) =
  NextMode(normal, w_ok, st_ok, pumps_ok, qa_1, qa_2, qc_1, qc_2)

NextMode(rescue, w_ok, st_ok, pumps_ok, qa_1, qa_2, qc_1, qc_2) =
  if *w_ok* **then** *NextMode(normal, w_ok, st_ok, pumps_ok, qa_1, qa_2, qc_1, qc_2)*
  else (**if** *not(st_ok)* **and** *pumps_ok*) **or** *water_crisis(qa_1, qa_2, qc_1, qc_2)*
    then *emergency* **else** *rescue*)

NextMode(emergency, ...) = *emergency.*

The function *water_crisis* is *true* if the water level is risking to reach one of the limit values M_1 or M_2. We will not give a maximal solution, but use the suggestions given in the (extended) problem description, i.e.

water_crisis$(qa_1, qa_2, qc_1, qc_2) = (qa_1 \leq M_1$ **or** $qa_2 \geq M_2$ **or** $qc_1 \leq M_1$ **or** $qc_2 \geq M_2).$

To compute the next mode of operation of the steam-boiler, the controller receives the necessary values from the other objects. We write (*all_msg*) as an abbreviation for $(w\_stat : w\_ok, qa_1, qa_2, qc_1, qc_2)(st\_stat : st\_ok)(p_1\_stat : p_1\_ok)\ldots(p_4\_stat : p_4\_ok)$. At the same time, the value of *stop_v* is reset to zero if no *stop* message has been received in this round.

$$\langle c|timer: timer(0), state: emergency\rangle \xrightarrow{0} \langle c|timer: timer(\Delta t)\rangle(mode(emergency))$$

$(all\_msg)\langle c|timer : timer(0), state : startup, stop\_v : v, stoprec : b\rangle \xrightarrow{0}$
$\langle c|timer : timer(\Delta t), state : startup, stop\_v :$ **if** b **then** v **else** $0, stoprec : false\rangle$
$(mode(init))$

sorts	*brokenstate, devicestate*	
subsort	*brokenstate* $\leq$ *devicestate*	
ops	$ok: \to devicestate$	
	$fail, sign, ack: \to brokenstate$	
var	*broken : brokenstate*	
class	*steam*	
atts	*timer : Timer*	– stops at time $n\Delta t$.
	state : devicestate	– *ok, fail, sign* and *ack* states.
	$va_1, va_2 : nat$	– current min and max steam output.
	strec : bool	– *steam(v)* message received in current round?
initially	$timer := timer(\Delta t), state := ok, va_1 := 0, va_2 := 0, strec := false.$	

Table 3. *Declaration of the class* steam.

$(all\_msg)\langle c|timer : timer(0), state : StBR, stop\_v : v, stoprec : b\rangle \xrightarrow{0}$
$\langle c|timer : timer(\Delta t), state : ch\_w\_s, stop\_v : \text{if } b \text{ then } v \text{ else } 0, stoprec : false\rangle$
$(mode(init))\ dly(\text{to } w\ init, \Delta t)\ dly(\text{to } st\ init, \Delta t)$

Note that the *water* and *steam* objects expect the init messages in the *next* round, i. e. Δt time units after the current time.

$(all\_msg)\langle c|timer : timer(0), state : PrReady, stop\_v : v, stoprec : b\rangle \xrightarrow{0}$
$\langle c|timer : timer(\Delta t), stop\_v : \text{if } b \text{ then } v \text{ else } 0, stoprec : false\rangle$
$(mode(init))(ProgramReady)$

$(all\_msg)\langle c|timer : timer(0), state : PUReady, stop\_v : v, stoprec : b\rangle \xrightarrow{0}$
$\langle c|timer : timer(\Delta t), state : newmode, stop\_v : \text{if } b \text{ then } v \text{ else } 0, stoprec : false\rangle$
$(mode(newmode))$
if $newmode = (\text{if } w\_ok \text{ then}$
$NextMode(normal, w\_ok, st\_ok, p_1\_ok$ and ... and $p_4\_ok, qa_1, qa_2, qc_1, qc_2)$
else *emergency*)

$(all\_msg)\langle c|timer : timer(0), state : boiling, stop\_v : v, stoprec : b\rangle \xrightarrow{0}$
$\langle c|timer : timer(\Delta t), state : newmode, stop\_v : \text{if } b \text{ then } v \text{ else } 0, stoprec : false\rangle$
$(mode(newmode))$
if $newmode = NextMode(boiling, ...)$.

Steam object. During the cycle the steam object *st*, whose class is defined in Table 3, receives the possible *steam repaired, steam failure acknowledged*, and *steam(v)* messages from the plant. Based on this information, it decides whether the steam device is working correctly. The values va_1 and va_2 denote at each moment the current estimated minimal and maximal levels of steam output.

Receive:	Send:	Abbreviation for:	From/To:
(st_ack)		steam failure acknowledged	environment
(st_rep)		steam device repaired	environment
$(steam(v))$		measured steam output	environment
	(st_fail)	steam device failure detected	environment
	(st_rep_ack)	steam repair acknowledged	environment
	(to ps st is va_1 to va_2)	steam output interval	pump system
	$(st\_stat : st\_ok)$	steam device status	controller
	(to w st is va_1 to va_2)	steam output interval	water
(to st $init$)		init request	controller
	(st is x)	reply to init request	controller
	$(trans\_fail)$	transmission failure detected	controller

Table 4. *Messages received or sent by the steam object.*

These values are updated continuously by the global rule. A received steam output value not within this interval indicates a device failure.

The steam object has four states for the physical condition of the steam output measurement device: *ok* if the steam measurement device is assumed to work well, *fail* indicates that a failure has been detected but not yet signalled, *sign* means that the device is broken and signalled but the failure is not yet acknowledged by the plant, and, finally, *ack* means the device is broken and the failure is acknowledged (but not repaired) by the plant.

The messages from the environment are treated in the following way:

$$(st\_ack)\langle st|state : sign\rangle \xrightarrow{0} \langle st|state : ack\rangle$$

$$(st\_ack)\langle st|state : s\rangle \xrightarrow{0} \langle st\rangle(trans\_fail) \text{ if } not(eq(s, sign))$$

$$(st\_rep)\langle st|timer : timer(r), state : ack\rangle \xrightarrow{0} \langle st|state : ok\rangle \; dly((st\_rep\_ack), r)$$

$$(st\_rep)\langle st|state : s\rangle \xrightarrow{0} \langle st\rangle(trans\_fail) \text{ if } not(eq(s, ack))$$

$$(steam(v))\langle st|state : ok, va_1 : a_1, va_2 : a_2, strec : false\rangle \xrightarrow{0}$$
$$\langle st|va_1 : v, va_2 : v, strec : true\rangle \text{ if } v \leq W \text{ and } a_1 \leq v \leq a_2$$

$$(steam(v))\langle st|state : ok, va_1 : a_1, va_2 : a_2, strec : false\rangle \xrightarrow{0}$$
$$\langle st|state : fail, strec : true\rangle \text{ if } not(v \leq W \text{ and } a_1 \leq v \leq a_2)$$

$$(steam(v))\langle st|state : broken, strec : false\rangle \xrightarrow{0} \langle st|strec : true\rangle.$$

At time $n\Delta t$, after having received all the messages from the environment, the object checks, provoked by a timer, whether the mandatory *steam(v)* message was received in the cycle. At the same time, the steam object sends its values va_1 and va_2 to the pump system object (which might need them when deciding what to do with the pumps) and to the water object (which needs them to calculate qc_1 and qc_2). It also sends a message to the controller object informing about the device status. If in *fail* or *sign* state, a *st_fail* message must also be sent:

$$\langle st|timer : timer(0), strec : false\rangle \xrightarrow{0} \langle st|timer : timer(\Delta t)\rangle(trans\_fail)$$

Receive:	Send:	Abbreviation for:	From/To:
(w_ack)		level failure acknowledged	environment
(w_rep)		water level repaired	environment
$(level(x))$		measured water level	environment
	(w_fail)	water level failure detected	environment
	(w_rep_ack)	water repair acknowledged	environment
(to w st is va_1 to va_2)		steam output	steam
	$(w\_stat : w\_ok, qa_1, qa_2, qc_1, qc_2)$	water status	controller
(to w $init$)		init request	controller
	$(w$ is $x)$	reply to init request	controller
	$(valve)$	open valve	environment
	$(trans\_fail)$	transmission failure detected	controller

Table 5. *Messages defining the interface of the water object.*

$$\langle st|timer : timer(0), state : s, va_1 : a_1, va_2 : a_2, strec : true \rangle \xrightarrow{0}$$
$$\langle st|timer : timer(\Delta t), strec : false \rangle$$
$$(\text{to } ps \text{ } st \text{ is } a_1 \text{ to } a_2)(\text{to } w \text{ } st \text{ is } a_1 \text{ to } a_2) \text{ } (st\_stat : eq(s, ok))$$
$$\textbf{if } eq(s, ok) \textbf{ or } eq(s, ack)$$

$$\langle st|timer : timer(0), state : s, va_1 : a_1, va_2 : a_2, strec : true \rangle \xrightarrow{0}$$
$$\langle st|timer : timer(\Delta t), state : fail, strec : false \rangle$$
$$(\text{to } ps \text{ } st \text{ is } a_1 \text{ to } a_2)(\text{to } w \text{ } st \text{ is } a_1 \text{ to } a_2) \text{ } (st\_stat : false) \text{ } (st\_fail)$$
$$\textbf{if } eq(s, fail) \textbf{ or } eq(s, sign).$$

If the steam object receives an init-request from the controller object, it replies by sending a message with its estimated output (which must necessarily be 0) if the device is working correctly. In case the device is broken the constant *error* is sent:

$$(\text{to } st \text{ } init)\langle st|state : s \rangle \xrightarrow{0} \langle st \rangle (st \text{ is } (\textbf{if } eq(s, ok) \textbf{ then } 0 \textbf{ else } error)).$$

Water, Pump, and Pump System Objects. Due to lack of space, the specification of the water, pump, and pump system objects is deferred to CD-ROM Annex OKW.A. The interfaces of these objects to the rest of the program and the environment are given in Tables 5-7.

4.3 A Specification of the Environment

In this section we give a simplified specification of the environment which can be used when reasoning about the whole steam-boiler system.

To the control program, the environment appears as nothing but a device that sends messages arbitrarily, though no more than one message of each kind in a cycle. Furthermore, no message should arrive at time $n\Delta t$ for any n. The following specification satisfies these criteria:

Receive:	Send:	Abbreviation for:	From/To:
$(p_i\text{\_}ack)$		pump failure acknowledged	environment
$(p_i\text{\_}rep)$		pump i repaired	environment
$(p_i\text{\_}mode(v))$		pump mode	environment
	$(p_i\text{\_}fail)$	pump failure detected	environment
	$(p_i\text{\_}rep\text{\_}ack)$	pump repair acknowledged	environment
	$(p_i\text{\_}stat : p_i\text{\_}ok)$	pump i status	controller
	$(open(p_i))$	open pump i	environment
	$(close(p_i))$	close pump i	environment
	$(trans\text{\_}fail)$	transmission failure detected	controller

Table 6. *Interface of the pump object p_i.*

Receive:	Abbreviation for:	From:
(to ps st is va_1 to va_2)	steam interval	steam

Table 7. *Message defining the interface of the pump system object.*

class   *environment*
atts   *timer : Timer*
       stop_sent : bool
       StBW_sent : bool
       $\vdots$
       $p_4\text{\_}mode\text{\_}sent : bool$
initially $timer := timer(\Delta t), stop\text{\_}sent := false, \ldots, p_4\text{\_}mode\text{\_}sent := false.$

An attribute $m\text{\_}sent$ is true if a message of kind m has been sent from the environment in the current cycle. The timer reaches the value zero at the end of a cycle. During a cycle, any kind of message that is not sent in the current cycle can be sent:

$$\langle env | timer : timer(\epsilon), m\text{\_}sent : false \rangle \xrightarrow{0} \langle env | m\text{\_}sent : true \rangle (m) \text{ if } 0 < \epsilon < \Delta t$$

if m is in the set of messages sent from the environment. At the end of each cycle, the $m\text{\_}sent$ attributes are reset to *false* for all m:

$$\langle env | timer : timer(0), stop\text{\_}sent : b_1, \ldots, p_4\text{\_}mode\text{\_}sent : b_n \rangle \xrightarrow{0}$$
$$\langle env | timer : timer(\Delta t), stop\text{\_}sent : false, \ldots, p_4\text{\_}mode\text{\_}sent : false \rangle,$$

and messages from the control program are received:

$$(m')\langle env \rangle \xrightarrow{0} \langle env \rangle$$

for $(m') \in \{$messages to environment$\}$. Finally, to ensure that the environment object *env* proceeds in time, the reflexivity axiom

$$\langle env \rangle \xrightarrow{r} \langle env \rangle$$

must be added.

If messages take time δ to travel between the control program and the environment, every message m sent between the environment and the control program is replaced by the message $dly(m, \delta)$, and the timer in the environment object is initialized to $\Delta t - \delta$.

5 Validation of the Specification

In this section we give two examples of validation techniques for Timed Maude.

In Section 5.1 it is shown how an informal statement about the behavior of the steam-boiler can be deduced using timed rewriting logic. In particular, we prove that a certain configuration $conf_1$ can rewrite to another configuration $conf_2$ in time r.

On the other hand, it is often desirable to be able to prove that the specification is deadlock-free, that it excludes Zeno behavior, that $conf_1$ will never rewrite to a configuration $conf_3$, etc. Such properties cannot be proven directly in TRL, but it is possible to take advantage of the fact that the specification consists of equations and (timed) rewrite rules to reason about the specification outside the logic TRL. In Section 5.2 we use techniques for proving termination of sequential rewriting to show that an infinite number of zero-time transitions cannot occur in a row.

5.1 Validation of the Initial Behavior

In the informal specification it is stated that "as soon as the *steam-boiler waiting* message has been received the program checks whether the quantity of steam coming out of the steam-boiler is really zero. If the unit for detection of the level of steam is defective – that is, when [the steam value] v is not equal to zero – the program enters the *emergency stop* mode." In CD-ROM Annex OKW.D it is formally proved using TRL that whenever the environment sends messages $steam(v)$ ($v > 0$) and $(StBW)$ in the first cycle, the initial configuration can rewrite in time Δt to a configuration where the $mode(emergency)$ message occurs and the *state* attribute of the controller object has the value *emergency*. Informally, we see that when a $(StBW)$ message is sent, it triggers a rule

$$(StBW)\langle c|timer:timer(r), state:startup\rangle \xrightarrow{0} \langle c|state:ch\_w\_s\rangle dly((\text{to } w \text{ init}), r)$$
$$dly((\text{to } st \text{ init}), r)$$

which means that at time Δt, the (to *st init*) message is received by the steam object. The controller goes to ch_w_s state where it waits for the reply from *water* and/or *steam* objects. In the meantime, the steam object received the $steam(v)$ message, and since $v > 0$ and the attributes va_1 and va_2 are both 0 ("updated" by the synchronous rule in section 4.1), the rule

$$(steam(v))\langle st|state:ok, va_1:a_1, va_2:a_2, strec:false\rangle \xrightarrow{0}$$
$$\langle st|state:fail, strec:true\rangle \text{ if } not(v \leq W \text{ and } a_1 \leq v \leq a_2)$$

sets the *state* attribute of the steam object to *fail*. At time Δt, the rule

$$(\text{to } st \; init)\langle st|state:s\rangle \xrightarrow{0} \langle st\rangle(st \text{ is } (\textbf{if } eq(s,ok) \textbf{ then } 0 \textbf{ else } error))$$

is triggered which sends an (st is $error$) message, that is read by the controller object in the rule

$$(st \text{ is } error)\langle c|state:ch\_w\_s\rangle \xrightarrow{0} \langle c|state:emergency\rangle.$$

The rule

$$\langle c|timer:timer(0), state:emergency\rangle \xrightarrow{0} \langle c|timer:timer(\Delta t)\rangle(mode(emergency))$$

then sends the appropriate *mode(emergency)* message.

5.2 Absence of Zeno Behavior

In this section it is shown that only a finite number of zero-time state changes can occur in a row.

To prove this claim, one must consider all rules that can cause zero-time state changes that can be triggered at the same time. It is easy to see that the zero-time rules that can be used during the cycle (in the time interval $((n-1)\Delta t, n\Delta t)$), namely the rules for handling messages from the environment, will not lead to any infinite sequence of zero-time state changes. These rules only update some attributes or send a *trans_fail* message to the controller object.

The majority, and the most complicated, zero-time transitions occur at the end of each round. If it can be proven that an infinite zero-time computation cannot take place at time $n\Delta t$, then this must be true for any instant of time.

The zero-time rules that can be triggered at time $n\Delta t$ together with the general axiom $t \xrightarrow{0} t$ define an (untimed) rewriting logic theory. In [4] it is shown that each (non-trivial) transition in concurrent rewriting can be replaced by a finite number of sequential rewrite steps. This means that we can use techniques for proving termination of sequential rewriting (modulo a set of equations) to prove that any computation in time $n\Delta t$ involves only a finite number of state changes in our Timed Maude specification. This is done by considering the set of zero-time rules mentioned above as an ordinary (untimed) term rewrite system. Messages[7] and objects are treated as ordinary terms according to their interpretations in timed rewriting logic.

In CD-ROM Annex OKW.C, Theorem 1 below is proved using the termination ordering *multiset path ordering* (see e. g. [1]). This is done by proving that $t \succ_{mpo} t'$ whenever there is a zero-time state change from configuration t to t' taking place at the end of a cycle.

Theorem 1 *Let $conf_0$ be any configuration of the steam-boiler system. Then there is no infinite computation sequence of the form $conf_0 \xrightarrow{0} conf_1 \xrightarrow{0} \cdots \xrightarrow{0} conf_k \xrightarrow{0} \cdots$ such that $conf_i \neq conf_{i+1}$ for all i.*

---

[7] See CD-ROM Annex OKW.B for the formal definition of the messages.

6 Concluding Remarks

In this paper we have presented a Timed Maude solution to the specification problem posed in Chapter AS in this book. We believe that the fact that the solution consists of equations and timed rewrite rules, and that objects can be specified almost independently of each other, makes the specification easy to read and understand and enhances confidence in its correctness.

Methods for validating Timed Maude specifications were briefly presented. Properties of our solution were proved using both timed rewriting logic and existing tools for sequential rewriting. We proposed some programming principles for Timed Maude, but there is still a lot of uncovered ground in this area. Other interesting issues which will be investigated in the future include refinement notions for TRL and Timed Maude, and techniques for mapping Timed Maude specifications to parallel architectures.

7 Evaluation and Comparison

1. In this paper a formal specification of the control program is given in an object-oriented algebraic style. The specification is operational: it describes how to design a steam-boiler control program. For the environment only a simple specification is given.
2. Our specification is operational in nature and should be executable in either a mechanized version of TRL, or, through an adequate translation, in a rewriting logic interpreter.
3. As mentioned above, the specification presented is operational, algebraic, object-oriented, and quite modular. In this volume, the solutions in Chapters BBDGR, BCPR, GDK, and BW are examples of specifications in algebraic style, and are therefore at a similar level of abstraction somewhere between more abstract non-executable specifications and actual implementations. The solution BCPR is written in a classical algebraic style and focuses on computing the possible failure situations. Chapter GDK is more similar to ours. It is a state-based algebraic approach where the state is implicit (in contrast to our explicit treatment of state). Control is done by a kind of imperative control construct whereas in our paper control is made by pattern matching and timing. The BW specification uses objects but structures the system quite differently than we do, while the LM solution is very modular and organizes the system similarly to our solution. It is also worth noticing that timed rewriting logic naturally supports a high level of "true" concurrency, something that our solution tries to exploit by communicating through message passing during computations, while e. g. the BBDGR solution uses shared values extensively.

In contrast to many of the frameworks used (e. g. in Chapters BW, HW, AL), TRL provides a clear logical framework for describing many different aspects of a hybrid system. While e. g. Chapter BW uses object diagrams, statecharts, and the algebraic specification language Z for specifying the architecture, the reactive parts, and the functional parts of the system respectively, Timed Maude can be used very naturally for describing all of these three parts.

Most solutions take advantage of the fact that the steam-boiler system as stated in Chapter AS is merely a reactive one. It seems not entirely clear how many of the techniques applied could be used on a really hybrid system like the more general steam-boiler system treated in this paper. This is true both for the pure algebraic approaches such as in Chapter BCPR, and where more reactive frameworks are combined with static ones (e. g. in Chapters RW, AL, CD).

Using an operational framework as TRL, we naturally felt that for expressing more complex safety and liveness properties it would be helpful to have a more abstract way of expressing properties such as modal or temporal logic. Chapter LM provides a temporal logic specification which therefore complements our solution well. Being more abstract, such requirement specifications are less suitable as basis for a concrete implementation.

Our specification concentrates on design issues of the control program, but neither models or analyses more closely the physical properties of the steam-boiler, nor does it calculate the optimal solution regarding which pumps to open or close. Chapter LW focuses on these issues and hence complements our solution in this aspect.

4. Timed Maude, once understood, provides a nice framework for writing object-oriented specifications quite easily. The solution provided was written in a couple of man-weeks. We believe that once familiarity with elementary algebraic specification and rewriting techniques is obtained, *writing* similar specifications is probably easier in TRL than in most of the other frameworks presented in this book. As a rough estimate, we think that it would take the average programmer about 1 month to acquire the desired familiarity with the above mentioned techniques and TRL.

5. We believe that the specification is easy to *understand* even without detailed knowledge of timed rewriting logic. This claim is supported by the many helpful suggestions and comments on an earlier version of this paper received from a colleague with no prior knowledge of timed rewriting logic or rewriting logic.

Acknowledgement: We would like to thank Stephan Merz and José Meseguer for many valuable comments on earlier versions of this paper.

References

1. N. Dershowitz. Termination of rewriting. *Journal of Symbolic Computation*, 3:69–116, 1987.
2. P. Kosiuczenko and M. Wirsing. Timed rewriting logic, 1995. Working material for the 1995 Marktoberdorf International Summer School "Logic of Computation".
3. P. Kosiuczenko and M. Wirsing. Timed rewriting logic with an application to object-oriented specification. To appear in *Science of Computer Programming*, 1996.
4. J. Meseguer. Conditional rewriting logic as a unified model of concurrency. *Theoretical Computer Science*, 96:73–155, 1992.
5. J. Meseguer. A logical theory of concurrent objects and its realization in the Maude language. In Gul Agha, Peter Wegner, and Akinori Yonezawa, editors, *Research Directions in Concurrent Object-Oriented Programming*, pages 314–390. MIT Press, 1993.

Refinement from a Control Problem to Programs*

Michael Schenke[1] and Anders P. Ravn[2]

[1] FB Informatik, Universität Oldenburg
Postfach 2503, D-26111 Oldenburg, Germany
E-mail: schenke@arbi.informatik.uni-oldenburg.de
[2] Department of Information Technology
Technical University of Denmark, bldg. 344
DK-2800 Lyngby, Denmark

Abstract. Duration Calculus, a real-time interval logic, is used to specify requirements for a control task, exemplified by a steam boiler. The same formalism is used to refine requirements to a functional design. Functional designs use a subset of Duration Calculus formulas which through a suitable transformation links to an event and action based formalism. Finally, components in the resulting design for a distributed architecture are transformed to occam-like programs. The presented approach links formalisms from a top level requirements notation down to programs together in a mathematically coherent development trajectory.

1 Introduction

The Steam-Boiler control problem is a good example of a major application area for real-time computing systems. The overall task is to control a mechanical, chemical etc. plant or device under a variety of operating modes which are caused by changing operating conditions. Such plants are often safety critical and must therefore be developed using the highest standards of engineering. In particular, programs must be the result of a rationally documented design process.

The approach to the Steam-Boiler control problem demonstrated in this paper has been developed by the ESPRIT BRA "Provably Correct Systems" (ProCoS). The goal of the ProCoS project was to create a mathematical basis for the development of embedded real-time computer systems. A starting point in the development of such a system is to use a precise temporal logic, the Duration Calculus (DC) as requirements language. The requirements are then systematically transformed to occam-like programs. This part are in the following applied to the steam-boiler. ProCoS has furthermore developed correct compilation methods for such programs to conventional processors and to hardware. An overview of the ProCoS project is found in [1].

---

* This work is partially funded by the Commission of the European Communities (CEC) under the ESPRIT programme in the field of Basic Research Project No. 7071: "ProCoS II: Provably Correct Systems".

Overview. We want to demonstrate a deep development path with a reasonable level of detail. In order to do so without burdening the reader with repetitive constructions, we elide duplicated developments, e.g. we consider only the active phase of the steam boiler, not the initialisation, and we do not replicate fault handling components for every sensor or actuator. These do not contribute anything to the understanding of our method. The facets of the steam boiler that are covered or deliberately omitted are outlined below.

A formal specification of safety and liveness requirements to the overall, hybrid system consisting of continuous plant and the discrete control modes is presented in Section 2. The developed plant model is adapted from the presentation in [4], and the requirements are simplified versions of the ones found there. However, the functional design in that paper is very different, because it develops intelligent sensors and actuators to support a small main controller, whereas the current paper develops a global controller for unintelligent sensors. Performance is not considered in the control law that is shown to refine the requirements.

A functional design with specifications for sensors, actuators, and a control program is specified. The design concentrates on the operating mode and the fault detection and tolerance mechanisms. It is presented in Section 3. We omit initialisation, human operator interaction and multiple pumps. Fault handlers are illustrated by the most interesting one for the pump. For a version where other facets are included, see [4]. It is outlined, how it is formally shown that the design refines the requirements.

Section 4 contains a design for a distributed architecture with components that are specified in the Systems Language (SL) as communicating sequential processes using a network of point-to-point channels. The components include: main control, and devices with associated fault handlers. The component specifications are derived by syntax directed transformation rules from the global design.

In Section 5 occam-like real-time programs for the components are derived by another syntax directed transformation from the architectural SL-design. The programs have not been compiled and linked to a simulator. This is demonstrated in further work on correct compilation [1].

2 System and Requirements

Software requirements for complex control applications depends heavily on properties of the plant. These properties are usually investigated by control engineers that base their work on the theory of dynamical systems. The mathematical tool for this work is mathematical analysis, in particular the theory of differential equations.

Design and implementation of a hybrid controller means that control engineers and programmers need to cooperate from the early stage of requirements formulation. We do not consider it safe to rely on informal translations from the precise mathematical models of control to an informal software requirements document, as e.g. illustrated by the steam boiler problem statement for the Dagstuhl meeting.

In order to get a precise interface to control engineering, our language must include concepts of analysis like differential or integral calculus. A complex system is, however, characterised by having discrete changes in operation, caused by operator interventions, failing components, or changing environments. Therefore, the language must also be able to specify discrete state changes over time. The notation that we have selected is therefore DC, a real-time interval logic, which incorporates integral calculus.

As designer, one must consider further aspects like a module notation. The module notation used, is based on Z schemas because the logical operators on such schemas blends well with a logic language like DC.

2.1 Plant Model

The problem at hand is to control the level of water in a steam-boiler. The physical system consists of the following units:

- The steam-boiler.
- A water level indicator.
- Pumps to provide the steam-boiler with water.
- A steam flow indicator.

We are not treating initialisation, so a flush valve has been omitted.

Note that *heating* of the boiler is *not* controlled by the system, so the only way of coping with variations in the water level is by starting or stopping the pumps.

System Identification The first step in formalising requirements to a system is a *system identification*, where control engineers build a mathematical model of the system. We use the well-known *time-domain model* where a system is described by a collection of states that are functions of time, the non-negative real numbers: $Time \stackrel{\text{def}}{=} \mathbf{R}^+$.

For the steam boiler we use the states shown in the Z style schema:

─ *Boiler* ─────────────────────────
 BoilerParams
 $q : Time \to [0, C]$ (Water level (liters))
 $v : Time \to [0, W]$ (Steam (liters/sec))
 $p : Time \to [0, P]$ (Pump throughput (liters/sec))
──────────────────────────────────

where $[a, b] \stackrel{\text{def}}{=} \{x : \mathbf{R} \mid a \leq x \land x \leq b\}$ denote a closed interval of reals, and where the boiler parameters C, W, etc. are specified in the schema

```
┌─ BoilerParams ─────────────────────────────
│  C :        R⁺ (Max water level)
│  W :        R⁺ (Max quantity of steam)
│  P :        R⁺ (Max total throughput of pumps)
│  U₁, U₂ : R⁺ (Limits for steam gradient)
├────────────────────────────────────────────
│  U₁ > 0 ∧ U₂ > 0
└────────────────────────────────────────────
```

To illustrate the constraint part of a schema, the gradient limits are specified to be strictly positive.

Further properties of the system are *preservation of mass*: The change in the water level is equal to the throughput of the pumps minus the quantity of steam: $\forall t : Time \bullet \dot{q}(t) = p(t) - v(t)$. Here $\dot{q}$ is the time derivative of q. There are also limits on the change in the steam quantity: $\forall t : Time \bullet -U_2 \leq \dot{v}(t) \leq U_1$.

It turns out that these formulae are more useful when we integrate over an arbitrary time interval $[b, e]$ with $0 \leq b \leq e$. The first formula gives:

$$Mass \stackrel{\text{def}}{=} \forall b, e : Time \bullet q(e) - q(b) = \int_b^e p(t)\,dt - \int_b^e v(t)\,dt$$

The second formula gives $v(e) - v(b) \in [-(e-b)U_2, (e-b)U_1]$, and by a further integration

$$\int_b^e v(t)\,dt - (e-b)v(b) \in [-\tfrac{1}{2}(e-b)^2 U_2, \tfrac{1}{2}(e-b)^2 U_1]$$

If we have an estimate of $v(b)$, say $v(b) \in V = [v_1, v_2]$, then these formulae can be used to give estimates of $v(e)$ and $\int_b^e v(t)\,dt$:

$$Steam_1 \stackrel{\text{def}}{=} \forall b, e : Time, V : Interval \bullet$$
$$\qquad v(b) \in V \Rightarrow v(e) \in G(V, e - b)$$
$$Steam_2 \stackrel{\text{def}}{=} \forall b, e : Time, V : Interval \bullet$$
$$\qquad v(b) \in V \Rightarrow \int_b^e v(t)\,dt \in F(V, e - b)$$

where *Interval* denotes the set of closed subintervals of the real numbers and where $F : Interval \times \mathbf{R} \to Interval$ and $G : Interval \times \mathbf{R} \to Interval$ denote the functions

$$G([v_1, v_2], y) \stackrel{\text{def}}{=} [v_1 - yU_2, \, v_2 + yU_1]$$
$$F([v_1, v_2], y) \stackrel{\text{def}}{=} [y\,v_1 - \tfrac{1}{2}y^2 U_2, \, y\,v_2 + \tfrac{1}{2}y^2 U_1]$$

Duration Calculus As illustrated above it is possible to formalise properties of systems using predicate logic and mathematical analysis. However, formal reasoning is rather cumbersome when all formulae contain several quantifiers. This is one of the rationales for having a temporal logic: it will implicitly quantify over the time domain.

Syntax. The syntax of Duration Calculus distinguishes (*duration*) *terms*, each one associated with a certain type, and (*duration*) *formulae*. Terms are built from names of elementary states like q or v, *rigid variables* representing time independent logical variables and are closed under arithmetic and propositional operators. Examples of terms are $p = P$ and $d \neq 0$ of Boolean type and $p - d$ of real type.

Terms of type real are also called *state expressions* and terms of Boolean type are called *state assertions*. We use f for a typical state expression and P for a typical state assertion.

Duration terms are built from $\mathbf{b}.f$ and $\mathbf{e}.f$ denoting the *initial* and *final value* of f in a given interval, and $\int f$ denoting the *integral* of f in a given interval. For a state assertion P, the integral $\int P$ is called the *duration* because it measures the time P holds in the given interval.

Duration formulae are built from duration terms of Boolean type and closed under propositional connectives, the chop connective, and quantification over rigid variables and variables of duration terms. We use D for a typical duration formula.

Semantics. The semantics of Duration Calculus is based on an *interpretation* $\mathcal{I}$ that assigns a fixed meaning to each state name, type and operator symbol of the language, and a time interval $[b, e]$. For given $\mathcal{I}$ and $[b, e]$ the semantics defines what domain values duration terms and what truth values duration formulae denote. For example, $\int f$ denotes the integral $\int_b^e f(t)\,dt$, $\mathbf{b}.f$ denotes the limit from the right in b, and $\mathbf{e}.f$ denotes the limit from the left in e.

A duration formula D *holds* in $\mathcal{I}$ and $[b, e]$, abbreviated $\mathcal{I}, [b, e] \models D$, if it denotes the truth value *true* for $\mathcal{I}$ and $[b, e]$. A chopped formula $D_1 \,;\, D_2$ holds just when there exists a m, such that D_1 holds on $[b, m]$ and D_2 holds on $[m, e]$, cf. the following timing diagram:

$$\begin{array}{c}
\overset{D_1 \,;\, D_2}{\longleftrightarrow} \\
\overset{D_1}{\longleftrightarrow} \quad \overset{D_2}{\longleftrightarrow} \\
\text{Time } b \qquad m \qquad e
\end{array}$$

D is *true* in $\mathcal{I}$, abbreviated $\mathcal{I} \models D$, if $\mathcal{I}, [0, t] \models D$ for every $t \in \text{Time}$. A *model* of D is an interpretation $\mathcal{I}$ which makes D true, $\mathcal{I} \models D$. The formula D is *satisfiable* if there exists an interpretation $\mathcal{I}$ such that $\mathcal{I} \models D$.

Examples. The modalities $\Diamond$ (on some subinterval) and $\Box$ (on any subinterval) are defined by:

$\Diamond D \stackrel{\text{def}}{=} true \,;\, D \,;\, true$
$\Box D \stackrel{\text{def}}{=} \neg \Diamond (\neg D)$

so $\Diamond D$ is true on an interval if the interval can be chopped into three subintervals such that D is true on the middle interval, i.e. if there is *some subinterval* where D is true. A similar argument shows that $\Box D$ is true on an interval if D is true on *any subinterval*.

Preservation of mass is now the simple formula $\mathbf{e}.q - \mathbf{b}.q = \int p - \int v$. This should hold on *any* time interval $[b, e]$, i.e. for any subinterval of any interval

$[0, t]$, resulting in the schema:

$$Mass \,\widehat{=}\, [Boiler \mid \Box(\mathbf{e}.q - \mathbf{b}.q = \int p - \int v)]$$

The length of an interval, abbreviated ℓ, is the duration $\int 1$ of the constant function 1. The terms $e - b$ in the above $Steam_1$ and $Steam_2$ formulae denote the lengths. So the constraints result in the schema:

$$Steam_1 \,\widehat{=}\, [Boiler \mid \Box(\forall V : Interval \bullet \mathbf{b}.v \in V \Rightarrow \mathbf{e}.v \in G(V, \ell))]$$
$$Steam_2 \,\widehat{=}\, [Boiler \mid \Box(\forall V : Interval \bullet \mathbf{b}.v \in V \Rightarrow \int v \in F(V, \ell))]$$

This completes the system identification.

2.2 Requirements

The controller operates in two phases: an initial phase with no production of steam which ends when the water level is within some normal interval; and an *active* phase with production of steam and with the goal of operating the pumps such that the water level is kept inside a maximal interval $M = [M_1, M_2]$.

In the initial problem text these basic goals are intertwined with failure detection, failure reporting, and fault tolerant operation. These further features depend on a particular selection of components for a specific design of a steam boiler system. We state the basic requirements to the system without presenting a design by proceeding as follows:

> Requirements use the form: *Assumption* $\Rightarrow$ *Commitment*, where the formula *Commitment* is detailed from the start, while *Assumption* is left open for a design stage. It records assumptions that are made when certain components are selected in a design to satisfy the commitment.

The plant extends the identified system with a state *active* (normal operation) and various parameters

```
┌─ Plant ─────────────────────────────────────────────
│ Mass
│ Steam₁
│ Steam₂
│ active : Time → Bool
│ M :     Interval        (Limit interval for water level)
├─────────────────────────────────────────────────────
│ M ⊆ [0, C]
└─────────────────────────────────────────────────────
```

As mentioned above we only state the commitments $Commit_i$ of the requirements

$$Req_i \stackrel{def}{=} Assumpt_i \Rightarrow Commit_i \quad i = 1, \ldots 3$$

The assumptions $Assumpt_i$ are derived later, when the design is verified.

The first commitment states that we expect the controller to start in the *active* phase, i.e. we ignore initialisation.

For an arbitrary assertion P, it is obvious that it is *true* almost anywhere in an interval, just when the duration $\int P$ is equal to the length of the interval. Recalling that ℓ denotes the length of the interval, the property that P is *true* is thus given by the formula $\int P = \ell$. This holds trivially for a point interval, so we consider proper intervals of positive length. These two properties are combined in the abbreviation

$$\lceil P \rceil^r \stackrel{\text{def}}{=} \int P = \ell \wedge \ell = r$$

We use $\lceil P \rceil \stackrel{\text{def}}{=} \exists r > 0 \bullet \lceil P \rceil^r$ to denote the property above, and a point interval is denoted by $\lceil \, \rceil \stackrel{\text{def}}{=} \ell = 0$.

The first commitment for the controller is thus expressed by the formula

$$Commit_1 \stackrel{\text{def}}{=} \lceil \, \rceil \vee (\lceil active \rceil \; ; \; true)$$

It expresses that any time interval $[0, t]$ is either a point or can be chopped into two parts such that *active* is *true* on the first part.

The second commitment states that the controller stays *active*, i.e. liveness. In order to assure this it suffices to know that whenever *active* is *true* in a time interval $[b, m]$ then it will also be *true* in some succeeding interval $[m, m+\delta]$.

We introduce the abbreviation $D \longrightarrow \lceil P \rceil$ for this property for arbitrary duration formula D and assertion P. Formally a violation of a property $D \longrightarrow \lceil P \rceil$ is the existence of an interval with D followed by $\neg P$, i.e.

$$D \longrightarrow \lceil P \rceil \Leftrightarrow \neg \Diamond (D \; ; \; \lceil \neg P \rceil)$$

Within a timing diagram, $D \longrightarrow \lceil P \rceil$ means that an interval where D holds is immediately *followed by* an interval where P holds. It could also be phrased D leads immediately to P. The notation has many of the properties of normal implication, e.g., if $D' \Rightarrow (true \; ; \; D)$ and $P \Rightarrow P'$ then $D \longrightarrow \lceil P \rceil$ implies $D' \longrightarrow \lceil P \rceil$. It differs from implication by incorporating progress of time, e.g., the formula $\square (\lceil P \rceil \Rightarrow \lceil P \vee Q \rceil)$ is just a tautology, whereas $\lceil P \rceil \longrightarrow \lceil P \vee Q \rceil$ defines a transition from P to Q (the term $P \vee Q$ is needed to allow stuttering in state P).

Using this notation, the second commitment is

$$Commit_2 \stackrel{\text{def}}{=} \lceil active \rceil \longrightarrow \lceil active \rceil$$

The reader may wonder why we introduce *active*, because we seem committed to stay *active* forever. Recall though that we still need to specify under what assumptions this is the case. If these assumptions are violated, e.g. malfunctions in components, then we are free to leave *active*. This will help us to satisfy the safety commitment. The water level stays inside the M-interval in the *active* mode:

$$Commit_3 \stackrel{\text{def}}{=} \lceil active \rceil \longrightarrow \lceil q \in M \rceil$$

A control engineer may choose to say that this is Ljapunov stability.

We collect the requirements in the schema

$$Requirements \mathrel{\hat=} [Plant \mid Req_1 \wedge Req_2 \wedge Req_3]$$

Recall that we have yet to determine the assumptions under which the commitments are fulfilled.

3 Global Design

The purpose of the global design is to identify and describe suitable components that contribute to satisfying the requirements under acceptable assumptions on the overall system. In the context of complex control systems, we propose the following steps:

1. Definition of the control programs.
2. Definition of sensors.
3. Definition of actuators.

In the example we have two control programs: *Main* that controls the steam boiler, and *PFH* which is a simple failure handler. There are three sensors *PumpControl*, *Steam* and *Level* and one actuator *Pump*. The design is thus given by the schema that combines the components with the plant:

```
┌─ SB ──────────────────────────────
│ Plant
│ Main
│ Level
│ Steam
│ PumpControl
│ Pump
│ PFH
└───────────────────────────────────
```

A control program is a state machine that progresses through a number of *phases*. These are specified by formulae over phase control states. The formulae determine the conditions for a phase to be either stable or to progress to a new phase. For each phase there is a set of phase commitments that determine the interaction with sensors and actuators. They will in general observe or control a single component of the plant state for a specific phase. They constitute together with the phase sequencing constraints the *implementables* for a program design.

Implementable forms. In order to give implementables in a compact form, we introduce the shorthands:

$$D \xrightarrow{c} \lceil P \rceil \stackrel{\text{def}}{=} (D \wedge \ell = c) \longrightarrow \lceil P \rceil$$

This form is used to give an upper bound, c, on the time D holds before it is followed by P. It may be read as: 'D for c time units leads immediately to P'.

$$D \xrightarrow{\leq c} \lceil P \rceil \stackrel{\text{def}}{=} (D \wedge \ell \leq c) \longrightarrow \lceil P \rceil$$

This form constrains transitions for up to c time units. It is used to give lower bounds on transition times.

When $\pi, \pi_1, \ldots$ denote single *phase* states, and $\varphi, \varphi_1, \ldots$ denote combinations of plant states or phase states, the implementable forms are:

Sequencing: $\lceil \pi \rceil \longrightarrow \lceil \pi \vee \pi_1 \vee \cdots \vee \pi_n \rceil$, where $n = 0$ means that the program is stuck in phase π, while $n > 1$ means that there is a choice of a successor phase. Sequencing constraints specify non-deterministic transitions of an untimed state machine over the phases and plant states.

Progress: $\lceil \pi \wedge \varphi \rceil \xrightarrow{c} \lceil \neg \pi \rceil$, where the phase π is left at the latest when φ holds for c time units. A progress constraint specifies an upper bound on a transitions of the state machine. Note that c is an upper bound, the transition may be taken earlier, if selection allows it.

Progress may also express *active stability* of the phase: $\lceil \pi \wedge \varphi \rceil \xrightarrow{c} \lceil \pi \rceil$.

Selection: $\lceil \neg \pi \rceil \; ; \; \lceil \pi \wedge \varphi \rceil \xrightarrow{\leq c} \lceil \pi \vee \pi_1 \vee \ldots \vee \pi_n \rceil$, where the sequencing of phase π is constrained under the condition φ for c time units (or forever, if the time bound is omitted). If $n = 0$ the formula defines conditional, time bounded *stability* of the phase. Note that c is a lower bound, an implementation may keep the selection open for a longer time, but not for a shorter.

Synchronisation: $\lceil \pi_1 \vee \ldots \vee \pi_n \rceil \xrightarrow{c} \lceil \varphi \rceil$, where the combined phase $\pi_1 \vee \ldots \vee \pi_n$ will cause φ to hold after c time units. Note that c is an upper bound, an implementation is allowed to cause φ sooner but not later.

Framing: $\lceil \neg \pi \rceil \; ; \; \lceil \pi \wedge \varphi \rceil \xrightarrow{\leq c} \lceil \varphi \rceil$, is dual to phase stability. It is a commitment that the state φ will remain stable for c time units if it is present when the phase π is entered.

These forms define a state machine which observes (read) plant states or states of other machines through the φ conditions in progress and selection forms, and which controls the phases and certain plant states through synchronisation and framing. Although not used in the following, we make the remark that if controlled states are kept disjoint then conjunction of implementables corresponds to parallel composition of state machines.

3.1 Main control program

The main control program cycles through: A pump check phase (*getpc*), an update phase where the pump is controlled, and a wait phase where it waits for a full sampling period. An additional *emergency* phase is added, it is used to mark that the control is inactive. The *phase* state encodes the phases of the control cycle.

For later use in progress commitments we introduce an upper bound of ε for the time spent in the *getpc* and *update* phases. The sampling period is Δ which is used as a lower bound in the wait phase. We also make *emergency* a stuck phase, by having a sequencing commitment such that once entered it is never left.

MainSeq

$phase : Time \rightarrow \{getpc, update, wait, emergency\}$
$\Delta, \varepsilon : \mathbf{R}$

$0 < \varepsilon \wedge 0 < \Delta$
$\lceil phase = emergency \rceil \longrightarrow \lceil phase = emergency \rceil$

The main program records the sensor readings and computed results in additional states that model program variables. These are treated as part of the state machine phase, the π's of the implementables are thus assertions over *phase*, and the following states of *MainVar* and *PumpControl*, *Level* and *Steam*. The design is thus with one sequential program that incorporates sensors, actuators and control.

MainVar

MainSeq
$Q :\quad Time \rightarrow Interval$ Estimate of water level
$V :\quad Time \rightarrow Interval$ Estimate of the steam level
$pfail : Time \rightarrow \mathbf{Bool}\quad$ Pump failure indicator

---

The pump is obviously an actuator. We introduce one state *pon* to denote whether it is on or off. This state is framed such that it is only changed by the update phase.

Pump

MainSeq
$pon : Time \rightarrow \mathbf{Bool}\quad$ pump control signal

---
$\forall x \bullet$
$\lceil phase = update \rceil \;;\; \lceil phase \neq update \wedge pon = x \rceil \longrightarrow \lceil pon = x \rceil$

We shall later in the control strategy specify the value of *pon* at the end of an update phase.

The pump is working correctly, when it produces a flow within a period when turned on and blocks the flow almost immediately, when turned off. This is formalised in the synchronisation properties:

$$PumpOk \stackrel{\text{def}}{=} (\lceil pon \rceil \stackrel{\Delta}{\longrightarrow} \lceil p = P \rceil) \wedge (\lceil \neg pon \rceil \stackrel{\varepsilon}{\longrightarrow} \lceil p = 0 \rceil)$$

The pump control indicates in *flow*, whether there is water movement or not. Its value is framed over the corresponding *getpc* phase.

```
┌─ PumpControl ─────────────────────────────────────────┐
│ MainSeq                                               │
│ flow : Time → Bool                                    │
│                                                       │
│ ∀x•                                                   │
│ ⌈phase ≠ getpc⌉ ; ⌈phase = getpc ∧ flow = x⌉ ⟶ ⌈flow = x⌉ │
└───────────────────────────────────────────────────────┘
```

It works correctly when it within ε indicates the flow

$$PCOk \stackrel{\text{def}}{=} (\lceil p > 0 \rceil \stackrel{\varepsilon}{\longrightarrow} \lceil flow \rceil) \wedge (\lceil p = 0 \rceil \stackrel{\varepsilon}{\longrightarrow} \lceil \neg flow \rceil)$$

The water level sensor has a state $\hat{q}$, and the steam sensor has a state $\hat{v}$. Both are framed in the update phase.

```
┌─ Level ───────────────────────────────────────────────┐
│ MainSeq                                               │
│ q̂ : Time → R                                          │
│                                                       │
│ ∀x • ⌈phase ≠ update⌉ ; ⌈phase = update ∧ q̂ = x⌉ ⟶ ⌈q̂ = x⌉ │
└───────────────────────────────────────────────────────┘
```

The level sensor works correctly when the reading reflects the physical value at the start of the update phase

$$LevelOk \stackrel{\text{def}}{=} (\lceil phase \neq update \rceil \wedge \mathbf{e}.q = x) \longrightarrow \lceil phase \neq update \vee \hat{q} = x \rceil$$

```
┌─ Steam ───────────────────────────────────────────────┐
│ MainSeq                                               │
│ v̂ : Time → R                                          │
│                                                       │
│ ∀x • ⌈phase ≠ update⌉ ; ⌈phase = update ∧ v̂ = x⌉ ⟶ ⌈v̂ = x⌉ │
└───────────────────────────────────────────────────────┘
```

It works when

$$SteamOk \stackrel{\text{def}}{=} (\lceil phase \neq update \rceil \wedge \mathbf{e}.v = x) \longrightarrow \lceil phase \neq update \vee \hat{v} = x \rceil$$

The *Main* module consists of program commitments for the phases and an initialisation.

```
┌─ Main ───────┐
│ Getpc        │
│ Update       │
│ Wait         │
│ Init         │
└──────────────┘
```

Phase commitments The *getpc* phase checks that the pump command is consistent with the pump control indication. An inconsistency is indicated in *pfail*. Other program variables are framed. The phase lasts at most ε time units.

Getpc
MainVar
Pump
PumpControl

$\forall pf, pok \bullet$
$\lceil phase = getpc \land pfail = pf \land pok = (pon = flow) \rceil \longrightarrow$
$\lceil (phase = getpc \land pfail = pf) \lor (phase = update \land pfail = \neg pok) \rceil$
$\forall x \bullet \lceil phase \neq getpc \rceil \;;\; \lceil phase = getpc \land Q = x \rceil \longrightarrow \lceil Q = x \rceil$
$\forall x \bullet \lceil phase \neq getpc \rceil \;;\; \lceil phase = getpc \land V = x \rceil \longrightarrow \lceil V = x \rceil$
$\lceil phase = getpc \rceil \stackrel{\varepsilon}{\longrightarrow} \lceil phase \neq getpc \rceil$

The *update* phase reads the steam level and water level, to rigid variables q_0 and v_0, and check them for consistency using the previous estimates Q_0 and V_0. The check is integrated in calculating estimates of the next Q and V values, given by the expressions QE and VE, elaborated below.

It furthermore checks safety of the water level using the estimate of the level. The assertion *safe* is defined below. Finally, the next pump setting is calculated in the expression *nextp*, explained below.

The sequencing to either *wait* or *emergency* is determined by the value of *safe*. During the phase, the value of *pfail* is framed.

Update
MainVar
Level
Steam

$\forall q_0, Q_0, v_0, V_0 \bullet$
$\lceil\; phase = update \land Q = Q_0 \land \hat{q} = q_0 \land V = V_0 \land \hat{v} = v_0 \rceil \longrightarrow$
$\lceil\; (phase = update \land Q = Q_0 \land V = V_0) \lor$
$\quad (phase = wait \land safe \land Q = QE \land V = VE \land pon = nextp) \lor$
$\quad (phase = emergency \land \neg safe) \rceil$
$\forall x \bullet$
$\lceil phase \neq update \rceil \;;\; \lceil phase = update \land pfail = x \rceil \longrightarrow$
$\lceil pfail = x \rceil$
$\lceil phase = update \rceil \stackrel{\varepsilon}{\longrightarrow} \lceil phase \neq update \rceil$

Checks and estimates.

The steam estimate is calculated as $VE = G(V_1, T)$, where the current value V_1 is either the read value v_0, or the previous estimate V_0 when v_0 is outside this estimate.

$$(v_0 \in V_0 \wedge V_1 = [v_0, v_0]) \vee (v_0 \notin V_0 \wedge V_1 = V_0),$$

and the period is given by $T = \Delta + 3 \cdot \varepsilon$, the maximal cycle time plus ε for the next update.

The level estimate is given by a prediction $QE = H(Q_1, V_1, P_1, T)$, where the current level Q_1 is given by

$$(q_0 \in Q_0 \wedge Q_1 = [q_0, q_0]) \vee (q_0 \notin Q_0 \wedge Q_1 = Q_0),$$

and the pump output P_1 is a crude estimate, due to pump delays:

$$(nextp \wedge P_1 = [0, P]) \vee (\neg nextp \wedge P_1 = [0, 0])$$

A finer estimate would use the current *pon* value.

The function $H : Interval \times Interval \times Interval \times \mathbf{R} \to Interval$ computes the next level estimate according to the mass preservation law:

$$H([q_1, q_2], V, [p_1, p_2], c) \stackrel{\text{def}}{=} [q_1 - f_2 + p_1 \cdot c, q_2 - f_1 + p_2 \cdot c]$$

where $[f_1, f_2] = F(V, c)$ with F defined in the previous section.

Control strategy. The expression *safe* evaluates whether it is safe to continue for another period. It is given by

$$safe = QE \subseteq S$$

where the interval S denotes the worst case behaviour that stays within M for a period. It is defined implicitly by $M = H(S, [0, W], [0, P], T)$. One might think that as long as $H(Q_1, [0, W], [0, P], T) \subset M$ is sufficient for safety, but that assumes that Q_1 is correct. The current strategy is more robust it is sufficient that either Q_0 or Q_1 is correct.

If the level is guaranteed to be safe, we are interested in keeping it safe in the next interval. Thus we compute a normal interval N where we should not change the current setting of the pump. It is given implicitly as the worst behaviour over two periods that stays safe: $S = H(H(N, [0, W], [0, P], T), [0, W], [0, P], T)$.

We define the pump control *nextp* by the formula

$$(QE \subset N \wedge nextp = pon) \vee$$
$$(QE \not\subset N \wedge nextp = ((QE \cap [0, n_1]) \neq \{\}))$$

The pump is kept unchanged when the estimate stays normal. Otherwise is is turned on if the water is low and otherwise turned off. If there are several pumps this strategy should be made finer such that the change in pumped volume compensates the drift from normal.

The wait phase waits for the sampling period and progresses within ε of the expiration of the sampling period. The program variables are kept unchanged.

```
┌─ Wait ────────────────────────────────────────────────────
│  MainVar
│ ─────────────────────────────────────────────────────────
│  ⌈phase = wait⌉ ⟶ ⌈phase = wait ∨ phase = getpc⌉
│  ⌈phase ≠ wait⌉ ; ⌈phase = wait⌉ $\xrightarrow{\Delta}$ ⌈phase = wait⌉
│  ⌈phase = wait⌉ $\xrightarrow{\Delta+\varepsilon}$ ⌈phase ≠ wait⌉
│  ∀x • ⌈phase = wait ∧ pfail = x⌉ ⟶ ⌈pfail = x⌉
│  ∀x • ⌈phase = wait ∧ Q = x⌉ ⟶ ⌈Q = x⌉
│  ∀x • ⌈phase = wait ∧ V = x⌉ ⟶ ⌈V = x⌉
└───────────────────────────────────────────────────────────
```

Initialisation. The main program starts in a wait phase with a normal water level and an unknown steam level.

```
┌─ Init ────────────────────────────────────────────────────
│  Boiler
│  MainVar
│ ─────────────────────────────────────────────────────────
│  ⌈⌉ ∨
│  (⌈phase = wait ∧ (q ∈ Q ⊆ N) ∧ (V = [0, W]) ∧ ¬pfail⌉ ; true)
└───────────────────────────────────────────────────────────
```

3.2 Verification

In the following we outline the proof techniques that have been found useful in numerous case studies, detailed proofs are found in Annex RS.2.

The verification assumes that the *active* state is defined by to be a non-emergency phase:

$$Link \stackrel{def}{=} \Box (\lceil active \rceil \Leftrightarrow \lceil phase \neq emergency \rceil)$$

From this definition and the initialisation it follows by monotonicity and substitution that any behaviour has the form: $\lceil \rceil \vee (\lceil active \rceil ; \lceil true \rceil)$, thus $Commit_1$ is satisfied with $Assumpt_1 \stackrel{def}{=} Link$.

For the other commitments, we use a standard technique with proof by contradiction. If $Commit_2$ does *not* hold, we use that $\neg(D \longrightarrow \lceil P \rceil) \Leftrightarrow \Diamond(D ; \lceil \neg P \rceil)$. For $\neg Commit_2$ it leads us to consider:

$$true ; \lceil active \rceil ; \lceil \neg active \rceil$$

It may be seen as a complete history of a violation. The first formula *true* is the prehistory, while ⌈*active*⌉ is the state just before the violation by the transition to ⌈¬*active*⌉. We must prove that under the assumptions this history cannot occur, i.e., it is *false*.

We use *Link* to rewrite it into *phase* constraints. The sequencing constraints on *phase*, which only allows transitions to *emergency* from *update* when ¬*safe* holds, gives:

$true$; $\lceil phase = update \wedge \neg safe \rceil$; $\lceil \neg active \rceil$

Now it is time to analyse the prehistory. Initialisation has $q \in Q \subseteq N$, and due to the definition of N, the water level range cannot become unsafe for the first cycle, thus we have had at least two updates. And the estimate QE of the first one has been outside N and the estimate of the last one is outside S.

$\exists QE_1, QE_2 \bullet$
$\quad true$; $\lceil phase = update \wedge safe \wedge QE_1 \not\subset N \rceil$;
$\quad (\Delta < \ell \leq T)$; $\lceil phase = update \wedge QE_2 \not\subset S \rceil$; $true$

where $T = \Delta + 3 \cdot \varepsilon$.

This cannot happen if we have a water level measurement within N for one of the two updates. Thus,

$$Assumpt_2 \stackrel{\text{def}}{=} Link \wedge QOkread$$

with

$QOkread \stackrel{\text{def}}{=} \forall q_0, q_1 \bullet \Box \,($
$\quad \lceil phase = update \wedge \hat{q} = q_0 \rceil$;
$\quad \lceil phase \neq update \rceil$; $\lceil phase = update \wedge \hat{q} = q_1 \rceil$
$\quad \Rightarrow (q_0 \in N) \vee (q_1 \in N))$

If we make the N interval smaller by iterating K times S (here $K = 2$) we could improve robustness to one good reading within K updates.

The argument for $Commit_3$ is similar to the preceding one. It gives the assumption $Assumpt_3 \stackrel{\text{def}}{=} Link \wedge QOkread \wedge QOk$.

The formula QOk is similar to $QOkread$, but requires that one of the read level values is correct.

$QOk \stackrel{\text{def}}{=}$
$\quad \Box \,(\lceil phase = update \rceil$; $\lceil phase \neq update \rceil$; $\lceil phase = update \rceil$
$\quad \Rightarrow OneOK)$

with $OneOk \stackrel{\text{def}}{=} \Diamond \,(\lceil phase \neq update \rceil$; $\lceil phase = update \rceil \Rightarrow LevelOk)$.

Both $QOKread$ and QOK are rather strong assumptions that do not use the steam measurement or the control strategy. They can be replaced by an assumption: $PumpOk \wedge (P > W) \wedge MeasOK$ where

$MeasOk \stackrel{\text{def}}{=}$
$\quad \Box \,(\lceil phase = update \rceil$; ; $\lceil phase \neq update \rceil$; $\lceil phase = update \rceil$
$\quad \Rightarrow LevelOk \wedge (SteamOk \vee OneOk))$

It uses the control strategy to prove that the water level will return to N within two updates. This can be further weakened to within K updates by using a smaller N, if desired.

3.3 A Failure Handler

A completed system would also include some failure handlers that report faults to a human supervisor. We conclude this section with a simple pump failure handler *PFH* below. It is included in order to investigate refinement of asynchronous, state based interaction among two state machines in the next section.

This failure handler has only two states *normal* and its complement, where reporting takes place. The failure handler monitors asynchronously the *pfail* state of the main program. The normal phase is only left when *pfail* is observed, and an occurrence of *pfail* is framed by the reporting phase.

$$
\begin{array}{l}
\underline{\textit{PFH}} \\
\textit{MainVar} \\
\textit{normal} : \textit{Time} \to \textbf{Bool} \\
\hline
\lceil \textit{normal} \land \textit{pfail} \rceil \xrightarrow{\varepsilon} \lceil \neg \textit{normal} \rceil \\
\lceil \neg \textit{normal} \rceil \ ; \ \lceil \textit{normal} \land \neg \textit{pfail} \rceil \longrightarrow \lceil \textit{normal} \rceil \\
\lceil \neg \textit{pfail} \rceil \ ; \ \lceil \neg \textit{normal} \land \textit{pfail} \rceil \longrightarrow \lceil \textit{pfail} \rceil
\end{array}
$$

We do not elaborate the fault reporting in the ¬*normal* phase.

4 Network and Components

The aim of this section is to refine the global design of the previous section to a design for a network of occam programs that synchronise and communicate through channels.

The development is done in order to show the applicability of the approach to the development of provably correct systems as lined out in [2] and described in [6]. We shall not repeat the rules from these papers in full detail. Instead we shall give the intermediate forms of the specification at various stages. For the steam boiler we shall not need all steps of the development method. Hence there will be explained only the relevant steps. Most important will be the change from a state based view like the one from DC to an event based view. So at some stage we shall introduce events. Events may be names for state changes or for synchronisations. They will be introduced in the syntax of SL, a language designed particularly for the specification of real time systems in ProCoS. The idea of this part is to successively remove the implementables and replace them by pieces of SL syntax. This syntax will be explained along with the transformations. We shall exhibit the following steps:

1. Before the events are introduced, the phase description often is still too abstract. A technique called phase splitting enables us to describe a phase in terms of newly introduced subphases. In this paper phase splitting will only be used in connection with synchronisation.
2. Phase changes (events) which change only one observable do not involve internal synchronisation of different components. They express a communication with the environment.

3. Implementables that deal with synchronisation are removed by synchronising communications which due to the occam paradigm appear in the SL specifications of both components involved.
4. Already at this stage it is possible to argue within an SL specification. In general there are several possible replacements of SL syntax pieces for other such syntax pieces. Here we shall show only one such transformation, the introduction of so-called trace assertions.
5. The introduction of time restrictions finally makes the quantitative timing requirements dependent on events rather than on phases.

4.1 Phase Splitting

In accordance with the first step of the method lined out in [2] we fulfill the phase splitting step. Phase splitting often but not exclusively becomes necessary for more or less complicated communication protocols.

Phase splitting will become necessary for example for the observable *pfail*. The main program controls this state: If in the *getpc* phase it receives values which indicate that the pump and the pump control values disagree it sets *pfail*. This is observed by the pump failure. These implementables are condensed to

(R1) $\lceil phase = getpc \wedge (pon \neq flow) \rceil \xrightarrow{\varepsilon} \lceil pfail \rceil$

(R2) $\lceil normal \wedge pfail \rceil \xrightarrow{\varepsilon} \lceil \neg normal \rceil$

Hence we want to break the *pfail* phase into a subphase, where the PFH is not yet informed, called *pfailstart*, and one where the information has been transmitted, *pfailinfo*.
Additionally we have to take into account *pfail* stabilities

(R3) $\lceil \neg pfail \rceil \; ; \; \lceil pfail \wedge \neg normal \rceil \longrightarrow \lceil pfail \rceil$

(R4) $\lceil pfail \rceil \; ; \; \lceil \neg pfail \wedge (pon = flow) \rceil \longrightarrow \lceil \neg pfail \rceil$

because these are the only other commitments that control *pfail*. Since we do not want to split ¬*pfail* we need not consider commitments here which contain ¬*pfail* only. We shall reach our aim, the removal each positive occurrence of *pfail* in favour of the above mentioned subphases, automatically by the rules in [2]. A general framework for each application of the phase splitting rules is introduced by

(S1) $\square \, (\lceil pfail \rceil \Leftrightarrow \lceil pfailstart \vee pfailinfo \rceil)$

(S2) $\square \, \neg \lceil pfailstart \wedge pfailinfo \rceil$

(S3) $\lceil pfailstart \rceil \longrightarrow \lceil pfailstart \vee pfailinfo \rceil$

(S4) $\lceil pfailinfo \rceil \longrightarrow \lceil pfailinfo \vee \neg pfail \rceil$

(S5) $\lceil \neg pfail \rceil \longrightarrow \lceil \neg pfail \vee pfailstart \rceil$

Then by [2] it is sufficient to replace (R1) - (R4) by

(R1') $\lceil phase = getpc \wedge (pon \neq flow) \rceil \xrightarrow{\varepsilon} \lceil pfailstart \rceil$
(R2') $\lceil normal \wedge pfailstart \rceil \xrightarrow{\varepsilon} \lceil pfailinfo \wedge \neg normal \rceil$
(R3') $\lceil \neg pfail \rceil \,;\, \lceil pfailstart \wedge \neg normal \rceil \longrightarrow \lceil pfailstart \vee pfailinfo \rceil$
(R3'') $\lceil \neg pfail \rceil \,;\, \lceil pfailinfo \wedge \neg normal \rceil \longrightarrow \lceil pfailinfo \rceil$
(R4') $\lceil pfailinfo \rceil \,;\, \lceil \neg pfail \wedge (pon = flow) \rceil \longrightarrow \lceil \neg pfail \rceil$

For $j = 1, 4$ according to [2] we have

(*) $\bigwedge_i (S_i) \wedge R'_j \Rightarrow R_j$.

Similarly for $j = 3$ where we need $(R3')$ and $(R3'')$ as preconditions. In accordance with [2], for $j = 2$ we need the additional precondition

(R5') $\lceil \neg pfail \rceil \,;\, \lceil pfailstart \vee pfailinfo \wedge \neg normal \rceil \xrightarrow{\leq \varepsilon} \lceil pfailstart \vee pfailinfo \wedge \neg normal \rceil$.

The truth of $\bigwedge_i (S_i) \wedge R'_2 \wedge R'_5 \Rightarrow R_2$ follows from a case distinction. If $\lceil normal \wedge pfailstart \rceil$ lasts less than ε seconds we can argue with $(R5')$, otherwise with $(R2')$. In the absence of $(R5')$ we could not prevent a short $\lceil pfailstart \rceil$ phase (less than ε) followed by $\lceil pfailinfo \rceil$ without a change to $\lceil \neg normal \rceil$.

The introduction of $(R5')$ illustrates a principle reflection: This requirement cannot be proven, but must be claimed additionally. At first sight this might seem a disadvantage. It is justified, however, by the fact that on the most abstract level we describe a continuous system, whereas a computer is a discontinuous control mechanism. Hence we must rely on a not too chaotic behaviour of the environment. At some stage of the development this assumption must become relevant. A typical such situation is the introduction of (extremely short) intervals with an additional, previously unknown stability like $(R5')$.

4.2 Introduction of Events

Since the programming language does not speak about phases but about events, we must introduce events. They are the phase changes. So the next step is to name the phase changes. We shall use the names of the original tasks. But we have to introduce some more events, mainly internal synchronisation. The introduction of events happens as in the rules in [2] and [5]. In particular we need two sorts of events. In this section those events will be introduced which are communications with the environment. In the following section we shall deal with those events which involve synchronisation with other parallel components. For better readability we already here apply the notation for communication assertions from SL. Basically a communication assertion consists of several parts:

- the name of the communication (here the name of the event that changes the phase),
- a precondition which is necessary for the communication describing the phase before the change,
- a postcondition in Z-style showing the effect of the communication describing the phase after the change.

But we also have to declare the events as input or output. The decision which of the two must be chosen is simple. If a value is transmitted, this can be seen from the implementables, if no value is transmitted, the type of the communication is called *"signal"* which means that only some synchronisation is required. Then the decision is arbitrary.

For *Main* we get:

$$
\begin{array}{ll}
COM & Getpc \\
WHEN & phase = getpc \\
THEN & phase' = update \land pfail' = (pon \neq flow) \\
COM & Update \\
WHEN & phase = update \\
THEN & Q' = QE \land V' = VE \land pon' = nextp \\
& \land((safe \land phase' = wait) \lor (\neg safe \land phase' = emergency)) \\
COM & Wait \\
WHEN & phase = wait \\
THEN & phase' = getpc \\
COM & Emergency \\
WHEN & phase = emergency \\
THEN & phase' = emergency
\end{array}
$$

where VE, QE, $safe$ and $nextp$ are evaluated as explained in the previous section. The update phase has been split into two with a trailing continue action.

5 Removal of Synchronisation

In this section we deal with events that involve two different parallel components. As in the previous section the events will replace certain implementables. The difference here is that two components are involved. So the new event appears twice, in the specification of both components which communicate with each other. From the above mentioned schemas we have to transform the not yet mentioned synchronisation requirement for the failure handler. The interface of the main controller as far as this section is concerned is

```
input    pump_ok
output   pump_not_ok
```

The interface of the pump failure handler is correspondingly:

```
input    pump_not_ok
output   pump_ok
```

In the pump failure handler these events appear as

$$
\begin{array}{ll}
COM & pump\_not\_ok \\
WHEN & normal \\
THEN & \neg normal'
\end{array}
$$

COM   *pump_ok*
WHEN $\neg normal$
THEN $normal'$

In the main program they have the form

COM   *Getpc*
WHEN   *pfailinfo*
THEN $\neg pfailinfo'$
COM   *pump_not_ok*
WHEN $pfail \wedge \neg pfailstart \wedge \neg pfailinfo$
THEN $pfailstart'$
COM   *pump_ok*
WHEN   *pfailstart*
THEN $\neg pfailstart' \wedge pfailinfo'$

5.1 Introduction of Trace Assertions

When we introduced communication assertions we introduced the first pieces of SL syntax. What we have reached so far when describing events by their pre- and postconditions is a style which is close to action systems. Such a style is not appropriate for the transformation into a programming language which also contains constructs for a more structured programming. Without structuring the information we would end up in an unstructured WHILE loop. So we proceed as follows:

In many cases the way an observable attains different phase values is describable by a finite automaton. This information about the events can be encoded in a trace assertion, a regular expression which describes the automaton. This simplifies the already existing communication assertions and will help to give a structure to the full program. (For example "+" normally will turn into an alternative, "." into a sequential composition, etc.) Other information is retained in the communication assertions. In fact in many cases communication assertions must remain because the DC implementables can express more than regular languages. On the semantic level the interplay between trace assertions, communication assertions (and time conditions) is as follows: In an SL specification the trace assertions describe a superset of the set of possible sequences of the channel communications. At any point of time a communication via a channel ch is possible, iff three conditions hold.

- If the elapsed trace so far is tr, then $tr.ch$, the trace extended by ch, is in the language described by the trace assertions.
- The precondition of each communication assertion dealing with ch must be fulfilled.
- The communication must not be forbidden by the time conditions (see below).

For the main controller we decide to put most information into the traces using one trace assertion for the normal working. The communication assertions concerning the main controller from section 4.2 are shortened. Some of the event names are abbreviated.

TRACE
   pref $(Getpc.pinfo.Update.(Wait.repair + Emergency^*)^*$

with $pinfo = (<> + pump\_not\_ok)$ and $repair = (<> + pump\_ok)$.
The operator *pref* denotes the prefix closure and $<>$ the empty trace. The communication *pinfo* informs the failure handler about the state of the pump, if this is necessary. The communication *repair* can only be made during $phase = wait$ which is sensible because it is the only phase in which the main controller stays for more than a moment (ε time units). The failure handler is described by

TRACE pref $(pump\_not\_ok.pump\_ok)^*$

Information from the former events not concerning the normal working of the main controller remains in communication assertions:

COM $Getpc$
$WHEN$ $true$
$THEN$ $pfail' = (pon \neq flow) \wedge \neg pfailinfo'$
COM $pump\_not\_ok$
$WHEN$ $pfail \wedge \neg pfailstart$
$THEN$ $pfailstart'$
COM $Update$
$WHEN$ $true$
$THEN$ $Q' = QE \wedge V' = VE \wedge pon' = nextp$
         $\wedge((safe \wedge \neg emergency') \vee (\neg safe \wedge emergency'))$
COM $Wait$
$WHEN$ $\neg emergency$
COM $pump\_ok$
$WHEN$ $pfailstart$
$THEN$ $\neg pfailstart' \wedge pfailinfo'$
COM $Emergency$
$WHEN$ $emergency$

5.2 Introduction of Time Restrictions

In SL we distinguish between two forms of time restrictions: upper and lower bounds. The rules how to introduce them are complex, so we only describe the syntax: In both cases there is a condition given, when the restriction applies: A regular expression determines the set of those traces after which the restriction is valid. For example the following *lower bound* can be found in the specification of a timer, which causes a waiting period of about Δ time units. It is

AFTER $Wait.Wait^*$ WAIT $(Getpc, \Delta)$.

It applies when a trace from $Wait.Wait^*$ has elapsed and in this case prevents the event $GETPC$ for at least Δ, possibly for longer. In order to say that the timer must become ready for after $\Delta + \varepsilon$ we use the *upper bound* construct

AFTER $Wait.Wait^*$ READY $(Getpc, \Delta + \varepsilon)$.

The interplay between upper and lower bounds is particularly useful when communications shall not happen instantaneously.

The remaining time restrictions are trivial ones: After each event ev each follow up event which is possible by trace assertions and communication assertions must be ready immediately. In the semantics it is assumed that each channel is ready immediately, unless stated otherwise in the time restrictions. So these restrictions need not be stated explicitly in specifications.

6 Programs

The development of the final program can be done in two steps: the introduction of a recursion construct which is not part of the occam syntax and secondly the resolution of the recursion by means of standard techniques. Since the latter step does not contain anything new we skip it and simply show the complete final program. The semantics is the one from [6]. There, a set of rules is shown by which SL specifications can be transformed into programs. The following program is developed this way and therefore implies the SL specification. It is a parallel composition of sequential components of the form

```
SYSTEM Steam Boiler
  PAR
    Main
    PFH
END
```

The program for *Main* is

```
SYSTEM Main
  -- Interface
  CHANNEL OF signal
     pump_not_ok, pump_ok :

  VAR real
     Q,V,
     qmeas, vmeas :
  VAR BOOL
     emergency,
     pfail, pfailstart, pfailinfo,
     pon, flow :
  -- The variables qmeas, vmeas, pon, flow
  -- shall be PLACED on the I/O ports to
```

Applicative specifications. This style is similar to the one used in functional programming languages. Explicit functions describe the result of a function application.

A single specification can adopt different styles of specifications within the same module. Non functional requirements cannot be dealt directly within RSL: Requirements like timing and space constraints are instead taken into account during the RAISE development process.

The whole development process is supported by various tools, the most important one is a module editor used to create syntactic correct specifications. Furthermore it performs semantic checks (like type checking) and displays relevant error messages. Other available tools are:

- Justification tools: A collection of tools used for reasoning about specifications.
- Translators: Programming language code can be generated automatically from low-level designs for some languages (C++ and ADA).
- Document formatter: Printouts of all developed documents can be produced (by now only for typesetting using LaTeX).

This paper focuses mainly on creating a model based formal specification of the steam-boiler control in RSL notation.

2 Some Basic RSL Concepts

RSL allows to develop modular specifications, the main building blocks are modules. All declarations are made within class expressions, named class expressions are called schemes. Instantiations of class expressions, so called objects, can be created and they have at least all the properties of the scheme they are based on. References to other objects are made through contexts.

Each class expression contains several "slots" for defining different kind of entities:

Type. The concept of types in RSL is similar to other programming languages (except that RSL is higher order, that means that functions are treated like normal values). RSL has some atomic types (e.g. **Bool, Int, Real, Char**) and provides the possibility to define composite types to model more complex data structures. Examples of composite types are (for arbitrary types T1 and T2): Total functions (e.g. T1 $\rightarrow$ T2), finite sequences (e.g. T1*), sets (e.g. T1−**set**), variants (e.g. State == open | closed), ... Subtypes of existing types can be introduces as well, for example a set of all squared integers: **type** Sqr = $\{|\ i : \mathbf{Int} \bullet \exists\ j : \mathbf{Int} \bullet i = j * j\ |\}$.
But also abstract types without any further information can be defined by simply declaring their identifier (e.g. **type** Abs).

Value. Values of all types can be defined (atomic or composite). All types are treated in the same way, therefore also functions are defined as values. The user can choose if an implicit or explicit definition is appropriate.

Variable. A state can be introduced through the declaration of variables, normally variables are changed only by means of operations defined in the same module. Each operation has to declare read and write access for the part of the state it is using, all other parts are assumed to be unchanged. Although these kind of specifications are normally associated with imperative styles operations can also be defined implicitly or even algebraicly.

Axiom. Algebraic specifications of functions are normally used at an early stage of the development process together with abstract types. At this stage it is not necessary to introduce too many details of the used algorithms or data structures. Properties of operations are given in this slot.

Channel. Channels allow parallel processes to communicate with each other. A process can output a value on a channel, another process can read this value by accessing the same channel.

Scheme. RSL allows to create local class expressions within a class expressions and thus to build hierarchical specifications.

Object. Local instantiations of a class expression can be defined in this slot.

Class expressions can be parameterised by other class expressions. This allows generalised specifications and a flexible use of already existing modules (entities of existing modules can also be renamed or hidden).

Further explanations of the notation are given whenever it seems to be necessary during the following case study. A detailed description of all aspects of RSL is given in [1].

3 Formal Specification of the Boiler

Constructing a specification for the steam-boiler control means first of all understanding and modelling the problem domain. Therefore the first part of the specification deals with modelling the physical environment of the system: A mathematical model of the boiler and its environment is developed. Each component is examined, necessary parts of the state are introduced as variables.

Once this model of the system has been built it is then possible to formalise the dynamic behaviour of the steam-boiler.

This main text contains only some parts of the specification. Repetitive features that do not introduce new ideas are not shown. The whole specification showing all details is given in the appendix (see CD-ROM Annex S.B).

3.1 Static Components

In this part of the specification the physical environment of the steam-boiler is described. The whole system is split into various modules, these can later be combined to build the whole model.

The first scheme describes the boiler and all its physical characteristics together with predicates that have to be fulfilled by it. A subtype *Pump* that represents a means to identify an individual pump is also given. A single instantiation, the object *Boiler*, is declared based on this scheme. This object does not access entities declared in other objects (therefore an empty context):

```
1.0 context:
 .1
 .2 object
 .3   Boiler :
 .4     class
 .5       value
 .6         /* The maximal capacity of the boiler. */
 .7         C : Int,
 .8         /* The maximal steam−output. */
 .9         W : Int,
 .10        /* The allowed range of water in the boiler. */
 .11        N1, N2 : Int,
 .12        /* If the water level is in danger to leave one of these
 .13           ranges the program should enter the Emergency mode.
 .14        */
 .15        M1, M2 : Int,
 .16        /* Maximum gradient of increase/decrease of steam−output. */
 .17        U1, U2 : Int,
 .18        /* The cycle time. */
 .19        dt : Nat
 .20
 .21      axiom 0 < M1 ∧ M1 < N1 ∧ N1 < N2 ∧ N2 < M2 ∧ M2 < C
 .22
 .23      value
 .24        /* The number of pumps available. */
 .25        PumpCount : Nat
 .26
 .27      axiom
 .28        /* There must be at least one pump. */
 .29        PumpCount > 0
 .30
 .31      type
 .32        /* A new type for the identification of the pumps is
 .33           introduced.
 .34        */
 .35        Pump = {| i : Nat • i ∈ { 1 .. PumpCount } |}
 .36    end
```

The water level measurement device is one of the several gauges that take measurements of the physical environment. The device can also be defective: A variant type *ErrorState* has been defined to express this fact in the specification.

Variables that keep track of the state of the device are introduced (the current estimations of the water level and the current state). The object *WtrLvl* has to access entities that have been defined in the object *Boiler* and therefore has to list it in the context field. The entities defined in *Boiler* can now be accessed using dot notation.

```
2.0  context: Boiler
 .1
 .2  object
 .3    WtrLvl :
 .4      class
 .5        type
 .6          /* The water level measurement device can either be working,
 .7             broken or repairing.
 .8          */
 .9          ErrorState == working | broken | repairing
 .10
 .11       variable
 .12         /* The adjusted water level measurement. */
 .13         qa1 : Int := 0,
 .14         qa2 : Int := Boiler.C,
 .15         /* State of the water level measurement device. */
 .16         qst : ErrorState := working
 .17
 .18       axiom
 .19         /* Valid range of the adjusted water level measurements. */
 .20         qa1 ∈ { 0 .. Boiler.C } ∧ qa2 ∈ { 0 .. Boiler.C } ∧ qa1 ≤ qa2
 .21  end
```

The steam measurement device is modelled in a similar way to the water measurement device described above. Its formal description can be found in the appendix (see CD-ROM Annex S.B.1).

The state of the pumps are again modelled using variants, the state of each pump is modelled by a deterministic map from type *Pump* to type *PumpState*. For each pump information about its state has to be stored, the estimated amount of water poured into the boiler is recorded as well.

```
3.0  context: Boiler
 .1
 .2  object
 .3    Pump :
 .4      class
 .5        value
 .6          /* Nominal capacity of a pump. */
 .7          P : Int
```

```
.8
.9      type
.10        /* Possible states of a pump. */
.11        PumpState == open | closed | broken | repairing
.12
.13     variable
.14        /* State of the pumps. */
.15        pst : Boiler.Pump ↦ PumpState
.16           := [ p ↦ closed | p : Boiler.Pump ],
.17        /* Adjusted throughput of the pumps. */
.18        pa1 : Int := 0,
.19        pa2 : Int := P * Boiler.PumpCount
.20
.21     axiom
.22        /* For each pump exactly one state is stored. */
.23        ∀ p : Boiler.Pump • p ∈ dom (pst),
.24
.25        /* Valid range of the adjusted throughput of the pumps. */
.26        pa1 ∈ { 0 .. P * Boiler.PumpCount } ∧
.27        pa2 ∈ { 0 .. P * Boiler.PumpCount } ∧ pa1 ≤ pa2,
.28
.29        /* All non defective pumps are in the same state. */
.30        ∀ i, j : Boiler.Pump •
.31           pst(i) ∉ {broken, repairing} ∧ pst(j) ∉ {broken, repairing} ⇒
.32           pst(i) = pst(j)
.33
.34     value
.35        /* This function changes the state of a pump, which pump and
.36           the new state are given as arguments.
.37        */
.38        Set_Pump : Boiler.Pump × PumpState → write pst Unit
.39        Set_Pump(p, s) ≡ pst := pst † [p ↦ s]
.40     end
```

Pump control devices are modelled in a similar way to the approach used for the pumps above (see CD-ROM Annex S.B.1).

The informal specification describes that the steam-boiler can be in one of five modes: *Initialization, Normal, Degraded, Rescue,* or *Emergency*. During the initialisation phase the water level gets adjusted within the limits N_1 and N_2, several messages are also sent (e.g. *ProgramReady, PhysicalUnitsReady*). This makes the initialisation phase a complex task. To reduce the complexity four new modes (that replace the single *Initialization* mode) have been introduced: *Waiting, Emptying, Filling,* and *Ready*. To be consistent with the requirements expressed in the informal specification a separate type of modes for the communication with the environment has been defined (*MessageMode*). The actual state is stored in a variable:

```
4.0  context: Valve
 .1
 .2  object
 .3    Mode :
 .4      class
 .5        type
 .6          /* Extended modes which are used to describe the actual mode
 .7             of the system.
 .8          */
 .9          SystemMode ==
.10            Waiting |
.11            Emptying |
.12            Filling |
.13            Ready |
.14            Normal |
.15            Degraded |
.16            Rescue |
.17            Emergency,
.18          /* Modes as described in the informal specification. Only
.19             used for communication with the environment.
.20          */
.21          MessageMode ==
.22            Initialization | Normal | Degraded | Rescue | Emergency
.23
.24      variable
.25        /* The operating mode of the program. */
.26        mode : SystemMode := Waiting,
.27        /* The number of STOP−messages received so far in a row. */
.28        stop_count : Nat := 0
.29
.30      axiom
.31        /* If the program receives a STOP−message three times in a
.32           row the program enters the Emergency mode.
.33        */
.34        stop_count ≥ 3 ⇒ mode = Emergency,
.35
.36        /* The valve can only be opened during the initialisation of
.37           the system.
.38        */
.39        Valve.valve = Valve.open ⇒
.40          mode ∈ {Waiting, Emptying, Filling, Ready}
.41    end
```

The control program communicates with the environment via a message passing protocol. All messages that can be received by the program are described in the next scheme. The set of all messages received during the last cycle are stored, the function *Inmess_Ok* is responsible for detecting transmission failures:

```
5.0  context: WtrLvl, StmOut, Pump, PumpCtr, Mode
 .1
 .2  object
 .3    InMsg :
 .4      class
 .5        type
 .6          /* All possible messages which can be received by the program. */
 .7          InMessage ==
 .8            Stop |
 .9            SteamBoilerWaiting |
.10            PhysicalUnitsReady |
.11            PumpState
.12              (
.13                Boiler.Pump,
.14                {| s : Pump.PumpState • s ∈ {Pump.open, Pump.closed} |}
.15              ) |
.16            PumpCtrState
.17              (
.18                Boiler.Pump,
.19                {| s : PumpCtr.MonitorState •
.20                  s ∈ {PumpCtr.flow, PumpCtr.noflow}
.21                |}
.22              ) |
.23            Level(Int) |
.24            Steam(Int) |
.25            PumpRep(Boiler.Pump) |
.26            PumpCtrRep(Boiler.Pump) |
.27            LevelRep |
.28            SteamRep |
.29            PumpFlrAck(Boiler.Pump) |
.30            PumpCtrFlrAck(Boiler.Pump) |
.31            LevelFlrAck |
.32            SteamFlrAck,
.33          InMessages = InMessage−set
.34
.35        variable
.36          /* Set of messages received at the beginning of the actual
.37             cycle.
.38          */
.39          inmess : InMessages := {}
```

.40
.41   **value**
.42   /* This function is responsible for detecting transmission
.43   failures.
.44   */
.45   Inmess_Ok :
.46   **Unit** →
.47   **read** inmess, Mode.mode, Pump.pst, PumpCtr.mst,
.48   WtrLvl.qst, StmOut.vst
.49   **Bool**
.50   Inmess_Ok() ≡
.51   (
.52   ∀ p : Boiler.Pump •
.53   (∃! s : Pump.PumpState • PumpState(p, s) ∈ inmess) ∧
.54   (
.55   ∃! s : PumpCtr.MonitorState •
.56   PumpCtrState(p, s) ∈ inmess
.57   )
.58   ) ∧
.59   (∃! l : **Int** • Level(l) ∈ inmess) ∧
.60   **let** l : **Int** • Level(l) ∈ inmess **in** l ∈ { 0 .. Boiler.C } **end** ∧
.61   (∃! l : **Int** • Steam(l) ∈ inmess) ∧
.62   **let** l : **Int** • Steam(l) ∈ inmess **in** l ∈ { 0 .. Boiler.W } **end** ∧
.63   (SteamBoilerWaiting ∈ inmess ⇒ Mode.mode = Mode.Waiting) ∧
.64   (PhysicalUnitsReady ∈ inmess ⇒ Mode.mode = Mode.Ready) ∧
.65   (
.66   ∀ p : Boiler.Pump •
.67   (PumpRep(p) ∈ inmess ⇒ Pump.pst(p) = Pump.repairing) ∧
.68   (
.69   PumpCtrRep(p) ∈ inmess ⇒
.70   PumpCtr.mst(p) = PumpCtr.repairing
.71   ) ∧
.72   (PumpFlrAck(p) ∈ inmess ⇒ Pump.pst(p) = Pump.broken) ∧
.73   (
.74   PumpCtrFlrAck(p) ∈ inmess ⇒
.75   PumpCtr.mst(p) = PumpCtr.broken
.76   )
.77   ) ∧
.78   (LevelRep ∈ inmess ⇒ WtrLvl.qst = WtrLvl.repairing) ∧
.79   (SteamRep ∈ inmess ⇒ StmOut.vst = StmOut.repairing) ∧
.80   (LevelFlrAck ∈ inmess ⇒ WtrLvl.qst = WtrLvl.broken) ∧
.81   (SteamFlrAck ∈ inmess ⇒ StmOut.vst = StmOut.broken)
.82   **end**

Messages sent by the program are handled in the same way and the formal description is given in the appendix (see CD-ROM Annex S.B.1). The two remaining parts of the environment, the valve and the operator desk, are very simple and also only shown in the appendix (see CD-ROM Annex S.B.1).

Summary of the Static Components. So far all static components of the system introducing all necessary state variables have been described. Using this model the dynamic behaviour of the system can now be described.

3.2 Dynamic Behaviour

This part of the specification formalises all the actions that have to be taken to regulate the level of water in the steam-boiler. This contains functions for

- updating the water level,
- updating the steam output,
- updating the pump and pump controller states, and
- updating the program mode

as well as making a decision how to regulate the water level and how to control the whole system. All these operations are again specified in different RSL modules, at the end the main control operation for the steam-boiler is constructed from these modules.

As described in [2] new measurements are taken and transmitted to the control program by the gauges every dt seconds: Every dt seconds an update cycle of the whole model has to be performed. This cycle consists of

1. Reading the messages from the environment.
2. Makeing a decision which messages to send to the environment.
3. Sending messages to the environment.

According to [2] the time for transmitting the messages can be neglected. The same assumption has been made for the second task (this assumption seems to be reasonable but will have to be proved for a implementation of the control program).

Lets start by specifying the actions that have to be taken for updating the water level measurements: Based on the water level measured during the last cycle the program calculates an upper and lower bound on the new water level. This calculation is based on the physical behaviour of the system. If the program recognises that the received water level is not between these two boundaries the program considers the device to be broken. Therefore the state of the device is changed to *broken*, further calculations are based on the upper and lower bound calculated before. If the device is not considered to be broken (that means if the transmitted water level is between the boundaries) the program assumes that the device is working correctly and the transmitted water level is considered to be ok.

If the device recognises a failure of the water level measurement device an appropriate failure protocol has to be used (indeed this protocol is more or less the same for all the other devices and will be described only once):

- The state of the device is set to *broken*, the message *LevelFlrDet* will be sent to the physical units until the acknowledgement message *LevelFlrAck* has been received by the program.
- If the acknowledgement message has been received by the program the state of the device is set to *repaired*. The program assumes the device still to be broken until the message *LevelRep* has been received.
- If the message *LevelRep* has been received the program assumes that the device has been repaired and the transmitted water level is correct. An acknowledgement message will be sent to the physical units and the new state will be *working*.

This failure protocol must not be omitted by the physical units, therefore only the program can detect a failure of the water level measurement device.

In RSL the requirements above can be formalised as follows:

```
6.0   context: WtrLvl, InMsg, OutMsg, Functions
 .1
 .2   object
 .3     UWtrLvl :
 .4       class
 .5         value
 .6           /* Calculate new estimations for an upper and lower bound
 .7              of the boiler contents and detect possible errors of the
 .8              device.
 .9           */
.10           Update_Level :
.11             Unit →
.12               read WtrLvl.qa1, WtrLvl.qa2, WtrLvl.qst, InMsg.inmess,
.13                 StmOut.va1, StmOut.va2, Pump.pa1, Pump.pa2
.14               write WtrLvl.qa1, WtrLvl.qa2, WtrLvl.qst,
.15                 OutMsg.outmess_new
.16               Unit
.17           Update_Level() ≡
.18             let
.19               qc1 =
.20                 Functions.Max
.21                   (
.22                     {
.23                       0,
.24                       WtrLvl.qa1 − StmOut.va2 ∗ Boiler.dt −
.25                       1 / 2 ∗ Boiler.U1 ∗ Boiler.dt ∗ Boiler.dt +
.26                       Pump.pa1 ∗ Boiler.dt
.27                     }
.28                   ),
.29               qc2 =
.30                 Functions.Min
```

```
.31                  (
.32                    {
.33                      Boiler.C,
.34                      WtrLvl.qa2 − StmOut.val * Boiler.dt +
.35                      1 / 2 * Boiler.U2 * Boiler.dt * Boiler.dt +
.36                      Pump.pa2 * Boiler.dt
.37                    }
.38                  ),
.39            q : Int • InMsg.Level(q) ∈ InMsg.inmess
.40          in
.41            case WtrLvl.qst of
.42              WtrLvl.working →
.43                if q ∈ { qc1 .. qc2 } then
.44                  WtrLvl.qa1 := q ; WtrLvl.qa2 := q
.45                else
.46                  WtrLvl.qa1 := qc1 ;
.47                  WtrLvl.qa2 := qc2 ;
.48                  OutMsg.Add_Outmess(OutMsg.LevelFlrDet) ;
.49                  WtrLvl.qst := WtrLvl.broken
.50                end,
.51              WtrLvl.broken →
.52                if InMsg.LevelFlrAck ∈ InMsg.inmess then
.53                  WtrLvl.qst := WtrLvl.repairing
.54                else
.55                  OutMsg.Add_Outmess(OutMsg.LevelFlrDet)
.56                end ;
.57                WtrLvl.qa1 := qc1 ; WtrLvl.qa2 := qc2,
.58              WtrLvl.repairing →
.59                if InMsg.LevelRep ∈ InMsg.inmess then
.60                  OutMsg.Add_Outmess(OutMsg.LevelRepAck) ;
.61                  WtrLvl.qa1 := q ;
.62                  WtrLvl.qa2 := q ; WtrLvl.qst := WtrLvl.working
.63                else
.64                  WtrLvl.qa1 := qc1 ; WtrLvl.qa2 := qc2
.65                end
.66            end
.67          end
.68    end
```

Updating the state of the pumps can be done in a similar way. But this time the detection of a pump failure is more difficult because messages that have been sent during the last two cycles have to be analysed in order to find an incorrect behaviour of a pump. A pump failure is detected if at least one of the following events takes place:

- The message *OpenPump* has been sent to a pump during the last cycle but the pump does not indicate that its state is *open*.
- The message *ClosePump* has been sent to a pump during the last cycle but the pump does not report that its state is now *closed*.
- Neither the message *OpenPump* nor the message *ClosePump* has been sent to the pump during the last cycle but its state has changed spontaneously.

```
7.0  context: Pump, InMsg, OutMsg
 .1
 .2  object
 .3    UPump :
 .4      hide Update_Pump in
 .5        class
 .6          value
 .7            /* Detect if a single pump is still working correctly. */
 .8            Update_Pump :
 .9              Boiler.Pump →
.10                read Pump.pst, OutMsg.outmess, InMsg.inmess, OutMsg.log
.11                write Pump.pst, OutMsg.outmess_new
.12                Unit
.13            Update_Pump(p) ≡
.14              let
.15                s : Pump.PumpState •
.16                  InMsg.PumpState(p, s) ∈ InMsg.inmess
.17              in
.18                if
.19                  (
.20                    OutMsg.OpenPump(p) ∈ OutMsg.outmess ∧
.21                    InMsg.PumpState(p, Pump.open) ∉ InMsg.inmess
.22                  ) ∨
.23                  (
.24                    OutMsg.ClosePump(p) ∈ OutMsg.outmess ∧
.25                    InMsg.PumpState(p, Pump.closed) ∉ InMsg.inmess
.26                  ) ∨
.27                  (
.28                    OutMsg.ClosePump(p) ∉ OutMsg.outmess ∧
.29                    OutMsg.OpenPump(p) ∉ OutMsg.outmess ∧
.30                    s ≠ Pump.pst(p)
.31                  )
.32                then
.33                  Pump.Set_Pump(p, Pump.broken) ;
.34                  OutMsg.Add_Outmess(OutMsg.PumpFlrDet(p))
.35                else
.36                  Pump.Set_Pump(p, s)
.37              end
```

```
.38              end,
.39
.40        /* Update the state of each pump and perform if necessary
.41           the required failure protocol.
.42        */
.43        Update_Pumps :
.44          Unit →
.45            read Pump.pst, InMsg.inmess, OutMsg.outmess, OutMsg.log
.46            write Pump.pst, OutMsg.outmess_new
.47          Unit
.48        Update_Pumps() ≡
.49          for p in ⟨ 1 .. Boiler.PumpCount ⟩ do
.50            let
.51              s : Pump.PumpState •
.52                InMsg.PumpState(p, s) ∈ InMsg.inmess
.53            in
.54              case Pump.pst(p) of
.55                Pump.closed → Update_Pump(p),
.56                Pump.open → Update_Pump(p),
.57                Pump.broken →
.58                  if InMsg.PumpFlrAck(p) ∈ InMsg.inmess then
.59                    Pump.Set_Pump(p, Pump.repairing)
.60                  else
.61                    OutMsg.Add_Outmess(OutMsg.PumpFlrDet(p))
.62                  end,
.63                Pump.repairing →
.64                  if InMsg.PumpRep(p) ∈ InMsg.inmess then
.65                    Pump.Set_Pump(p, s) ;
.66                    OutMsg.Add_Outmess(OutMsg.PumpRepAck(p))
.67                  end
.68              end
.69            end
.70          end
.71        end
```

Similar update operations as described for the water level measurement device and the pumps have also to be performed for the other devices (the steam output measurement device and the pump controller). The RSL specification for these parts are given in the appendix (see CD-ROM Annex S.B.2).

Once all the state variables have been updated a regulation criteria has to be used to maintain the water level between the limits N_1 and N_2. This decision can be made based on the hints given in [2]. Formalising this strategy in RSL yields the following module:

8.0 **context**: WtrLvl, StmOut, Pump, OutMsg, Functions
.1
.2 **object**
.3   Regulate :
.4   **class**
.5   **value**
.6   /* Make a decision how the water level can be kept between
.7   the limits N1 and N2 (if possible).
.8   */
.9   Regulate_Level :
.10   **Unit** $\to$
.11   **read** WtrLvl.qa1, WtrLvl.qa2, StmOut.va1, StmOut.va2,
.12   Pump.pa1, Pump.pa2
.13   **write** Mode.mode, OutMsg.outmess_new
.14   **Unit**
.15   Regulate_Level() $\equiv$
.16   **let**
.17   qc1 =
.18   Functions.Max
.19   (
.20   {
.21   0,
.22   WtrLvl.qa1 $-$ StmOut.va2 $*$ Boiler.dt $-$
.23   1 / 2 $*$ Boiler.U1 $*$ Boiler.dt $*$ Boiler.dt $+$
.24   Pump.pa1 $*$ Boiler.dt
.25   }
.26   ),
.27   qc2 =
.28   Functions.Min
.29   (
.30   {
.31   Boiler.C,
.32   WtrLvl.qa2 $-$ StmOut.va1 $*$ Boiler.dt $+$
.33   1 / 2 $*$ Boiler.U2 $*$ Boiler.dt $*$ Boiler.dt $+$
.34   Pump.pa2 $*$ Boiler.dt
.35   }
.36   )
.37   **in**
.38   **if**
.39   WtrLvl.qa1 $\leq$ Boiler.M1 $\vee$
.40   WtrLvl.qa2 $\geq$ Boiler.M2 $\vee$
.41   qc1 $\leq$ Boiler.M1 $\vee$ qc2 $\geq$ Boiler.M2
.42   **then**
.43   Mode.mode := Mode.Emergency ;
.44   OutMsg.Add_Outmess(OutMsg.State(Mode.Emergency))
.45   **else**

```
.46            let
.47              low = { 0 .. Boiler.N1 − 1 },
.48              ok = { Boiler.N1 .. Boiler.N2 },
.49              high = { Boiler.N2 + 1 .. Boiler.C }
.50            in
.51              if WtrLvl.qa1 ∈ low ∧ WtrLvl.qa2 ∈ low then
.52                OutMsg.Open_Pumps()
.53              elsif WtrLvl.qa1 ∈ low ∧ WtrLvl.qa2 ∈ ok then
.54                OutMsg.Open_Pumps()
.55              elsif WtrLvl.qa1 ∈ low ∧ WtrLvl.qa2 ∈ high then
.56                Mode.mode := Mode.Emergency ;
.57                OutMsg.Add_Outmess(OutMsg.State(Mode.Emergency))
.58              elsif WtrLvl.qa1 ∈ ok ∧ WtrLvl.qa2 ∈ ok then
.59                skip
.60              elsif WtrLvl.qa1 ∈ ok ∧ WtrLvl.qa2 ∈ high then
.61                OutMsg.Close_Pumps()
.62              else
.63                OutMsg.Close_Pumps()
.64              end
.65            end
.66          end
.67        end
.68  end
```

The next scheme summarises the overall control of the program. First the function is responsible for detecting major failures of the system which cause the program to enter the *Emergency* mode. This can occur due to one of the following reasons:

- The program receives three *Stop* messages in a row.
- The function responsible for detecting transmission failures (described during the specification of the possible messages received by the program) reports that a failure during the transmission occurred.
- Due to intervention from the operator desk the program should terminate (immediate shutdown at the beginning of the next cycle).

If the program enters the *Emergency* mode because of one of these conditions the function terminates, otherwise dependent on the actual operating mode appropriate control actions are taken. One of the following events can occur:

- The program is in one of the initialization modes and has to initialise the boiler according to the protocol described in the informal specification:
  - *Waiting*: The control program is waiting to receive the message *SteamBoilerWaiting* from the environment. As soon as this message has been received the program takes appropriate actions to get the water level between the limits N_1 and N_2 (either by opening the valve (new mode will be *Emptying*) or opening the pumps (new mode will be *Filling*)). If the

water level is already within these limits the control program changes into mode *Ready*.
- *Filling/Emptying*: Once the water level has reached an appropriate level between the two bounds the pumps are switched of (respectively the valve is closed) and the control program is now waiting to take over control over the environment. The control program changes into the mode *Ready*.
- *Ready*: In this mode the control problem is waiting for the message *PhysicalUnitsReady* to be sent from the environment. Once this message has been received the control program takes over control over the environment and starts maintaining the water lever between the limits N_1 and N_2.
- The control program is in one of the control states (*Normal, Rescue* or *Degraded*) and tries to maintain the water level between the limits N_1 and N_2: All variables for the physical devices have to be updated according to the received messages, a decision how to regulate the water level has to be made, and the new operating mode has to be calculated.

In the following modules only some parts of the specification are shown. The formal descriptions of the functions *Control_Filling, Control_Emptying*, and *Control_Ready* are only given in the appendix (see CD-ROM Annex S.B.2):

```
9.0   context: Valve, OpDesk, UWtrLvl, UStmOut, UPump, UPumpCtr,
 .1           UThrough, Regulate, UState
 .2
 .3   object
 .4     Control :
 .5     hide
 .6        Control_Normal, Control_Waiting, Control_Filling,
 .7           Control_Emptying, Control_Ready
 .8     in
 .9     class
.10        value
.11         Control_Waiting :
.12           Unit →
.13             read WtrLvl.qa1, WtrLvl.qa2, StmOut.va1, StmOut.va2,
.14               Pump.pa1, Pump.pa2, WtrLvl.qst, StmOut.vst, Pump.pst,
.15               PumpCtr.mst, OutMsg.outmess, OutMsg.log, Mode.mode,
.16               InMsg.inmess
.17             write WtrLvl.qa1, WtrLvl.qa2, StmOut.va1, StmOut.va2,
.18               Pump.pa1, Pump.pa2, WtrLvl.qst, StmOut.vst, Pump.pst,
.19               PumpCtr.mst, Mode.mode, OutMsg.outmess_new,
.20               Valve.valve
.21             Unit
.22         Control_Waiting() ≡
.23           let
```

```
.24            q : Int • InMsg.Level(q) ∈ InMsg.inmess,
.25            v : Int • InMsg.Steam(v) ∈ InMsg.inmess
.26         in
.27            UPump.Update_Pumps() ;
.28            UPumpCtr.Update_Pump_Controls() ;
.29            UThrough.Update_Throughput() ;
.30            if InMsg.SteamBoilerWaiting ∈ InMsg.inmess then
.31               if v ≠ 0 then
.32                  StmOut.vst := StmOut.broken ;
.33                  OutMsg.Add_Outmess(OutMsg.SteamFlrDet) ;
.34                  Mode.mode := Mode.Emergency ;
.35                  OutMsg.Add_Outmess(OutMsg.State(Mode.Emergency))
.36               else
.37                  StmOut.va1 := v ;
.38                  StmOut.va2 := v ;
.39                  if q > Boiler.N2 then
.40                     Valve.valve := Valve.open ;
.41                     OutMsg.Add_Outmess(OutMsg.Valve) ;
.42                     Mode.mode := Mode.Emptying
.43                  elsif q < Boiler.N1 then
.44                     OutMsg.Open_Pumps() ; Mode.mode := Mode.Filling
.45                  else
.46                     Mode.mode := Mode.Ready
.47                  end ;
.48                  OutMsg.Add_Outmess
.49                     (OutMsg.State(Mode.Initialization)) ;
.50                  WtrLvl.qa1 := q ; WtrLvl.qa2 := q
.51               end
.52            end
.53         end,
.54
.55      Control_Filling :
.56         Unit →
.57            read WtrLvl.qa1, WtrLvl.qa2, StmOut.va1, StmOut.va2,
.58               Pump.pa1, Pump.pa2, WtrLvl.qst, StmOut.vst, Pump.pst,
.59               PumpCtr.mst, OutMsg.outmess, OutMsg.log, Mode.mode,
.60               InMsg.inmess
.61            write WtrLvl.qa1, WtrLvl.qa2, StmOut.va1, StmOut.va2,
.62               Pump.pa1, Pump.pa2, WtrLvl.qst, StmOut.vst, Pump.pst,
.63               PumpCtr.mst, Mode.mode, OutMsg.outmess_new,
.64               Valve.valve
.65            Unit,
.66
.67      Control_Emptying :
.68         Unit →
```

```
.69         read WtrLvl.qa1, WtrLvl.qa2, StmOut.va1, StmOut.va2,
.70             Pump.pa1, Pump.pa2, WtrLvl.qst, StmOut.vst, Pump.pst,
.71             PumpCtr.mst, OutMsg.outmess, OutMsg.log, Mode.mode,
.72             InMsg.inmess
.73         write WtrLvl.qa1, WtrLvl.qa2, StmOut.va1, StmOut.va2,
.74             Pump.pa1, Pump.pa2, WtrLvl.qst, StmOut.vst, Pump.pst,
.75             PumpCtr.mst, Mode.mode, OutMsg.outmess_new,
.76             Valve.valve
.77         Unit,
.78
.79       Control_Ready :
.80         Unit →
.81           read Pump.pa1, Pump.pa2, Pump.pst, PumpCtr.mst,
.82             WtrLvl.qst, StmOut.vst, OutMsg.outmess, OutMsg.log,
.83             Mode.mode, InMsg.inmess
.84           write Pump.pa1, Pump.pa2, Pump.pst, PumpCtr.mst,
.85             Mode.mode, OutMsg.outmess_new
.86         Unit,
.87
.88       Control_Normal :
.89         Unit →
.90           read WtrLvl.qa1, WtrLvl.qa2, StmOut.va1, StmOut.va2,
.91             Pump.pa1, Pump.pa2, WtrLvl.qst, StmOut.vst, Pump.pst,
.92             PumpCtr.mst, InMsg.inmess, OutMsg.outmess, OutMsg.log
.93           write WtrLvl.qa1, WtrLvl.qa2, StmOut.va1, StmOut.va2,
.94             Pump.pa1, Pump.pa2, WtrLvl.qst, StmOut.vst, Pump.pst,
.95             PumpCtr.mst, Mode.mode, OutMsg.outmess_new
.96         Unit
.97       Control_Normal() ≡
.98         UWtrLvl.Update_Level() ;
.99         UStmOut.Update_Steam() ;
.100        UPump.Update_Pumps() ;
.101        UPumpCtr.Update_Pump_Controls() ;
.102        UThrough.Update_Throughput() ;
.103        Regulate.Regulate_Level() ; UState.Update_State(),
.104
.105      Control :
.106        Unit →
.107          read WtrLvl.qa1, WtrLvl.qa2, StmOut.va1, StmOut.va2,
.108            Pump.pa1, Pump.pa2, WtrLvl.qst, StmOut.vst, Pump.pst,
.109            PumpCtr.mst, OutMsg.outmess, OutMsg.log, Mode.mode,
.110            InMsg.inmess, Mode.stop_count, OpDesk.shutdown
.111          write WtrLvl.qa1, WtrLvl.qa2, StmOut.va1, StmOut.va2,
.112            Pump.pa1, Pump.pa2, WtrLvl.qst, StmOut.vst, Pump.pst,
.113            PumpCtr.mst, Mode.mode, OutMsg.outmess_new,
```

```
.114                    Valve.valve, Mode.stop_count
.115                 Unit
.116              Control() ≡
.117                 if InMsg.Stop ∈ InMsg.inmess then
.118                    Mode.stop_count := Mode.stop_count + 1
.119                 else
.120                    Mode.stop_count := 0
.121                 end ;
.122                 if
.123                    OpDesk.shutdown ∨
.124                    Mode.stop_count = 3 ∨ ∼ InMsg.Inmess_Ok()
.125                 then
.126                    Mode.mode := Mode.Emergency ;
.127                    OutMsg.Add_Outmess(OutMsg.State(Mode.Emergency))
.128                 else
.129                    case Mode.mode of
.130                       Mode.Waiting → Control_Waiting(),
.131                       Mode.Filling → Control_Filling(),
.132                       Mode.Emptying → Control_Emptying(),
.133                       Mode.Ready → Control_Ready(),
.134                       _ → Control_Normal()
.135                    end
.136                 end
.137           end
```

Now all necessary modules have been defined in order to construct the whole control program. But still some parts have to be considered more carefully.

The program exchanges messages with the physical units with the help of two channels, namely *input* and *output*. These channels have the abstract types *Input* and *Output*. How messages are transmitted across these channels (how the set of messages are converted to be transmitted across a channel) will not be further discussed – this should be made during the implementation of the program. Only the signatures of two functions *Transform_In* and *Transform_Out* are given for this purpose.

Each cycle therefore consists of the following actions:

– Read new messages from the physical units.
– Based on the received messages handle all possible failures and regulate the water level.
– Send new messages to the physical units (do not forget to remember the two last sent message sets).
– Wait until the start of the next cycle.

These actions are repeatedly executed until the program enters the mode *Emergency*. After the mode *Emergency* occurred the physical devices are left on their own and the program terminates.

10.0 **context**: Control
.1
.2 **object**
.3   SteamBoiler :
.4   **class**
.5   **type**
.6   /* An abstract type for receiving information over a channel. */
.7   Input,
.8   /* An abstract type for sending information over a channel. */
.9   Output
.10
.11   **channel**
.12   /* A channel for receiving information from the environment. */
.13   input : Input,
.14   /* A channel for sending information to the environment. */
.15   output : Output
.16
.17   **value**
.18   /* This (not further specified) function converts the
.19   information from the environment (received over a channel)
.20   to an appropriate set of messages. It also returns a flag:
.21   true if the conversion has been successfully, otherwise
.22   false.
.23   */
.24   Transform_In : Input $\to$ InMsg.InMessages $\times$ **Bool**,
.25
.26   /* This (not further specified) function converts the
.27   information stored in a message−set to an appropriate
.28   representation to be sent over a channel.
.29   */
.30   Transform_Out : OutMsg.OutMessages $\to$ Output,
.31
.32   /* This (also not further specified) function waits for the
.33   amount of seconds given as an argument.
.34   */
.35   Wait_Time : **Nat** $\to$ **Unit**,
.36
.37   /* This function reads information from the input−channel and
.38   stores the information in the set of received messages. It
.39   returns true if the transmission has been successful,
.40   otherwise false.
.41   */
.42   Read_Input : **Unit** $\to$ **write** InMsg.inmess **in** input **Bool**
.43   Read_Input() $\equiv$
.44   **let**

```
.45          new_input = input?,
.46          (new_mess, state) = Transform_In(new_input)
.47        in
.48          InMsg.inmess := new_mess ; state
.49        end,
.50
.51     /* This function is responsible for sending information to the
.52        environment, it also maintains the history of the messages
.53        sent so far.
.54     */
.55     Write_Output :
.56        Unit →
.57          write OutMsg.log, OutMsg.outmess, OutMsg.outmess_new
.58          out output
.59          Unit
.60     Write_Output() ≡
.61        let new_out = Transform_Out(OutMsg.outmess_new) in
.62          output ! new_out ;
.63          OutMsg.log := OutMsg.outmess ;
.64          OutMsg.outmess := OutMsg.outmess_new ;
.65          OutMsg.outmess_new := {}
.66        end,
.67
.68     /* This function represents the main function of the system.
.69        The function always calls itself recursively until the
.70        Emergency mode is entered.
.71     */
.72     Steam_Boiler :
.73        Unit →
.74          read WtrLvl.qa1, WtrLvl.qa2, StmOut.va1, StmOut.va2,
.75             Pump.pa1, Pump.pa2, WtrLvl.qst, StmOut.vst, Pump.pst,
.76             PumpCtr.mst, OutMsg.outmess, OutMsg.log, Mode.mode,
.77             InMsg.inmess, Mode.stop_count, OpDesk.shutdown
.78          write WtrLvl.qa1, WtrLvl.qa2, StmOut.va1, StmOut.va2,
.79             Pump.pa1, Pump.pa2, WtrLvl.qst, StmOut.vst, Pump.pst,
.80             PumpCtr.mst, Mode.mode, OutMsg.outmess_new,
.81             OutMsg.outmess, OutMsg.log, InMsg.inmess, Valve.valve,
.82             Mode.stop_count
.83          in input
.84          out output
.85          Unit
.86     Steam_Boiler() ≡
.87        if ∼ Read_Input() then
.88          Mode.mode := Mode.Emergency ;
.89          OutMsg.outmess_new := {OutMsg.State(Mode.Emergency)} ;
```

```
.90             Write_Output()
.91          else
.92             Control.Control() ;
.93             Write_Output() ;
.94             if Mode.mode = Mode.Emergency then
.95                skip
.96             else
.97                Wait_Time(Boiler.dt) ; Steam_Boiler()
.98             end
.99          end
.100      end
```

3.3 Loose Ends and Discussion

Throughout the development of the specification various question concerning information given in [2] have been raised. In this section I will describe assumptions which have been made in order to develop this specification and aspects of the informal specification which could lead to major problems during the execution of the control program:

- The initial values of the components of the system have not been described in [2]. Therefore states that seem to be plausible have been chosen.
- It is not described in which quantities the measures are taken. In the specification all values are stored in integer variables. A sufficient mapping to the real representation can be used to receive a necessary accuracy.
- [2] gives no information concerning the regulation strategy that should be used. In the specification the most simple way has been chosen: Either all pumps are closed or open. Other more sophisticated rules (for example using the actual amount of steam exiting the boiler) are also possible.
- "Three *Stop* messages in a row" are assumed to be received if the message sets received during three consecutive cycles each contain a *Stop* signal.
- It is assumed that the stop signal from the operator desk should be handled in the same way as normal messages received by the system. Therefore only at the beginning of a cycle the program can enter the *Emergency* mode.
- The informal specification describes that "The program follows a cycle and a priori does not terminate." and "Once the program has reached the *Emergency* mode [...] the program terminates.". Therefore the following actions could be taken once the program enters the *Emergency* mode:
    1. The program sends the message *Emergency* to the environment and terminates.
    2. The program does not terminate and sends the message *Emergency* every dt seconds.

  Both strategies lead to different behaviour of the control program, in this specification I have chosen to implement the first strategy.
- What should be done if necessary acknowledgement messages for the failure protocol are not received by the program? This can for example happen because of one of the following reasons:

1. The failure message sent by the program has been lost: This is a transmission failure!
2. The repairing of the device takes a long time and is not finished yet.

Which action should the program perform? Is a timeout mechanism appropriate?
- After a device has been repaired, in which state will the device continue to work?
- How can transmission failures during writing to the environment be detected?
- Should messages be sent to defective pumps?
- Is the delay of the pumps and the cycle time the same?

4 Conclusion

This contribution describes one way of handling specifications using RSL notation. It consists of a formal model of the environment together with a detailed description of the dynamic behaviour of the system. It has been shown in an appendix that RSL is also capable of handling specifications in a more abstract way then presented in the case study. But due to the given problem it has been very difficult for me to build a more abstract one for the steam-boiler.

5 Evaluation and Comparison

1. This contribution represents a detailed formal requirements specification of all parts of the steam-boiler control specification problem. The resulting model reflects the operational view of the informal specification at almost the same abstraction level. This has the drawback that it is difficult to proof properties of the resulting model.
2. Due to lack of time there is no executable implementation of the control program available.
3. It is difficult to find comparable solutions to this work. Almost all other solutions use a higher level of abstraction to specify the behaviour of the control program.
   Alternative solutions that are also based on a model based approach are [3], [4] and [7]. The specification languages used in these solutions are VDM-SL and Z, they focus more on creating an abstract specification of the system.
4. This work has been developed during summer and early autumn of 1995. It has been produced by a single person and I would estimate the effort spent with one person month.
   An average programmer will need training in a number of areas: It will not be sufficient to teach just the used notation. He also needs to get knowledge about the used tools and time to get some experience in applying the set of notation to new problems. A training course should at least last four to six weeks.

5. In order to understand the presented solution a knowledge of the used formalism is essential and enough. Therefore an average programmer without any experience in formal methods (and especially the notation used in this paper) will not be able to understand it. He will need a training period to gain the necessary skills – one or two weeks should be enough for this purpose.

References

1. The RAISE Language Group: The RAISE Specification Language, Prentice Hall, 1992
2. Jean-Raymond Abrial: Steam-boiler control specification problem, August 1994 (see Chapter AS, this book)
3. Pascal Bernard: A Z specification of the boiler, January 1996 (available online at http://www.informatik.uni-kiel.de/~procos/dag9523/bernard-fulltext.ps.Z)
4. Robert Buessow and Matthias Weber: A Steam-Boiler Control Specification with Statecharts and Z (see Chapter BW, this book)
5. Zoë Hellinger: RAISE Frequently Asked Questions, February 1995, (available online at http://dream.dai.ed.ac.uk/raise/faq.html)
6. Cliff B. Jones: Systematic Software Development using VDM, second edition, Prentice Hall, 1990
7. Yves Ledru and Marie-Laure Potet: A VDM specification of the steam-boiler problem (see Chapter LP, this book)
8. J.G. Turner and T.L. McCluskey: The Construction of Formal Specifications – An Introduction to the Model-Based and Algebraic Approaches, McGraw Hill, 1994

Assertional Specification and Verification Using PVS of the Steam Boiler Control System

Jan Vitt[1] and Jozef Hooman[2]

[1] Institut für Informatik und Praktische Mathematik
Christian-Albrechts-Universität zu Kiel, Germany
e-mail: jav@informatik.uni-kiel.d400.de
[2] Dept. of Mathematics and Computing Science
Eindhoven University of Technology, The Netherlands
e-mail: wsinjh@win.tue.nl

Abstract. An implementation of the steam boiler control system has been derived using a formal method based on assumption/commitment pairs. Intermediate stages of top-down design are represented in a mixed formalism where programs and assertional specifications are combined in a single framework. Design steps can be verified by means of compositional proof rules. This framework has been defined in the specification language of the verification system PVS. By the interactive proof checker of PVS, the correctness of each refinement step has been checked mechanically.

1 Introduction

The steam boiler control system, as described in chapter AS of this book, has been designed in an assertional framework. That is, the system and its components are described by listing their properties in a certain logic. The formalism used here is based on Hoare logic (precondition, program, postcondition), which has been extended and modified to deal with distributed real-time systems. Verification is supported by compositional proof rules, i.e. parts of the system can be considered as black boxes and only their specifications are used to verify design steps. To represent the intermediate stages of top-down design, we use a mixed formalism in which one can freely combine assertional specifications and constructs of the programming language.

The application of a formal method in the design process leads to a large number of verification conditions, including many trivial proof obligations. Therefore the tool PVS[3] (Prototype Verification System) [ORS92, ORSvH95] has been used to support the design. The approach presented here is based on [Hoo94] where our mixed framework has been formulated in terms of the PVS specification language, a strongly-typed higher-order logic. This enables us to verify the correctness of each refinement step by means of the interactive proof checker of PVS. Using the powerful decision procedures of PVS, simple verification conditions can be discharged automatically, allowing the user to concentrate on the essential structure of a proof.

---

[3] PVS is free available, see WWW page http://www.csl.sri.com/sri-csl-pvs.html

For simplicity, in this paper the pump system is reduced to a single pump and one pump controller. Furthermore, we assume that the initialization phase has been finished. The resulting steam boiler control system has been specified and an implementation has been derived formally, verifying all refinement steps by means of the PVS proof checker.

Our design of the control system comprises three stages, where the design in a later stage is based on the ones in earlier stages. In the first stage, only real-time aspects of the steam boiler control system are considered, assuming no failures occur. In the second stage, failures of communication channels are allowed and a control component that is able to detect such failures is derived. It is an extension of the real-time controller derived before. In the last stage, also failures of physical components are taken into account.

The first stage is described in Section 2 where a real-time control program for the steam boiler system is developed. The steam boiler system is a typical example of a hybrid system [GNRR93] with discrete and continuous components such as program controlled digital computers and continuous processes. Hence we could follow an approach which has been applied before [Hoo93, Hoo94, Hoo96] to derive a real-time control program for such systems.

The next stages, where failures of channels or components are allowed, are less straightforward, since we had not much experience with fault-tolerance in our framework. First, in Section 3, we consider failing channels that might loose messages. A communication axiom is modified and the design of the previous stage is adapted, making assumptions about the correctness of channels explicit. It is shown that a detectable failure of a channel leads to an alarm message.

Section 4 concerns failing components, formulating a failure hypothesis about the effect of faults and making all assumptions about the correctness of components explicit. A steam sensor is introduced to deal with a failing water level sensor.

An evaluation and comparison can be found in Section 5. In general, only the main outline of the PVS theories is described; see CD-ROM Annex VH for more details.

2 Real-Time Part of the Steam Boiler Control System

Section 2.1 contains a brief introduction on the basic parts of the mixed formalism in the notation of the PVS specification language. Then an implementation of a real-time control program is derived following the standard approach of earlier work [Hoo93, Hoo94, Hoo96].

- In Section 2.2 we give a formal description of the top-level specification of the complete steam boiler control system, i.e., control system plus continuous processes (such as inflow of water and outflow of steam). Further, relevant properties of the physical processes to be controlled are formalized. Next we describe a control strategy in terms of a continuous interface and prove that it leads to the top-level specification.

- In Section 2.3 we introduce a communication mechanism for parallel programs using asynchronous channels.
- This communication mechanism is used in Section 2.4 to refine the control system by introducing a pump system and a water sensor. They transform the continuous interface into discrete events. This leads to a specification of a discrete control program. The correctness of these design steps is also proved by means of PVS, using only the specifications of the components in terms of their real-time communication interface.
- A real-time programming language and a compositional proof system are presented in Section 2.5.
- In Section 2.6, the control component is implemented in this language.

2.1 Basic Framework

We give a brief introduction on the assertional framework in terms of the PVS specification language which is indicated by **typewriter** font.

Values, Time, and Programs Consider a value domain **Value**, which equals the built-in type of the real numbers, with subtype **PositiveValue**. Time domain **Time** equals **PositiveValue**, since the (positive) real numbers are convenient to model hybrid systems. Time intervals are represented as sets of time points, **setof[Time]**, which is equivalent to functions from **Time** to **bool**, **[Time -> bool]**. We use **co** for left-closed right-open intervals, etc. The standard PVS operators =, <=, ... on **real** are overloaded and also defined on predicates over **Time**. A predicate **pred[T]** of an arbitrary type **T** is a function **[T -> bool]**. We define when a time predicate **P** holds *inside* or *during* an interval **I**. Note the declaration of **t** as a variable of type **Time**. Without showing the declaration, we also use variables **v0,v1** of type **Value**, interval **I**, and predicate **P**.

```
Value          : TYPE = real
PositiveValue  : TYPE = { posval: Value | posval >= 0 }
Time           : TYPE = { r: real | r >= 0 }
Interval       : TYPE = setof[Time]

t              : VAR Time
co(v0,v1)      : Interval = { t | v0 <= t AND t < v1 }
cc(v0,v1)      : Interval = { t | v0 <= t AND t <= v1 }

inside(P,I)    : bool = (EXISTS t: I(t) AND P(t))
dur(P,I)       : bool = (FORALL t: I(t) IMPLIES P(t))
```

A real-time program is simply a relation on states, i.e., a function from pairs of states to **bool**. A state **s** of type **State** is a record, denoted by **[# ... #]**, consisting of three fields: **val(s)** gives the values of program variables, **now(s)** records the current time, and **term(s)** indicates termination. Let **Vars** denote a set of program variables (in fact, this set is a parameter of the theory). A parallel composition operator on programs is introduced.

```
State   : TYPE = [# val: [Vars -> Value], now: Time, term: bool #]
program: TYPE = pred[State, State]

prog1, prog2: VAR program
par(prog1, prog2) : program              % parallel composition
```

Specifications Assertions, used to specify real-time systems, are predicates over states and a notion of validity is defined. To support the mixed approach, a specification is also considered as a program, i.e., a relation on states. A specification is a pair (A,C) with the meaning that if the initial state satisfies assumption A then the final state should satisfy commitment C.

```
A, C         : VAR pred[State]
Valid(A)   : bool    = (FORALL s: A(s))
spec(A,C) : program = (LAMBDA s0,s1: A(s0) IMPLIES C(s1))
```

For refinement the infix operator => is overloaded and shown to be transitive. E.g., **prog** => **spec(A,C)** expresses that program **prog** satisfies **spec(A,C)**.

```
=> : [program, program -> bool] = (LAMBDA prog1,prog2:
            (FORALL s0,s1: prog1(s0,s1) IMPLIES prog2(s0,s1) ))

reftrans: THEOREM (prog1 => prog2) IFF
            (EXISTS prog: (prog1 => prog) AND (prog => prog2))
```

Proof System The proof system contains the following consequence rule that reflects the classical consequence rule of Hoare logic [Hoa69].

```
rulecons: THEOREM Valid(A IMPLIES A0) AND Valid(C0 IMPLIES C)
            IMPLIES (spec(A0,C0) => spec(A,C))
```

The rule for parallel composition of processes is given as an axiom rather than a theorem. Also, for simplicity, syntactic constraints which require that the assertions of one process do not refer to observables of the other process are omitted. Additionally, it is assumed that neither **now** nor **term** occur in the commitments. We refer to [Hoo91] for more details and a soundness proof of the parallel composition rule. Here we concentrate on the use of this rule during top-down program design of distributed systems.

```
rulepar: AXIOM  par(spec(A1,C1), spec(A2,C2))
            => spec(A1 AND A2, C1 AND C2)
```

2.2 Example Steam Boiler Control System

Top-Level Specification The steam boiler contains a vessel filled with water. The water level is dangerous if it is below M1 or above M2. Hence, the top-level specification spec(A,CTL) of the steam boiler system requires that the water level, denoted by q, always stays in the range [M1,M2].

```
q     : [Time -> PositiveValue]  % quantity of water in liters
M1, M2 : PositiveValue
A     : pred[State] = (LAMBDA s: now(s) = 0 AND term(s))
CTL   : pred[State] = (LAMBDA s: (FORALL t: M1 <= q(t) AND
                                             q(t) <= M2))
```

Properties of the Vessel Since accumulated outflow of steam and inflow of water can be expressed using integrals, function integral(t1,t2,f) is introduced to denote $\int_{t1}^{t2} f(t)dt$. We do not give a mathematical definition of integrals, but formulate a few simple properties. They are given as AXIOMs, since proofs can be found in many textbooks and are not in the scope of this paper.

First the physical properties of the water level in the vessel are specified. Suppose the water level in the Vessel is the result of the initial water level plus the amount of water added (the accumulated inflow) minus the amount of water removed (the accumulated outflow). The outflow of steam (vapor) and the throughput of the pump system are denoted by, respectively, v (in liters/second) and p (in liters/second), as illustrated in Figure 1. Outflow is bounded by W.

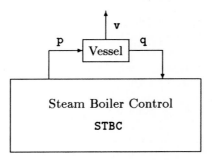

Fig. 1. Steam Boiler System

```
p, v  : [Time -> PositiveValue]   % liters/second
W     : PositiveValue             % liters/second
CV1   : pred[State] = (LAMBDA s: (FORALL t:
                    q(t) = q(0) + integral(0,t,p) - integral(0,t,v) ))
CV2   : pred[State] = (LAMBDA s: (FORALL t: v(t) <= W))
CV    : pred[State] = CV1 AND CV2
```

Control Strategy Given the properties of the water level in the vessel, we formulate a strategy for a steam boiler control system STBC (STeam Boiler Control). The strategy describes the relation between the water level and the required inflow. During regular operation the system tries to maintain the water level between normal values N1 and N2, to cope with reaction times of the pump and the water measuring unit. Commitment CSTBC1 specifies that there should be an inflow of P as long as the water level is at most N1. We allow a reaction time of DelayOpen time units. Similarly, CSTBC2 (not shown here) expresses that there should be no inflow within `DelayClose` time units as long as the water level is at least `N2`. Note that opening and closing of the pump can take different amounts of time, so we have chosen two different timing constants. Finally, `CSTBC3` asserts that inflow `p` is bounded by the capacity `P` of the pump.

```
N1, N2, P              : PositiveValue
DelayOpen, DelayClose  : Time

CSTBC1: pred[State] = (LAMBDA s:
   (FORALL t1,t2: t1 + DelayOpen <= t2 AND dur( q <= N1, cc(t1,t2))
                     IMPLIES dur( p = P, cc(t1 + DelayOpen, t2))))

CSTBC3: pred[State] = (LAMBDA s: (FORALL t: p(t) <= P))

CSTBC:  pred[State] = CSTBC1 AND CSTBC2 AND CSTBC3
```

The correctness of this design step can be proved using requirement `TLreq` which is the conjunction of four constraints:

```
P_W         : bool = P >= W AND W > 0
init_quant  : bool = N1 < q(0) AND q(0) < N2
DiffN1M1    : bool = M1 <= N1 - DelayOpen * W
DiffN2M2    : bool = M2 >= N2 + DelayClose * P
TLreq       : bool = P_W AND init_quant AND DiffN1M1 AND DiffN2M2
```

E.g., `DiffN2M2` expresses that the difference between the normal limit `N2` and the absolute limit `M2` should be large enough to close the pump system, letting `p` become 0, without risking to exceed `M2` even if there is no outflow of steam.

Then we can prove lemma `correctCTL` using some properties of integrals and the intermediate value property of continuous functions. Theorem `correctTL` is easily proved by rule `rulepar` and lemma `correctCTL`.

```
correctCTL: LEMMA TLreq IMPLIES Valid(CV AND CSTBC IMPLIES CTL)
correctTL : THEOREM TLreq IMPLIES
              par(spec(A,CV), spec(A,CSTBC)) => spec(A,CTL)
```

2.3 Asynchronous Communication

We consider parallel components that communicate via message passing along unidirectional channels. Suppose communication is asynchronous, i.e., a sender does not wait for synchronization but sends the message immediately. A receiver waits until a message is available. There is no buffering of messages; a message gets lost if there is no receiver.

Let `Chan` be a set of channel names. To describe asynchronous communication, we define primitives expressing when a process is waiting to receive a message and when it starts receiving, resp., sending a value. Sometimes we abstract from the value transmitted, using channel variable `ch`.

```
waitrec    : [Chan -> pred[Time]]
recv, sendv : [Chan,Value -> pred[Time]]
send(ch)(t) : bool = (EXISTS val: sendv(ch,val)(t))
```

The following axioms characterize the communication mechanism.

```
valueax: AXIOM sendv(ch,v0)(t) AND sendv(ch,v1)(t) IMPLIES v0 = v1
minwait: AXIOM NOT(send(ch)(t) AND waitrec(ch)(t))
recsend: AXIOM recv(ch,val)(t) IMPLIES sendv(ch,val)(t)
```

The first axiom guarantees that at any point of time at most one value is transmitted on any channel. The second axiom expresses that no process waits to receive along channel ch if a message is transmitted (and hence available) on ch. The third one expresses that a received message has been sent. Let predicate awaitrec express that a process starts waiting to receive input along ch at time t until it receives input, allowing the possibility of infinite waiting.

Although we consider an asynchronous communication mechanism in which messages can easily get lost, we are able to show that under certain conditions every message sent is received. If the sender does not send before a certain time Start and with a distance of *at least* Period time units (expressed by maxsend(ch,Start,Period)), and if the receiver is ready to receive after Start and with a distance of *at most* Period time units (expressed by minawait(ch,Start,Period)), then each output is received (see CD-ROM Annex VH, lemma sendrecv of theory asyn).

Sometimes it is required that at least one communication takes place each period of T time units. If the sender sends regularly (sendperiod(ch,T1)(t), for all t) and if the receiver is ready to receive regularly (waitperiod(ch,T2)(t), for all t) then we can prove that a communication occurs at least once every period of T1+T2 time units (commperiod(ch, T1+T2)(t), for all t).

2.4 Refinement of the Control Strategy – Introducing a Pump System and a Water Sensor

Pump System PS To implement spec(A,CSTBC) we use a pump system PS to provide the steam boiler with water (see Figure 2). Suppose the pump system

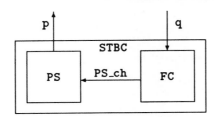

Fig. 2. Pump System and Flow Control

modifies the inflow of water into the steam boiler according to messages it receives along channel PS_ch. By CPS1 we specify that the pump system is ready to receive input along `PS_ch` periodically, using timing constants `Init` and `Repeat`. Commitment `CPS2` asserts that it responds to message `OPEN_PUMP` by increasing `p` up to its full capacity `P` in at most `DelayOn` time units. Similarly, `CPS3` expresses that it reacts to message `CLOSE_PUMP` by decreasing `p` down to 0 within `DelayOff` time units. Henceforth we will show only one of similar specifications and, hence, `CPS3` is not shown (see CD-ROM Annex VH for more details). As expressed by `CSTBC3`, the capacity of the pump is `P`.

```
DelayOn, DelayOff, Init, Repeat : Time
OPEN_PUMP, CLOSE_PUMP            : Value
axOPENCLOSE : AXIOM NOT(OPEN_PUMP = CLOSE_PUMP)

CPS1: pred[State] = (LAMBDA s:
                     (FORALL t: minawait(PS_ch,Init,Repeat)(t)))
CPS2: pred[State] = (LAMBDA s:
 (FORALL t1,t2: t1 + DelayOn <= t2 AND recv(PS_ch,OPEN_PUMP)(t1)
    AND dur( NOT recv(PS_ch,CLOSE_PUMP), oc(t1,t2))
       IMPLIES dur( p = P, cc(t1 + DelayOn, t2)) ))
CPS: pred[State] = CPS1 AND CPS2 AND CPS3 AND CSTBC3
```

In parallel with pump system `PS` we design a flow control component `FC` that sends messages to the pump system along `PS_ch`. In view of `CPS1`, we guarantee that no message gets lost by specifying in `CFC1` the maximal frequency with which it sends messages along `PS_ch`. `CFC2` expresses that `FC` sends message `OPEN_PUMP` within `DelayFC` time units if it recognizes that the water level `q` is less or equal `N1`. Similarly, `CFC3` asserts that it sends message `CLOSE_PUMP` after at most `DelayFC` time units if the water level is at least `N2`.

```
CFC1: pred[State] = (LAMBDA s:
                     (FORALL t: maxsend(PS_ch,Init,Repeat)(t)))
CFC2: pred[State] = (LAMBDA s:
  (FORALL t1,t2: t1 + DelayFC <= t2 AND dur( q <= N1, cc(t1,t2))
     IMPLIES (EXISTS t0: t1 <= t0 AND t0 < t1 + DelayFC
           AND sendv(PS_ch,OPEN_PUMP)(t0)
               AND dur( NOT sendv(PS_ch,CLOSE_PUMP), oc(t0,t2))) ))
CFC: pred[State] = CFC1 AND CFC2 AND CFC3
```

Commitments `CPS1` and `CFC1` imply (by lemma `sendrecv`) that no message along `PS_ch` gets lost. Then this design step can be proved correct, assuming two constraints on timing constants.

```
DelayOpen_On    : bool = DelayOpen >= DelayOn + DelayFC
DelayClose_Off  : bool = DelayClose >= DelayOff + DelayFC
STBCreq         : bool = DelayOpen_On AND DelayClose_Off
correctSTBC: THEOREM STBCreq IMPLIES
                par(spec(A,CPS),spec(A,CFC)) => spec(A,CSTBC)
```

Refinement of the Pump System In this section pump system PS is refined into a pump **Pump** and a pump controller **PC**. They communicate via channel P_ch, as illustrated in Figure 3. The pump is turned on or off, depending on the

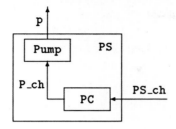

Fig. 3. Pump and Pump Control

messages it receives. Recall that control program Contr sends message OPEN_PUMP when the received water level is at most N1 + WSDev. Usually many OPEN_PUMP messages are sent although the pump is already on. The main task of PC is to record the state of the pump and to forward only those messages from Contr that change the pump state.

First we specify the pump by spec(S, CPump). Suppose Pump is ready to receive input along P_ch periodically within RepeatPC time units and initially before time InitPC. Further it should respond to message START by increasing p up to its capacity P within DelayP_On time units.

```
DelayP_On, DelayP_Off, InitPC, RepeatPC: Time
CPump1: pred[State] = (LAMBDA s:
                  (FORALL t: minawait(P_ch,InitPC,RepeatPC)(t) ))
CPump2: pred[State]  = (LAMBDA s:
   (FORALL t1,t2: t1 + DelayP_On <= t2 AND recv(P_ch,START)(t1)
                  AND dur( NOT sendv(P_ch,STOP), oc(t1,t2))
      IMPLIES dur( p = P, cc(t1 + DelayP_On, t2)) ))
```

Similarly, CPump3 expresses that it should respond to STOP by decreasing p down to 0 after at most DelayP_Off time units. Again, DelayP_On and DelayP_Off might represent different values since opening and closing the pump can take different amounts of time. Inflow p is still bounded by P, as expressed by CSTBC3.

Next we specify the pump controller PC. Suppose PC is ready to receive input along PS_ch with the same frequency as PS, i.e. it satisfies CPS1. To guarantee that no message gets lost on P_ch we specify, in view of CPump1, that it sends along P_ch with a frequency of at most RepeatPC and not before InitPC.

```
DelayPC : Time
CPC1 : pred[State] = (LAMBDA s:
                  (FORALL t: maxsend(P_ch,InitPC,RepeatPC)(t) ))
```

CPC2 expresses that if PC receives the first OPEN_PUMP after a series of CLOSE_PUMP messages, it sends message START along P_ch in at most DelayPC time units.

Similarly CPC3 asserts that it sends STOP after receiving the first CLOSE_PUMP that follows a series of OPEN_PUMP messages. Further PC should send START and STOP only if it has received messages OPEN_PUMP and CLOSE_PUMP, respectively, earlier. Then CPC: pred[State] = CPS1 AND CPC1 AND CPC2 AND CPC3 AND CPC4 AND CPC5 leads to spec(A,CPC). Observe that CPC1 and CPump1 imply (by lemma sendrecv) that no message gets lost on P_ch. Then the correctness of this refinement step, as expressed by theorem correctPS, can be proved assuming certain constraints on timing constants, expressed by PSreq.

correctPS: THEOREM PSreq IMPLIES
                (par(spec(A,CPC),spec(A,CPump)) => spec(A,CPS))

Water Sensor WS Next we refine flow control FC by a water sensor WS and a control component, as shown in Figure 4. Assume given a water sensor WS that

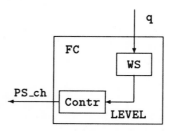

Fig. 4. Water Sensor and Control Program

measures the water level and sends the measured values along channel LEVEL at least once every DelayWS time units. These values do not deviate more than WSDev from the real water level.

CWS1: pred[State] = (LAMBDA s:
                      (FORALL t: sendperiod(LEVEL,DelayWS)(t)))
CWS2: pred[State] = (LAMBDA s:
      (FORALL t,val: sendv(LEVEL,val)(t)
            IMPLIES val - WSDev <= q(t) AND q(t) <= val + WSDev))
CWS: pred[State] = CWS1 AND CWS2

In parallel with WS we design a control component Contr which is ready to receive input from the water sensor along LEVEL at least once every DelayReadWS time units. It sends messages along channel PS_ch with the same maximal frequency as FC, as specified by CFC1.

DelayReadWS, DelayContr: Time
CContr1: pred[State] = (LAMBDA s:
                  (FORALL t: waitperiod(LEVEL,DelayReadWS)(t)))

Since the values that the control component receives from the water sensor do not necessarily equal the real water level but can deviate from it up to WSDev, we have to decide when the pump system should be opened or closed. A message should be sent along PS_ch if there is a possibility that the water level is outside the range [N1,N2]. Hence, if a value below N1 + WSDev is received, then message OPEN_PUMP is transmitted within a reaction time of DelayContr time units. Further, no CLOSE_PUMP is transmitted as long as only values below N2 - WSDev are received. Similarly, a CLOSE_PUMP message is transmitted if a value above N2 - WSDev is received and as long as only values above N1 + WSDev are being received. The control component is specified by spec(A,CContr).

By CWS1 and CContr1 we obtain that there is at least one communication along LEVEL each DelayWS + DelayReadWS time units. Assuming requirement CFCreq, theorem correctFC can be proved.

```
DelayFC_Contr: bool = DelayFC >= DelayWS+DelayReadWS+DelayContr
Min_MaxNorm   : bool = N1 + 2 * WSDev < N2
CFCreq        : bool = DelayFC_Contr AND Min_MaxNorm
correctFC:    THEOREM CFCreq IMPLIES
                      par(spec(A,CWS),spec(A,CContr)) => spec(A,CFC)
```

2.5 Programming Language

A simple real-time programming language and its semantics are presented briefly. Recall that programs are defined as a relation between initial state s0 and final state s1. Let Ta represent the execution time of an assignment.

```
vvar : VAR Vars
exp  : VAR [State -> Value]
Ta   : Time
assign(vvar,exp): program = (LAMBDA s0,s1:
  term(s0) IMPLIES term(s1) AND now(s1) = now(s0) + Ta AND
                  val(s1) = val(s0) WITH [(vvar):= exp(s0)])

seq(prog1,prog2): program = (LAMBDA s0,s1:
  term(s0) IMPLIES (EXISTS s: prog1(s0,s) AND prog2(s,s1)) )
```

Similarly, ifthenelse(b,prog1,prog2), ifthen(b,prog), and while(b,prog) have been defined. Assume that the evaluation of a boolean takes Tb time units.

Input and output statements, input(ch,vvar) and output(ch,exp), are defined using the communication primitives of Section 2.3. Note that an input statement might have to wait before it starts receiving a message and that it need not terminate because it might have to wait forever. Each communication takes Tc time units.

We introduce a select statement select(ch,vvar,prog1,exp,prog2). Basically the select statement waits for input along channel ch. If it receives a message within exp time units then program prog1 is executed. Otherwise, delay exp expires and program prog2 is executed.

Proof Rules of the Programming Language Proof rules of the proof system are formulated as theorems and can be proved by means of their semantic definitions. Here we only show the sequential composition rule.

```
ruleseq: THEOREM seq(spec(A,B),spec(B,C)) => spec(A,C)
```

2.6 Steam Boiler Control System – Final Implementation

Next a program `Control_prog` that satisfies specification `spec(A,CContr)` is presented. The parameter `Vars`, representing the set of program variables as introduced in Section 2.1, is now fixed to the set $\{x\}$.

```
low(s)  : bool = val(s)(x) <= N1 + WSDev
high(s) : bool = val(s)(x) >= N2 - WSDev
analysis: program =
        ifthenelse(low, output(PS_ch, (LAMBDA s: OPEN_PUMP)),
            ifthen(high, output(PS_ch, (LAMBDA s: CLOSE_PUMP))))

Control_prog: program = while(TRUE, seq(input(LEVEL,x),analysis))
```

Requirements on timing constants `DelayReadWS`, `DelayContr Init`, and `Repeat` are expressed by `Control_progreq`, leading to program correctness.

```
corImpl: THEOREM Control_progreq IMPLIES
                        Control_prog => spec(A,CContr)
```

Conclusion Steam Boiler Control System The following monotonicity rule `monopar` allows us to express that the parallel execution of the physical components and the control program satisfies the given top-level specification. Similar to the parallel composition rule, `monopar` is formulated as an axiom here.

```
monopar: AXIOM (prog3 => prog1) AND (prog4 => prog2)
            IMPLIES (par(prog3,prog4) => par(prog1,prog2))
```

Assuming all constraints on constants that have evolved during refinement steps, represented by `totalreq`, we can prove the correctness of the total system.

```
totalreq: bool = TLreq AND Control_progreq AND
 DelayOpen >= DelayPC+DelayPOn+DelayWS+DelayReadWS+DelayContr AND
 DelayClose >= DelayPC+DelayPOff+DelayWS+DelayReadWS+DelayContr

correcttotal: THEOREM totalreq IMPLIES
  par(spec(A,CV), par(spec(A,CPC), par(spec(A,CPump),
                par(spec(A,WS), Control_prog)))) => spec(A,CTL)
```

Hence we have obtained a system which implements the top-level specification `spec(A,CTL)` assuming specifications of the physical properties of the vessel, the pump system, and the water sensor, and provided the timing constants meet certain requirements.

3 Failures of Communication Channels

3.1 Introduction

Observe that the correctness of the system derived in the previous section depends on the implicit assumption that all physical components and all communication channels work correctly. In reality, however, some components or communication channels fail during execution of the system. In this section failures of communication channels are considered.

Suppose messages can get lost on a failing channel, i.e., although a receiver waits for messages along a certain channel and a sender sends a message along this channel, this message never reaches the receiver in case of a channel failure. For simplicity, we assume that this is the only effect of a failing channel. Thus, e.g., messages are neither inserted nor corrupted.

3.2 Detection of Channel Failures

Observe that in Section 2 asynchronous communication has been used in two different ways. This also leads to two ways of detecting failures.

For channels PS_ch and P_ch the sending process should not send faster than certain frequencies. If the receiving process is ready to receive before messages are sent, no message gets lost. Detecting failures along such channels can be done by the usual transmission protocols. Here we do not consider failures on these channels, but refer to the literature on communication protocols.

For channel LEVEL we have specified that the water sensor WS sends at least one value every DelayWS time units. The control component, which is ready to receive every DelayReadWS time units, need not receive every value sent, but there is at least one communication along LEVEL every interval of length DelayWS + DelayReadWS. If the control component is ready to receive a message along LEVEL during DelayWS and no message and no communication occurs, then the channel is considered broken (since the water sensor does not fail here).

Thus, an additional task of the control program in this section is to detect failures of LEVEL and to send an appropriate alarm message. Then the environment should take an appropriate action, e.g., shutting down the system.

3.3 Adaptation of the Framework to deal with Faulty Channels

To distinguish correctly working communication channels from those that are broken, we introduce a primitive correct. Chan is the set of all channel names.

correct: [Chan -> pred[Time]]

Next consider the three communication axioms formulated in Section 2.3. Since we do not allow insertion or corruption of messages, valueax and recsend are still valid. But messages can get lost, so minwait need not hold for a failing channel. Suppose, e.g., that a process starts sending along channel ch at time t. If ch is broken at t, then there can be another process that still waits to receive along `ch` at `t` since the message sent is lost on the channel. Therefore, axiom `minwait` is replaced by `FT_minwait` which expresses that the minimal waiting property holds for correct channels.

```
FT_minwait: AXIOM correct(ch)(t) IMPLIES
                  NOT( send(ch)(t) AND waitrec(ch)(t) )
```

Of course, all communication lemmas that depend on this axiom are modified in a similar fashion. In general, we have to mention explicitly that the channels involved should be correct. A predicate `awaitrecterm(T,ch)(t)` is defined to indicate that a process is ready to receive input along channel `ch` at time `t`. It differs from `awaitrec(ch)(t)` of Section 2.3 by requiring the process to wait for at least `T` time units, instead of infinite waiting.

3.4 Example Steam Boiler Control System – Channel Failures

Consider again the steam boiler system as specified in Section 2.2. In this section we aim at an implementation of a steam boiler control program that behaves similar to `Control_prog` of Section 2.6, keeping the water level inside the vessel between the critical values if all communication channels work correctly. Moreover, it should detect failures of channel `LEVEL` and send an alarm message to the environment if it has detected an error.

We start with the top-level specification, the vessel properties, and the control strategy of Section 2.2, and consider the design steps of Section 2 in view of failing channels.

Introducing a Pump System To refine the control strategy, again a pump system `PS` and a flow control `FC` are introduced. Both have the same specifications as in Section 2.4. To prove that `PS` and `FC` indeed refine the control strategy, channel `PS_ch` that connects these two components should not fail.

```
FT_correctSTBC: THEOREM (FORALL t: correct(PS_ch)(t)) AND STBCreq
       IMPLIES (par( spec(A,CPS), spec(A,CFC)) => spec(A,CSTBC))
```

Introducing a Water Sensor Flow control component `FC` is refined further by introducing water sensor `WS`, which is specified as in Section 2.4. In parallel with `WS` we introduce a Fault-Tolerant control component `FT_Contr` which should behave similar to `Contr` of Section 2.4. First we require that `FT_Contr` is ready to receive input along `LEVEL` for at least `DelayWS` time units.

```
FT_CContr1: pred[State] = (LAMBDA s: (FORALL t:
       inside( awaitrecterm(DelayWS,LEVEL), co(t, t+DelayReadWS))
```

We copy the commitments `CContr2`, `CContr3`, and `CContr4`. To detect failures of channel `LEVEL`, observe that the specification of the water sensor asserts that `WS` sends at least one message every `DelayWS` time units. Hence, if `FT_Contr` has waited for input along `LEVEL` for at least `DelayWS` consecutive time units without receiving any message, then `LEVEL` must be broken. Recall that in this section all physical components are assumed to work correctly, so `WS` does indeed send regularly. We specify that `FT_Contr` sends message `LEVEL_failure` along channel `ALARM` iff it has detected a failure of `LEVEL`. This results in specification `spec(A,FT_CContr)`. Then we show that `WS` and `FT_Contr` refine flow control `FC`.

```
FT_correctFC: THEOREM (FORALL t: correct(LEVEL)(t)) AND CFCreq
    IMPLIES (par(spec(A,CWS), spec(A,FT_CContr)) => spec(A,CFC))
```

Since we concentrate on failures of channel `LEVEL`, we do not consider a refinement of pump system `PS` here. One could choose the same specifications for the pump and the pump controller as in Section 2.4 and then require the correctness of the channel connecting them explicitly.

3.5 Implementation of the Control Component

Similar to Section 2.6 we derive an implementation of the control component `FT_Contr`. To implement `spec(A,FT_CContr)`, define `FT_Control_prog` using program **analysis** of Section 2.6 and a program **alarm**.

```
alarm: program = output(ALARM, (LAMBDA s: LEVEL_failure))
FT_Control_prog: program =
    while(TRUE,
          select(LEVEL, x, analysis, (LAMBDA s: DelayWS), alarm))
```

For the correctness proof, we define requirement `FT_Control_progreq`, consisting of timing constraints and the failure hypothesis `incorrect_LEVEL`, defined by

```
incorrect_LEVEL(t): bool = NOT correct(LEVEL)(t) IMPLIES
        (t >= DelayWS AND dur(waitrec(LEVEL), co(t - DelayWS,t)))
```

Using this requirement it has been proved that `FT_Control_prog` satisfies specification `spec(A,FT_CContr)`.

Conclusion Steam Boiler Control System Collecting all constraints on constants that have evolved during refinement steps, we can prove a theorem about the whole system. This theorem is similar to `correcttotal` of Section 2.6, now with `FT_Control_prog` instead of `Control_prog`. Additionally, explicit assumptions about the correctness of all channels are required.

We can also give a final conclusion about the system behavior if a failure of channel `LEVEL` is detected by the control program `FT_Control_prog`.

```
Failure: pred[State] = (LAMBDA s:
                    (EXISTS t: sendv(ALARM, LEVEL_failure)(t)) )
correctFT_failure: THEOREM FT_Control_progreq AND
  (EXISTS t: NOT correct(LEVEL)(t) AND incorrect_LEVEL(t))
    IMPLIES par(spec(A,CV), par(spec(A,CPS),
          par(spec(A,CWS), FT_Control_prog ))) => spec(A,Failure)
```

4 Failures of Physical Units

4.1 Introduction

Whereas Section 3 concerns the detection of errors on a certain communication channel, here failures of physical units are considered. The physical units of the steam boiler are the water sensor and the pump of the pump system. Computer controlled units such as the pump controller or the control component itself are not classified as physical units here and are assumed to be free of failures.

Failures of the pump might have the effect that it does not react to messages from the pump controller or that it changes the inflow of water spontaneously. By calculations that take into account the assumed pump state, the water level, and the outflow of steam, these failures can be corrected by the system. Of course, this method requires correctly working sensors.

Here we consider only failures of the water sensor and implicitly assume that the pump works correctly. Suppose the only effect of a water sensor failure is sending values that do not reflect the current water level. More specifically, assume a broken sensor sends values that are either out of the interval [0,C], where C is the boiler capacity, or that are impossible related to values received before. The latter effect can be detected by calculations that take into account the inflow of water, the outflow of steam, and the dynamics of the system.

If the water sensor is considered broken, the control component should immediately send an alarm message that causes the environment to take some action, like shutting down the whole system. Hence we do not consider the possibility to let the system run further until there is a possibility that the water level is about to reach one of the limit values M1 or M2.

4.2 Extension of the Framework

To deal with failures of physical units, function **correct** is defined for elements of type **Component**. All physical units that might fail are listed in the enumeration type **Component**. Component WLS represents a water level sensor which is introduced in the next section.

```
Component : TYPE = { WS, WLS }
correct   : [Component -> pred[Time]]
```

4.3 Example Steam Boiler Control System – Failures of Physical Units

The pump and all communication channels are assumed to function correctly. A sensor control component should detect failures of the water sensor by means of a steam sensor and send an alarm message to the environment if a failure occurs. This will be proved later, for the moment we start with the top-level specification **spec(A,CTL)** of Section 2.2. To refine the control strategy we introduce pump system PS and flow control FC. Again, both are specified exactly as in Section 2.4.

Introducing a Water Sensor Flow control FC is refined further by introducing water sensor WS and control component Contr. Contr is specified as in Section 2.4. In the specification of WS commitment CWS2 is changed by adding the requirement that WS should work correctly.

```
WSDev    : Value
FT_CWS2 : pred[State] = (LAMBDA s:
   (FORALL t,val: correct(WS)(t) AND sendv(LEVEL,val)(t)
             IMPLIES val - WSDev <= q(t) AND q(t) <= val + WSDev))
FT_CWS   : pred[State] = CWS1 AND FT_CWS2
```

Then the parallel composition of a correct water sensor WS and control component Contr indeed refines flow control FC provided conditions DelayFC_Contr and Min_MaxNorm, combined in CFCreq, of Section 2.4 hold.

```
FT_correctFC: THEOREM (FORALL t: correct(WS)(t)) AND CFCreq
     IMPLIES (par(spec(A,FT_CWS), spec(A,CContr)) => spec(A,CFC))
```

Introducing a Sensor Control Water sensor WS, specified by spec(A,FT_CWS), is refined further by introducing a water level sensor WLS and a sensor control Sens. Later also a steam sensor is introduced (see Figure 5).

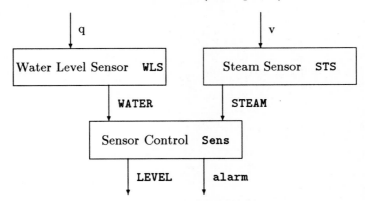

Fig. 5. Refinement of the Water Sensor

The main task of Sens is to forward values that it receives from WLS along channel WATER if it considers WLS working correctly. If Sens detects a failure of WLS, it sends an alarm message. The water level sensor WLS behaves similar to WS and is specified by spec(A, CWLS).

To specify Sens we first postulate a predicate broken_WLS(t,val) which represents that WLS is considered broken at time t. The second parameter val denotes the water level value that Sens receives at t. Based on val, Sens decides whether WLS is broken or not. The exact definition is given in the next section when the steam sensor is introduced.

At least once every DelayReadWLS time units, Sens is ready to receive input from WLS along channel WATER. If it receives a value and WLS is considered correct, then this value is simply forwarded to the control component `Contr` along channel `LEVEL` after a delay of at most `DelaySens` time units. If a failure of `WLS` is detected, then alarm message `WLS_failure` is sent to the environment along channel `ALARM` and dummy value `fail_value` is sent to `Contr` within the same reaction time. This is specified by `spec(A, CSens)`.

Next a correct water level sensor is related with `broken_WLS` and the correctness of `WLS`.

```
correct_WLS: bool = (FORALL t: correct(WLS)(t) IFF
                                (FORALL val: NOT broken_WLS(t,val)))
correct_WS: bool = (FORALL t,val:
   correct(WS)(t) AND sendv(LEVEL,val)(t)
         IMPLIES (EXISTS t1: t1 <= t AND t1 > t - DelaySens AND
                  sendv(WATER,val)(t1) AND correct(WLS)(t1)) )
```

To prove that `WLS` and sensor control `Sens` refine the water sensor `WS` we first define three properties reflecting the physical behavior of the steam boiler. Also constraints on timing constants are defined. Constraint `P_W` has been defined in Section 2.2. Further predicate `correct_WS` should hold.

```
q_prop: bool = (FORALL t: q(t) =
                          q(0) + integral(0,t,p) - integral(0,t,v))
p_prop: bool = (FORALL t: p(t) <= P)
v_prop: bool = (FORALL t: v(t) <= W)
DelayWS_Sens : bool = DelayWS >= DelayWLS+DelayReadWLS+DelaySens
DevWS_WLS    : bool = WSDev >= WLS + P * DelaySens
CWSreq       : bool = DelayWS_Sens AND DevWS_WLS AND P_W AND
                      q_prop AND p_prop AND v_prop
FT_correctWS: THEOREM correct_WS AND CWSreq IMPLIES
              (par(spec(A,CWLS), spec(A,CSens)) => spec(A,FT_CWS))
```

We can also prove that `Sens` sends an alarm message along channel `ALARM` if a failure occurs in `WLS`, as expressed by commitment `Alarm`.

```
WLS_failure: Value
Alarm: pred[State] = (LAMBDA s:
                     (EXISTS t: sendv(ALARM,WLS_failure)(t) ))
FT_incorrectWLS: THEOREM NOT correct(WLS)(t) AND correct_WLS
                 IMPLIES Valid(CSens IMPLIES Alarm)
```

Introducing a Steam Sensor To define predicate `broken_WLS`, we introduce a steam sensor `STS`. The steam sensor is specified similar to `WS` and `WLS` by `spec(A, CSTS)`. It regularly sends values that reflect the outflow of steam along channel `STEAM`. The values do not deviate more than `STSDev` from the real outflow of steam. Note that we assume that the steam sensor always works correctly.

In parallel with the steam sensor we specify a modified sensor control `SC` that behaves similar to `Sens` but, additionally, receives messages from `STS`. This leads to specification `spec(A, CSC)`.

Now **broken_WLS** can be defined by taking into account the steam values that **SC** receives from **STS**. First the minimal and the maximal amount of steam that can exit the steam boiler between two measurements of the water level is calculated, based on the maximal increase of steam **U1**, the maximal decrease of steam **U2**, and the maximal outflow of steam **W**.
Assuming a timing constraint, the correctness of this step can be proved.

```
DelaySens_SC: bool = DelaySens >= DelaySTS+DelayReadSTS+DelaySC
correctSens: THEOREM DelaySens_SC IMPLIES
             (par(spec(A,CSTS), spec(A,CSC)) => spec(A,CSens))
```

5 Evaluation and Comparison

1) The system configuration is somewhat reduced from the original problem description; the operator desk is not modeled and only a single pump and corresponding pump controller are considered. Moreover, only normal mode is fully dealt with. In particular, proper initialization is assumed. The aspects that are included are completely formal, that is, there is a formal specification of the components (control program, water level sensor, steam sensor, pump actuator, pumps) and a formal derivation of the control program.

2) The resulting control program has been written in a pseudo programming language, modeled in the PVS language. It has not been linked to the FZI simulator and no experimentation has been done.

3) Close to our approach is the solution given by chapter RS of this book, since it also starts from a formal specification (expressed in the Duration Calculus) and then derives an implementation (in Occam) by means of a number of design steps. Similarly, chapter CW1 is strongly related, also because of the use of a theorem prover. An alternative approach to program refinement is given in chapter BSS using action systems. Comparable are further solutions that express properties of the system in some form of logic, such as TRIO, TLA, and VDM in chapters GM, LM, and LP, respectively.

Complementary are solutions that put more emphasis on requirements engineering. For instance, chapter BCPR contains an analysis of the informal requirements. Interesting is that validation has been supported by the Larch Prover to prove properties from the specification. Also complementary are approaches that concentrate more on simulation and the design of an executable program. For instance, chapter AT considers deductive synthesis of programs. Finally, we would mention chapter LW which is complementary since it deals with performance and optimization.

4) The work described here has been done as the Master's thesis work of the first author and took about nine months. The first part, in which no faults are considered, was rather straightforward since we could build on a previous control example verified in PVS [Hoo94]. The most complicated part was the proof of the first refinement step, since this concerns the correctness of a control strategy in terms of continuous quantities. The proof depends on mathematical analysis, such as properties of continuous functions and integration, which had to be incorporated in PVS. Also the correctness proof of the final control program was rather intricate and time consuming. Later stages of the design, considering failures, took more time due to our limited experience in this field. Note, however, that PVS has been used extensively for fault-tolerant applications [ORSvH95].

5) According to our experience, it requires some basic knowledge of first-order logic (the higher-order aspects of PVS seem easy understandable) to understand PVS specifications. Given this background, usually a few days of introduction and hands-on experience are sufficient to start working with the PVS system. Of course, it takes some time to get acquainted with the interactive proof checker of PVS and to complete non-trivial proofs. The basic principles of compositional methods for distributed (real-time) systems can be learned in a course of fifteen hours.

Acknowledgments

Many thanks go to Kai Engelhardt for his frequent and valuable advice to the first author during the work on the Master's thesis that forms basis of this paper. We are greatly indebted to Willem-Paul de Roever for establishing our collaboration and valuable suggestions. An anonymous reviewer is thanked for useful suggestions and a valuable contribution to the above evaluation.

References

[GNRR93] R. Grossman, A. Nerode, A. Ravn, and H. Rischel, editors. *Hybrid Systems*, volume 736 of *LNCS*. Springer-Verlag, 1993.

[Hoa69] C.A.R. Hoare. An axiomatic basis for computer programming. *Communications of the ACM*, 12(10):576–580, 583, October 1969.

[Hoo91] J. Hooman. *Specification and Compositional Verification of Real-Time Systems*, volume 558 of *LNCS*. Springer-Verlag, 1991.

[Hoo93] J. Hooman. A compositional approach to the design of hybrid systems. In Grossman et al. [GNRR93], pages 121–148.

[Hoo94] J. Hooman. Correctness of real-time systems by construction. In H. Langmaack, W.P. de Roever, and J. Vytopil, editors, *Formal Techniques in Real-Time and Fault-Tolerant Systems*, volume 863 of *LNCS*, pages 19–40. Springer-Verlag, 1994.

[Hoo96] J. Hooman. Assertional specification and verification. In M. Joseph, editor, *Real-time Systems: Specification, Verification and Analysis*, chapter 5, pages 97–146. Prentice Hall, 1996.

[ORS92] S. Owre, J. Rushby, and N. Shankar. PVS: A prototype verification system. In *Conference on Automated Deduction*, volume 607 of *Lecture Notes in Artificial Intelligence*, pages 748–752. Springer-Verlag, 1992.

[ORSvH95] S. Owre, J. Rushby, N. Shankar, and F. von Henke. Formal verification for fault-tolerant architectures: Prolegomena to the design of PVS. *IEEE Transactions on Software Engineering*, 21(2):107–125, 1995.

```
  specification SteamBoiler_System [Fault,Repaired, Actions_In_EmergencyStop,
2                                   System_Stop, Explosion]: noexit ...
  behaviour
4 hide tsg_ps_in, tsg_ps_out, tsg_cu_in,tsg_cu_out in
  ( PS [tsg_ps_in, tsg_ps_out,Fault,Repaired, Explosion]
6      |||
    CU [tsg_cu_in, tsg_cu_out,Actions_In_EmergencyStop,System_stop] )
8 |[tsg_ps_in, tsg_ps_out, tsg_cu_in,tsg_cu_out]|
    TM[tsg_ps_in, tsg_ps_out, tsg_cu_in,tsg_cu_out]
```

Fig. 2. The System Behavior

system and the control unit to be independent of each other, except of their common communication over the transmission medium (line 8-9).

```
  process PS [tsg_in, tsg_out, Fault, Repaired, Explosion]: noexit :=
2 tsg_out !SteamBoiler!STEAM_BOILER_WAITING !0 of TParameter;
  ( ( GetMode [..] ||| Valve [..] |||
4     Measurer [..](Steam) ||| Measurer [..](Water)
    ||| ( hide sb_notify in TheSteamboiler[..] )
6                          |[sb_notify]|
                           Pump [..] (Pump1) )
8   |||
    ( tsg_in !Controller !PROGRAM_READY !0 of TParameter;
10  (      DeviceFailMsgs [..] (Pump1) ||| DeviceFailMsgs [..] (Steam)
        ||| DeviceFailMsgs [..] (Water)
12      ||| tsg_out !SteamBoiler!PHYSICAL_UNITS_READY !0 of TParameter; stop ) )
    [> ( tsg_in !Controller !MODE !ModeEmergencyStop; stop ) )
14 where ... endproc (*PS*)
```

Fig. 3. The Physical System

The Physical System The specification of the physical system is given in Fig. 3. The physical system PS has an input and output port, the gates for device faults and repairs, and the gate of explosion (line 1). When the startup message STEAM_BOILER_WAITING is sent to the control unit (line 2), all physical devices and the GetMode process are instantiated independently of each other (line 4-12).

The pump communicates with the steam boiler in the internal gate sb_notify in order to inform the steam boiler about the real opening or closing of the pump (line 5-7). Using this information, the steam boiler can evaluate the real level of water.

In parallel, the physical system waits for the PROGRAM_READY message from the control unit (line 9). Once this messages is received, the processes for handling failure detection messages from the control unit are instantiated (line 10-11). In addition, the PHYSICAL_UNITS_READY message is sent to the control unit, after which it can leave the initialization mode (line 12). Whenever a ModeEmergencyStop message is received from the control unit, the whole physical system is interrupted and stops its behavior completely (line 13).

```
  process CU [tsg_in, tsg_out, Actions_In_EmergencyStop, System_stop]:
2 noexit :=
    hide  EmergencyStop, Passive_EmergencyStop in
4  ( ( hide iga in
        (      StopScanner [..]
6        ||| ( iga !SteamBoiler !PHYSICAL_UNITS_READY !0 of TParameter;
                ( SyncScanner [..] ||| AsyncScanner [..] ) ) )
8        |[iga]|
          ( tsg_in !SteamBoiler !STEAM_BOILER_WAITING !0 of TParameter;
10            ( ModeInitialization[..](..) ||| ThreeTimesStop [..] ) ) )
        [>] Passive_EmergencyStop; stop  )
12  |[EmergencyStop, Passive_EmergencyStop]|
      ( EmergencyStop; Passive_EmergencyStop; ModeEmergencyStop[..] )
14  where ... endproc (*CU*)
```

Fig. 4. The Control Unit

The Control Unit Fig. 4 gives the specification of the control unit. It has an input and output port. The Actions_In_EmergencyStop gate informs the operator about the emergency stop mode, while System_stop is used to stop the system externally (line 1).

The internal action EmergencyStop is used by the control unit components to inform the others about an exceptional situation, i.e. about the violation of a safety requirement (line 3). The internal action Passive_EmergencyStop is only needed technically for the interruption of all control unit components once the EmergencyStop has been issued by one of the components (line 5, 7, 10). The interruption is caused by the disabling expression (line 11), which allows us to specify exception handling elegantly.

The control unit CU works in a cyclic manner. This cycle takes place every 5 sec and consists of acquiring the measurement values (i.e. level of steam, etc.), processing the values, and if necessary, to open or close the pumps[5].

---

[5] In the functional specification the timing constraints are not specified.

Roughly speaking, gathering all incoming messages is executed by the scanning processes, while the automaton of the control unit is responsible for the decision making according to the current situation.

The SyncScanner is responsible for getting all synchronous messages from the physical system (line 7). This process can only work when the physical system is ready, i.e. it is instantiated after the PHYSICAL_UNITS_READY message (line 6). Besides the synchronous cycle, CU asynchronously exchanges messages with the physical system. Asynchronous messages are caused by a detected failure of a physical device. Asynchronous messages are scanned by the AsyncScanner process (line 7). The StopScanner takes care of the external gate System_stop.

Once all incoming messages of the control unit are scanned, they are offered at the internal communication gate iga for further processing (line 8). If none of the control unit components is able to process a message that is offered at iga within one cycle, a failure situation is assumed and EmergencyStop is issued. This ensures that any unexpected external communication leads to the emergency stop mode.

The control units contains five processes — ModeInitialization, ModeNormal, ModeDegraded, ModeRescue, ModeEmergencyStop, which are also graphically represented in Fig. 1. At any time, only one of them is active and represents the current mode of the control unit. If the current mode of the control unit changes, the current mode process instantiates one of the others. Initially, the control unit is in ModeInitialization (line 10). This process is instantiated when the physical system awaits the control unit, i.e. when it issues the STEAM_BOILER_WAITING message (line 9).

The mode processes define the automaton of the steam boiler system and constitute the main behavior of the control unit. They decide on proper reactions on incoming synchronous and asynchronous messages (for further details see below).

Finally, the control unit contains a process to observe the occurrence of System_Stop from outside in three subsequent periods. This is the ThreeTimesStop process (line 10).

The Transmission Medium The bidirectional transmission medium TM that is specified in Fig. 5, consists of two dedicated unidirectional lines Line between the physical system and the control unit and vice versa. Each message that is transmitted contains the physical device (a value of TDevice), a specific message type (a value of TMessage), and possibly a certain value (a value of TParameter) [6]. For example, "tsg_ps !Water !LEVEL !4" represents a message that is sent by the physical system, in particular by the water measurer. The actual level of water is 4. The transmission medium may properly transfer (line 5), generate (line 6)[7] or loose messages (line 7), what is specified as nondeterministic choices with the internal action i.

---

[6] All data types are explained in Section 4

[7] A random data generation in LOTOS is expressed as a communication between two actions, where both of them request a value of a given type, e.g. a ?x: Int; stop |[AS]| a ?y: Int; stop generates a random integer value.

```
  process TM[tsg_ps_in, tsg_ps_out, tsg_cu_in,tsg_cu_out]: noexit:=
2   Line[tsg_ps_out,tsg_cu_in] ||| Line[tsg_cu_out,tsg_ps_in]
    where process Line[a,b]: noexit:=
4         a ?d: TDevice ?t: TMessage ?p: TParameter;
          (    i; b !d !t !p; Line[a,b]
6         [] i; b ?d: TDevice ?t: TMessage ?p: TParameter; Line[a,b]
          [] i; Line[a,b] )
8      endproc (*Line*)
  endproc (*TM*)
```

Fig. 5. The Transmission Medium

Some Purely Functional Processes This section offers the specification of some processes in the physical system that have a purely functional behavior, i.e. without any time constraints.

```
  process DeviceFailMsgs [Repaired, tsg_in, tsg_out] (d: TDevice):noexit :=
2   tsg_in !d !FAILURE_DETECTION  !0 of TParameter;
    tsg_out !d !FAILURE_ACK !0 of TParameter;
4   Repaired !d; tsg_out !d !REPAIRED !0 of TParameter;
    tsg_in !d !REPAIRED_ACK !0 of TParameter; DeviceFailMsgs[..](..)
6 endproc (*DeviceFailMsgs*)
```

Fig. 6. The Device Failure Process

Failure Messages The process `DeviceFailMsgs` that is specified in Fig. 6, is responsible for the message handling in the case of device failure. As a simplification, a defective device is assumed to restrain from doing anything. Hence, the message exchange about failure detection, repair, and the respective acknowledgments is executed by a separate process. For any physical device exists one `DeviceFailMsgs` process.

Whenever the control unit detects a device failure, it sends a `FAILURE_DETECTION` message to the physical system. After receiving this message (line 2), `DeviceFailMsgs` issues an acknowledgment message, i.e. `FAILURE_ACK`. Once the device has been repaired by the operator (line 4), a `REPAIRED` message is send to the control unit (line 4), which answers with `REPAIRED_ACK` (line 5). This finishes the message handling for a device failure and the `DeviceFailMsgs` is re-instantiated (line 5).

```
  process Measurer [Fault, Repaired, tsg_in, tsg_out](d: TDevice):noexit :=
2   ( tsg_in !d !LEVEL_REQ !(Null); tsg_out !d !LEVEL ?p:TParameter; exit
    [> Fault !d; Repaired !d; exit )
4   >> Measurer[..](..)
    endproc (*Measurer*)
```

Fig. 7. The Measurer

Measurers The specification of the measurer process is given in Fig. 7[8]. Whenever the control unit requests the current level with a LEVEL_REQ message (line 2), the measurer answers with a random value (line 2). This reflects the possibility to send completely wrong values to the control unit. This is the reason why the control unit estimates the expected values (see below). The measurer stops to send level messages when it is faulty (line 3). Only after the repair it can send measurement values again.

2.2 Functional Verification

The formal semantics of LOTOS offers a wide variety of methods for formal validation and verification. In particular, we used simulation, validation, and testing of the specification.

The Simulation We applied LITE — the LOTOS Integrated Tool Environment [pE93] — to validate the steam boiler specification by the use of SMILE [Eer94]. This LOTOS simulator offers means

- to execute the specification interactively, i.e. the simulator determines all possible actions in a given behavior expression, under which one of them is chosen for further execution
- to generate automatically all possible executions of a given specification, and
- to generate automatically the underlying extended finite state machine (EFSM).

The simulator was very helpful in removing errors in the first versions of the specification [9]. However, a simulator cannot be used to verify the correctness.

The Verification Therefore, after validating the specification, we applied CADP — the Caesar/Aldebaran Distribution Package [Gar95] — for verifying the functional correctness. We had to restrict the specification in its size due to

---

[8] Let us note that the LEVEL message is both used to give the level of water or the level of steam. The messages are differentiated with the TDevice parameter, i.e. Steam or Water.

[9] The annex contains the functional specification for simulation purposes

restrictions that were imposed by CADP. We considered a specification of the steam boiler system where no faults of physical devices can occur and where transmission errors are neglected.

CADP uses full state space exploration techniques for the generation of the EFSM. In this sense, it is similar to SMILE, however it is more powerful with respect to the size of the specification that can be analyzed. In addition, it is capable of minimizing the generated EFSM and to automatically detect deadlocks. Moreover, once the EFSM has been generated, model checking methods (the CADP evaluator) can be applied in order to verify requirements that are described as temporal logic formulae.

The good news of the verification with CADP is that only one class of deadlocks has been detected. These are exactly the deadlocks that occur when the control unit enters the emergency stop mode. The bad news were that the EFSM generation took several days and was therefore interrupted before it finished.

The Testing Approach In order to overcome these problems, we decided to use in addition a testing approach. This allows us to define the requirements on the steam boiler system as separate LOTOS behavior expressions and to verify whether they are met by the specification or not. A tester defines the requirements to be tested as a LOTOS behavior expression, which possibly ends with a dedicated action, the so called Success action. The tester process is composed in parallel with the system to be tested and synchronizes in those externally visible gates of the system, that are sensitive to the test. There is no synchronization at the Success gate

$$\text{Tester}[\text{<gates>}, \text{Success}] \ |[\text{<gates>}]| \ \text{System}[\text{<gates>}, \ldots] \quad (1)$$

The parallel composition (1) of the Tester with the System to be tested enforces the System

- to execute only those behaviors that are dictated by the Tester until Success is reached,
- to end in a deadlock, i.e. the System does not contain the specified test sequence and Success is not reached, or
- to end in a livelock, i.e. the System ends in an infinite cycle with internal actions and Success is not reached.

If one considers all executions of the parallel composition (1), it is possible to distinguish three cases:

- The system *fails* the test, if there is no execution that leads to the Success action, i.e. the requirement is not fulfilled by the system.
- The system *may pass* the test, if there exists an execution where the Success action occurs and an execution where the Success action does not occur. Hence, the requirement is met in some executions.
- The system *must pass* the test, if the Success action occurs in all executions of the system, i.e. the requirement is always met.

Two important classes of tests are *acceptance* and *rejection* tests. In the first case, the system is tested to accept a given sequence of interactions with the environment. In the latter case, the system is tested to reject sequences of interactions that are not allowed to occur. Since tests reflects only a subset of requirements, which should of course be sufficiently large and should reflect all essential requirements, care has to be taken by the selection of the requirements.

The safety requirements of the steam boiler system require that the steam boiler system has to enter the emergency stop mode under all faulty conditions. These requirements can be represented by acceptance tests. Even more, since classical LOTOS has no means to enforce the execution of the internal action EmergencyStop, i.e. there has to be the possibility to enter the emergency stop mode but not the necessity[10], the tests are only may tests. We successfully tested for example whether

- a proper start up of the steam boiler system (may test) is possible
- the messages STEAM_BOILER_WAITING and PHYSICAL_UNITS_READY are each awaited only once by the control unit
- the message STEAM_FAILURE_DETECTED from the control unit is correctly acknowledged by the physical system
- the system enters the EmergencyStop mode after three times System_Stop
- whether the CU has to enter the EmergencyStop mode, in the case that the water and steam measurer are defective

For more details on the results of the testing approach, please refer to the annex.

3 The Time-Extended Specification in TE-LOTOS

3.1 The Timed Behavior of the Steam Boiler System

The Pumps We discuss the specification of the timed behavior of the steam boiler system in TE-LOTOS exemplarily for those processes with severe timing constraints. Firstly, we discuss the specification of a pump (given in Fig. 8) as a representative of the physical system. A pump is identified by its device identifier (a value of type TDevice). The rules for its behavior are the following:

- A pump is either open or closed (the initial state). The state information is offered to the control unit whenever needed without any delay (line 12-13).
- A pump can handle OPEN_PUMP and CLOSE_PUMP messages, where a closed pump ignores CLOSE_PUMP and an open pump ignores OPEN_PUMP.
- An open pump is closed immediately after receiving CLOSE_PUMP (line 9, 15-19).

---

[10] This is in difference to the timed specification of the steam boiler system, where the emergency stop mode has to be entered immediately due to the time concept in TE-LOTOS.

```
   process Pump[Fault,Repaired,tsg_in,tsg_out,sb_notify] (d:TDevice): noexit :=
2  hide req_open, req_close, notify in
   ( pmp [..] (d, PumpIsClosed) |[req_open, req_close, notify]| pumpswitch [..] )
4  [> ( Fault !d; Repaired !d; Pump[..](..)
   where
6  process pmp [req_open, req_close, notify, tsg_in, tsg_out]
                (d: TDevice, state : TPumpState): noexit :=
8       tsg_in !d !OPEN_PUMP !0 of TParameter; req_open; pmp [..](d, state)
     [] tsg_in !d !CLOSE_PUMP !0 of TParameter; req_close;
10      notify ! PumpIsClosed {0}; pmp [..](d, PumpIsClosed)
     [] notify ?new_state: TPumpState; pmp [..] (d, new_state)
12   [] tsg_in !d !PUMP_STATE_REQ !0 of TParameter;
        tsg_out !d !PUMP_STATE !state; pmp [..](d, state)
14 endproc (*pmp*)
   process pumpswitch[req_open,req_close,notify,sb_notify]:noexit :=
16    ( req_open; Ticker[req_open, req_close, notify, sb_notify] (0.0)
        [> req_close; notify !PumpIsClosed; sb_notify !PumpIsClosed; exit )
18    >> pumpswitch[..]
   endproc (*pumpswitch*)
20 process Ticker[req_open,req_close,notify,sb_notify](eps: Time): exit:=
        Wait (period−eps); notify !PumpIsOpen; sb_notify !PumpIsOpen; exit
22   [] req_open @t; Ticker[..](eps + t)
   endproc (*ticker*)
24 endproc (*Pump*)
```

Fig. 8. The Pumps

- A closed pump is opened within 5 sec after receiving OPEN_PUMP. An additional OPEN_PUMP message arriving within this time interval has no effect. A CLOSE_PUMP message arriving within this time makes the OPEN_PUMP message ineffective, i.e. the pump remains closed (line 8, 15-23).
- A pump can break down and stay completely inactive until it is repaired (line 4).

The subprocess pmp is responsible for the message exchange with the control unit. The subprocess pumpswitch specifies the behavior to open a closed pump. During that period additional open and close pump requests are possible. The Ticker counts the time interval of 5 sec.

The Scanners The process Sync_Scanner that is given in Fig. 9 scans every 5 sec the state information of physical devices such as the water level, the steam level, and the pump states. Scanning messages is realized by a pair of request and answer message. For example, tsg_out !Steam !LEVEL_REQ !(Null) {0} requests the level of steam from the steam measurer (line 2) and awaits the answer immediately with tsg_in !Steam !LEVEL ?qs: TParameter {0} (line 3).

Simulation of a Steam-Boiler

Annette Lötzbeyer

Forschungszentrum Informatik, Karlsruhe

Abstract. In this paper, we describe the implementation and usage of a simulation with graphical visualization for a steam-boiler [Abr94] (see Chapter AS, this book).
The simulation imitates the behaviour of the steam-boiler and can be controlled using a simple ASCII protocol. It reacts to defined commands and sends information about its state every five seconds.
Furthermore, we describe the options available, the protocol which functions as an interface between the simulation and the controller, and how one can simulate component failures within the boiler. We give information about the simulation and how to install it.

1 Goal and Motivation

The aim of the steam-boiler case study is to examine various formal methods in relation to the steam-boiler specification problem. In the case study, different solutions to the steam-boiler specification problem are compared. One aspect of this comparison is to investigate the ability of a method to create or to develop an executable program. Our aim is to provide a graphical visualization of a steam-boiler simulation, which can be used to show the validity of a control program. Using this simulation, one connects a program to it and tests the program. To limit the effort needed for connecting the simulation and the program, we have chosen a simple ASCII protocol to interface the simulation.

2 Architecture and Behaviour of the Simulation

The simulation imitates the behaviour of the steam-boiler. In Figure 1, a view of the running simulation is shown. On the left side, there are the four pumps with four pump controllers which determine the water influx. On the right side, the boiler with a valve, a scale, the water level display (the black label) and the steam measurement device are shown.

Every five seconds, the simulation sends messages about the status of its devices (physical units). Devices are defined as: pumps, pump controllers, water level display, and the steam measurement device. According to the task description, the simulation has to stop after sending messages and must continue to imitate the behaviour of the steam-boiler, only when it receives the commands from the program. In a realistic situation in a plant, this synchronous assumption cannot be satisfied. Therefore, we decided not to implement a synchronous simulation. Instead, our simulation keeps running after sending the status messages and executes the commands, as soon as they are received. A delay may occur due to the time needed for transmission.

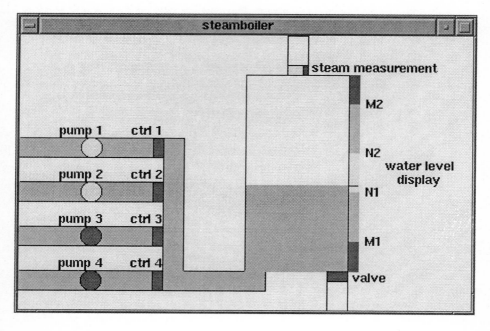

Fig. 1. View of the steam-boiler simulation

3 Installing the Simulation

To implement the simulation, we used the Tool Command Language (Tcl) and the widget Toolkit (Tk). Tcl/Tk can be easily used to integrate tools and to construct graphical user interfaces. The simulation runs on the UNIX operation system with X-windows.

In order to use the simulation, a running Tcl/Tk installation must exist. We have tested the simulation with the following versions of Tcl/Tk:

Tcl 7.3
Extended Tcl 7.3b
Tk 3.6

These versions are available on our WWW server

http://www.fzi.de/divisions/prost/
    projects/steam_boiler/steam_boiler.html

or via ftp:

ftp://ftp.fzi.de/pub/korso/steam_boiler/

The source code of the simulation is located at the same place (see also CD-ROM Annex L). After getting the file simulator-2.1a.tar.gz, the file must be de-compressed and un-tared in a directory of your choice. The README file contains a list of files, which can be used to check whether the installation has been completed. The file **startsimu** must be changed by entering the correct path and name for the wish interpreter. These changes must be made in the first and the fifth line of the file. Now the steam boiler simulation should run.

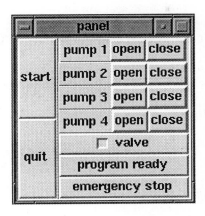

Fig. 2. The control panel

4 Starting the Simulation

There are two possibilities for running the simulation. First, the simulation can run with a control panel, so that no controller is needed (Fig. 2). Using this panel, the simulation can be controlled manually. Starting the simulation together with the control panel is done by entering:

    startsimu

Secondly, the simulation can be connected to an separate control program:

    startsimu -con *name_of_controller*

name_of_controller is the name of the controller. In this case, the simulation sends messages about its devices to the controller and executes the commands received from the control program.

In both cases, an additional window appears, which shows a failure panel. Using this panel, one can simulate failures and repairs of the physical devices.

The option

   -fail *name_of_failure_file*

suppresses the failure panel. Instead, the failures are generated by commands written in the failure file. A more detailed description of how to simulate failures can be found in Section 5.

There are several extra options available:

   -dbg   Puts the simulation in a debug mode. All commands sent by the simulation are displayed.
   -mon   Puts the simulation in a monitoring mode. All commands received by the simulation are displayed.
   -end   This option causes the simulation to end each message list with the additional message END_OF_TRANSMISSION (instead of a blank line).
   -bw   Puts the simulation in a black-and-white mode.

5 The Protocol to control the Simulation

The simulation and the controller are connected via UNIX-pipes. Therefore, the control program only has to write ASCII messages to *stdout* and read the messages from the simulation from *stdin*, in order to control the simulation. The messages sent to the simulation must be separated by a carriage return. It is important to flush the pipe, since otherwise the messages will be buffered and their effects will be delayed. In the following section, the messages are described in detail.

5.1 Messages sent to the simulation

The messages sent to the simulation can be divided into three classes: control commands, failure detection messages and acknowledgment messages. The control commands are sent from the control program to the simulation to steer the steam boiler. In our simulation, the commands are executed as soon as they are received by the simulation. The other two classes contain messages which do not result in any changes in the behaviour of the system. The following messages are available:

Control commands:

   MODE m If the program sends the command MODE stop, the system will stop. All other modes will have no effect on the system.
   PROGRAM_READY   Starts the boiler.
   VALVE   Opens/closes the valve. Initially the valve is closed. The first VALVE command opens the valve, the next closes it again.

OPEN_PUMP *n*	Opens pump n.
CLOSE_PUMP *n*	Closes pump n.
system_quit	Quits the system. This and the system_start commands are additional commands which are not included in the task description.
system_start	Re-starts the system if it has been stopped.

Failure detection messages:

If the control program detects that a device is broken, it sends a failure detection message. After that the simulation will send the corresponding failure acknowledgement message.

```
PUMP_FAILURE_DETECTION n
PUMP_CONTROL_FAILURE_DETECTION n
LEVEL_FAILURE_DETECTION
STEAM_OUTCOME_DETECTION
```

Repair acknowledgement messages:

If a device has been repaired, the simulation sends the corresponding repair message to the control program. The acknowledgement messages, which should be sent back to the simulation, are accepted, but have no effect on system functioning.

```
PUMP_REPAIRED_ACKNOWLEDGEMENT n
PUMP_CONTROL_REPAIRED_ACKNOWLEDGEMENT n
LEVEL_REPAIRED_ACKNOWLEDGEMENT
STEAM_OUTCOME_REPAIRED_ACKNOWLEDGEMENT
```

5.2 Messages sent by the simulation

In each cycle, the simulation sends a list of messages, which contain information about the boiler devices. There are certain other messages, which do not appear in every cycle. These are messages sent in the initialization mode and messages sent in response to incoming messages.

Messages sent in the initialization mode:

```
STEAM_BOILER_WAITING
PHYSICAL_UNITS_READY
```

Information messages:

The following messages are sent each cycle:

PUMP_STATE(n,b)	State of pump n. PUMP_STATE(1,0) means: pump 1 is closed.
PUMP_CONTROL_STATE(n,b)	State of pump n. PUMP_CONTROL_STATE(1,0) means, that there is no water flow at pump controller 1.
LEVEL(v)	Indicates the water level (in litres). v is a real number.
STEAM(v)	Indicates the steam output (in litre/sec). v is a real number.

Repair messages:

PUMP_REPAIRED(n)	Pump n has been repaired.
PUMP_CONTROLLER_REPAIRED(n)	Pump controller n has been repaired.
LEVEL_REPAIRED	The water level measuring unit has been repaired.
STEAM_REPAIRED	The steam output measuring unit has been repaired.

Failure acknowledgement messages:

PUMP_FAILURE_ACKNOWLEDGEMENT(n)
PUMP_CONTROL_FAILURE_ACKNOWLEDGEMENT(n)
LEVEL_FAILURE_ACKNOWLEDGEMENT
STEAM_OUTCOME_FAILURE_ACKNOWLEDGEMENT

Stop message:

STOP   Every use of the emergency stop button will cause one STOP message. The emergency stop button is on the failure panel (Section 6).

6 How to simulate Failures

There are two ways to simulate failures and repairs to the devices. When starting the simulation, a failure panel appears (Fig. 3). The device shows a failure, when the corresponding button is pressed. After this, the indicator button turns red. To repair a faulty device, the corresponding button must be pressed again. Then, the indicator button becomes green. The stop button functions in the same way as the emergency button in the task description. The quit button can be used to end the simulation.

Instead of using the failure panel, the simulation can be used with the option -fail *name_of_failure_file*. As a result, the simulation reads the commands of the specified failure file, which creates the failures and the repairs. Some example files are provided for the simulation in the subdirectory `failurefiles`. It is easy to write failure files, after taking a look at the example files.

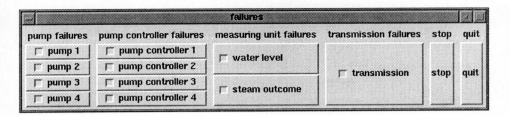

Fig. 3. The failure panel

7 Physical constants

The steam boiler contains several physical constants for which are no fixed numbers were given in the task description (e.g., the maximum capacity of the steam boiler). Consequently, we had to give the symbolic constants real values. In our configuration, at least two working pumps are needed to control the water level. All those constants are defined in the file system_const.tcl. They can be changed according to the needs of the user.

References

[Abr94] Jean-Raimond Abrial. Steam-boiler control specification problem. Technical report, F-75014 Paris, 26 rue des Plantes, 1994.

Steam-Boiler Control Specification Problem

Jean-Raymond Abrial

1 Problem Statement

1.1 Introduction

This text constitutes an informal specification of a program which serves to control the level of water in a steam-boiler. It is important that the program works correctly because the quantity of water present when the steam-boiler is working has to be neither too low nor to high; otherwise the steam-boiler or the turbine sitting in front of it might be seriously affected.

The proposed specification is derived from an original text that has been written by LtCol. J.C. Bauer for the Institute for Risk Research of the University of Waterloo, Ontario, Canada. The original text has been submitted as a competition problem to be solved by the participants of the International Software Safety Symposium organized by the Institute for Risk Research. It has been given to us by the Institut de Protection et de Sureté Nucléaire, Fontenay-aux-Roses, France. We would like to thank the author, the Institute for Risk Research and the Institut de Protection et de Sureté Nucléaire for their kind permission to use their text.

The text to follow is severly biased to a particular implementation. This is very often the case with industrial specifications that are rarely independent from a certain implementation people have in mind. In that sense, this specification is realistic. Your first formalization steps could be *much more abstract* if that seems important to you (in particular if your formalism allows you to do so). In other words, you are encouraged to *structure* your specification in a way that is not necessarily the same as the one proposed in what follows. But in any case, you are asked to demonstrate that your specification can be refined to an implementation that is close enough to the functional requirements of the "specification" proposed below.

You might also judge that the specification contains some loose ends and inconsistencies. Do not hesitate to point them out and to take yourself some appropriate decisions. The idea, however, is that such inconsistencies should be solely within the *organization* of the system and *not within its physical properties*.

We are aware of the fact that the text to follow does not propose any precise model of the physical evolution of the system, only elementary suggestions. As a consequence, you may have to take some simple, even simplistic, abstract decisions concerning such a physical model.

1.2 Physical environment

The system comprises the following units

- the steam-boiler
- a device to measure the quantity of water in the steam-boiler
- four pumps to provide the steam-boiler with water
- four devices to supervise the pumps (one controller for each pump)
- a device to measure the quantity of steam which comes out of the steam-boiler
- an operator desk
- a message transmission system

1.3 The steam-boiler

The steam-boiler is characterized by the following elements:

- A valve for evacuation of water. It serves only to empty the steam-boiler in its initial phase.
- Its total capacity C (indicated in litres).
- The minimal limit quantity M_1 of water (in litres). Below M_1 the steam-boiler would be in danger after five seconds, if the steam continued to come out at its maximum quantity without supply of water from the pumps.
- The maximal limit quantity M_2 of waters (in litres). Above M_2 the steam-boiler would be in danger after five seconds, if the pumps continued to supply the steam-boiler with water without possibility to evacuate the steam.
- The minimal normal quantity N_1 of water in litres to be maintained in the steam-boiler during regular operation ($M_1 < N_1$).
- The maximal normal quantity N_2 of water (in litres) to be maintained in the steam-boiler during regular operation ($N_2 < M_2$).
- The maximum quantity W of steam (in litres/sec) at the exit of the steam-boiler.
- The maximum gradient U_1 of increase of the quantity of steam (in litres/sec/sec).
- The maximum gradient U_2 of decrease of the quantity of steam (in litres/sec/sec).

1.4 The water level measurement device

The device to measure the level of water in the steam-boiler provides the following information

- the quantity q (in litres) of water in the steam-boiler.

1.5 The pumps

Each pump is characterized by the following elements

- Its capacity P (in litres/sec)
- Its functioning mode: on or off
- it's being started: after having been switched on the pump needs five seconds to start pouring water into the boiler (this is due to the fact that the pump does not balance instantaneously the pressure of the steam-boiler).
- it's being stopped: with instantaneous effect

1.6 The pump control device

Each pump controller provides the following information:

- the water circulates from the pump to the steam-boiler or, in the contrary, it does not circulate.

1.7 The steam measurement device

The device to measure the quantity of steam which comes out of the steam-boiler provides the following information:

- a quantity of steam v (in litres/sec).

1.8 Summary of constants and variables

The following table summerizes the various constants or physical variables of the system:

	Unit	Comment
		Quantity of water in the steam-boiler
C	litre	Maximal capacity
M_1	litre	Minimal limit
M_2	litre	Maximal limit
N_1	litre	Minimal normal
N_2	litre	Maximal normal
		Outcome of steam at the exit of the steam-boiler
W	litre/sec	Maximal quantity
U_1	litre/sec/sec	Maximum gradient of increase
U_2	litre/sec/sec	Maximum gradient of decrease
		Capacity of each pump
P	litre/sec	Nominal capacity
		Current measures
q	litre	Quantity of water in the steam-boiler
p	litre/sec	Throughput of the pumps
v	litre/sec	Quantity of steam exiting the steam-boiler

1.9 The overall operation of the program

The program communicates with the physical units through messages which are transmitted over a number of dedicated lines connecting each physical unit with the control unit. In first approximation, the time for transmission can be neglected.

The program follows a cycle and a priori does not terminate. This cycle takes place each five seconds and consists of the following actions:

- Reception of messages coming from the physical units.
- Analysis of informations which have been received.
- Transmission of messages to the physical units.

To simplify matters, and in first approximation, all messages coming from (or going to) the physical units are supposed to be received (emitted) *simultaneously* by the program at each cycle.

1.10 Operation modes of the program

The program operates in different modes, namely: *initialization, normal, degraded, rescue, emergency stop*.

1.11 *Initialization* mode

The *initialization* mode is the mode to start with. The program enters a state in which it waits for the message STEAM-BOILER_WAITING to come from the physical units. As soon as this message has been received the program checks whether the quantity of steam coming out of the steam-boiler is really zero. If the unit for detection of the level of steam is defective—that is, when v is not equal to zero—the program enters the *emergency stop* mode. If the quantity of water in the steam-boiler is above N_2 the program activates the valve of the steam-boiler in order to empty it. If the quantity of water in the steam-boiler is below N_1 then the program activates a pump to fill the steam-boiler. If the program realizes a failure of the water level detection unit it enters the *emergency stop* mode. As soon as a level of water between N_1 and N_2 has been reached the program can send continuously the signal PROGRAM_READY to the physical units until it receives the signal PHYSICAL_UNITS_READY which must necessarily be emitted by the physical units. As soon as this signal has been received, the program enters either the mode *normal* if all the physical units operate correctly or the mode *degraded* if any physical unit is defective. A transmission failure puts the program into the mode *emergency stop*.

1.12 *Normal* mode

The normal mode is the standard operating mode in which the program tries to maintain the water level in the steam-boiler between N_1 and N_2 with all physical units operating correctly. As soon as the water level is below N_1 or above N_2 the level can be adjusted by the program by switching the pumps on or off. The corresponding decision is taken on the basis of the information which has been received from the physical units. As soon as the program recognizes a failure of the water level measuring unit it goes into *rescue* mode. Failure of any other physical unit puts the program into *degraded* mode. If the water level is risking to reach one of the limit values M_1 or M_2 the program enters the mode *emergency stop*. This risk is evaluated on the basis of a maximal behaviour of the physical units. A transmission failure puts the program into *emergency stop* mode.

1.13 *Degraded* mode

The *degraded* mode is the mode in which the program tries to maintain a satisfactory water level despite of the presence of failure of some physical unit. It is assumed however that the water level measuring unit in the steam-boiler is working correctly. The functionality is the same as in the preceding case. Once all the units which were defective have been repaired, the program comes back to *normal* mode. As soon as the program sees that the water level measuring unit has a failure, the program goes into mode *rescue*. If the water level is risking to reach one of the limit values M_1 or M_2 the program enters the mode *emergency stop*. A transmission failure puts the program into *emergency stop* mode.

1.14 *Rescue* mode

The *rescue* mode is the mode in which the program tries to maintain a satisfactory water level despite of the failure of the water level measuring unit. The water level is then estimated by a computation which is done taking into account the maximum dynamics of the quantity of steam coming out of the steam-boiler. For the sake of simplicity, this calculation can suppose that exactly n liters of water, supplied by the pumps, do account for exactly the same amount of boiler contents (no thermal expansion). This calculation can however be done only if the unit which measures the quantity of steam is itself working and if one can rely upon the information which comes from the units for controlling the pumps. As soon as the water measuring unit is repaired, the program returns into mode *degraded* or into mode *normal*. The program goes into *emergency stop* mode if it realizes that one of the following cases holds: the unit which measures the outcome of steam has a failure, or the units which control the pumps have a failure, or the water level risks to reach one of the two limit values. A transmission failure puts the program into *emergency stop* mode.

1.15 *Emergency stop* mode

The *emergency stop* mode is the mode into which the program has to go, as we have seen already, when either the vital units have a failure or when the water level risks to reach one of its two limit values. This mode can also be reached after detection of an erroneous transmission between the program and the physical units. This mode can also be set directly from outside. Once the program has reached the *Emergency stop* mode, the physical environmente is then responsible to take approrpiate actions, and the program stops.

1.16 Messages sent by the program

The following messages can be sent by the program:

– MODE(m): The program sends, at each cycle, its current mode of operation to the physical units.

- PROGRAM_READY: In *initialization* mode, as soon as the program assumes to be ready, this message is continuously sent until the message PHYSICAL_UNITS_READY coming from the physical units has been received.
- VALVE: In *initialization* mode this message is sent to the physical units to request opening and then closure of the valve for evacuation of water from the steam-boiler.
- OPEN_PUMP(n): This message is sent to the physical units to activate a pump.
- CLOSE_PUMP(n): This message is sent to the physical units to stop a pump.
- PUMP_FAILURE_DETECTION(n): This message is sent (until receipt of the corresponding acknowledgement) to indicate to the physical units that the program has detected a pump failure.
- PUMP_CONTROL_FAILURE_DETECTION(n): This message is sent (until receipt of the corresponding acknowledgement) to indicate to the physical units that the program has detected a failure of the physical unit which controls a pump.
- LEVEL_FAILURE_DETECTION: This message is sent (until receipt of the corresponding acknowledgement) to indicate to the physical units that the program has detected a failure of the water level measuring unit.
- STEAM_FAILURE_DETECTION: This message is sent (until receipt of the corresponding acknowledgement) to indicate to the physical units that the program has detected a failure of the physical unit which measures the outcome of steam.
- PUMP_REPAIRED_ACKNOWLEDGEMENT(n): This message is sent by the program to acknowledge a message coming from the physical units and indicating that the corresponding pump has been repaired.
- PUMP_CONTROL_REPAIRED_ACKNOWLEDGEMENT(n): This message is sent by the program to acknowledge a message coming from the physical units and indicating that the corresponding physical control unit has been repaired.
- LEVEL_REPAIRED_ACKNOWLEDGEMENT: This message is sent by the program to acknowledge a message coming from the physical units and indicating that the water level measuring unit has been repaired.
- STEAM_REPAIRED_ACKNOWLEDGEMENT: This message is sent by the program to acknowledge a message coming from the physical units and indicating that the unit which measures the outcome of steam has been repaired.

1.17 Messages received by the program

The following messages can be received by the program:

- STOP: When the message has been received three times in a row by the program, the program must go into *emergency stop*.
- STEAM_BOILER_WAITING: When this message is received in *initialization* mode it triggers the effective start of the program.

- PHYSICAL_UNITS_READY: This message when received in *initialization* mode acknowledges the message PROGRAM_READY which has been sent previously by the program.
- PUMP_STATE(n, b): This message indicates the state of pump n (open or closed). This message must be present during each transmission.
- PUMP_CONTROL_STATE(n, b): This message gives the information which comes from the control unit of pump n (there is flow of water or there is no flow of water). This message must be present during each transmission.
- LEVEL(v): This message contains the information which comes from the water level measuring unit. This message must be present during each transmission.
- STEAM(v): This message contains the information which comes from the unit which measures the outcome of steam. This message must be present during each transmission.
- PUMP_REPAIRED(n): This message indicates that the corresponding pump has been repaired. It is sent by the physical units until a corresponding acknowledgement message has been sent by the program and received by the physical units.
- PUMP_CONTROL_REPAIRED(n): This message indicates that the corresponding control unit has been repaired. It is sent by the physical units until a corresponding acknowledgement message has been sent by the program and received by the physical units.
- LEVEL_REPAIRED: This message indicates that the water level measuring unit has been repaired. It is sent by the physical units until a corresponding acknowledgement message has been sent by the program and received by the physical units.
- STEAM_REPAIRED: This message indicates that the unit which measures the outcome of steam has been repaired. It is sent by the physical units until a corresponding acknowledgement message has been sent by the program and received by the physical units.
- PUMP_FAILURE_ACKNOWLEDGEMENT(n): By this message the physical units acknowledge the receipt of the corresponding failure detection message which has been emitted previously by the program.
- PUMP_CONTROL_FAILURE_ACKNOWLEDGEMENT(n): By this message the physical units acknowledge the receipt of the corresponding failure detection message which has been emitted previously by the program.
- LEVEL_FAILURE_ACKNOWLEDGEMENT: By this message the physical units acknowledge the receipt of the corresponding failure detection message which has been emitted previously by the program.
- STEAM_OUTCOME_FAILURE_ACKNOWLEDGEMENT: By this message the physical units acknowledge the receipt of the corresponding failure detection message which has been emitted previously by the program.

1.18 Detection of equipment failures

The following erroneous kinds of behaviour are distinguished to decide whether certain physical units have a failure:

- PUMP: (1) Assume that the program has sent a start or stop message to a pump. The program detects that during the following transmission that pump does not indicate its having effectively been started or stopped. (2) The program detects that the pump changes its state spontaneously.
- PUMP_CONTROLLER: (1) Assume that the program has sent a start or stop message to a pump. The program detects that during the second transmission after the start or stop message the pump does not indicate that the water is flowing or is not flowing; this despite of the fact that the program knows from elsewhere that the pump is working correctly. (2) The program detects that the unit changes its state spontaneously.
- WATER_LEVEL_MEASURING_UNIT: (1) The program detects that the unit indicates a value which is out of the valid static limits–i.e. between 0 and C. (2) The program detects that the unit indicates a value which is incompatible with the dynamics of the system.
- STEAM_LEVEL_MEASURING_UNIT: (1) The program detects that the unit indicates a value which is out of the valid static limits—i.e. between 0 and W. (2) The program detects that the unit indicates a value which is incompatible with the dynamics of the system.
- TRANSMISSION: (1) The program receives a message whose presence is aberrant. (2) The program does not receive a message whose presence is indispensable.

2 Additional Information Concerning the Physical Behaviour of the Steam Boiler

In this section, we propose some additional information about a possible model of the boiler system. Such information can be taken into account in the construction of your own model. Besides the raw measures q, p, and v, we shall consider the following quantities that are called the adjusted values

qa_1, qa_2 minimal and maximal adjusted quantity of water
pa_1, pa_2 minimal and maximal adjusted throughput of the pumps
va_1, va_2 minimal and maximal adjusted quantity of exiting steam

Such adjusted quantities are defined to be either the *raw* values effectively delivered in the messages or the *calculated* values estimated from previous cycle. The raw values are chosen in case the corresponding equipment is considered to be not broken. Otherwise, the calculated quantities are chosen.

The calculated quantities are denoted as follows:

qc_1, qc_2 minimal and maximal calculated quantity of water
pc_1, pc_2 minimal and maximal calculated throughput of the pumps
vc_1, vc_2 minimal and maximal calculated quantity of exiting steam

We have thus:

$qa_1 = qc_1$ if the water level equipment is considered broken, and
$qa_1 = q$ otherwise.

$qa_2 = qc_2$ if the water level equipment is considered broken, and
$qa_2 = q$ otherwise.

Similar definitions hold for the other quantities.

The calculated quantities can be determined from the adjusted quantities as follows:

$$qc_1 = qa_1 - va_2 \Delta t - \frac{1}{2} U_1 \Delta t^2 + pa_1$$

$$qc_2 = qa_2 - va_1 \Delta t + \frac{1}{2} U_2 \Delta t^2 + pa_2$$

$$rc_1 = ra_1 - U_2 \Delta t$$

$$rc_2 = ra_2 + U_1 \Delta t$$

$$pc_1 = \sum_{i=1}^{4} pc_{1,i}$$

$$pc_2 = \sum_{i=1}^{4} pc_{2,i}$$

where t is the cycle time and where $pc_{1,i}$ and $pc_{2,i}$ are the minimal and maximal throughputs of each individual pump/monitor. Such quantities are defined as follows:

$pc_{1,i} = 0$ if C_i holds, and
$pc_{1,i} = P$ otherwise.
$pc_{2,i} = P$ if D_i holds, and
$pc_{2,i} = 0$ otherwise.

When the condition C_i holds, we *consider* that the water is not flowing through the i-th pump. This is the case when the i-th pump/monitor is broken (since, in that case, we have *no information* and thus we estimate the worst *minimal* throughput to be 0). This is also the case when either the order to close the pump has just been given or when we know that the pump was already closed.

When the condition D_i holds, we *consider* that the water is flowing through the i-th pump. This is the case when the i-th pump/monitor is broken (since, in that case, we have *no information* and thus we estimate the worst *maximal* throughput to be P). This is also the case when either the order to open the pump has just been given or when we know that the pump was already open. Clearly the previous conditions C_i and D_i could be made more elaborate.

Note that an equipment (that is not already considered broken) becomes broken when the corresponding raw quantity is not a member of the interval of quantities *calculated* at the previous cycle.

To determine whether the water level is too low (opening the pumps) or too high (closing the pumps), we have to use now the interval (qa_1, qa_2). There are thus 6 cases to consider according to the following diagrams:

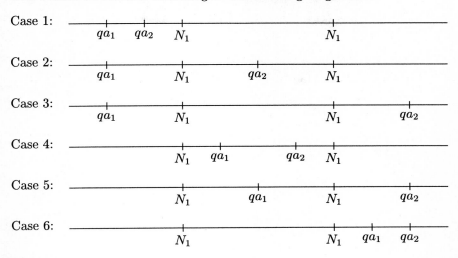

We might consider the following decisions:

Case 1	opening pumps
Case 2	opening pumps
Case 3	?
Case 4	do nothing
Case 5	closing pumps
Case 6	closing pumps

Case 3 poses a problem. We might decide to do nothing, as in case 4. The decision concerning shutdown is simpler. We might decide to shut the system down when at least one of the following condition holds

$$qa_1 \leq M_1$$
$$qa_2 \geq M_2$$
$$qc_1 \leq M_1$$
$$qc_2 \geq M_2$$

Here qc_1 and qc_2 are the calculated values for next cycle.

Author Index

Abrial, J.-R., 1, 500
Addibpour, M., 13
Andriessens, C., 35

Beierle, C., 52
Bidoit, M., 79
Börger, E., 1, 52
Büssow, R., 109
Butler, M., 129

Cattel, T., 149, 203
Chevenier, C., 79
Cuéllar, J., 165, 184

Dauchy, P., 233
Đurđanović, I., 52
Duval, G., 149, 203

Gargantini, A., 218
Gaudel, M.-C., 233
Glässer, U., 52

Henzinger, T.A., 265
Hooman, J., 453
Howard, W.-T., 265

Khoury, C., 233
Kosiuczenko, P., 379

Langmaack, H., 1
Ledru, Y., 283
Leeb, G., 318
Lesske, F., 339

Li, X.S., 359
Lindner, T., 35
Lötzbeyer, A., 493
Lynch, N., 318

Merz, S., 339
Morzenti, A., 218

Ölveczky, P.C., 379

Pellen, C., 79
Potet, M.-L., 283

Ravn, A.P., 403
Riccobene, E., 52
Ryckbosch, J., 79

Schenke, M., 403
Schieferdecker, I., 473
Schinagl, C.P., 428
Sekerinski, E., 129
Sere, K., 129

Tyugu, E., 13

Vitt, J., 453

Wang, J.A., 359
Weber, M., 109
Wildgruber, I., 165, 184
Willig, A., 473
Wirsing, M., 379

Springer and the environment

At Springer we firmly believe that an international science publisher has a special obligation to the environment, and our corporate policies consistently reflect this conviction.

We also expect our business partners – paper mills, printers, packaging manufacturers, etc. – to commit themselves to using materials and production processes that do not harm the environment. The paper in this book is made from low- or no-chlorine pulp and is acid free, in conformance with international standards for paper permanency.

Lecture Notes in Computer Science

For information about Vols. 1–1091

please contact your bookseller or Springer-Verlag

Vol. 1092: H. Kleine Büning (Ed.), Computer Science Logic. Proceedings, 1995. VIII, 487 pages. 1996.

Vol. 1093: L. Dorst, M. van Lambalgen, F. Voorbraak (Eds.), Reasoning with Uncertainty in Robotics. Proceedings, 1995. VIII, 387 pages. 1996. (Subseries LNAI).

Vol. 1094: R. Morrison, J. Kennedy (Eds.), Advances in Databases. Proceedings, 1996. XI, 234 pages. 1996.

Vol. 1095: W. McCune, R. Padmanabhan, Automated Deduction in Equational Logic and Cubic Curves. X, 231 pages. 1996. (Subseries LNAI).

Vol. 1096: T. Schäl, Workflow Management Systems for Process Organisations. XII, 200 pages. 1996.

Vol. 1097: R. Karlsson, A. Lingas (Eds.), Algorithm Theory – SWAT '96. Proceedings, 1996. IX, 453 pages. 1996.

Vol. 1098: P. Cointe (Ed.), ECOOP '96 – Object-Oriented Programming. Proceedings, 1996. XI, 502 pages. 1996.

Vol. 1099: F. Meyer auf der Heide, B. Monien (Eds.), Automata, Languages and Programming. Proceedings, 1996. XII, 681 pages. 1996.

Vol. 1100: B. Pfitzmann, Digital Signature Schemes. XVI, 396 pages. 1996.

Vol. 1101: M. Wirsing, M. Nivat (Eds.), Algebraic Methodology and Software Technology. Proceedings, 1996. XII, 641 pages. 1996.

Vol. 1102: R. Alur, T.A. Henzinger (Eds.), Computer Aided Verification. Proceedings, 1996. XII, 472 pages. 1996.

Vol. 1103: H. Ganzinger (Ed.), Rewriting Techniques and Applications. Proceedings, 1996. XI, 437 pages. 1996.

Vol. 1104: M.A. McRobbie, J.K. Slaney (Eds.), Automated Deduction – CADE-13. Proceedings, 1996. XV, 764 pages. 1996. (Subseries LNAI).

Vol. 1105: T.I. Ören, G.J. Klir (Eds.), Computer Aided Systems Theory – CAST '94. Proceedings, 1994. IX, 439 pages. 1996.

Vol. 1106: M. Jampel, E. Freuder, M. Maher (Eds.), Over-Constrained Systems. X, 309 pages. 1996.

Vol. 1107: J.-P. Briot, J.-M. Geib, A. Yonezawa (Eds.), Object-Based Parallel and Distributed Computation. Proceedings, 1995. X, 349 pages. 1996.

Vol. 1108: A. Díaz de Ilarraza Sánchez, I. Fernández de Castro (Eds.), Computer Aided Learning and Instruction in Science and Engineering. Proceedings, 1996. XIV, 480 pages. 1996.

Vol. 1109: N. Koblitz (Ed.), Advances in Cryptology – Crypto '96. Proceedings, 1996. XII, 417 pages. 1996.

Vol. 1110: O. Danvy, R. Glück, P. Thiemann (Eds.), Partial Evaluation. Proceedings, 1996. XII, 514 pages. 1996.

Vol. 1111: J.J. Alferes, L. Moniz Pereira, Reasoning with Logic Programming. XXI, 326 pages. 1996. (Subseries LNAI).

Vol. 1112: C. von der Malsburg, W. von Seelen, J.C. Vorbrüggen, B. Sendhoff (Eds.), Artificial Neural Networks – ICANN 96. Proceedings, 1996. XXV, 922 pages. 1996.

Vol. 1113: W. Penczek, A. Szałas (Eds.), Mathematical Foundations of Computer Science 1996. Proceedings, 1996. X, 592 pages. 1996.

Vol. 1114: N. Foo, R. Goebel (Eds.), PRICAI'96: Topics in Artificial Intelligence. Proceedings, 1996. XXI, 658 pages. 1996. (Subseries LNAI).

Vol. 1115: P.W. Eklund, G. Ellis, G. Mann (Eds.), Conceptual Structures: Knowledge Representation as Interlingua. Proceedings, 1996. XIII, 321 pages. 1996. (Subseries LNAI).

Vol. 1116: J. Hall (Ed.), Management of Telecommunication Systems and Services. XXI, 229 pages. 1996.

Vol. 1117: A. Ferreira, J. Rolim, Y. Saad, T. Yang (Eds.), Parallel Algorithms for Irregularly Structured Problems. Proceedings, 1996. IX, 358 pages. 1996.

Vol. 1118: E.C. Freuder (Ed.), Principles and Practice of Constraint Programming — CP 96. Proceedings, 1996. XIX, 574 pages. 1996.

Vol. 1119: U. Montanari, V. Sassone (Eds.), CONCUR '96: Concurrency Theory. Proceedings, 1996. XII, 751 pages. 1996.

Vol. 1120: M. Deza. R. Euler, I. Manoussakis (Eds.), Combinatorics and Computer Science. Proceedings, 1995. IX, 415 pages. 1996.

Vol. 1121: P. Perner, P. Wang, A. Rosenfeld (Eds.), Advances in Structural and Syntactical Pattern Recognition. Proceedings, 1996. X, 393 pages. 1996.

Vol. 1122: H. Cohen (Ed.), Algorithmic Number Theory. Proceedings, 1996. IX, 405 pages. 1996.

Vol. 1123: L. Bougé, P. Fraigniaud, A. Mignotte, Y. Robert (Eds.), Euro-Par'96. Parallel Processing. Proceedings, 1996, Vol. I. XXXIII, 842 pages. 1996.

Vol. 1124: L. Bougé, P. Fraigniaud, A. Mignotte, Y. Robert (Eds.), Euro-Par'96. Parallel Processing. Proceedings, 1996, Vol. II. XXXIII, 926 pages. 1996.

Vol. 1125: J. von Wright, J. Grundy, J. Harrison (Eds.), Theorem Proving in Higher Order Logics. Proceedings, 1996. VIII, 447 pages. 1996.

Vol. 1126: J.J. Alferes, L. Moniz Pereira, E. Orlowska (Eds.), Logics in Artificial Intelligence. Proceedings, 1996. IX, 417 pages. 1996. (Subseries LNAI).

Vol. 1127: L. Böszörményi (Ed.), Parallel Computation. Proceedings, 1996. XI, 235 pages. 1996.

Vol. 1128: J. Calmet, C. Limongelli (Eds.), Design and Implementation of Symbolic Computation Systems. Proceedings, 1996. IX, 356 pages. 1996.

Vol. 1129: J. Launchbury, E. Meijer, T. Sheard (Eds.), Advanced Functional Programming. Proceedings, 1996. VII, 238 pages. 1996.

Vol. 1130: M. Haveraaen, O. Owe, O.-J. Dahl (Eds.), Recent Trends in Data Type Specification. Proceedings, 1995. VIII, 551 pages. 1996.

Vol. 1131: K.H. Höhne, R. Kikinis (Eds.), Visualization in Biomedical Computing. Proceedings, 1996. XII, 610 pages. 1996.

Vol. 1132: G.-R. Perrin, A. Darte (Eds.), The Data Parallel Programming Model. XV, 284 pages. 1996.

Vol. 1133: J.-Y. Chouinard, P. Fortier, T.A. Gulliver (Eds.), Information Theory and Applications II. Proceedings, 1995. XII, 309 pages. 1996.

Vol. 1134: R. Wagner, H. Thoma (Eds.), Database and Expert Systems Applications. Proceedings, 1996. XV, 921 pages. 1996.

Vol. 1135: B. Jonsson, J. Parrow (Eds.), Formal Techniques in Real-Time and Fault-Tolerant Systems. Proceedings, 1996. X, 479 pages. 1996.

Vol. 1136: J. Diaz, M. Serna (Eds.), Algorithms – ESA '96. Proceedings, 1996. XII, 566 pages. 1996.

Vol. 1137: G. Görz, S. Hölldobler (Eds.), KI-96: Advances in Artificial Intelligence. Proceedings, 1996. XI, 387 pages. 1996. (Subseries LNAI).

Vol. 1138: J. Calmet, J.A. Campbell, J. Pfalzgraf (Eds.), Artificial Intelligence and Symbolic Mathematical Computation. Proceedings, 1996. VIII, 381 pages. 1996.

Vol. 1139: M. Hanus, M. Rogriguez-Artalejo (Eds.), Algebraic and Logic Programming. Proceedings, 1996. VIII, 345 pages. 1996.

Vol. 1140: H. Kuchen, S. Doaitse Swierstra (Eds.), Programming Languages: Implementations, Logics, and Programs. Proceedings, 1996. XI, 479 pages. 1996.

Vol. 1141: H.-M. Voigt, W. Ebeling, I. Rechenberg, H.-P. Schwefel (Eds.), Parallel Problem Solving from Nature – PPSN IV. Proceedings, 1996. XVII, 1.050 pages. 1996.

Vol. 1142: R.W. Hartenstein, M. Glesner (Eds.), Field-Programmable Logic. Proceedings, 1996. X, 432 pages. 1996.

Vol. 1143: T.C. Fogarty (Ed.), Evolutionary Computing. Proceedings, 1996. VIII, 305 pages. 1996.

Vol. 1144: J. Ponce, A. Zisserman, M. Hebert (Eds.), Object Representation in Computer Vision. Proceedings, 1996. VIII, 403 pages. 1996.

Vol. 1145: R. Cousot, D.A. Schmidt (Eds.), Static Analysis. Proceedings, 1996. IX, 389 pages. 1996.

Vol. 1146: E. Bertino, H. Kurth, G. Martella, E. Montolivo (Eds.), Computer Security – ESORICS 96. Proceedings, 1996. X, 365 pages. 1996.

Vol. 1147: L. Miclet, C. de la Higuera (Eds.), Grammatical Inference: Learning Syntax from Sentences. Proceedings, 1996. VIII, 327 pages. 1996. (Subseries LNAI).

Vol. 1148: M.C. Lin, D. Manocha (Eds.), Applied Computational Geometry. Proceedings, 1996. VIII, 223 pages. 1996.

Vol. 1149: C. Montangero (Ed.), Software Process Technology. Proceedings, 1996. IX, 291 pages. 1996.

Vol. 1150: A. Hlawiczka, J.G. Silva, L. Simoncini (Eds.), Dependable Computing – EDCC-2. Proceedings, 1996. XVI, 440 pages. 1996.

Vol. 1151: Ö. Babaoğlu, K. Marzullo (Eds.), Distributed Algorithms. Proceedings, 1996. VIII, 381 pages. 1996.

Vol. 1152: T. Furuhashi, Y. Uchikawa (Eds.), Fuzzy Logic, Neural Networks, and Evolutionary Computation. Proceedings, 1995. VIII, 243 pages. 1996. (Subseries LNAI).

Vol. 1153: E. Burke, P. Ross (Eds.), Practice and Theory of Automated Timetabling. Proceedings, 1995. XIII, 381 pages. 1996.

Vol. 1154: D. Pedreschi, C. Zaniolo (Eds.), Logic in Databases. Proceedings, 1996. X, 497 pages. 1996.

Vol. 1155: J. Roberts, U. Mocci, J. Virtamo (Eds.), Broadbank Network Teletraffic. XXII, 584 pages. 1996.

Vol. 1156: A. Bode, J. Dongarra, T. Ludwig, V. Sunderam (Eds.), Parallel Virtual Machine – EuroPVM '96. Proceedings, 1996. XIV, 362 pages. 1996.

Vol. 1157: B. Thalheim (Ed.), Conceptual Modeling – ER '96. Proceedings, 1996. XII, 489 pages. 1996.

Vol. 1158: S. Berardi, M. Coppo (Eds.), Types for Proofs and Programs. Proceedings, 1995. X, 296 pages. 1996.

Vol. 1159: D.L. Borges, C.A.A. Kaestner (Eds.), Advances in Artificial Intelligence. Proceedings, 1996. XI, 243 pages. (Subseries LNAI).

Vol. 1160: S. Arikawa, A.K. Sharma (Eds.), Algorithmic Learning Theory. Proceedings, 1996. XVII, 337 pages. 1996. (Subseries LNAI).

Vol. 1161: O. Spaniol, C. Linnhoff-Popien, B. Meyer (Eds.), Trends in Distributed Systems. Proceedings, 1996. VIII, 289 pages. 1996.

Vol. 1162: D.G. Feitelson, L. Rudolph (Eds.), Job Scheduling Strategies for Parallel Processing. Proceedings, 1996. VIII, 291 pages. 1996.

Vol. 1163: K. Kim, T. Matsumoto (Eds.), Advances in Cryptology – ASIACRYPT '96. Proceedings, 1996. XII, 395 pages. 1996.

Vol. 1165: J.-R. Abrial, E. Börger, H. Langmaack (Eds.), Formal Methods for Industrial Applications. VIII, 511 pages. 1996.

Vol. 1166: M. Srivas, A. Camilleri (Eds.), Formal Methods in Computer-Aided Design. Proceedings, 1996. IX, 470 pages. 1996.

Vol. 1167: I. Sommerville (Ed.), Software Configuration Management. VII, 291 pages. 1996.

Vol. 1168: I. Smith, B. Faltings (Eds.), Advances in Case-Based Reasoning. Proceedings, 1996. IX, 531 pages. 1996. (Subseries LNAI).

Vol. 1169: M. Broy, S. Merz, K. Spies (Eds.), Formal Systems Verification. XXIII, 541 pages. 1996.

Vol. 1173: W. Rucklidge, Efficient Visual Recognition Using the Hausdorff Distance. XIII, 178 pages. 1996.

Vol. 1175: K.G. Jeffery, J. Král, M. Bartošek (Eds.), SOFSEM'96: Theory and Practice of Informatics. Proceedings, 1996. XII, 491 pages. 1996.

Lecture Notes in Computer Science

Edited by G. Goos, J. Hartmanis and J. van Leeuwen

Advisory Board: W. Brauer   D. Gries   J. Stoer

Springer
*Berlin
Heidelberg
New York
Barcelona
Budapest
Hong Kong
London
Milan
Paris
Santa Clara
Singapore
Tokyo*